Sustainable Development Goals Series

The **Sustainable Development Goals Series** is Springer Nature's inaugural cross-imprint book series that addresses and supports the United Nations' seventeen Sustainable Development Goals. The series fosters comprehensive research focused on these global targets and endeavours to address some of society's greatest grand challenges. The SDGs are inherently multidisciplinary, and they bring people working across different fields together and working towards a common goal. In this spirit, the Sustainable Development Goals series is the first at Springer Nature to publish books under both the Springer and Palgrave Macmillan imprints, bringing the strengths of our imprints together.

The Sustainable Development Goals Series is organized into eighteen subseries: one subseries based around each of the seventeen respective Sustainable Development Goals, and an eighteenth subseries, "Connecting the Goals", which serves as a home for volumes addressing multiple goals or studying the SDGs as a whole. Each subseries is guided by an expert Subseries Advisor with years or decades of experience studying and addressing core components of their respective Goal.

The SDG Series has a remit as broad as the SDGs themselves, and contributions are welcome from scientists, academics, policymakers, and researchers working in fields related to any of the seventeen goals. If you are interested in contributing a monograph or curated volume to the series, please contact the Publishers: Zachary Romano [Springer; zachary.romano@springer.com] and Rachael Ballard [Palgrave Macmillan; rachael.ballard@palgrave.com].

Florence D. Hudson
Editor

The COVID Information Commons

Research Insights from the Coronavirus

Springer

Editor
Florence D. Hudson
Columbia University
New York City, NY, USA

FDHint, LLC
New York City, NY, USA

Co-editors Lauren G. Close and Emily Rothenberg

ISSN 2523-3084 ISSN 2523-3092 (electronic)
Sustainable Development Goals Series
ISBN 978-3-032-06124-9 ISBN 978-3-032-06125-6 (eBook)
https://doi.org/10.1007/978-3-032-06125-6

This Springer imprint is published by the registered company Springer Nature Switzerland AG
The registered company address is: Gewerbestrasse 11, 6330 Cham, Switzerland

We dedicate this book to everyone affected by the coronavirus around the world.
We grieve the 7 million individuals who lost their lives during the COVID-19 pandemic.
We thank the brave healthcare workers who cared for patients, risking their own lives by doing so.
We respect and applaud the researchers who continue to study the impact of the coronavirus and share their insights for the betterment of humanity.

Preface

In March of 2020, the U.S. National Science Foundation (NSF) contacted the NSF-funded Northeast Big Data Innovation Hub (NEBDHub) at Columbia University in New York City to request a proposal to develop an open, accessible portal to enable researchers, government officials and citizens to easily discover government-funded coronavirus research and inspire scientific collaboration to mitigate the societal impacts of the COVID-19 pandemic. The result was the development and delivery of the COVID Information Commons (CIC), which was initially funded by NSF in May 2020 and delivered as an open portal in July 2020. This book, *The COVID Information Commons—Research Insights from the Coronavirus*, was created to communicate scientific insights from COVID-19 researchers, with the goal to inform and enable scientific researchers, healthcare providers, government leaders, and citizens to mitigate the societal impacts of COVID-19 and future pandemics.

In March 2020, when the CIC project was first envisioned, there was very little information about prior research on the coronavirus, which could be leveraged to inform new research and actions to combat the novel coronavirus sweeping the world. COVID-19 was first identified in late 2019 (Lu et al. 2020; Sarda et al. 2020) as a novel form of the coronavirus discovered by Dr. June Almeida in 1964 at her laboratory in St. Thomas' Hospital in London, England (Brocklehurst 2020; Combs 2020). The NEBDHub team decided to ensure that the research done in the era of the COVID-19 pandemic would be findable, accessible, interoperable, and reusable, in line with the FAIR open data principles. We accomplished this through the CIC portal and community (https://covidinfocommons.datascience.columbia.edu/). The number of U.S. NSF-funded COVID-19 research awards in March 2020 was 32, and by 2025, there were nearly 14,000 NSF- and National Institutes of Health (NIH)-funded COVID-19-related research grants in the CIC. As a next step to provide lasting access to research from the COVID 19 pandemic and the resulting insights, the CIC website content was archived in the Dryad repository, and this book was commissioned with Springer Nature. The chapters in this book are authored by the researchers in the CIC community funded by the government to combat the COVID-19 pandemic. We are committed to making our research available for posterity, to inform and enable future pandemic mitigation efforts. COVID 19 would eventually cause over seven million deaths globally (https://www.worldometers.info/coronavirus/). We were convinced we needed to understand and share what happened so that we could prevent it from happening again (Morens et al. 2020).

COVID-19 researchers who presented a research lightning talk in the CIC webinar series between 2020 and 2024 were invited to publish their research results in this book. This resulted in 95 authors contributing 24 chapters, led by researchers supported by 50 research grants from the NSF, NIH, United States Department of Agriculture (USDA), Office of the Director of National Intelligence (ODNI), Intelligence Advanced Research Projects Activity (IARPA), and Michigan Alzheimer's Disease Research Center. The researchers are grateful for the funding that supported their research, and the opportunity to share the results and insights from their research to benefit humanity in the future.

We would like to acknowledge the teamwork and support from many people and organizations, beginning with the funding agencies and institutes. None of this research would have been possible without their generous funding. We also want to thank the Columbia University Information Technology and Libraries teams, and the CIC System Architect, who designed, developed, deployed, and curated the CIC portal. The collaboration of the broader NSF community, including the Midwest, South, and West Big Data Innovation Hub teams, was pivotal in the broader engagement, reaching thousands of users and participants in the CIC programs from the research lightning talks to the Student Paper Challenges. We are especially grateful for the staff and students on the Columbia University Northeast Big Data Innovation Hub team who played a valuable role in the development and delivery of the content and collaborative innovation community realized through the COVID Information Commons.

We hope this book and the related scientific insights shared by the 95 researchers enable the mitigation of the long-term effects of the COVID-19 pandemic and future pandemic challenges.

New York City, USA Florence D. Hudson

References

Lu R et al (2020) Genomic characterisation and epidemiology of 2019 novel coronavirus: implications for virus origins and receptor binding. Lancet 395(10224):565–574. https://www.thelancet.com/journals/lancet/article/PIIS0140-6736(20)30251-8/fulltext

Sarda S, Tekale SU, Kótai L, Domb A, Pawar R (2020) COVID-19: a global pandemic. Eur Chem Bull 9:266. https://pdfs.semanticscholar.org/4c4a/97b9f2b84e642b6c93cffb71b025e2caceff.pdf

Brocklehurst S (2020) The woman who discovered the first coronavirus. BBC Scotland News. https://www.bbc.com/news/uk-scotland-52278716

Combs S (2020) She discovered coronaviruses decades ago—but got little recognition—scientific pioneer June Almeida is finally being acknowledged for virology breakthroughs she made a half century ago. National Geographic. https://www.nationalgeographic.com/history/article/june-almeida-discovered-coronaviruses-decades-ago-little-recognition

COVID Information Commons, https://covidinfocommons.datascience.columbia.edu/

Worldometer, NOTE: As of April 13, 2024, the Coronavirus Tracker was no longer being updated due to the unfeasibility of providing statistically valid global totals, as the majority of countries had stopped reporting, https://www.worldometers.info/coronavirus/

Morens DM, Breman JG, Calisher CH, Doherty PC, Hahn BH, Keusch GT, Kramer LD, LeDuc JW, Monath TP, Taubenberger JK (2020) The origin of COVID-19 and why it matters. Am J Trop Med Hyg 103(3):955–959. DOI: 10.4269/ajtmh.20-0849. PMID: 32700664; PMCID: PMC7470595

Contents

Editors and Contributors

Editor in Chief

Florence D. Hudson is the Founder and CEO of FDHint, LLC, a global advanced technology consulting firm, and Executive Director and Principal Investigator (PI) for the Northeast Big Data Innovation Hub (NEBDHub) in the Data Science Institute at Columbia University in the City of New York. She has led over $10 million in funded projects from the U.S. National Science Foundation (NSF), National Institutes of Health (NIH), and the Department of Transportation Federal Highway Administration. She is PI for the COVID Information Commons (CIC), which is the basis of this book. She has also served as a PI for the NSF-funded Prototype Open Knowledge Network, Cybersecurity Transition to Practice (TTP) Acceleration Program, Cybersecurity Risk Conference, and End-to-End Trust and Security for the Internet of Things (IoT) workshop. She is Vice-Chair for the IEEE Engineering Medicine and Biology Society, and leads the development of IEEE standards to improve Trust, Identity, Privacy, Protection, Safety, and Security (TIPPSS) in connected healthcare and cyberphysical systems globally. As Chair of the IEEE/UL P2933 Clinical Internet of Things Data and Device Interoperability with TIPPSS working group of over 300 individuals in 33 countries and six continents, she led the development of the first TIPPSS standard, which was honored with the 2024 IEEE Standards Association Emerging Technology Award. A former IBM Vice President and Chief Technology Officer, Internet2 Senior Vice President and Chief Innovation Officer, and Special Advisor for the NSF Cybersecurity Center of Excellence, she began her career as an aerospace engineer at the NASA Jet Propulsion Lab and Grumman Aerospace Corporation. She has served on industry, academic, and nonprofit boards. She is a globally renowned speaker, author, editor, board director, and trustee. She earned her Mechanical and Aerospace Engineering degree from Princeton University and executive education certificates from Harvard Business School and Columbia University.

Other Springer Books by Florence D. Hudson:

- Women Securing the Future with TIPPSS for IoT—Trust, Identity, Privacy, Protection, Safety, Security for the Internet of Things (2019). https://doi.org/10.1007/978-3-030-15705-0

- Women Securing the Future with TIPPSS for Connected Healthcare—Trust, Identity, Privacy, Protection, Safety, Security (2022). https://doi.org/10.1007/978-3-030-93592-4

Assistant Editors

Lauren G. Close is the Senior Communications & Outreach Manager at Columbia University's Center for Teaching and Learning, and a former Operations & Communications Manager at the Northeast Big Data Innovation Hub in the Data Science Institute at Columbia University in the City of New York. She holds an MS in Strategic Communications from Columbia University and an MA in Art History and Archaeology from New York University's Institute of Fine Arts. Lauren specializes in digital strategy and communications, with expertise in analytics-driven content creation and building educational communities.

Emily Rothenberg is the Program Manager of the National Student Data Corps (NSDC), a global program hosted by the Northeast Big Data Innovation Hub (NEBDHub), headquartered in the Data Science Institute at Columbia University in the City of New York. Her passion lies in using mathematics, data science, and logic to solve real-world problems, with a focus on supporting missions that leverage data for social good. With a degree in Mathematical Finance, Information Technology, and Data Analytics, she works to create data science and STEM resources, projects, and programs for learners everywhere, and hopes to continue to broaden the community for data enthusiasts.

Student Editorial Assistants

Yonara Milagros Anastacio Cubas is an undergraduate student at Columbia University, pursuing a degree in Computer Science, and an Undergraduate Student Assistant at the Northeast Big Data Innovation Hub. A passionate advocate for equitable STEM education and the amplification of underrepresented voices in technology, Yonara is committed to using her expertise to foster inclusivity and social impact. Currently, she is deeply interested in Artificial Intelligence (AI), Machine Learning (ML), and independent research, exploring how these fields intersect with Public Health to drive innovation and solve real-world challenges. At the Northeast Big Data Innovation Hub, she collaborated on the COVID Information Commons, the Data Essentials and Hypothesis Building Project, and the Transportation Data Science Project.

Krishen Chetankumar Jagani is a graduate student in the M.S. in Data Science program at Columbia University and a Graduate Student Assistant at the Northeast Big Data Innovation Hub. He has a strong foundation in machine learning, deep learning, and statistical analysis, and enjoys applying data-driven thinking to a wide range of real-world problems. He's especially interested in the intersection of data science with finance, healthcare, and social impact. Krishen values collaboration and curiosity, and is passionate about making data understandable and impactful.

Sstuti Deepak Mehra is an M.S. in Data Science student at Columbia University and a Graduate Student Assistant at the Northeast Big Data Innovation Hub (NEBDHub). Driven by a passion for data and community impact, she actively contributes through research and collaboration. Her key interests lie in machine learning and data analytics.

Shermaine Tay received her B.A. in Political Science from Columbia University. She was a Student Assistant at the Northeast Big Data Innovation Hub (NEBDHub), where she supported projects including the COVID Information Commons, National Student Data Corps, Student Portfolio and Network Building Group, and the Transportation Data Science Project, while also assisting with operations and communications at the NEBDHub. Her interests lie at the intersection of data science, public health, and the environment, exploring how data-driven insights can inform policy and address pressing societal challenges.

Authors

The biographies for the 95 authors who contributed to this book, *The COVID Information Commons—Research Insights from the Coronavirus*, are included in their chapters. The authors were supported by 50 research awards, including 37 research grants focused on mitigating the impacts of the COVID-19 pandemic and 13 additional grants for Centers and other supporting efforts. These grants were sponsored by the NSF, NIH, United States Department of Agriculture (USDA), Intelligence Advanced Research Projects Activity (IARPA) within the Office of the Director of National Intelligence, and the Michigan Alzheimer's Disease Research Center. Many of these projects are highlighted in the CIC portal, and each project in this book was presented on a CIC Research Lightning Talk Webinar between 2020 and 2024. The research lightning talk videos and transcripts for all the COVID research projects in this book are available through the Columbia University Academic Commons (https://academiccommons.columbia.edu/search?f%5Bfeatured_search%5D%5B%5D=cic).

Links to the individual research project CIC video(s) and transcripts in written English, Spanish, French, and Hindi are included in each chapter.

We are grateful to the 24 Principal Investigators of the research projects highlighted in this book for stepping up to share their COVID-19 scientific insights and recommendations to inform the mitigation of future pandemics.

Contributors

Matthew Albrecht University of Florida, Gainesville, FL, USA

Lilian Azer University of California, Irvine, Irvine, CA, USA

Mina Becker University of Alabama at Birmingham, Birmingham, AL, USA

Archna Bhatia Florida Institute of Human and Machine Cognition, Pensacola, FL, USA

Megan E. Bounds University of Colorado Anschutz, Aurora, CO, USA

Sarah E. J. Bowman Department of Biochemistry, Jacobs School of Medicine and Biomedical Sciences, University at Buffalo State University of New York, Buffalo, NY, USA

Scott Burg Rockman Et Al Cooperative, Inc., San Francisco, CA, USA

Yifeng Troy Cai Brown University, Providence, RI, USA

Kathleen M. Carley Carnegie Mellon University, Pittsburgh, PA, USA

Katrina Carr University of Pittsburgh, Pittsburgh, PA, USA

Craig Carter Arizona State University, Tempe, AZ, USA

Chi-Ning Chang Virginia Commonwealth University, Richmond, VA, USA

Michael Chertkov University of Arizona, Tucson, AZ, USA

Lauren G. Close Columbia University, New York, NY, USA

J. Coffey University of West Florida, Pensacola, FL, USA

Susan R. Compton Yale University School of Medicine, New Haven, CT, USA

Yvette Conley University of Pittsburgh, Pittsburgh, PA, USA

Cathleen M. Connell University of Michigan, Ann Arbor, MI, USA

Kathryn Corvey University of Alabama at Birmingham, Birmingham, AL, USA

Ryan Culbertson Texas Tech University, Lubbock, TX, USA

Elizabeth Davis University of California, Riverside, Riverside, CA, USA

Natalie M. Davoodi Johns Hopkins Bloomberg School of Public Health, Baltimore, MD, USA

Isadora Farias Lopes de Queiroz University of California, Riverside, Riverside, CA, USA

Kathy DiRanna WestEd, San Francisco, CA, USA

Bonnie J. Dorr University of Florida, Gainesville, FL, USA

Kelly Dunning University of Wyoming, Laramie, WY, USA

Alon Efrat University of Arizona, Tucson, AZ, USA

Amir Mohammad Esmaieeli Sikaroudi University of Arizona, Tucson, AZ, USA

Ann-Margaret Esnard Georgia State University, Atlanta, GA, USA

Maria Chang Fay San Francisco Exploratorium, San Francisco, CA, USA

Leah Ferguson University of California, Riverside, Riverside, CA, USA

Jason R. Gantenberg Brown University, Providence, RI, USA

Niu Gao American Institutes for Research, Sacramento, CA, USA

Laura N. Gitlin AgeWell Collaboratory, and Johns Hopkins School of Nursing Drexel University, Philadelphia, PA, USA

Elizabeth M. Goldberg University of Colorado Anschutz, Aurora, CO, USA

R. Goodenough University of West Florida, Pensacola, FL, USA

Jennifer Hamilton NORC, University of Chicago, Chicago, IL, USA

Joshua K. Hartshorne Communication Sciences and Disorders, MGH Institute of Health Professions, Boston, MA, USA

Florence D. Hudson FDHint, LLC and Columbia University, New York, NY, USA

John Hui Virginia Commonwealth University, Richmond, VA, USA

Sylvester Ibemgbo Yale University School of Medicine, New Haven, CT, USA

Esther M. Jackson Columbia University, New York, NY, USA

Mengguo Jing Crane Center for Early Childhood Research and Policy, Ohio State University, Columbus, OH, USA

Branden B. Johnson Decision Research Center, Oregon Research Institute, (formerly at Decision Science Research Institute), Springfield, OR, USA

Jeffrey C. Johnson University of Florida, Gainesville, FL, USA

Cammie Justus-Smith Virginia Commonwealth University, Richmond, VA, USA

Helen C. Kales University of California, Davis, Sacramento, CA, USA

Aditya Khanna Brown University School of Public Health, Providence, RI, USA

Debbie Kim NORC, University of Chicago, Chicago, IL, USA

Esra Kurum University of California, Riverside, Riverside, CA, USA

Yena Kyeong National Taiwan Normal University, Taipei, Taiwan

Wenhua Lai Wayne State University and University of Michigan, Detroit, MI, USA

Asheley R. Landrum Arizona State University, Tempe, AZ, USA

Christian Lebiere Carnegie Mellon University, Pittsburgh, PA, USA

Amanda N. Leggett Wayne State University and University of Michigan, Detroit, MI, USA

Bryn C. Loftness Biobe and University of Vermont, Burlington, VT, USA

Peyton Luiz Brown University School of Public Health, Providence, RI, USA

Mark N. Lurie Brown University School of Public Health, Providence, RI, USA

Miranda L. Lynch Department of Structural Biology, Jacobs School of Medicine and Biomedical Sciences, University at Buffalo State University of New York, Buffalo, NY, USA

Katherine A. Mason Brown University, Providence, RI, USA

Brodie Mather Florida Institute of Human and Machine Cognition, Pensacola, FL, USA

Konstantinos Mitsopoulos Florida Institute of Human and Machine Cognition, Pensacola, FL, USA

Diana Mitsova Florida Atlantic University, Boca Raton, FL, USA

Don Morrison Carnegie Mellon University, Pittsburgh, PA, USA

Sarah Mullin Roswell Park Comprehensive Cancer Center, Buffalo, NY, USA

S. Namilae Embry-Riddle Aeronautical University, Daytona Beach, FL, USA

Natasha Nemmers Wayne State University and University of Michigan, Detroit, MI, USA

Mark Orr Florida Institute of Human and Machine Cognition, Pensacola, FL, USA

R. Pahle New York University, New York, NY, USA

George Perrett New York University, New York, NY, USA

Peter Pirolli Florida Institute of Human and Machine Cognition, Pensacola, FL, USA

Rebecca L. R. Powell Department of Medicine, Division of Infectious Diseases, Icahn School of Medicine at Mount Sinai, New York, NY, USA

Tania M. Rodriguez University of California, Riverside, Riverside, CA, USA

Samuel F. Rosenblatt University of Vermont, Burlington, VT, USA

Emily Rothenberg Columbia University, New York, NY, USA

Alka Sapat Florida Atlantic University, Boca Raton, FL, USA

Guan Saw Claremont Graduate University, Claremont, CA, USA

Ryan Scherle Dryad, Durham, NC, USA

Susan M. Sereika University of Pittsburgh, Pittsburgh, PA, USA

A. Srinivasan University of West Florida, Pensacola, FL, USA

Carla M. Strickland-Hughes Colorado School of Mines, Golden, CO, USA

Natasha Strydhorst Morgridge Institute for Research, University of Wisconsin-Madison, Madison, WI, USA

Tomek Strzalkowski Rensselaer Polytechnic University, Troy, NY, USA

Karen D. Sweeting University of Rhode Island, South Kingstown, RI, USA

Choh Man Teng Florida Institute of Human and Machine Cognition, Pensacola, FL, USA

Jeremiah Trinidad-Christensen Columbia University, New York, NY, USA

Tracy Van Holt New York University, New York, NY, USA

Guixing Wei Brown University, Providence, RI, USA

Susan W. Wesmiller University of Pittsburgh, Pittsburgh, PA, USA

Ryan D. Williamson University of Wyoming, Laramie, WY, USA

Janna R. Willoughby Auburn University, Auburn, AL, USA

Rachel Wu University of California, Riverside, Riverside, CA, USA

Y. Wu Embry-Riddle Aeronautical University, Daytona Beach, FL, USA

John Yin University of Wisconsin-Madison, Madison, WI, USA

Caroline Zeiss Yale University School of Medicine, New Haven, CT, USA

Abbreviations

AAAI	Association for the Advancement of Artificial Intelligence
AAAS	American Association for the Advancement of Science
Ab	Antibody
ABCD	Adolescent Brain and Cognitive Development
ABM	Agent-Based Model
AC	Columbia (University) Academic Commons
ACE2	Angiotensin-Converting Enzyme 2
ACL	Association for Computational Linguistics
ACM	Association for Computing Machinery
ACS	American Chemical Society
ACT-R	Adaptive Control of Thought—Rational
ADI	Area Deprivation Index
AERJ	American Educational Research Journal
AI	Artificial Intelligence
AIAA	American Institute of Aeronautics and Astronautics
AIC	Agent-In-Cell
AIDS	Acquired immunodeficiency syndrome
AIR	American Institutes for Research
AISL	Advancing Informal STEM Learning
ALGOSENSORS14	Algorithms and Experiments for Sensor Systems, Wireless Networks, and Distributed Robotics
ALS	Amyotrophic Lateral Sclerosis
ANOVA	Analysis of Variance
ANTLR	ANother Tool for Language Recognition
AP	Advanced Placement
APA	American Psychological Association
APHIS	Animal and Plant Health Inspection Service
API	Application Programming Interface
AP-NORC	Associated Press-NORC Center for Public Affairs Research
ARI	Army Research Institute
ASA	American Society of Anesthesiologists
ASA	American Sociological Association
ASA	American STEM Alliance

AST	Abstract Syntax Tree
AVD	Average Visit Duration
BA	Bachelor of Arts
BCS	NSF Division of Behavioral and Cognitive Sciences
BEPI	Business and Economics Polling Initiative (Florida Atlantic University)
BLE	Bluetooth Low-Energy
BOSE	Board on Science Education
BP	Base Pair
BS	Bachelor of Science
BSL-3	Biosafety Level 3
BSN	Bachelor of Science in Nursing
BT	Bluetooth
BWell	Brown University's Health Promotion Office
CAB	Community Advisory Board
CALLA	Cognitive Agility across the Lifespan via Learning and Attention laboratory
CALM	Constrained Linear Movement Model
CAPS	Center for Advanced Procurement Strategy
CARES Act	Coronavirus Aid, Relief, and Economic Security Act
CASE	California Science Educator Association
CAST	California Science Test
CBET	NSF Division of Chemical, Bioengineering, Environmental, and Transport Systems
CBG	Census Block Group
CBGVD	Census Block Group-based Voronoi Diagram
CCGRID	Cluster, Cloud and Grid Computing
CCL28	C-C Motif Chemokine Ligand 28
CCR10	C-C Chemokine Receptor 10
CCV	Center for Computation and Visualization (Brown University)
CDC	Centers for Disease Control and Prevention
CD-RISC 10	Connor-Davidson Resilience Scale-10
CEO	Chief Executive Officer
CFI	Comparative Fit Index
CHARM	dynamiC Hospital wARd Management
CHIPS and Science Act	Creating Helpful Incentives to Produce Semiconductors Act
CI	Confidence Interval
CI	Cyberinfrastructure
CIC SPC	COVID Information Commons Student Paper Challenge
CIC	COVID Information Commons
CIC-E award	COVID Information Commons Extension award
CLC	Certified Lactation Counselor
CMU	Carnegie Mellon University

CNI	Coalition for Networked Information
CNN	Cable News Network
COVID-19	Coronavirus Disease 2019
CPU	Central Processing Unit
Cryo-EM	Cryo-Electron Microscopy
CSV	Comma Separated Values
CU Libraries	Columbia University Libraries
CUIT	Columbia University Information Technology
DARPA	Defense Advanced Research Projects Agency
DHS	Department of Homeland Security
DIPs	Defective Interfering Particles
DMEM	Dulbecco's Modified Eagle Medium
DOAJ	Directory of Open Access Journals
DOI	Digital Object Identifier
DOT	Department of Transportation
DPE	Days Post Exposure
DPI	Days Post-Infection
DRCN	International Conference on the Design of Reliable Communication Networks
DRL	NSF Division of Research on Learning in Formal and Informal Setting
DrPH	Doctor of Public Health
DSL	Domain Specific Language
DVM	Doctor of Veterinary Medicine
EAGER	EArly-Concept Grants for Exploratory Research
EdD	Doctor of Education
EL	English Learner
ELA	English Language Arts
ELISA	Enzyme-Linked Immunosorbent Assay
EP&C	Environmental Principles and Concepts
ER	Estrogen Receptor
ES	Elementary School
ESG	Environmental, Social, and Governance
FAAN	Fellow of the American Academy of Nursing
FACS	Flu And Coronavirus Simulator
FAIR	Findable, Accessible, Interoperable, Reusable
FASEB	Federation of American Societies for Experimental Biology
FAST	Functional Assessment Staging Tool
Fc	Fragment crystallizable
FcRs	Fragment crystallizable Receptors
FDA	Food and Drug Administration
FEMA	Federal Emergency Management Agency
FGDs	Focus Group Discussions
FGSA	Fellow of the Gerontological Society of America
FIML	Full Information Maximum Likelihood
FOCS	Foundations of Computer Science
GALT	Gut-Associated Lymphoid Tissue

GAPcare	Geriatric Acute and Post-Acute Fall Prevention Intervention
GB	Gigabyte
GDP	Gross Domestic Product
GDPR	General Data Protection Regulation
GE R&D	General Electric Research and Development
GEOINT 2018	United States Geospatial Intelligence Foundation Conference 2018
GI	Gastrointestinal
GIF	Graphics Interchange Format
GIS	Geographic Information Systems
GNU	GNU's Not Unix
GPT-4	Generative Pre-trained Transformer 4
GUI	Graphical User Interface
HCoV-229E	Human Coronavirus 229E
HCoV-HKU1	Human Coronavirus HKU1
HCoV-OC43	Human Coronavirus OC43
HER-2	Human Epidermal Growth Factor Receptor 2
HHMI	Howard Hughes Medical Institute
HIV	Human Immunodeficiency Virus
HKU2	Rhinolophus bat coronavirus HKU2
HPC	High Performance Computing
HS	High School
IARPA	Intelligence Advanced Research Projects Activity
iBAQ	Intensity-Based Absolute Quantification
IBL	Instance-Based Learning
IBM	International Business machines
IC50	Half maximal inhibitory concentration
ICERM	Institute for Computational and Experimental Research in Mathematics
ICGEB-JNU	International Centre for Genetic Engineering and Biotechnology—Jawaharlal Nehru University
ICU	Intensive Care Unit
ID	Identification
IDeaS	Director of the Center for Informed Democracy and Social CyberSecurity
IDP	Individual Development Plan
IEEE	Institute of Electrical and Electronics Engineers
IES	Institute of Education Sciences
IgA	Immunoglobulin A
IgG	Immunoglobulin G
IHMC	Institute for Human and Machine Cognition
IIBR	Infrastructure Innovation Biological Research
IJCGA	International Journal of Computational Geometry and Applications
ILS	Institute for Informatics, Logics, and Security Studies

IN	Intranasal
INFOCOM	IEEE Conference on Computer Communications
INSNA	International Network for Social Network Analysis
IO	Input/Output
IPA	Intergovernmental Personnel Act (IPA)
IPR	Infection-to-Population Ratio
IQR	Interquartile Range
IRB	Institutional Review Board
IT	Information Technology
IT	Intratracheal
IUCrJ	International Union of Crystallography Journal
JCHAIN	Joining Chain
JD	Juris Doctor
JDA	Journal of Discrete Algorithms
JSON	JavaScript Object Notation
K-8	Kindergarten through 8th Grade (approximately ages 5–14)
K-12	Kindergarten through 12th Grade (approximately ages 5–18)
LCAPs	Local Control Accountability Plans
LCTG-002	Lactiga 002
LEAs	Local Education Agencies
LLC	Limited Liability Company
LTCFs	Long-Term Care Facilities
M protein	Membrane Protein
MA	Master of Arts
mAb	Monoclonal Antibody
MadCAM	Mucosal Addressin Cell Adhesion Molecule-1
MAPPS	Mobility Analysis for Pandemic Prevention Strategies
MARCO	MAchine Recognition of Crystallization Outcomes
MAT	Master of Arts in Teaching
MCB	NSF Division of Molecular and Cellular Biosciences
MD	Molecular Dynamics
MERS	Middle East Respiratory Syndrome
MFI	Median Fluorescence Intensity
MGH	Massachusetts General Hospital
MILCOM	IEEE Military Communications Conference
MIPO model	Mentoring Input-Process-Outcome model
ML	Machine Learning
MLS	Master of Library Science
MPH	Master of Public Health
Mpro	Main viral protease
mRNA	Messenger RNA
MS	Master of Science

MS	Middle School
MVP	Minimum viable product
MVPOI	Most Visited Point of Interest
NAICS	North American Industry Classification System
NASA	National Aeronautics and Space Administration
NASDAQ	National Association of Securities Dealers Automated Quotations
NASEM	National Academies of Sciences, Engineering, and Medicine
NCI	National Cancer Institute
NEBDHub	Northeast Big Data Innovation Hub
NERSC	National Energy Research Scientific Computing Center
NGO	Non-Governmental Organization
NGSS	Next Generation Science Standards
NIA	National Institute on Aging
NIAID	National Institute of Allergy and Infectious Diseases
NIGMS	National Institute of General Medical Sciences
NIH NRSA	National Institutes of Health National Research Service Award
NIH	National Institutes of Health
NIJ	National Institute of Justice
NINR	National Institute of Nursing Research
NLP	Natural Language Processing
NLP&C	Natural Language Processing and Culture Laboratory
NMR	Nuclear Magnetic Resonance
NOAA	National Oceanic and Atmospheric Administration
NOV	Number of Visits
NPIs	Non-Pharmaceutical Interventions
NPR	National Public Radio
NREA	National Rural Education Association
NSDC	National Student Data Corps
NSF CAREER	National Science Foundation Faculty Early Career Development
NSF	National Science Foundation
NSP	Non-Structural proteins
NT	No Tessellation
NUCES	Nuclear Cardiology Expert System
NUMA	Non-Uniform Memory Access
NYC	New York City
NYS	New York State
NYU	New York University
OD	Optical Density
ODE	Ordinary Differential Equation
ODNI	Office of the Director of National Intelligence

OoA	Office on Aging (Riverside County)
OPV	Sabin Oral Polio Vaccine
OR	Odds Ratio
ORA	Organization Risk Analyzer
ORCID	Open Researcher and Contributor ID
ORF	Open Reading Frame
OS	Operating System
OSM	OpenStreetMap
PADM	Protective Action Decision Model
PBS	Phosphate-Buffered Saline
PBS	Public Broadcasting System
PCP	Primary Care Physicians
PCR	Polymerase Chain Reaction
PCV	Probability of Co-Visiting
PD	Professional Development
PDB	Protein Data Bank
PDF	Portable Document Format
PECASE	Presidential Early Career Award for Scientists and Engineers
PFU	Plaque Forming Units
Ph.D.	Doctor of Philosophy
PI	Post-Infection
PIDs	Persistent Identifiers
pIgR	Polymeric Immunoglobulin Receptor
PIPP	Predictive Intelligence for Pandemic Prevention
PIs	Principal Investigators
PJP	Pandemic Journaling Project
PNB Group	Portfolio and Network Building Group
PO	Per os, by mouth, Oral Administration
POIs	Points-Of-Interest
PPA	Program for Positive Aging (University of Michigan)
PPE	Personal Protective Equipment
PR	Progesterone Receptor
PROMIS®	Patient Reported Outcomes Measurement System®
PTSD	Post-Traumatic Stress Disorder
P-value	Probability Value
PVAs	Psychologically Valid Agents
qRT-PCR	Quantitative Reverse Transcription Polymerase Chain Reaction
QSES	Qualitative Socio-Economic Status
RA	Research Assistant
RAPID	Rapid Response Research
RBD	Receptor-Binding Domain
RBD-ACE2	Receptor-Binding Domain Angiotensin-Converting Enzyme 2
RCV	Parker's Rat Coronavirus

Re	Reproduction Number
RESTful API	REpresentational State Transfer Application Programming Interface
RMB	Renminbi (Chinese Yuan Currency)
RMCBG	Randomly Merged Census Block Groups
RMSEA	Root Mean Square Error of Approximation
ROR	Research Organization Registry (identifiers)
RPI	Rensselaer Polytechnic Institute
RPP	Research Practice Partnership
R-PVAs	Regional Psychologically Valid Agents
RSSI	Received Signal Strength Indicator
RSV	Respiratory Syncytial Virus
RT-PCR	Reverse Transcription Polymerase Chain Reaction
RUI	Research in Undergraduate Institutions
S4	Spatial Structures in the Social Sciences
SA	Super-Agent
sAb	Secretory Antibody
SARS-CoV-1	Severe Acute Respiratory Syndrome Coronavirus 1
SARS-CoV-2	Severe Acute Respiratory Syndrome Coronavirus 2
SAS	Statistical Analysis System
SB	Bachelor of Science (alternate form)
SC	Secretory Component
SD	Standard Deviation
SDAV	Sialodacryoadenitis Virus
SDS-PAGE	Sodium Dodecyl Sulfate–Polyacrylamide Gel Electrophoresis
SEIR model	Susceptible-Exposed-Infectious-Recovered Susceptible model
SEL	Social-Emotional Learning
SEM	Structural Equation Modeling
SES	Socio-Economic Status
SHG	Second Harmonic Generation
SIG	(U.S. Department of Health and Human Services) SARS-CoV-2 Interagency Group
sIgA	Secretory Immunoglobulin A
SIGSPATIAL	ACM SIGSPATIAL International Conference on Advances in Geographic Information Systems
SIR	Susceptible–Infectious–Recovered
SLATE	Senior Learning and Accessibility Technology Evaluation
SoCG	International Symposium on Computational Geometry
SoCG'14	Thirtieth Annual Symposium on Computational Geometry
SPH	School of Public Health (Brown University)

SRMR	Standardized Root Mean Squared Error
STEM	Science, Technology, Engineering and Mathematics
STEMM	Science, Technology, Engineering, Mathematics, and Medicine
STI	Sexually Transmitted Infection
STRING	Search Tool for the Retrieval of Interacting Genes/Proteins
STSG	DataCite Services and Technology Steering Group
SUNY	State University of New York
SYNS	Society of Young Network Scientists
TCID50	50% Tissue Culture Infectious Dose
TIPPSS	Trust, Identity, Privacy, Protection, Safety and Security
TIPs	Therapeutic Interfering Particles
TLI	Tucker-Lewis Index
TMPRSS2	Transmembrane Protease, Serine 2
TOSA	Teachers On Special Assignment
TV	Television
U.S.	United States
U.S.A.	United States of America
UAB	University of Alabama at Birmingham
UB HWI	University at Buffalo Hauptman-Woodward Research Institute
UB	University at Buffalo
UCR	University of California, Riverside
UCSB	University of California, Santa Barbara
UDL	Universal Design for Learning
UFL	University of Florida
UI	User Interface
UK	United Kingdom
UL	Underwriters Laboratories
UM	University of Michigan
UPMC	University of Pittsburgh Medical Center
USD	United States Dollar
USDA	United States Department of Agriculture
USGIF	United States Geospatial Intelligence Foundation
UV-TPEF	UltraViolet-Two Photon Excited Fluorescence
UWF	University of West Florida
UX	User Experience
VA	Veterans Affairs
VAPP	Vaccine-Derived Polio-Virus strains
VBM	Variants Being Monitored
VCU	Virginia Commonwealth University
VD	Voronoi Diagram
VDPV	Vaccine-Derived Polio-Virus
VIPRA	Viral Infection Propagation during Air travel

VOC	Variants of Concern
VOHC	Variants of High Consequence
VOI	Variants of Interest
VPN	Virtual Private Network
VRS	VIPRA Recommender System (see VIPRA)
VSV	Vesicular Stomatitis Virus
VTT	Video Text Track
WHO	World Health Organization
Wi-Fi	Wireless Fidelity (wireless networking technology)
wwPDB	Worldwide Protein Data Bank
XML	eXtensible Markup Language
YAML	YAML Ain't Markup Language

Part I
The COVID Information Commons

The COVID Information Commons—A Portal and Community for Researcher Collaboration

1

Florence D. Hudson, Lauren G. Close, Esther M. Jackson, Emily Rothenberg, Ryan Scherle, and Jeremiah Trinidad-Christensen

Abstract

The COVID Information Commons (CIC) was funded by the National Science Foundation (NSF) in 2020 to develop a public access portal to facilitate the discovery of COVID-related research. The goal was to enable knowledge sharing and collaboration across NSF-funded COVID-related projects to accelerate research and mitigate the effects of the COVID-19 pandemic. The success of the CIC as a portal and community, as well as the growth in COVID-19 pandemic-related research, led to the approval of a CIC-Extension (CIC-E) grant in 2021. With this extension, additional COVID-related research projects from NSF- and National Institutes of Health (NIH)-funded researchers were added to the CIC, enabling cross-agency collaboration. The CIC has served as a valuable resource for thousands of researchers, students, and decision-makers in government, academia, nonprofits, and industry to collaborate and accelerate the most promising COVID research. This chapter describes the goal, process, and impact of the CIC as a research portal and collaborative community, presenting it as a role model for future societal challenges that can be addressed with scientific research and researcher engagement.

Research Funded by

NSF Award #2028999—RAPID (Rapid Response Research Award): COVID Information Commons (CIC); NSF Award #2139391—CIC-E: COVID Information Commons Extension for Pandemic Recovery.

COVID Information Commons Video Presentations

The COVID Information Commons (CIC) Research Lightning Talk video presentations compendium (Columbia Academic Commons, https://academiccommons.columbia.edu/search?f%5Bseries_ssim%5D%5B%5D=COVID-19+Research+Lightning+Talks&search_field=all_fields) and individual lightning talk videos with multilingual transcripts for this project can be found in the Columbia Academic Commons (AC) (Hudson 2022, 2024).

F. D. Hudson (✉)
FDHint, LLC and Columbia University, New York, NY, USA
e-mail: florence.distefano.hudson@gmail.com

L. G. Close · E. M. Jackson · E. Rothenberg · J. Trinidad-Christensen
Columbia University, New York, NY, USA
e-mail: lc3460@columbia.edu

E. M. Jackson
e-mail: ej2432@columbia.edu

E. Rothenberg
e-mail: er3101@columbia.edu

J. Trinidad-Christensen
e-mail: jt2118@columbia.edu

R. Scherle
Dryad, Durham, NC, USA
e-mail: rs4256@columbia.edu

F. D. Hudson (ed.), *The COVID Information Commons*, Sustainable Development Goals Series,
https://doi.org/10.1007/978-3-032-06125-6_1

Keywords

COVID-19 · Pandemic · Research · Collaboration · Libraries

1.1 Introduction

The COVID Information Commons (CIC) was funded by the National Science Foundation (NSF) in 2020 to develop an open access portal to facilitate discovery of COVID-related research (U.S. National Science Foundation, https://www.nsf.gov/awardsearch/showAward?AWD_ID=2028999&HistoricalAwards=false). The goal was to enable knowledge sharing and collaboration across NSF-funded COVID-related projects to accelerate research and mitigate the effects of the COVID-19 pandemic.

The success of the CIC as a portal and community, as well as the growth in research related to the COVID-19 pandemic, led to the approval of a CIC-Extension grant (CIC-E) in 2021. The CIC has served as a valuable resource for thousands of researchers, students, and decision-makers in government, academia, nonprofits, and industry to enable collaboration to accelerate government-funded research into practical application. The CIC serves as a model for integrated knowledge sharing and collaboration for public health challenges today and into the future.

The CIC grew from a portal aimed at providing quick discovery of hundreds of NSF Rapid Response Research (RAPID) grants at its launch in July 2020 to a dynamic portal of over 14,000 NSF- and National Institutes of Health (NIH)-funded COVID-related research projects by 2025. The NSF funding vehicle known as a Rapid Response Research (RAPID) Proposal is used when there is a severe urgency with regard to availability of, or access to, data, facilities, or specialized equipment, including quick-response research on natural or anthropogenic events and similar unanticipated occurrences, such as the COVID-19 pandemic. The CIC search portal was developed with FAIR—Findable, Accessible, Interoperable, Reusable—data and research principles to encourage cross-domain research, spanning projects related to biology, education, public health, human behavior, and more. As of October 2025, there were over 11,800 individuals in the CIC community, including researchers, professionals, students, and citizens based in 1,450+ academic, government, industry, and nonprofit institutions. From 2020 through early 2025, the CIC's open online content attracted over 63,000 viewers of data visualizations, research lightning talk videos and transcripts, datasets, data science projects, and research websites.

This chapter discusses the CIC conceptualization, the evolution of the project from a minimum viable product (MVP) in 2020 to the development of a dynamic CIC portal and community, as well as the CIC's ongoing value and legacy. *The COVID Information Commons—Research Insights from the Coronavirus* book published by Springer Nature in 2025, in which this chapter is included, provides a glimpse into 50 of the 14,000+ research projects in the CIC which have enabled the development of science-based insights and recommendations to address the COVID-19 pandemic. The CIC book was envisioned with the intention to communicate critical insights from COVID-19 research to the scientific community and to inform the mitigation of future pandemics. It is hoped that this resource provides value to society for many years to come.

1.2 The COVID Information Commons (CIC) Journey

The CIC journey began when NSF contacted the Northeast Big Data Innovation Hub (NEBDHub) at Columbia University (funded by NSF #1916585 (U.S. National Science Foundation, https://www.nsf.gov/awardsearch/showAward?AWD_ID=1916585&HistoricalAwards=false) and #1550284 (U.S. National Science Foundation, https://www.nsf.gov/awardsearch/showAward?AWD_ID=1550284)) in March 2020 to request their leadership in developing a proposal to create the COVID Information Commons, https://covidinfocommons.datascience.columbia.edu/) as a findable, searchable, interoperable virtual portal for researchers, students, citizens, and decision-makers in government, academia, nonprofits, and industry. This portal was envisioned as a means to enable research discovery and collaboration in order to mitigate the impact of the COVID-19 pandemic. The CIC proposal was submitted to NSF in March 2020 and funded in May 2020 by a RAPID grant (NSF #2028999) (U.S. National Science Foundation, https://www.nsf.gov/awardsearch/showAward?AWD_ID=2028999&HistoricalAwards=false) under the Coronavirus Aid, Relief, and Economic Security (CARES) Act. The CIC was developed by the NEBDHub, in collaboration with the Midwest Big Data Innovation Hub (https://midwestbigdatahub.org/), South Big Data Innovation Hub (https://southbigdatahub.org/), and West Big Data Innovation Hub (www.westbigdatahub.org). The development of the CIC's model of integrated information and knowledge sharing was led by staff from the NEBDHub, Columbia University Libraries (CU Libraries), and Columbia University Information Technology (CUIT) teams.

When Phase I launched in 2020, the CIC functioned as an open, centralized portal for easy access to the hundreds of COVID RAPID projects funded by NSF. As the COVID-19 pandemic stretched into 2021, the CIC-E award (NSF #2139391) (U.S. National Science Foundation, https://www.nsf.gov/awardsearch/showAward?AWD_ID=2139391&HistoricalAwards=false) provided an additional $2,000,000 to fund Phase II of the CIC. This new award supported the continued expansion of the CIC portal to include thousands of projects funded by federal grants from NSF and NIH. As of 2025, the CIC Awards & Researcher Database included over 14,000 NSF- and NIH-funded research awards related to the COVID-19 pandemic and its broad societal impacts.

The CIC project provided support for researcher collaboration through webinars, working groups, and interactive online programs. These programs included a Research Lightning Talk Webinar Series which began in July 2020 with two presentations from NSF-funded Principal Investigators (PIs). The PIs discussed their COVID-19-related research to an attentive audience, inspiring interest from over 40 other COVID researchers who wanted to present their research projects as well. This dynamic spawned the CIC webinar series, which began as a monthly set of lightning talks on new research and then transitioned into a quarterly webinar series where presenters discussed their research results and insights. Recordings of these presentations were shared on the CIC website, YouTube, and the Columbia University Academic Commons (AC) digital repository. With the use of curated playlists, the CIC video series enabled ongoing public discovery and access to this important research content.

The CIC project team also created programs that supported the professional development of the students and early-career researchers within the broader CIC community. These emerging public health leaders were encouraged to participate in collaborative projects to explore current and future pandemic research and recovery efforts. Among these programs were the CIC Student Working Group, student research projects, and Student Paper Challenges. To support these initiatives, additional data and metadata were added to the CIC portal, including listings of COVID funding opportunities, publications, datasets, and more.

Looking to the future of the CIC beyond 2025, the team decided it was important to ensure ongoing access to the CIC database, the project website and content. In 2024, the CIC content was archived with Dryad, an open data publishing platform and community committed to the open availability and routine re-use of research data (Hudson et al. 2024). This enables the CIC's rich data and metadata compendium to serve as a FAIR role model for collaborative research and learning opportunities today and in the future.

1.3 The CIC as a Collaborative Innovation Community

The CIC project was collaborative from its inception. The CIC Leadership Team webpage on the project website included details for all four NSF-funded regional Big Data Innovation Hubs (The Big Data Innovation Hubs, https://bigdatahubs.org/) whose network of researchers and scholars provided the CIC programs with a national research and support network.

The collaborative culture of the CIC also enabled teamwork across Columbia University and multiple associated organizations. The NEBDHub led the CIC project from its position within Columbia University's Data Science Institute in coordination with Columbia University Information Technology (CUIT) who created and hosted the website, as well as the Columbia University Libraries (CU Libraries). The collaboration with CU Libraries was very beneficial in developing the information management, organization, and UX/UI (user experience/user interface) design for CIC users. The website development, hosting, and scaling from the original CIC MVP to the broader CIC-E database was led and hosted by CUIT in collaboration with the CIC-E system architect.

1.3.1 Collaboration Between the Northeast Big Data Innovation Hub and Columbia University Libraries

The collaboration between the CU Libraries and the NEBDHub began with a casual conversation and an immediate connection between high-level administrators at a campus event. At the start of the COVID-19 pandemic, this connection continued through email and the NEBDHub invited CU Libraries to review a data management plan from the NEBDHub, including possible collaboration opportunities. Little did the participants know that they were modeling a lasting culture of communication, setting the tone for future collaborations at the very initial planning stages of this project.

Although the initial invitation of participation to the CU Libraries team was centered on data curation, it quickly became clear that the CU Libraries staff had valuable expertise to share on other topics, including metadata practices, information architecture, and information access. The success of the relationship between the NEBDHub and the CU Libraries can be attributed to a sense of mutual respect and to a general team ethos of curiosity. The CU Libraries team was initially tasked with providing recommendations for the metadata standards and architecture for the site and suggestions about the general design of the public website interface. In later stages, the team also provided recommendations about how internal website pages accessed by staff members would look and operate, as well as workflows and recommendations for staff adding new grant records to the database. Further, the CU Libraries team agreed to publish the videos and transcripts of the CIC Research Lightning Talks in the Columbia University Academic Commons (AC), providing persistent digital object identifiers (DOIs) for over 140 CIC research lightning talks with English transcripts and multilingual translations of each video. This ensured that CIC content would remain findable, accessible, interoperable, and reusable into the future, abiding by the FAIR open science principles.

1.3.1.1 CIC Website Design

A key aspect of the way that the CIC team worked was to keep an open mind and thoughtfully weigh the various needs and desires of all website visitors. It was important that the website be navigable for users coming from different academic backgrounds or professional knowledge of COVID-19 research and with different devices.

When the CIC production environment MVP was deployed in July 2020, just two months after the NEBDHub received a RAPID award from NSF in May 2020, the website included two online search mechanisms available through the homepage (Fig. 1.1). The first search mechanism was a customized search into the NSF Simple Search database for each NSF Directorate, including: Biological Sciences; Computer and Information Science and Engineering; Education and Human Resources; Engineering; Geosciences; Mathematical and Physical Sciences; Social, Behavioral and Economic Sciences; and the Office of the Director. This highlights how multi-disciplinary COVID-related research was from the beginning of the pandemic, and underscores the rare opportunity for multidimensional collaboration to assess and mitigate broad and interrelated societal impacts that the CIC provided. The second search

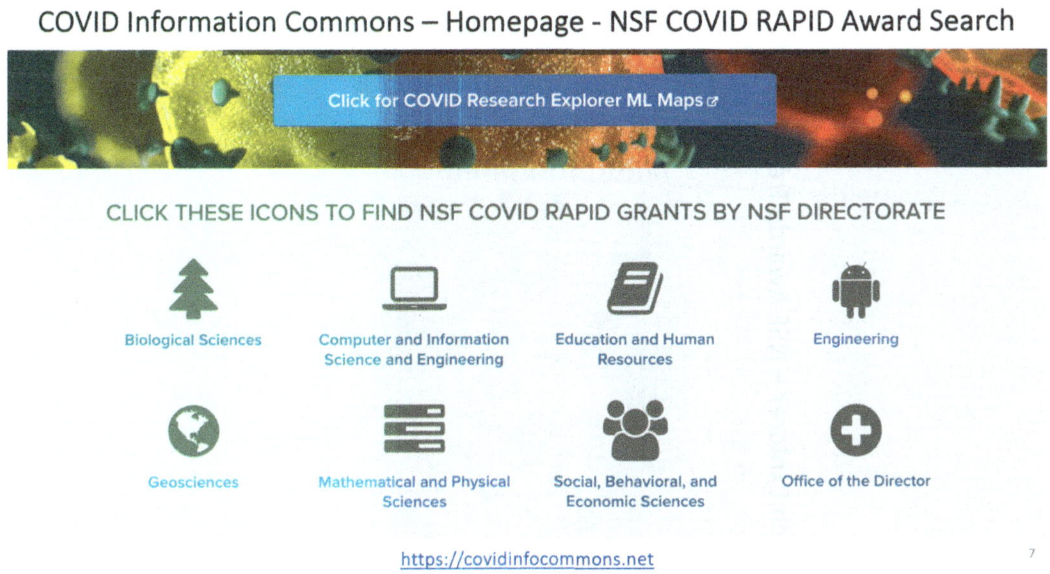

Fig. 1.1 COVID Information Commons (CIC) homepage—launched in July 2020

mechanism was a machine learning (ML)-enabled clustering tool, named the COVID Research Explorer ML Maps. Built on the Lingo4G clustering and visualization engine from CarrotSearch (https://carrotsearch.com/lingo4g/), the ML Maps tool provided a colorful and visually-structured faceted search of all the COVID awards in the CIC database. Awards were presented in topographical maps by keyword (Fig. 1.2a) and in tree maps with kaleidoscope-like visualizations of research projects within topical clusters (Fig. 1.2b). The colors could be customized to show awards associated with various parameters (Fig. 1.3a) (e.g., by institutions in a given U.S. state, awardee organization, PI, or funding organization) (Fig. 1.3b). These visualizations provided a FAIR mechanism where users could identify potential scholars and research projects in a given topical area, geography, institution, and more. This tool enabled researchers from around the world to find collaborators, meeting the mission of the CIC to enable scientific cooperation to mitigate the impact of COVID-19. A prime example of the broad collaboration which was enabled through the CIC was when an NSF-funded COVID researcher located in Buffalo, New York was discovered through the CIC portal in 2020 by a student researcher in South Korea, who engaged a researcher in the United Kingdom (UK) to enable collaboration when there was so much uncertainty and learning in countries around the world related to COVID-19. The FAIR capabilities of the CIC enabled and fueled collaborative innovation and partnerships.

The CU Libraries team was an invaluable collaborator as the CIC website and UX/UI was designed, as libraries are often publishers and maintainers of content-rich websites. Web designers in libraries must regularly balance the need to present precise and detailed information for sites that should be browsable and not overwhelming to users in search of specific resources. One recommendation that the CU Libraries team members made early on was to provide users with multiple ways of viewing information about COVID-19 awards. This included the website search results page, a detailed view of the specific grant with links back to more comprehensive records on the NSF website, and the ability to download search results as a CSV (comma separated values) file, a feature that was not available on the NSF website at the time in 2020.

In the CIC-Extension phase of the project, CU Libraries team members provided feedback on how NIH grant data could be seamlessly inte-

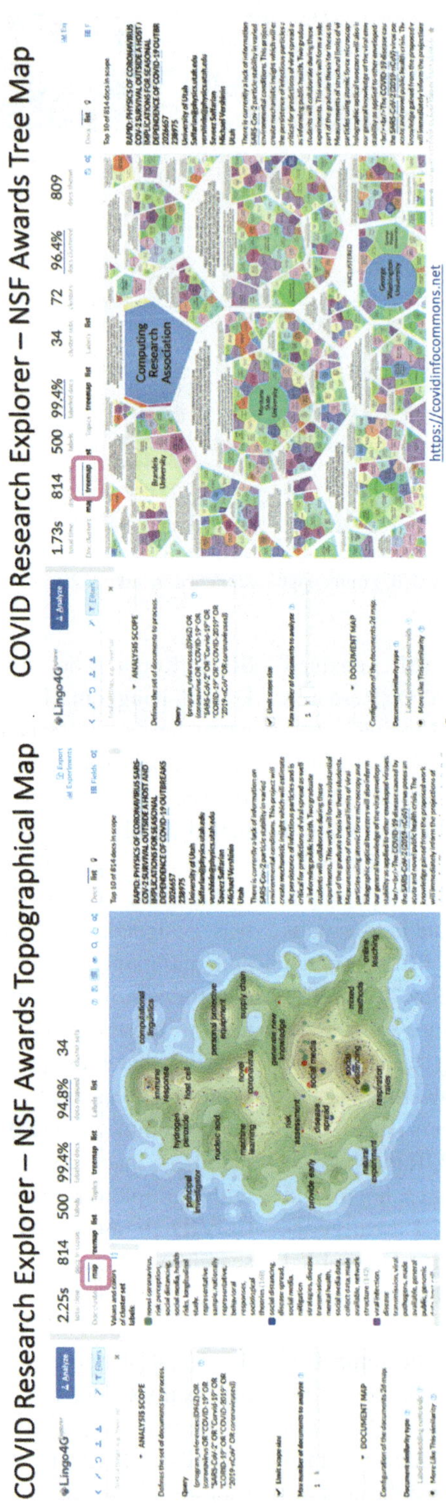

Fig. 1.2 CIC COVID research explorer ML maps tool, **a** Topographical map, **b** Tree map

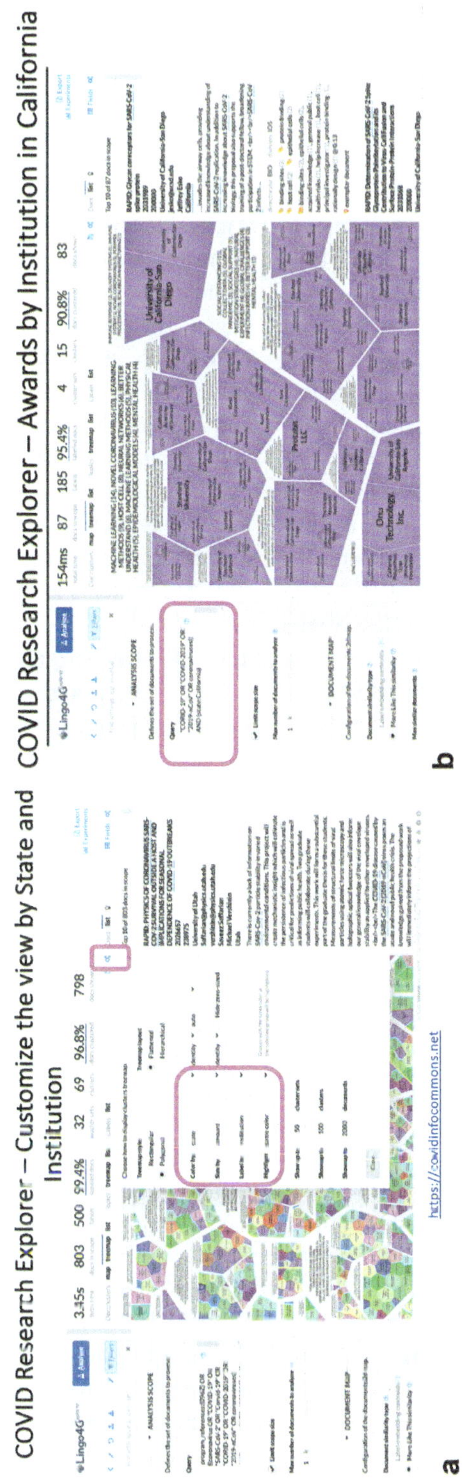

Fig. 1.3 CIC COVID research explorer customized view, **a** view options, **b** view by U.S. State

grated into the public view, as well as recommendations for the behavior that users expect when filtering and faceting search results.

1.3.1.2 Metadata Mapping Recommendations

CU Libraries team members made initial recommendations about what content would be most useful to harvest from the NSF Application Programming Interface (API) for the CIC. There were some early challenges with the data, including number codes for entities that were not translated into their corresponding text values. The CUIT team members were able to review and address this by building a custom dictionary of terms for the CIC database. CU Libraries team members also made suggestions about the frequency of the API call to NSF, and made recommendations for the desired CIC database behavior if there were differences in the metadata between the sources (live NSF JSON and static NSF XML) and local CIC database records.

The CU Libraries team also provided suggestions about how the metadata from NSF could be supplemented with persistent identifiers (PIDs) including researcher-supplied ORCID iDs. As a part of the larger project team designing the CIC researcher survey to capture information for CIC PI profiles, they also recommended collecting DOIs for related works from researchers for the purpose of later linking their awards data to their publication data. Libraries team members also recommended using Research Organization Registry (ROR) identifiers which were incorporated into Phase II of the CIC. Wikidata identifiers were also considered, but were not adopted in favor of RORs. At the time of the metadata architecture work, there was not an industry-recommended PID to use for grant IDs, and so external grant identifiers (other than the NSF, and later NIH, grant IDs) were not included.

1.3.1.3 Database Management Recommendations and Workflows

CU Libraries team members advocated for internal website tools that could be used by NEBDHub students and staff to add, remediate, and enhance metadata for CIC records. With web projects, it is common for the internal pages of a web application to receive less attention, especially when so much attention is being paid to the accessibility of the public pages. However, staff workflows and efficiencies are heavily reliant on tools that function, layout that is intuitive, and helper language and documentation that is clear. Alongside the development of these tools, CU Libraries team members made training recommendations for staff who would use the internal pages. The goal was to build capacity within the NEBDHub and CIC teams—of both staff and students—which would allow for the maintenance and eventual growth of the resource to be internally sustainable. The process of designing and finalizing the internal tools took time, but was ultimately beneficial and saved the time of staff in the longer term, including time required to orient new staff and students to the system.

1.3.1.4 Raising the Visibility of CIC Research

The CU Libraries team members included staff who managed the Columbia University Academic Commons (AC) (https://academiccommons.columbia.edu/), the university's digital repository. When the CIC Research Lightning Talk series began, it seemed natural to include these works in the repository. NEBDHub team members gathered video files and video text track (VTT) caption files of the talks while generating English transcripts as Microsoft Word and PDF documents. Those materials were packaged and uploaded to the AC, where they could be viewed and downloaded from the repository. Each transcript was given a digital object identifier (DOI) enabling ongoing findability of the content (https://academiccommons.columbia.edu/search?f%5Bseries_ssim%5D%5B%5D=COVID-19+Research+Lightning+Talks&search_field=all_fields).

The CU Libraries and NEBDHub staff worked closely to develop a workflow which leaned on existing digital translation tools while refining the content with an eye toward scientific specificity and accuracy, with a multilingual approach to raise the visibility and leverage of the CIC research. The technical workflow used was described in the blog post "The COVID Information Commons & Columbia University Libraries – using translation & transcription to increase accessibility to NSF-funded research," which was published through CU Libraries in 2025 (Close et al. 2023).

1.3.2 Enabling Researcher and Student Collaboration Across the CIC Community

1.3.2.1 The CIC Website Portal and Collaboration Tools

The CIC website was developed as an open portal providing easy access to many researcher assets, as well as multiple research and researcher discovery and collaboration tools. Beyond the original CIC project search mechanisms shown in Fig. 1.1, a new COVID Awards and PI Database was developed and deployed in October 2020. The new database interface included a faceted search of hundreds and eventually thousands of COVID awards, originally from NSF with the later addition of NIH when the CIC-E funding was committed in 2021. By simply providing a keyword (Fig. 1.4.a), PI name (Fig. 1.4.b), institution, or topical area, one could find thousands of research grants with metadata about the research, researcher, institution, funder, funding amount and grant duration.

A CIC PI Profile feature was developed as well which was populated by CIC researchers filling out an online survey form where they were asked to provide information about their COVID research and their ORCID iD in order to link to their overall research profile. ORCID is a free, unique, PID for individuals to use to disambiguate their work as they engage in research, scholarship, and innovation activities (https://orcid.org/). The CIC PI Profiles also included links to the PI's CIC Research Lightning Talk video(s), keywords for their research, links to their NSF or NIH abstract on the agency websites, and links to the PI's research websites for ease of data and research discovery. A sample CIC PI profile is provided in Fig. 1.5.

In addition to the COVID Awards & Researcher Database, the CIC team developed several other resources to serve the CIC's cross-disciplinary community of scientists, students, and researchers. These online resources included lists and links for:

- Open datasets related to the diverse impacts of the pandemic including health models, vaccination data, genetic sequences, and mental health insights from many countries and regions around the globe
- U.S. and international research funding opportunities
- Career and professional development resources and opportunities including jobs, internships, fellowships, presentations, publications, and events
- Groups and guides featuring COVID-19-related organizations and networks
- The CIC Student Working Group which focused on COVID-related research projects
- CIC Student Paper Challenge opportunities for undergraduate, community college, and graduate students, and recent graduates
- CIC user tutorial videos
- A video library of all CIC videos beginning with the kickoff webinar in July 2020, plus 30 additional CIC Research Lightning Talk Webinars, and over 140 separate CIC Research Lightning talk videos and transcripts in multiple languages.

1.3.2.2 CIC Working Groups—A Collaboration Engine

In late 2022, the CIC team developed a project plan to spearhead a Working Group specifically designed to support undergraduate and graduate students with an interest in public health research (https://covidinfocommons.datascience.columbia.edu/content/student-working-group). The first Working Group session was held in January 2023 and hosted 20+ students from schools across the USA. The CIC team continued to hold monthly meetings with the students, and by September 2023, Working Group members came to a consensus that student-led research projects should be a priority for their community.

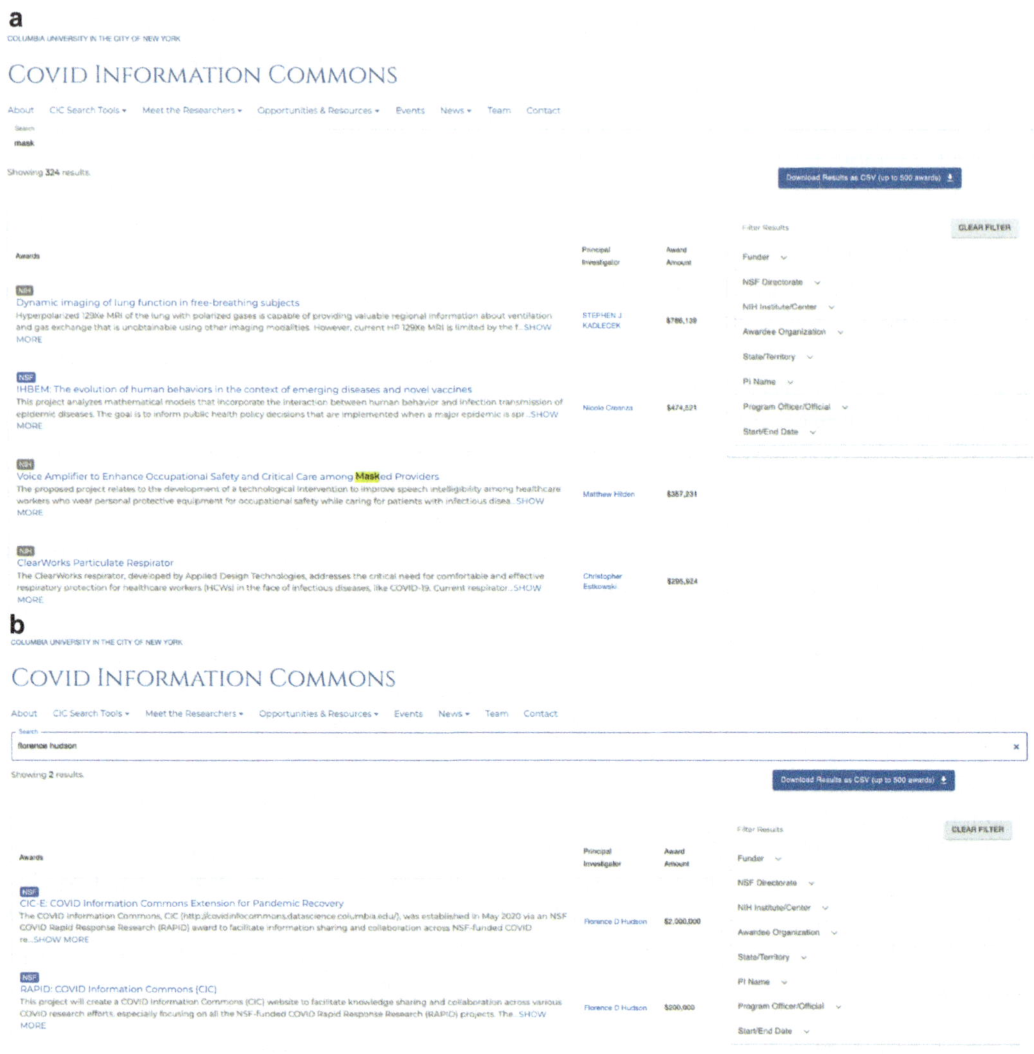

Fig. 1.4 CIC Phase II COVID awards & PI database, **a** search by keyword, **b** search by name

With help from part-time student employees at Columbia University, the CIC Student Working Group launched its first collaborative research project, the COVID-19 Data Visualization Project, in 2023 (https://nebigdatahub.org/cic-data-viz-project/). The CIC Student Working Group's membership increased exponentially after this new project was launched, growing to over 2,400 members based at 574 organizations by October 2025. The CIC Student Working Group hosted three additional independent research projects suitable for individuals and groups of learners alike–the Analyzing Pandemic Responses Project, the COVID-19 Data Dashboard Project, and the Data & Research Methods Project. These four projects engaged over 2,500 learners based in nearly 500 academic, nonprofit, government, and professional institutions (https://nebigdatahub.org/cic-data-viz-project/; https://nebigdatahub.org/cic-apr-project/; https://nebigdatahub.org/covid-dashboard/; https://nebigdatahub.org/drm-project/). Students who completed these projects received a Certificate of Completion from the CIC team which could be featured on their resume or LinkedIn profile. These students were

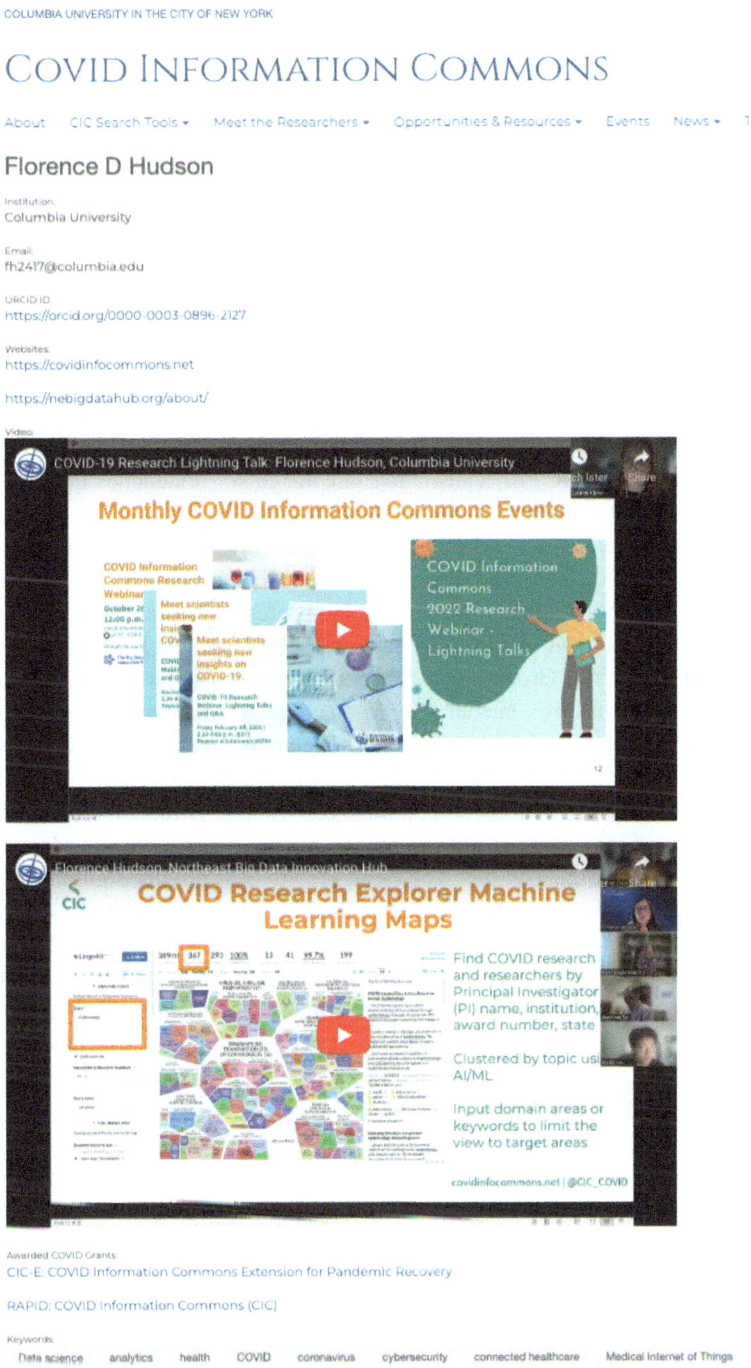

Fig. 1.5 Sample CIC PI Profile

also encouraged to mentor others who were in earlier stages of their projects, fostering a collaborative and supportive environment along the way. Faculty at multiple universities approached the CIC team to discuss how these open source projects could be added to their STEM (Science, Technology, Engineering and Mathematics) curriculum—an exciting opportunity for broader impact and community engagement.

With input from members of the CIC Student Working Group, the CIC Project team also developed a new program to facilitate student networking and professional development to support their research journeys in May 2024 called the Portfolio & Network Building (PNB) Group (https://nebigdatahub.org/pnb-group/). CIC Student Working Group members initially requested a space for direct student-to-student networking, and quickly realized that they would also benefit from project portfolio reviews and peer resume reviews. Additional requests for the CIC Project team to share insights and best practices on leadership styles and professional communications led to the development of a five-session workshop series designed to support STEM student learning, engagement, and professional development. A LinkedIn page was developed for the PNB Group and a cohort of 15 data science mentors was encouraged to share their data science career insights with new members in that space (https://www.linkedin.com/groups/13009480/). As of October 2025, the PNB Group had over 3,000 members based at 713 schools and 153 other organizations. The rapid growth of these student-facing programs was indicative of the CIC's educational and professional development value as well as the community's broader interest in communal research and collaboration.

1.3.2.3 CIC Student Paper Challenge

When the COVID-19 pandemic sparked societal challenges, the CIC team saw an opportunity to engage students in meaningful research, highlighting the importance of broad collaboration and knowledge sharing. In December 2020, the CIC Student Paper Challenge (CIC SPC) (https://covidinfocommons.datascience.columbia.edu/content/2024-cic-student-paper-challenge) was launched to encourage current and recently-graduated undergraduate students to contribute to the scientific community by exploring COVID-19 research topics, and to develop research skills and insights to spur innovation in mitigating current and future pandemics. The CIC-E funding in 2021 supported further development of the annual CIC SPC, which was expanded to include community college and graduate students. It also funded a modest incentive for the winners of the undergraduate/community college cohort and the graduate student cohort. Winners received certificates and were invited to present on the CIC Research Lightning Talk webinars, alongside professional NSF- and NIH-funded researchers, providing professional development opportunities for student researchers who may lead pandemic and public health research in the future. The CIC SPC quickly expanded, with annual relaunches in 2022, 2023, 2024, and 2025. Each year, students from institutions around the world were invited to submit papers addressing COVID recovery, health and scientific insights, potential future research discoveries, and innovations.

Students were encouraged to develop their own research questions, leveraging the CIC's resources to explore topics that aligned with their interests, such as the pandemic's impact on education, or public policy differences regionally. Through virtual mentorship calls, they were guided by data science, STEM, and healthcare professionals who provided guidance for students to complete their compelling research.

From 2020 through 2025, the CIC SPC attracted participation from 804 students and 65 mentors, representing 150+ institutions, offering students a unique opportunity to contribute to pandemic response research and health innovation. Judges, including academic, industry, and nonprofit professionals, evaluated the SPC submissions, ensuring students received feedback from various disciplines. Winning papers were judged based on intellectual merit and broader impact, two criteria used by NSF to evaluate project proposals. Winning papers were published on the CIC website and presented during CIC Research Lightning Talks webinars, CIC Student Working Group Meetings, and Northeast Big Data Innovation Hub Symposiums. Community feedback was overwhelmingly positive. One of the student presenters was offered research funding and a potential internship by webinar attendees.

The CIC SPC successfully engaged students across multiple disciplines, from data science and finance to public policy and public health. The challenge successfully fostered an environment where students could contribute to the ongoing pandemic response while also gaining access to professional development opportunities.

1.4 Evolving the CIC for Growth

1.4.1 CIC Portal and Community Evolution

When the original CIC portal was designed, it was expected to house no more than a few thousand research awards. This was based on optimism that the COVID-19 pandemic might be short lived and contained. The reality was that the effects of COVID-19 would persist for years, with a much broader impact from a timeframe and societal perspective than originally hoped. When it was decided to include all NSF- and NIH-funded COVID-related research in the CIC to extend researcher collaboration and research evolution, the CIC needed to be redesigned to scale to many more awards. The CIC-E system architect took on this challenge including multiple elements of design as follows:

- Redesigned the database to scale for many more thousands of awards
- Deployed new software design approaches for scale which included:
  - API-centric design
  - Serverless deployment
  - Keeping up with updates to the NSF and NIH APIs
  - Addressing potential cybersecurity concerns
- Expanded the data model to encompass more types of content which included:
  - Added support for publications and datasets
  - A common model to manage award information from NSF, NIH, and other funders
  - An administrative interface for manual corrections to content
- Built automated tools to import and refine content
- Developed a process to periodically archive content in Dryad for persistence beyond the life of the CIC itself (Hudson et al. 2024)
- Designed and built a next-generation search system to allow searching over all of the content simultaneously and navigating between different types of content including research awards, datasets, and publications

1.4.2 CIC Community Communications Evolution

Communications was a central tenet of the CIC's growth and outreach efforts from its inception in 2020, in part because of the remote-only work and research environment in which the CIC was launched during COVID-19 quarantine. Additionally, the CIC team understood early on that the sweeping impacts of the pandemic meant that every academic discipline could benefit from the scientific conversations fostered through the CIC ecosystem.

Building upon an audience engagement strategy designed for the NEBDHub, the CIC team identified three key channels for reaching the target audience. The first was the CIC website (COVID Information Commons, https://covidinfocommons.datascience.columbia.edu/), launched in July 2020, designed, and maintained with support from CUIT. Website traffic was bolstered by frequent updates to the site, including the addition of new CIC Student Working Group projects, webinar videos, and COVID-related resources. In 2025, the CIC website continued to attract an average of 2,000 visitors quarterly, the majority based in the USA. This demonstrated the ongoing value of the CIC as the world navigated beyond the original COVID-19 situation.

Another crucial communications channel for the CIC was the email listserv used to communicate the CIC newsletter and invite professionals and students to participate in CIC programs and webinars. The CIC newsletter began as a monthly communication in 2020 and was transitioned to a quarterly CIC newsletter in January 2025. The newsletter audience grew to 2,100+ subscribers by October 2025 who engaged with the COVID and multidisciplinary research-related news curated by the CIC staff and student contributors. Another 8,200+ NSF and NIH researchers were invited to participate in CIC research and collaboration opportunities via a separate listserv. In 2024, the CIC team spearheaded a redesign of the CIC newsletter, streamlining the content design to maximize engagement and drive traffic to the CIC website.

The third leg of the CIC communications strategy was social media. With targeted campaigns on X (formerly Twitter), Instagram, LinkedIn, YouTube, and Slack, the CIC team reached thousands of followers with news about new research challenges, event announcements, and networking opportunities (https://x.com/NEbigdatahub; https://www.instagram.com/nebigdatahub/; https://www.linkedin.com/company/northeast-big-data-innovation-hub; https://www.youtube.com/channel/UCaHjFSdvLnxWeYski3QjU_g; https://join.slack.com/t/covidcommons/shared_invite/zt-2rkr2h7x1-sx9eSXJUsJ6x_nfdGcae6A). As of October 2025, the CIC reached over 4,500 subscribers across social media platforms, with the greatest growth taking place via the YouTube and LinkedIn channels. Whereas the CIC website and newsletters were vehicles for sharing news and updates, the social media platforms were key in fostering and maintaining long-term community engagement with the CIC. The CIC team developed longitudinal campaigns that celebrated the CIC Student Working Group's progress with new research projects and sparked peer conversations. In combination, the CIC website, newsletter, and social media platforms were an effective integrated communications toolkit that the CIC team used to sustain the momentum of scientific research and collaboration across the community.

1.5 Outcomes and Broader Impact

The culture of collaborative innovation engendered in the CIC project, community, and website from its inception set the proper tone to enable researchers and students to leverage each other's research to develop valuable multidisciplinary insights to benefit COVID research and related scientific opportunities. The tools and content developed and delivered through the CIC website enabled researcher discovery around the world.

The thousands of researchers who engaged through the CIC shared their experience, research and learnings to provide broader impact and value to other academic and scientific projects starting in 2020 and continuing throughout the project. Project team members presented together at conferences such as the Coalition for Networked Information (CNI) in November 2020 where CIC PI Hudson and the CU Libraries leader Trinidad-Christensen presented their CIC collaboration and the value of the NEBDHub and Libraries partnership (Hudson et al. 2020), with another CIC presentation in August 2023 to provide an update on the growth and broader impact of the CIC. Ten of the PIs whose research was in the CIC portal presented their research virtually at the International FAIR Convergence Symposium in Paris in November 2020 to increase the international collaboration opportunities for the COVID researchers. The CIC was also presented at National Cancer Institute (NCI) conferences on Biomedical Digital Twins in 2023, 2024 and 2025, to inspire the idea to develop a CIC-like portal to share knowledge, insight, research, videos, and inspire student engagement in biomedical digital twins to cure cancer.

The CIC website continually evolved with the introduction of new programs in response to valuable input from the community. By March 2025, the CIC website (Fig. 1.6) provided easy access to over 14,000 research grants funded by NSF and NIH through the COVID Awards & PI database and the COVID Research Explorer ML Maps Tool. To increase ease of use for new users, the project team recorded a CIC User Tutorial video (An Introduction to the COVID Information Commons video, https://youtu.be/X1rMZPlOH50). The links to the Student Working Group, and Student Projects and Challenges, inspired student participation and collaboration.

1.6 COVID Information Commons Learnings for Mitigation of Future Pandemics

The COVID Information Commons project proved to be a pivotal collaboration mechanism for thousands of researchers, students, and professionals around the world who shared a common mission to mitigate the societal impacts from the COVID-19 pandemic. Together, they developed and discovered complementary research to inform future research, and to inform recommendations for their communities. The ability to provide data discovery of over 14,000 NSF- and NIH-funded grants, the associated researchers, datasets, and networks, all through one portal, was very valuable for the scientific, government, academia, and student community. The advanced ML clustering tools provided clear and compelling visualizations of the research topics which advanced critical thinking for leverage of various research approaches and insights for valuable collaborations. The virtual programming through synchronous research lighting talks and question and answer webinars and easy asynchronous access to over 140 researcher videos online provided broader impact, including translations of the lightning talk transcripts for English, Spanish, French, and Hindi, increasing the FAIR nature of the CIC output. All of these mechanisms were easy to create and highly impactful. Future challenges, such as those presented by the COVID-19 pandemic, could benefit from the development of similar mechanisms and communities. The CIC team began working with NIH and the NCI in 2023 to discuss how a CIC-like portal and community for researcher and student engagement could enable more data and insight sharing to accelerate precision medicine and cures for cancer. Curing cancer would be a great silver lining to the results of the COVID-19 pandemic, which claimed over seven million lives around the world (Worldometer, https://www.worldometers.info/coronavirus/).

The collaborative and supportive culture developed at the onset of the CIC project, between the Big Data Innovation Hubs across the USA, across disciplines from the Data Science Institute, to the Columbia University Information Technology and Columbia Libraries teams, and for researchers, professionals, and students from all scientific domains, supported the art of listening, learning, and collaborating. The CIC team itself had a culture of responsiveness, flexibility, and the ability to move quickly, which engendered the same in the CIC community. Building an open, listen-first culture allowed increased understanding across boundaries and built a foundation for future collaborative innovation opportunities. Engaging students in understanding and addressing the pandemic enabled them to learn how to think about challenges they have never encountered before and helped prepare them for future challenges, as they grow to be the leaders of tomorrow.

The coming together of nearly 100 COVID-19 researchers in the CIC book in 2025, five years after the COVID-19 pandemic began and the CIC portal and community was created, shows the deep and long-lasting camaraderie created for individuals focused on a mission to continue to reduce the impact of the COVID-19 pandemic. The thousands of students engaged through the CIC focused not just on the present situation, but also on what they may be dealing with in future pandemics and public health emergencies.

The CIC teamwork has the potential to continue for years to come, as COVID-19 continues to evolve and penetrate the world, with the hope that it inspires similar collaborative innovation communities and technology-enabled portals and knowledge sharing to combat future societal challenges.

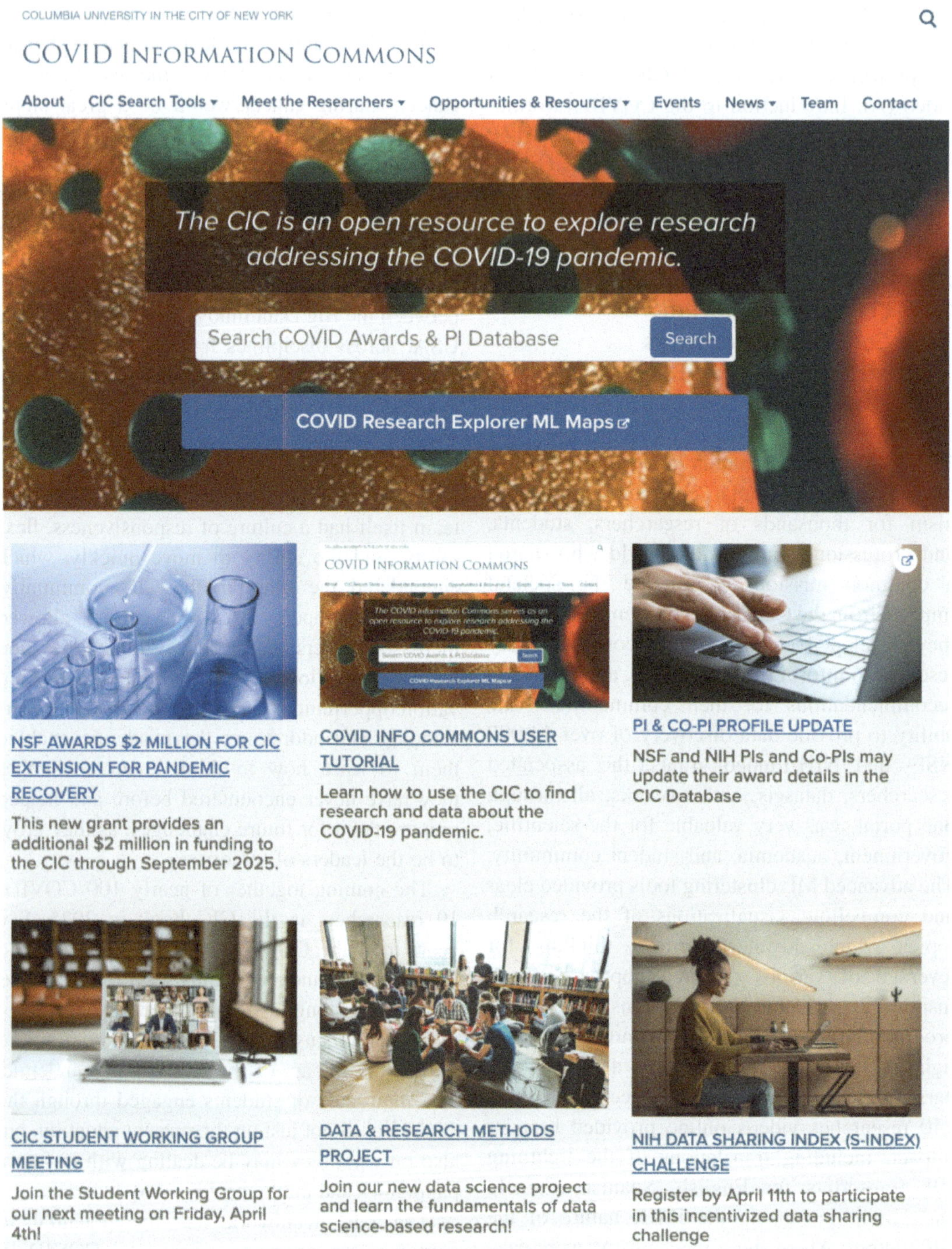

Fig. 1.6 CIC homepage—March 2025

References

An Introduction to the COVID Information Commons video, https://youtu.be/X1rMZPlOH50

Close L, Teran L, Jackson E, Hudson F, Moujabber M, Graham-Martinez I, Mercurio J (2023) The Covid information commons & Columbia University libraries—using translation & transcription to increase accessibility to NSF-funded research. Retrieved 30 April 2025, from: https://blogs.library.columbia.edu/the-tinker/2025/04/30/the-covid-information-commons-columbia-university-libraries-using-translation-transcription-to-increase-accessibility-to-nsf-funded-research-2/

Columbia Academic Commons, Academic Unit: Northeast Big Data Innovation Hub, Series: COVID-19 Research Lightning Talks. https://academiccommons.columbia.edu/search?f%5Bseries_ssim%5D%5B%5D=COVID-19+Research+Lightning+Talks&search_field=all_fields

COVID Information Commons, https://covidinfocommons.datascience.columbia.edu/

https://carrotsearch.com/lingo4g/

https://academiccommons.columbia.edu/

https://academiccommons.columbia.edu/search?f%5Bseries_ssim%5D%5B%5D=COVID-19+Research+Lightning+Talks&search_field=all_fields

https://orcid.org/

https://covidinfocommons.datascience.columbia.edu/content/student-working-group

https://nebigdatahub.org/cic-data-viz-project/

https://nebigdatahub.org/cic-apr-project/

https://nebigdatahub.org/covid-dashboard/

https://nebigdatahub.org/drm-project/

https://nebigdatahub.org/pnb-group/

https://www.linkedin.com/groups/13009480/

https://covidinfocommons.datascience.columbia.edu/content/2024-cic-student-paper-challenge

https://x.com/NEbigdatahub; https://www.instagram.com/nebigdatahub/; https://www.linkedin.com/company/northeast-big-data-innovation-hub; https://www.youtube.com/channel/UCaHjFSdvLnxWeYski3QjU_g; https://join.slack.com/t/covidcommons/shared_invite/zt-2rkr2h7x1-sx9eSXJUsJ6x_nfdGcae6A

Hudson F (2022) Florence Hudson: CIC-E: COVID information commons extension for pandemic recovery. Columbia Academic Commons, Academic Unit: Northeast Big Data Innovation Hub, Series: COVID-19 Research Lightning Talks. https://doi.org/10.7916/bxaz-rw47

Hudson F (2024) Florence Hudson: CIC-E: COVID information commons extension for pandemic recovery. Columbia Academic Commons, Academic Unit: Northeast Big Data Innovation Hub, Series: COVID-19 Research Lightning Talks. https://doi.org/10.7916/m75q-hm85

Hudson FD, Trinidad-Christensen J, Garza N, Bowman S (2020) The coalition for networked information (CNI) virtual membership meeting, Nov 2020. https://www.cni.org/topics/economic-models/the-covid-information-commons

Hudson F, Scherle R, Close L et al (2024) COVID information commons archive. Dryad. https://doi.org/10.5061/dryad.37pvmcvqp

Midwest Big Data Innovation Hub, https://midwestbigdatahub.org/

South Big Data Innovation Hub, https://southbigdatahub.org/

The Big Data Innovation Hubs, https://bigdatahubs.org/

U.S. National Science Foundation, Award Abstract # 2028999, RAPID: COVID Information Commons (CIC), https://www.nsf.gov/awardsearch/showAward?AWD_ID=2028999&HistoricalAwards=false

U.S. National Science Foundation, Award Abstract # 1916585, BD Hubs: NORTHEAST: The Northeast Big Data Innovation Hub, https://www.nsf.gov/awardsearch/showAward?AWD_ID=1916585&HistoricalAwards=false

U.S. National Science Foundation, Award Abstract # 1550284, BD Hubs: NORTHEAST: The Northeast Big Data Innovation Hub, https://www.nsf.gov/awardsearch/showAward?AWD_ID=1550284

U.S. National Science Foundation, Award Abstract # 2139391, CIC-E: COVID Information Commons Extension for Pandemic Recovery, https://www.nsf.gov/awardsearch/showAward?AWD_ID=2139391&HistoricalAwards=false

West Big Data Innovation Hub, https://www.westbigdatahub.org/

Worldometer, NOTE: As of April 13, 2024, the Coronavirus Tracker was no longer being updated due to the unfeasibility of providing statistically valid global totals, as the majority of countries had stopped reporting, https://www.worldometers.info/coronavirus/

Florence D. Hudson is Executive Director of the Northeast Big Data Innovation Hub in the Data Science Institute at Columbia University in the City of New York, leading over $10 million in projects funded by the U.S. National Science Foundation (NSF), National Institutes of Health (NIH), Department of Transportation Federal Highway Administration (US DOT FHWA), and Defense Advanced Research Projects Agency (DARPA). She is also Founder and CEO of FDHint, LLC, an advanced technology consulting firm. She is a former IBM Vice President and Chief Technology Officer, Internet2 Senior Vice President and Chief Innovation Officer, Special Advisor for the NSF Cybersecurity Center of Excellence, and aerospace engineer at the NASA Jet Propulsion Lab and Grumman Aerospace Corporation. She is an Editor in Chief and Author for Springer, Elsevier, Wiley, Institute of Electrical and Electronics Engineers (IEEE), and other publications. She leads the development of IEEE and Underwriters Laboratories (UL) standards to increase Trust, Identity, Privacy, Protection, Safety, and Security (TIPPSS) for connected healthcare data and devices, and cyber-physical systems, and is Vice Chair of the IEEE Engineering Medicine and Biology Society Standards Committee. She has served on over 15 academic, nonprofit, startup, and NASDAQ-listed boards. She earned her Mechanical and Aerospace Engineering degree from Princeton University, and executive education certificates from Harvard Business School and Columbia University.

Lauren G. Close is the Senior Communications & Outreach Manager at Columbia University's Center for Teaching and Learning, and a former Operations & Communications Manager at the Northeast Big Data Innovation Hub in the Data Science Institute at Columbia University in the City of New York. She holds an MS in Strategic Communications from Columbia University and an MA in Art History and Archaeology from New York University's Institute of Fine Arts. Lauren specializes in digital strategy and communications, with expertise in analytics-driven content creation and building educational communities.

Esther M. Jackson is the Head of Open Scholarship at Columbia University Libraries where she oversees digital publishing and digital pedagogy activities, and consults with researchers on open access and open infrastructure topics. In addition, she serves as a committee member for the Wikimedia Foundation's Regional Fund for North America, on the OpenRefine Advisory Committee, on the DataCite Services and Technology Steering Group (STSG), and on the Directory of Open Access Journals (DOAJ) Editorial Policy Advisory Group.

Emily Rothenberg is the Program Manager of the National Student Data Corps (NSDC), a program hosted within the Northeast Big Data Innovation Hub (NEBDHub) at the Data Science Institute of Columbia University in the City of New York. Her work centers on advancing the use of mathematics, data science, and logic to address real-world challenges, with a particular focus on initiatives that leverage data for social good. She has managed and contributed to projects funded by the U.S. National Science Foundation (NSF), National Institutes of Health (NIH), Department of Transportation Federal Highway Administration (US DOT FHWA), and the Defense Advanced Research Projects Agency (DARPA). Emily holds a degree in Mathematical Finance, Information Technology, and Data Analytics, and is dedicated to developing data science programs, resources, and collaborations that expand engagement across learning communities.

Ryan Scherle is the System Architect for the COVID Information Commons and the Head of Platform Development for the Dryad open data publishing platform. Ryan has over 30 years of experience designing and developing systems to manage collections of digital objects. He holds a Ph.D. in Computer Science and Cognitive Science from Indiana University. He believes that the best technology strikes a balance between anticipating what a user needs and allowing the user to remain in control.

Jeremiah Trinidad-Christensen has been with Columbia University Libraries for more than 20 years and has worn many hats. Currently, he is the Head of Research Data Services, which is responsible for the numeric and geospatial data collections within the library, and provides data and data management-related consultations and services. He started as the first Geographic Information Systems (GIS) librarian at Columbia, and still holds a strong passion for geospatial-related research and technologies.

Part II

COVID and Biology

Using Animal Models to Inform Pandemic to Endemic Transition of COVID-19 in Humans

2

Sarah Mullin, Susan R. Compton, Sylvester Ibemgbo, and Caroline Zeiss

Abstract

Population-wide coronaviruses are common in animals, providing a rich source of epidemiologic data that can be applied to disease modeling of coronaviral infection in humans. We modeled heterogeneous SARS-CoV-2 exposure in a naturally infected human population using a range of high-risk (inoculation and direct contact) and low-risk (inanimate fomite) exposures in a rat model of SARS-CoV-2 transmission. Using these data, we developed deterministic compartmental models in humans with the goal of predicting pandemic to endemic transition dynamics of COVID-19 in the United States. After accounting for translationally altered parameters between rat-derived data and human SARS-CoV-2 transmission, and after introducing vaccination, we arrived at a median time to endemic stability of 1437 days (interquartile range (IQR) = 749.25) with a median 15.4% of the population remaining susceptible. We extended the models to introduce successive variants with increasing transmissibility and included the effect of varying duration of immunity. Our models predicted approximate pandemic to endemic transition of COVID-19 occurring in December 2023 and were consistent with highest mean national peaks occurring in the late summer and winter. Test positivity rates in early summer and fall were lower, however reporting declined due to home testing vs. reported testing with a healthcare provider. Rates of vaccine uptake declined; however, our models indicated that vaccination combined with natural immunity maintained a lower prevalence of infection than natural infection.

Research funded by

USDA-APHIS (United States Department of Agriculture (USDA) Animal and Plant Health Inspection Service (APHIS))-10025-OA000000-23-0001 (APP-22174); NSF Award #2031960—RAPID (Rapid Response Research Award): An in vivo driven SEIRS (susceptible-exposed-infectious-recovered susceptible) model of coronaviral infection.

COVID Information Commons Video Presentation

The COVID Information Commons (CIC) Research Lightning Talk video presentation and individual multilingual transcripts for this project can be found in the Columbia Academic Commons (AC) (Zeiss 2021).

S. Mullin
Roswell Park Comprehensive Cancer Center, Buffalo, NY, USA
e-mail: sarah.falls@roswellpark.org

S. R. Compton · S. Ibemgbo · C. Zeiss (✉)
Yale University School of Medicine, New Haven, CT, USA
e-mail: caroline.zeiss@yale.edu

S. R. Compton
e-mail: susan.compton@yale.edu

S. Ibemgbo
e-mail: sylvester.ibemgbo@yale.edu

F. D. Hudson (ed.), *The COVID Information Commons*, Sustainable Development Goals Series,
https://doi.org/10.1007/978-3-032-06125-6_2

Keywords

COVID-19 · Animal models · Vaccines · Immunity · Pandemic to endemic transition · Infection

2.1 Introduction

In early 2020, it became apparent that SARS-CoV-2 was spreading rapidly through a naïve human population on a global scale. By March of that year, COVID-19 was declared a pandemic. In contrast to familiar and relatively innocuous common cold coronaviruses, the behavior of SARS-CoV-2 presented a shocking distinction. Its genetic fluidity, relentless generation of increasingly transmissible variants and relatively high pathogenicity gave no clear indication of how the pandemic would resolve. For the human species, the experience was novel and outside of living memory. However, in animals, coronaviruses have been very much a presence in domestic, laboratory and wild populations (Kenney et al. 2021; Bhatt et al. 1972; Franzo et al. 2019; Turlewicz-Podbielska and Pomorska-Mól 2021; Percy et al. 1990; Lee 2015) for decades, thus providing insight into the potential trajectory of SARS-CoV-2 in humans. A characteristic element of coronaviruses is their tendency to cross the species barrier (Zhou et al. 2018; Vijgen et al. 2005). Entry into a new naïve host species results in a pandemic-to-endemic infection trajectory shaped by non-sterilizing immunity (Turlewicz-Podbielska and Pomorska-Mól 2021; Percy et al. 1990; Bhatt and Jacoby 1985), generally declining disease severity (Bhatt and Jacoby 1985; Pensaert et al. 1993), immune evasion following vaccination (Franzo et al. 2019; Bhuiyan et al. 2021), and periodic resurgence associated with new variants (Lee 2015; Lin and Chen 2017; Fan et al. 2020). With these realities in mind, we undertook studies to determine how long it would take for the COVID-19 pandemic to reach a stable endemic state.

2.2 Goal of the Research

Epidemiologic models interrogate and predict transmission dynamics to inform policies on mitigation strategies, vaccination roll-out, and healthcare settings (Giordano et al. 2021; Usherwood et al. 2021; Dashtbali and Mirzaie 2021). However, due to their observational nature, the resulting mathematical models may not yield accurate insights of disease dynamics (Massonis et al. 2021). Therefore, we informed parameter estimates for the model using in vivo animal model transmission data. First, we generated controlled reinfection data after high- and low-risk natural exposure or heterologous vaccination to a respiratory betacoronavirus in rats (Zeiss et al. 2021). Next, after accounting for translationally altered parameters between rat-derived data and human SARS-CoV-2 transmission, we applied these data to develop deterministic compartmental models in humans with the goal of predicting pandemic to endemic transition dynamics in the United States. Components of the model included increasing transmission rates associated with successive waves of variants, variable duration of immunity, and the effect of vaccination on patterns of viral transmission.

Because we were modeling viral transmission in populations, our selection of an animal model was based on characteristics defining viral infection, shedding, transmission, and immunity rather than disease pathogenesis. We selected sialodacryoadenitis virus (SDAV), a highly infectious rat betacoronavirus (Bhatt et al. 1972; Bhatt and Jacoby 1985; Regina et al. 1992) that is closely related to two human upper respiratory pathogens, human coronavirus HKU1 (HCoV-HKU1) and human coronavirus OC43 (HCoV-OC43) (Vijgen et al. 2005; Yoo et al. 2000; Hu et al. 2021). SDAV infects the upper respiratory tract (Bihun and Percy 1995), lacrimal and salivary glands (Eisenbrandt et al. 1982; Percy et al. 1984), and lung (Percy et al. 1984; Wojcinski and Percy 1986; Funk et al. 2009) of rats. SDAV infection results in a two-week course of asymptomatic to mild respiratory disease (Wojcinski and Percy 1986; Funk et al. 2009). Like SARS-CoV-2 (Johansson et al. 2021; Arons et al. 2020; Zhang et al. 2020a; Kraay et al. 2021; Hu et al. 1533), SDAV infection can be transmitted by asymptomatically infected individuals (Bhatt and Jacoby 1985) via airborne, direct contact, or fomite routes (Regina et al. 1992). Both SARS-CoV-2 and SDAV can persist on hard surfaces for two to 28 days (Kraay et al. 2021; Hu et al. 1533; Doremalen et al. 2020) and two days (Gaertner et al. 1993), respectively. Like human immunity to seasonal coronavirus (Edridge et al. 2020), and SARS-CoV-2 natural infection (Spicer et al. 2022) or vaccination (Goldberg et al. 2021; Levin et al. 2021; Eyre et al. 2022), immunity in rats elicited by SDAV infection (Bihun and Percy 1994; Weir et al. 1990) is temporary.

2.3 Using Rats to Model Natural Infection, Immune Heterogeneity, Reinfection and Vaccination

Mixing interactions characterizing infectious disease transmission in human populations are complex and dynamic, resulting in extremely heterogeneous exposure and immunity patterns (Zhang et al. 2020b; Mistry et al. 2021). Because the full spectrum of these interactions cannot be feasibly modeled in animals, we modeled heterogeneous viral exposure in a naturally infected population using a range of high-risk (inoculation and direct contact) and low-risk (fomite) exposures (Zeiss et al. 2021). The exposure paradigm is illustrated in Fig. 2.1. Animals in the high-risk transmission group experienced short-range viral exposure through direct physical contact, aerosol, or droplet routes (Zhang et al. 2020a; Hu et al. 1533; Leung 2021), whereas animals in the low-risk transmission group experienced long-range viral exposure via fomite exposure (Regina et al. 1992; Kraay et al. 2021; Leung 2021). These exposures occurred 48 h after inoculating the source rat, at a time when viral shedding and upper respiratory tract infection are established (Bhatt and Jacoby 1985; Funk et al. 2009).

2.3.1 Initial Infection with SDAV ($n = 105$)

Naïve animals ($n = 19$) were inoculated intranasally with $2X10^4$ pfu SDAV. After 48 h, inoculated animals or their dirty cages were used to expose naïve animals to the virus. For direct contact exposure, one inoculated animal was placed in a clean cage with naïve animals ($n = 31$) for 24 h. For fomite exposure, naïve animals were placed in a dirty cage that had been inhabited by an inoculated rat for 24 h. After 24 h, all rats were relocated to new clean cages to constitute four groups: inoculated rats, direct exposure rats, and two fomite exposure groups—a fomite cohabitation group ($n = 30$) constituting two to three animals, and a fomite single group ($n = 25$) with only one animal. Oral swabs were taken on Day 2, 3, 4, 7, and 10 days post-infection (dpi). Infection is defined as inoculation or infection via transmission on all animals, and serology performed five to six weeks later. A control group (not shown) was inoculated with Dulbecco's Modified Eagle Medium (DMEM) alone, and similarly swabbed and bled. After initial exposure, animals assumed either seronegative ($n = 40$) or seropositive status ($n = 66$). All groups were evenly split by sex.

2.3.2 Reinfection with SDAV

Naïve seronegative rats ($n = 18$) were inoculated intranasally with 2X10e4 pfu SDAV to provide a source of infection, and to represent the putative most susceptible group. Rats that had originally received intranasal infection with SDAV were re-infected intranasally again with the same viral dose to constitute the putative least susceptible group. Seropositive and seronegative animals from the initial infection experiment were randomly assigned direct contact ($n = 28$), fomite contact-cohabitation

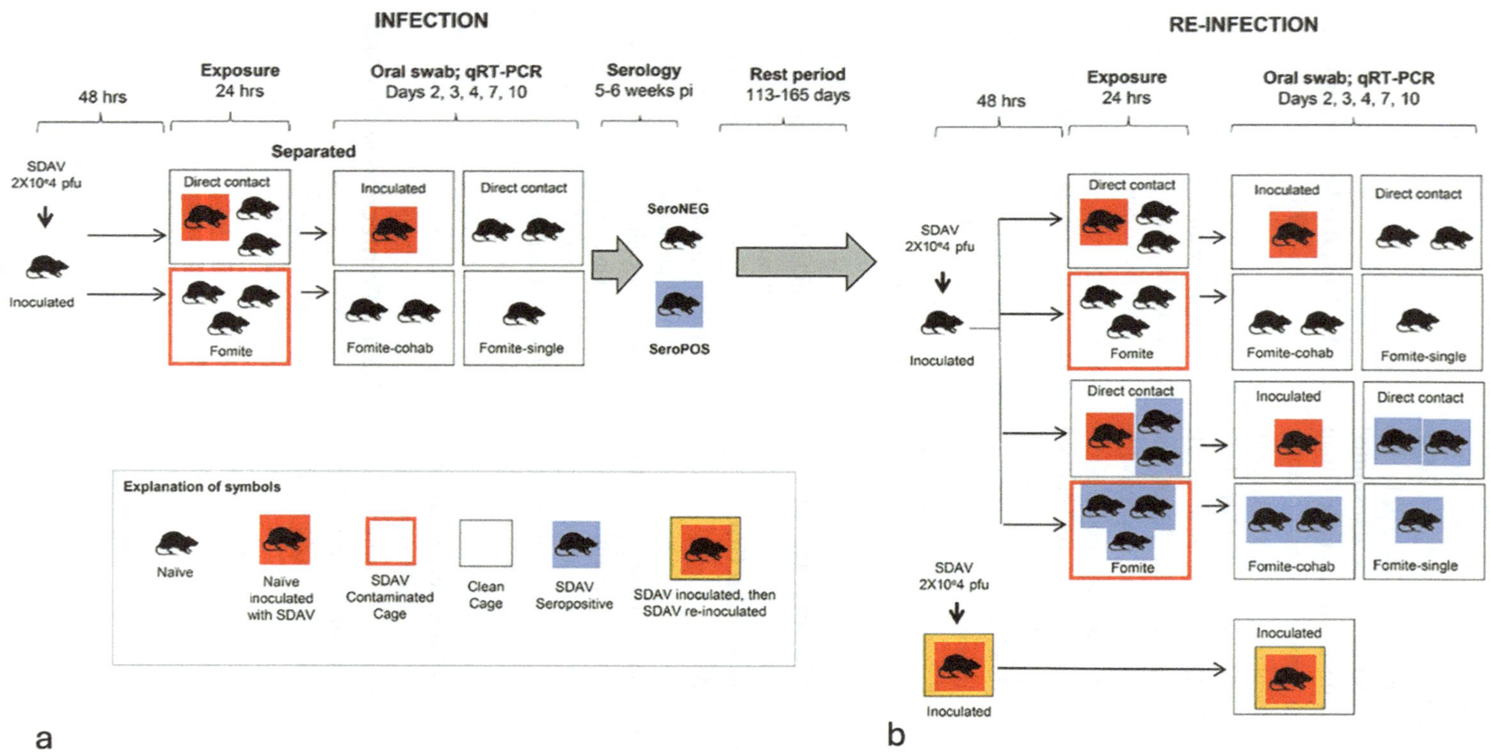

Fig. 2.1 **a** SDAV infection paradigm, **b** SDAV re-infection paradigm (Note: pfu = plaque forming units; pi = post-infection (inoculation or transmission))

(n = 26), and fomite contact-singly housed (n = 16) contact groups for their second exposure. Time between initial and second exposure ranged from 113 to 165 days. Oral swabs were taken on Day 2, 3, 4, 7, and 10 days post-exposure (dpe) on all animals. Animals were sacrificed at 10 dpe and assessed for seroconversion. All groups were evenly split by sex.

Results of these experiments have been previously reported (18). Briefly, on initial exposure, 74% of rats shed virus across all exposure groups. Amount and duration of viral shedding and seroconversion following initial natural infection was heterogeneous and influenced by route of exposure. Those animals that had been inoculated, or experienced high-risk close contact transmission, shed more virus for a longer period of time and seroconverted more reliably. Of 105 rats used to model heterogeneous natural infection and immunity, 66 were seropositive, based on their initial high- or low-risk exposures. After 113–165 days, rats were randomly assigned to high- and low-risk groups for their second natural exposure. Subsequent testing paradigms were identical to those described for initial reinfection. As expected, reinfection of rats that emerged without seroconverting after the initial infection occurred at similar rates to the original infection. Importantly, a majority (59%) of naturally infected seropositive rats shed virus on re-exposure after 3.7–5.5 months, consistent with waning immunity.

Data from naturally SDAV-infected and re-exposed animals were compared to those from animals exposed to SDAV after infection with Parker's Rat Coronavirus (RCV), a closely related coronavirus (Weir et al. 1990; Barker et al. 1994). RCV elicits protective cross-immunity to SDAV (Bihun and Percy 1994) and was used in this context as a heterologous vaccine for SDAV. To create the vaccinated group, naïve rats were inoculated intranasally with RCV, and seroconversion confirmed one month later. RCV-vaccinated rats were challenged by intranasal inoculation with SDAV six weeks later, and viral shedding assessed as described above. After vaccination, shedding occurred in a third of animals following SDAV exposure. In both groups exposed to SDAV (after either initial SDAV exposure and recovery, or RCV vaccination), viral shedding occurred at lower levels and for shorter duration following reinfection.

Next, these data were used to generate estimates for compartmental models (Mullin et al. 2022).

2.4 Compartmental Models for Pandemic to Endemic Transition of COVID-19

Using deterministic compartmental models, our overall goal was to model the combined effects of variable transmission rates, variable duration of immunity, successive waves of variants, and vaccination on patterns of SARS-CoV-2 transmission in humans. We focused on extending the basic SIR (susceptible infectious–recovered) model using ordinary differential equations (Hethcote 1989). We chose to model this deterministically, as opposed to stochastically, due to the sheer volume of the human population impacted by SARS-CoV-2. To mirror SARS-CoV-2 in humans, we added an exposed, E, compartment to the SIR model given an incubation period between a susceptible person becoming infected and shedding the virus (Rai et al. 2021). Epidemiologic compartmental models divide the population into compartments of disjoint classes, which change within a time t, represented by a day in our models (Hethcote 1989). We developed two models in which we focused first on effects of direct and indirect transmission, waning immunity, and seroconversion (Model 1). Next, we included the effects of vaccination (Model 2). For simplicity, model assumptions were the following: (1) no demographic effects, meaning the population is constant, (2) homogeneous mixing, and (3) the population size assumed constant without birth or natural death (89). The compartmental model diagrams for Models 1 and 2, and parameter definitions are given in Mullin et al. (2022).

2.4.1 Translational Considerations When Determining Parameter Estimates

When using experimental rat-derived estimates to model SARS-CoV-2 transmission in humans, we accounted for similarities and differences between transmission of SDAV and SARS-CoV-2. As in our experimental SDAV transmission paradigm, humans experience high-risk (H) and low-risk (L) exposure (Leung 2021) to SARS-CoV-2. However, in contrast to SARS-CoV-2 propagation in human populations, our entire rat population was exposed within a narrow time-period (two months) during initial infection, followed by re-exposure of the entire population approximately four months later. As in our rats, seroconversion following exposure is variable in humans (Masiá et al. 2021; Feldman and Mishra 2019). Timed exposure and repeated testing using a sensitive method (quantitative reverse transcription polymerase chain reaction (qRT-PCR)) after SDAV infection (Zeiss et al. 2021) allowed us to detect viral shedding rates on re-exposure that were much higher than those reported in humans (Hansen et al. 2021) in which substantial infection was undetected (Li et al. 2020).

Waning immunity and reinfection have been widely reported for SARS-CoV-2 naturally acquired and vaccination immunity (Goldberg et al. 2021; Levin et al. 2021), and are used as markers of epidemic emergence and persistence (Feldman and Mishra 2019; Hamami et al. 2017). Like SARS-CoV-2 infection (Spicer et al. 2022) or vaccination (Goldberg et al. 2021; Levin et al. 2021; Eyre et al. 2022), prior SDAV infection or RCV heterologous vaccination (Bihun and Percy 1994) did not result in sterilizing immunity. Our assumption was that immunity achieved by RCV was close to that achieved by natural SDAV infection, and that immunity declined at a similar rate (Bihun and Percy 1994). We incorporated our vaccination compartment similarly to previous models (Ghostine et al. 2021). Waning immunity following vaccination was assumed to occur at the same rate as that imparted by natural infection.

The incubation period between exposure and viral shedding differed slightly between SDAV (approximately two days) (Zeiss et al. 2021) and SARS-CoV-2 variants. In the latter, incubation times ranged from three to four days with Omicron and Delta variants, respectively (Brandal et al. 2021), and five to six days with prior variants (Rai et al. 2021; Wassie et al. 2020). Duration of viral shedding from the respiratory tract was 3–10 days in SDAV (Zeiss et al. 2021); in COVID-19, this varied from an average of 8–10 days in naïve individuals post onset of symptoms (Cevik et al. 2021) to 5.5 days in vaccinated people (Kissler et al. 2021). Because the incubation period for SDAV is short (2 days), the parameter δ (duration between first exposure and first positive PCR) was increased from the observed estimate in the rat study (3.477) to up to six days (Rai et al. 2021; Wassie et al. 2020; Alene et al. 2021; Grant et al. 2022). Similarly, because viral shedding occurs for a longer period in SARS-CoV-2 patients, the parameter ψ (duration of recovery period) was increased from the observed estimate in the rat study (2.477) to 10 and 17 days (Cevik et al. 2021; Kampen et al. 2021).

We modeled transmission only, not disease; thus, modeling viral propagation in healthy often asymptomatic humans, a common means of SARS-CoV-2 transmission (Johansson et al. 2021). No deaths occurred in our rats; consequently, the population size was assumed to be constant without birth or natural death. Model parameters using transmission estimates obtained from in vivo rat data and from translationally adapted (SDAV to COVID-19) estimate ranges are given in Mullin et al., 2022 (Mullin et al. 2022).

2.4.2 Core Compartmental Curves Modeling Natural Infection (Model 1) and Vaccination (Model 2)

The compartmental curves using estimates obtained from rat and human-adapted data are shown in Fig. 2.2. Using rat-derived estimates (Model 1, natural infection, Fig. 2.2.a),

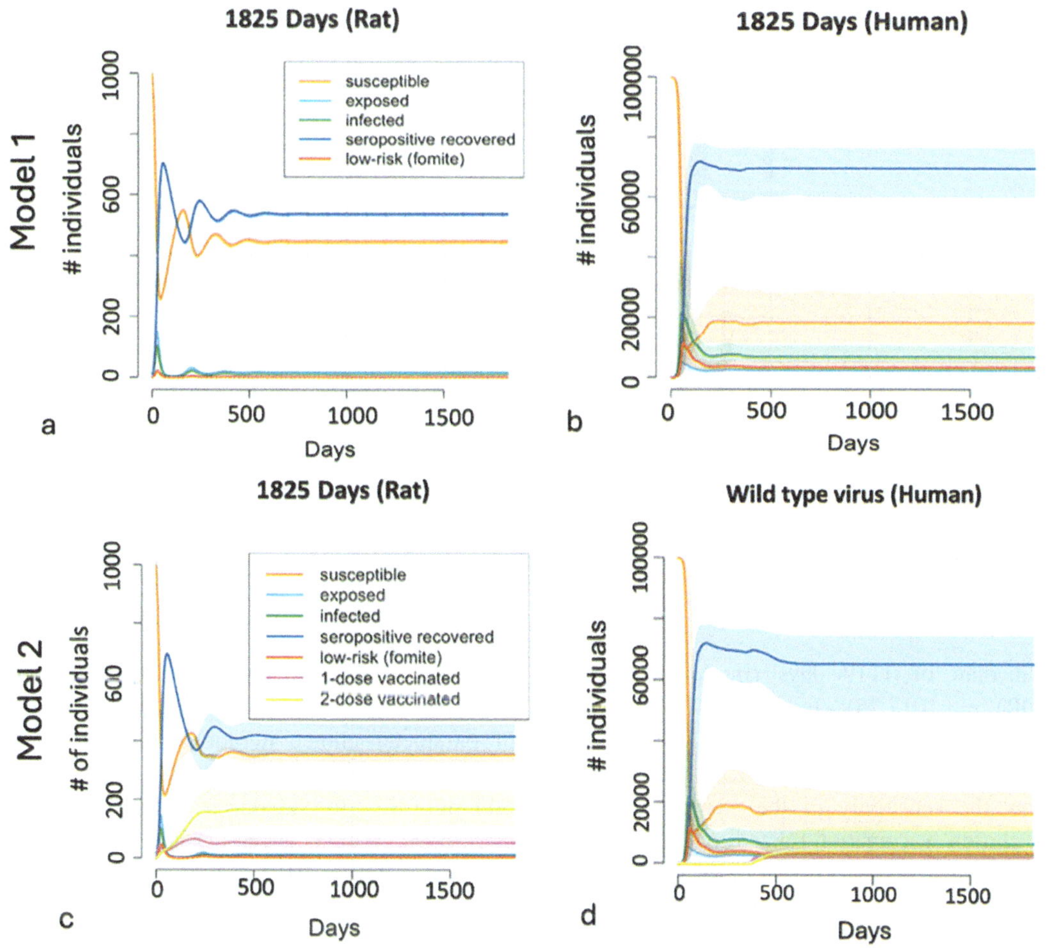

Fig. 2.2 SARS-CoV-2 Disease Compartment Curves for Model 1 (natural infection: **a, b**), and Model 2 (with vaccination: **c, d**) with predominantly in vivo rat study estimates **a, c**, and human-adapted estimates **b, d**

an endemic state achieved by natural infection occurred after approximately 724 days with approximately 41.3% of the population remaining susceptible to reinfection. Compared to SDAV-infected rats, initial spread of SARS-CoV-2 in humans prior to vaccination occurred over a longer period and involved a lower percentage of the population engaged in heterogeneous and complex mixing patterns. To account for this, we repeated the model using an infection duration of 10 to 17 days to accommodate longer reported shedding periods for SARS-CoV-2 than for SDAV, and a range of estimates for the transmission parameters β_H and β_L. This produced an estimated R0 (R-naught) ranging from 1.05 to 9.27.

Using predominantly rat data (Fig. 2.3.a), iterated for 1,825 days, Model 1 stabilizes after approximately 724 days at an endemic state with approximately 41% of the population susceptible to reinfection. Using translational human estimates (Fig. 2.3.b), iterated for 1,825 days, curves in Model 1 illustrate the median at time t (25th and 75th percentile) to show the sensitivity of the curves to our set of chosen parameters. When an analysis of endemic stability across all parameter choices is performed, endemic stability is reached at

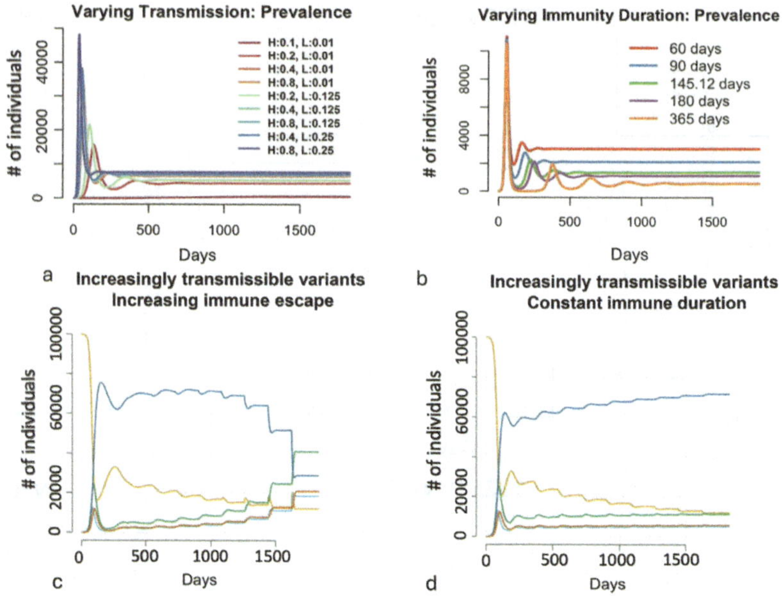

Fig. 2.3 Effects of varying transmission (**a**), immune duration (**b**) and their combinations (**c, d**)

a median of 1119 days (interquartile range (IQR) = 1017.75). In Model 2, using rat-derived data (Fig. 2.3.c), curves illustrate the median at time t (25th and 75th percentile) to show the sensitivity of the curves to our set of chosen vaccination-related parameters iterated for 1,825 days. Lower rates of infections over time are observed following introduction of vaccination, with time to endemic status varied from approximately 600 to 800 days. Applying translationally adapted parameters to Model 2 in humans (Fig. 2.3.d), iterative analysis across all parameter choices predicted a slightly longer median time to reach endemic stability of 1437 days (IQR = 49.25) with vaccination, with 15.4% of the population remaining susceptible. All models assume a single ancestral variant. The compartments and associated curves are denoted in the legend. Population size (closed with no births or deaths) is given as number (#) of individuals on the Y-axis.

To provide a sensitivity analysis of our model with human translationally alternated analysis, Fig. 2.2b illustrates the variability in the compartments. This plot denotes the median at time t for each compartment bounded by the 25th and 75th percentiles based on a grid-search of human-adapted parameters, which resulted in 320 plausible combinations. A total of 85.6% of parameter choice iterations reach endemic stability before 5 years with 318 days as the minimum time to endemic stability ($\beta_H = 0.8$, $m^+ = 90$, $\delta = 3$, $\psi = 17$) and a median of 1119 days (IQR = 1017.75).

2.4.3 Assessing the Effects of Vaccination

We modeled both one- and two-dose vaccination strategies using estimates obtained from SARS-CoV-2 vaccination reports in humans (Mathieu et al. 2021; Kriss et al. 2021). When these estimates were included in Model 2 using predominantly rat data (Fig. 2.2c), lower rates of infections over time were observed following introduction of vaccination. The susceptible compartment sharply decreased and only rose slightly due to immune state conferred by vaccination and through natural infection with seropositivity. The cyclic nature of infection was tempered by vaccination with lower peaks of infection across time. Addition of heterologous vaccination using RCV (Model 2) attenuated infection peaks. Time to endemic status varied from approximately 600 to 800 days, influenced by vaccine uptake and efficacy. Applying translationally adapted parameters to Model 2 in humans (Fig. 2.2d), iterative analysis across all parameter choices (including successive variants) predicted a slightly longer median time to reach endemic stability of 1437 days (IQR = 749.25) with vaccination, with 15.4% of the population remaining susceptible. This model used vaccination as an intervention to reduce the chance of infection and reduce transmission (flatten the curve), leading to a median difference of 2.16 (IQR = 2.32) between the basic reproduction number and the effective reproduction number (Re). However, immunity is modeled as transient and repeated reinfection can occur. Therefore, endemic stability is pushed farther out with a lower percentage (%) infected at the time of endemic stability for the model with vaccination (Model 2: 6.75% (IQR 6.3%), Model 1: 7.56% (IQR 5.65%)).

2.4.4 Introducing Successive Variants and Variable Immunity

In humans, variants of concern such as Alpha, Beta, Gamma, Delta, and Omicron emerged with the spread of these variants having a combination of enhanced transmissibility and evasion of immunity elicited by prior variants (Fan et al. 2022; Callaway 2021).

First, we assessed the effects of increased transmissibility and immune evasion separately (Fig. 2.3a, b). Greater transmissibility shortens the time to endemic stability, however at the cost of higher ongoing transmission. Immunity following infection (Kojima and Klausner 2022) or vaccination (Ferdinands et al. 2022) for SARS-CoV-2 persists for a variable period thus rendering a proportion of individuals susceptible to reinfection. In Fig. 2.3b, we provide different estimates of immunity duration. Our model shows that the longer infection-induced immunity lasts, the lower and more spread out the peaks of infection remain. Our longest modeled duration of immunity (365 days) generated a seasonal pattern similar to that seen with endemic coronaviruses causing seasonal colds (Kissler et al. 2020). When duration of immunity was varied, incremental accumulation of immunity occurred even with relatively shorter duration of immunity. Immunity lasting less than 365 days resulted in 15–20% of the population remaining susceptible to reinfection once the endemic state was reached.

Next, we extended human models to include repeated waves of reinfection by introducing a more transmissible virus every six months, progressively increasing β_H and β_L. Because successive variants demonstrate partial immune escape (Davis et al. 2021; Ma et al. 2022), we combined each introduction of a more transmissible virus with decreasing number of days of natural immunity. This resulted in reversal of infected and seropositive populations (Fig. 2.3c). In Fig. 2.3d, we presented a model which also accommodated increased viral transmissibility, but in which immune duration was stable. In this scenario, accumulation of a seropositive recovered population eventually resulted in endemic stability.

Figure 2.3 graphs a, b depict assessing prevalence estimates with varying combinations of high-risk (H) and low-risk (L) transmission rates (Fig. 2.3a), and with varying immunity duration above and below rat-derived estimate of 145.12 days (Fig. 2.3b).

Figure 2.3 graphs c, d depict assessing an introduction of variants into our SARS-CoV-2 Disease Compartment Curves with human estimates and introducing increasingly transmissible variants at 180-day intervals. In (Fig. 2.3c), we modeled decreasing length of natural immune protection from reinfection, whereas in (Fig. 2.3d), immune protection was held constant. The compartments and associated curves were denoted as follows: susceptible (yellow; S), exposed (light blue; E), infected (green; I), seropositive recovered (dark blue; R),and low-risk/fomite (orange; L). Population size (closed with no births or deaths) is given as the number (#) of individuals on the Y-axis.

2.5 Comparison of Model Predictions with 2024 SARS-CoV-2 Infection Patterns

After accounting for translationally altered parameters between rat-derived data and human SARS-CoV-2 transmission, and after introducing vaccination, we arrived at a median time to endemic stability of 1437 days (IQR = 749.25) from the beginning of the pandemic with a median of 15.4% of the population remaining susceptible. Starting from January 11, 2020 (Fig. 2.4), this would position approximate pandemic to endemic transition to December 2023. In the United States (Centers for Disease Control 2024), test positivity assumed a biphasic pattern in 2022–2024 (Wiemken et al. 2023; Cohen 2024). The Centers for Disease Control and Prevention (CDC) (Centers for Disease Control 2024) recorded highest mean national peaks occurring in the summer (July and August, 13–14% positivity) and a second slightly lower peak in the winter (December and January, 10–13% positivity). There is regional variation with recent late summer peaks up to 18.5% (Iowa, Kansas, Missouri, and Nebraska). These data are in line with our predictions, although test positivity rates in early summer (April to June, 3–4% positivity) and early fall (October and November, 6.5–9%) are lower. Due to broad adoption of home testing and lack of mandatory reporting, official COVID-19 case counts increasingly underestimate the number of true infections (Park et al. 2023).

Applying translationally adapted parameters to Model 2 in humans predicted a median time to reach endemic stability of 1,437 days (IQR = 749.25) with vaccination (18, 2023, indicated by the red bar), with 15.4% of the population remaining susceptible (Adapted from Centers for Disease Control (2024)).

With the emergence of a large number of variants, the World Health Organization (WHO) and U.S. Department of Health and Human Services SARS-CoV-2 Interagency Group (SIG) established a classification system for distinguishing the emerging variants of SARS-CoV-2. The SIG divides these into variants of high consequence (VOHC), variants of concern (VOC), variants of interest (VOI), and variants being monitored (VBM) based on decreasing transmissibility, vaccine evasion, or pathogenicity (Centers for Disease Control 2025). Current circulating strains in 2024 fall into the VOI and VBM categories, although assignment of individual viruses to one or the other of these categories may vary across the CDC (Centers for Disease Control 2025) and WHO (World Health Organization 2024). The lack of increasing infection surges such as that seen with the introduction of the Omicron variant is consistent with our model in which duration of population immunity was held stable in the face of increasingly transmissible variants. This

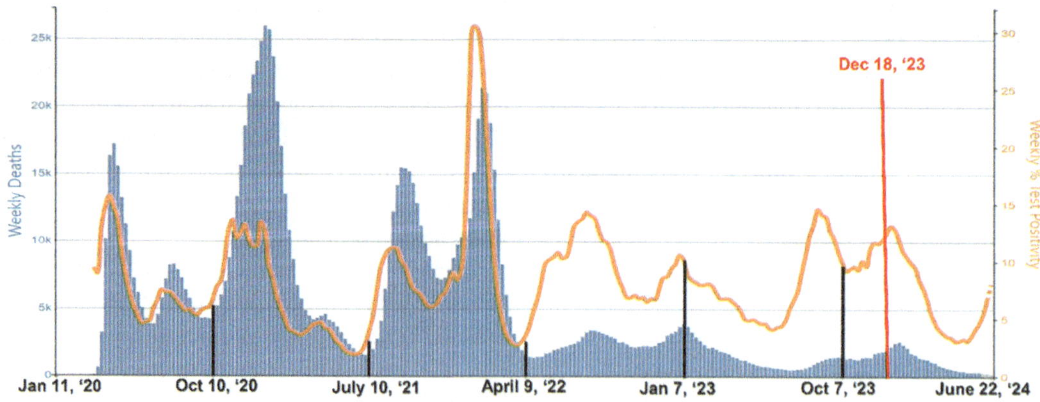

Fig. 2.4 Weekly COVID-19 death (blue) and test positivity (orange) rates in the United States

model is more consistent with an emerging pattern of actual infection rates (Fig. 2.4) and implies that transmission of new variants can be blunted by relatively stable retention of heterogeneous population immunity created by vaccination and cross-protective immunity accompanying successive waves of natural infection (Bushman et al. 2021; Hall et al. 2022; Wang et al. 2024).

Vaccine uptake declined as of 2024, particularly among younger people in the United States (Higgins et al. 2025). However, because most Americans have developed some immunity to the virus (Carreño et al. 2024), COVID-19 deaths have also greatly declined (Cohen 2024) (Fig. 2.4), thus resulting in the perception that COVID-19 presents a lower threat now than during the pandemic. Anecdotal reports of vaccine associated adverse events are quite widespread, but appear relatively rare in controlled studies (Elias et al. 2023; Lamprinou et al. 2023; Wise 2024).

2.6 Recommendations

The perception that COVID-19 presents a lower threat in 2024 than during the pandemic has begged the question of whether COVID-19 vaccines are still needed (Nohynek and Wilder-Smith 2022; Gong et al. 2023; Moghadas et al. 2021). We believe that vaccination and surveillance remain essential defenses against SARS-CoV-2. Vaccination greatly reduced morbidity and mortality associated with transition to global vaccine and natural infection-induced herd immunity to SARS-CoV-2 (Macchia et al. 2021; Pušnik et al. 2023). While deaths have declined, reinfection rates remain quite robust even at endemic equilibrium. Our variant scenarios indicate that at endemic equilibrium, combined vaccination and natural immunity maintains a much lower infection prevalence compared to ineffective or absent vaccination. Breakthrough infections by new variants can variably augment vaccine immunity (Bormann et al. 2023; Stein et al. 2023; Hornsby et al. 2023), thus creating an uneasy balance between population immunity and the constant threat of an emerging variant that may result in a transmission surge or more severe disease. As immunity due to vaccination and previous infection with SARS-CoV-2 wanes, there is a risk of reinfection in an increasing proportion of susceptible individuals. This is particularly germane to younger individuals in whom vaccine uptake is the lowest. The accumulation of SARS-CoV-2 naive infants could increase the proportion of susceptible individuals and further impact immunity at the population level (Li and Zhu 2024).

Whether COVID-19 is fated to become a relatively benign seasonal respiratory infection similar to seasonal cold viruses such as HCoV-OC43 remains to be seen. Historical precedent for this evolutionary trajectory has been proposed for three of the four globally endemic common cold viruses, HCoV-229E (Corman et al. 2018), HCoV-OC43 (Vijgen et al. 2005), and HCoV-HKU1 (Forni et al. 2021). The observed phenomenon of increased transmissibility, reduced morbidity, and altered tissue tropism demonstrated by the Omicron variant (Bushman et al. 2021; Lundberg et al. 2022) has been previously noted in other species. In pigs, transmissible gastroenteritis virus, an endemic gastrointestinal coronavirus discovered in 1946 (Doyle and Hutchings 1946), underwent a 600 base pair (bp) deletion in its S gene (Wesley et al. 1991) to generate a new, more transmissible but milder variant with respiratory tropism (porcine respiratory coronavirus) that elicited cross-protective immunity to the original virus (Pensaert et al. 1993). However, alternative insights can be gained from observations in longstanding endemic domestic animal coronaviral disease (Turlewicz-Podbielska and Pomorska-Mól 2021; Xiong et al. 2024; Rani et al. 2024; Cuadrado et al. 2024; Attipa et al. 2023; Meir et al. Sep 2004; Read et al. 2015). Non-sterilizing host immunity, viral genomic plasticity and immune escape (Franzo et al. 2019; Rani et al. 2024; Cuadrado et al. 2024) are at the root of persistent circulation of these viruses within their hosts, thus creating a dynamic array of co-circulating strains. These may manifest clinically as outbreaks of more severe disease (Attipa et al. 2023), or variant disease due to altered tissue tropism (Meir et al. 2004). Recently, in 2023, an outbreak of feline infectious peritonitis was associated with emergence of a novel highly pathogenic variant due to a recombination event between endemic canine and feline coronavirus strains (Attipa et al. 2023). A long history of vaccine development against infectious bronchitis in chickens provides insight into the effect of vaccination on viral evolution (Franzo et al. 2019; Read et al. 2015). As likely occurred with SARS-CoV-2, domestic animals are also susceptible to the propensity of coronaviruses for cross-species transmission. In 2018, a novel HKU2-related bat coronavirus with broad in vitro host tropism (Yang et al. 2019) resulted in a fatal swine enteric syndrome in China (Zhou et al. 2018).

These examples, and the broad range of animal species infected by SARS-CoV-2 (Nerpel et al. 2022), underscore widespread cross-species virus circulation and genomic plasticity characteristics of the vertebrate coronaviral landscape. Consequently, coordinated surveillance and ongoing collection of sequence data of coronaviruses in humans and non-human animals (both domestic and wild) will be an essential element for early detection of emergent variants with pandemic potential. Because surveillance is resource intensive and inevitably non-representative of an entire landscape, epidemiologic modeling becomes another essential tool to predict potential animal–human transmission. Such modeling relies upon surveillance data and can be supplemented by translationally adapted experimental data derived from other animal species, as shown in this study.

References

Alene M, Yismaw L, Assemie MA, Ketema DB, Gietaneh W, Birhan TY (2021) Serial interval and incubation period of COVID-19: a systematic review and meta-analysis. BMC Infect Dis 21:257. https://doi.org/10.1186/s12879-021-05950-x

Arons MM, Hatfield KM, Reddy SC, Kimball A, James A, Jacobs JR et al (2020) Presymptomatic SARS-CoV-2 infections and transmission in a skilled nursing facility. N Engl J Med. https://doi.org/10.1056/NEJMoa2008457

Attipa C, Warr AS, Epaminondas D, O'Shea M, Fletcher S, Malbon A et al (2023) Emergence and spread of feline infection peritonitis due to a highly pathogenic canine/feline recombinant coronavirus. bioRxiv

Barker MG, Percy DH, Hovland DJ, MacInnes JI (1994) Preliminary characterization of the structural proteins of the coronaviruses, sialodacryoadenitis virus and Parker's rat coronavirus. Can J Vet Res 58:99–103

Bhatt PN, Jacoby RO (1985) Epizootiological observations of natural and experimental infection with sialodacryoadenitis virus in rats. Lab Anim Sci 35:129–134

Bhatt PN, Percy DH, Jonas AM (1972) Characterization of the virus of sialodacryoadenitis of rats: a member of the coronavirus group. J Infect Dis 126:123–130. https://doi.org/10.1093/infdis/126.2.123

Bhuiyan MSA, Amin Z, Rodrigues KF, Saallah S, Shaarani SM, Sarker S et al (2021) Infectious bronchitis virus (gammacoronavirus) in poultry farming: vaccination, immune response and measures for mitigation. Vet Sci 8. https://doi.org/10.3390/vetsci8110273

Bihun CG, Percy DH (1994) Coronavirus infections in the laboratory rat: degree of cross protection following immunization with a heterologous strain. Can J Vet Res 58:224–229

Bihun CG, Percy DH (1995) Morphologic changes in the nasal cavity associated with sialodacryoadenitis virus infection in the Wistar rat. Vet Pathol 32:1–10. https://doi.org/10.1177/030098589503200101

Bormann M, Brochhagen L, Alt M, Otte M, Thümmler L, van de Sand L et al (2023) Immune responses in COVID-19 patients during breakthrough infection with SARS-CoV-2 variants Delta, Omicron-BA.1 and Omicron-BA.5. Front Immunol 14:1150667. https://doi.org/10.3389/fimmu.2023.1150667

Brandal LT, MacDonald E, Veneti E, Ravlo T, Lange H, Naseer U et al (2021) Outbreak caused by the SARS-CoV-2 omicron variant in Norway, November to December 2021. Euro Surveill 26. https://doi.org/10.2807/1560-7917.ES.2021.26.50.2101147

Bushman M, Kahn R, Taylor BP, Lipsitch M, Hanage WP (2021) Population impact of SARS-CoV-2 variants with enhanced transmissibility and/or partial immune escape. Cell 184:6229–6242. https://doi.org/10.1016/j.cell.2021.11.026

Callaway E (2021) Beyond Omicron: what's next for COVID's viral evolution. Nature 600:204–207. https://doi.org/10.1038/d41586-021-03619-8

Carreño JM, Wagner AL, Monahan B, Singh G, Floda D, Gonzalez-Reiche AS et al (2024) SARS-CoV-2 serosurvey across multiple waves of the COVID-19 pandemic in New York City between 2020–2023. Nat Commun 15:5847. https://doi.org/10.1038/s41467-024-50052-2

Centers for Disease Control. https://covid.cdc.gov/covid-data-tracker. Accessed 22 June 2024

Centers for Disease Control. https://www.cdc.gov/covid/php/interagency-group/index.html. Accessed 11 April 2025

Cevik M, Tate M, Lloyd O, Maraolo AE, Schafers J, Ho A (2021) SARS-CoV-2, SARS-CoV, and MERS-CoV viral load dynamics, duration of viral shedding,

and infectiousness: a systematic review and meta-analysis. Lancet Microbe 2:e13–e22. https://doi.org/10.1016/S2666-5247(20)30172-5
Cohen J (2024) COVID-19 is surging again-with far fewer serious cases. Science 385:814–815
Corman VM, Muth D, Niemeyer D, Drosten C (2018) Hosts and sources of endemic human coronaviruses. Adv Virus Res 100:163–188. https://doi.org/10.1016/bs.aivir.2018.01.001
Cuadrado C, Breedlove C, van Santen E, Joiner KS, van Santen VL, Toro H (2024) Protection against infectious bronchitis virus vaccine recombinants and chicken-selected vaccine subpopulations. Avian Dis 68:89–98. https://doi.org/10.1637/aviandiseases-D-23-00064
Dashtbali M, Mirzaie M (2021) A compartmental model that predicts the effect of social distancing and vaccination on controlling COVID-19. Sci Rep 11:8191. https://doi.org/10.1038/s41598-021-86873-0
Davis C, Logan N, Tyson G, Orton R, Harvey WT, Perkins JS et al (2021) Reduced neutralisation of the Delta (B.1.617.2) SARS-CoV-2 variant of concern following vaccination. PLoS Pathog 17:e1010022. https://doi.org/10.1371/journal.ppat.1010022
Doyle LP, Hutchings LM (1946) A transmissible gastroenteritis in pigs. J Am Vet Med Assoc 108:257–259
Edridge AWD, Kaczorowska J, Hoste ACR, Bakker M, Klein M, Loens K et al (2020) Seasonal coronavirus protective immunity is short-lasting. Nat Med 26:1691–1693. https://doi.org/10.1038/s41591-020-1083-1
Eisenbrandt DL, Hubbard GB, Schmidt RE (1982) A subclinical epizootic of sialodacryoadenitis in rats. Lab Anim Sci 32:655–659
Elias MD, Truong DT, Oster ME, Trachtenberg FL, Mu X, Jone PN et al (2023) Examination of adverse reactions after COVID-19 vaccination among patients with a history of multisystem inflammatory syndrome in children. JAMA Netw Open 6:e2248987. https://doi.org/10.1001/jamanetworkopen.2022.48987
Eyre DW, Taylor D, Purver M, Chapman D, Fowler T, Pouwels KB et al (2022) Effect of Covid-19 vaccination on transmission of alpha and delta variants. N Engl J Med. https://doi.org/10.1056/NEJMoa2116597
Fan B, Jiao D, Zhang R, Zhou J, Guo R, Yu Z et al (2020) Origin and epidemic status of porcine epidemic diarrhea virus variants in China. Transbound Emerg Dis 67:1364–1370. https://doi.org/10.1111/tbed.13444
Fan Y, Li X, Zhang L, Wan S, Zhang L, Zhou F (2022) SARS-CoV-2 Omicron variant: recent progress and future perspectives. Signal Transduct Targeted Therapy 7:141. https://doi.org/10.1038/s41392-022-00997-x
Feldman J, Mishra S (2019) What could re-infection tell us about R(0)? A modeling case-study of syphilis transmission. Infect Dis Model 4:257–264. https://doi.org/10.1016/j.idm.2019.09.002
Ferdinands JM, Rao S, Dixon BE, Mitchell PK, DeSilva MB, Irving SA et al (2022) Waning 2-dose and 3-dose effectiveness of mRNA vaccines against COVID-19-associated emergency department and urgent care encounters and hospitalizations among adults during periods of delta and omicron variant predominance—VISION Network, 10 States, August 2021-January 2022. MMWR Morb Mortal Wkly Rep 71:255–263. https://doi.org/10.15585/mmwr.mm7107e2
Forni D, Cagliani R, Arrigoni F, Benvenuti M, Mozzi A, Pozzoli U et al (2021) Adaptation of the endemic coronaviruses HCoV-OC43 and HCoV-229E to the human host. Virus Evol 7:veab061. https://doi.org/10.1093/ve/veab061
Franzo G, Legnardi M, Tucciarone CM, Drigo M, Martini M, Cecchinato M (2019) Evolution of infectious bronch0itis virus in the field after homologous vaccination introduction. Vet Res 50:92. https://doi.org/10.1186/s13567-019-0713-4
Funk CJ, Manzer R, Miura TA, Groshong SD, Ito Y, Travanty EA et al (2009) Rat respiratory coronavirus infection: replication in airway and alveolar epithelial cells and the innate immune response. J Gen Virol 90:2956–2964. https://doi.org/10.1099/vir.0.014282-0
Gaertner DJ, Compton SR, Winograd DF (1993) Environmental stability of rat coronaviruses (RCVs). Lab Anim Sci 43:403–404
Ghostine R, Gharamti M, Hassrouny S, Hoteit I (2021) An extended SEIR model with vaccination for forecasting the COVID-19 pandemic in Saudi Arabia using an ensemble Kalman filter. Mathematics 9:636. https://doi.org/10.3390/math9060636
Giordano G, Colaneri M, Di Filippo A, Blanchini F, Bolzern P, De Nicolao G et al (2021) Modeling vaccination rollouts, SARS-CoV-2 variants and the requirement for non-pharmaceutical interventions in Italy. Nat Med 27:993–998. https://doi.org/10.1038/s41591-021-01334-5
Goldberg Y, Mandel M, Bar-On YM, Bodenheimer O, Freedman L, Haas EJ et al (2021) Waning immunity after the BNT162b2 vaccine in Israel. N Engl J Med 385:e85. https://doi.org/10.1056/NEJMoa2114228
Gong W, Parkkila S, Wu X, Aspatwar A (2023) SARS-CoV-2 variants and COVID-19 vaccines: current challenges and future strategies. Int Rev Immunol 42:393–414. https://doi.org/10.1080/08830185.2022.2079642
Grant R, Charmet T, Schaeffer L, Galmiche S, Madec Y, Von Platen C et al (2022) Impact of SARS-CoV-2 Delta variant on incubation, transmission settings and vaccine effectiveness: results from a nationwide case-control study in France. Lancet Reg Health Eur 13:100278. https://doi.org/10.1016/j.lanepe.2021.100278
Hall V, Foulkes S, Insalata F, Kirwan P, Saei A, Atti A et al (2022) Protection against SARS-CoV-2 after Covid-19 vaccination and previous infection. N Engl J Med. https://doi.org/10.1056/NEJMoa2118691
Hamami D, Cameron R, Pollock KG, Shankland C (2017) Waning immunity is associated with periodic large outbreaks of mumps: a mathematical modeling study of Scottish data. Front Physiol 8:233. https://doi.org/10.3389/fphys.2017.00233
Hansen CH, Michlmayr D, Gubbels SM, Mølbak K, Ethelberg S (2021) Assessment of protection against reinfection with SARS-CoV-2 among 4 million PCR-tested individuals in Denmark in 2020: a population-level

observational study. Lancet 397:1204–1212. https://doi.org/10.1016/S0140-6736(21)00575-4
Hethcote HW (1989) Three basic epidemiological models. In: Levin SA, Hallam TG, Gross LJ (eds) Applied mathematical ecology. Springer, Berlin, pp 119–144
Higgins DM, Bryan MA, O'Leary ST, Opel DJ (2025) The impact of the COVID-19 pandemic on parental childhood vaccine attitudes in the US. Acad Pediatr 102829. https://doi.org/10.1016/j.acap.2025.102829
Hornsby H, Nicols AR, Longet S, Liu C, Tomic A, Angyal A et al (2023) Omicron infection following vaccination enhances a broad spectrum of immune responses dependent on infection history. Nat Commun 14:5065. https://doi.org/10.1038/s41467-023-40592-4
Hu B, Guo H, Zhou P, Shi ZL (2021) Characteristics of SARS-CoV-2 and COVID-19. Nat Rev Microbiol 19:141–154. https://doi.org/10.1038/s41579-020-00459-7
Hu S, Wang W, Wang Y, Litvinova M, Luo K, Ren L et al (2021) Infectivity, susceptibility, and risk factors associated with SARS-CoV-2 transmission under intensive contact tracing in Hunan, China. Nat Commun 12:1533. https://doi.org/10.1038/s41467-021-21710-6
Johansson MA, Quandelacy TM, Kada S, Prasad PV, Steele M, Brooks JT et al (2021) SARS-CoV-2 transmission from people without COVID-19 symptoms. JAMA Netw Open 4:e2035057. https://doi.org/10.1001/jamanetworkopen.2020.35057
van Kampen JJA, van de Vijver D, Fraaij PLA, Haagmans BL, Lamers MM, Okba N et al (2021) Duration and key determinants of infectious virus shedding in hospitalized patients with coronavirus disease-2019 (COVID-19). Nat Commun 12:267. https://doi.org/10.1038/s41467-020-20568-4
Kenney SP, Wang Q, Vlasova A, Jung K, Saif L (2021) Naturally occurring animal coronaviruses as models for studying highly pathogenic human coronaviral disease. Vet Pathol 58:438–452. https://doi.org/10.1177/0300985820980842
Kissler SM, Tedijanto C, Goldstein E, Grad YH, Lipsitch M (2020) Projecting the transmission dynamics of SARS-CoV-2 through the postpandemic period. Science 368:860–868. https://doi.org/10.1126/science.abb5793
Kissler SM, Fauver JR, Mack C, Tai CG, Breban MI, Watkins AE et al (2021) Viral dynamics of SARS-CoV-2 variants in vaccinated and unvaccinated persons. N Engl J Med 385:2489–2491. https://doi.org/10.1056/NEJMc2102507
Kojima N, Klausner JD (2022) Protective immunity after recovery from SARS-CoV-2 infection. Lancet Infect Dis 22:12–14. https://doi.org/10.1016/S1473-3099(21)00676-9
Kraay ANM, Hayashi MAL, Berendes DM, Sobolik JS, Leon JS, Lopman BA (2021) Risk for fomite-mediated transmission of SARS-CoV-2 in child daycares, schools, nursing homes, and offices. Emerg Infect Dis 27:1229–1231. https://doi.org/10.3201/eid2704.203631
Kriss JL, Reynolds LE, Wang A, Stokley S, Cole MM, Harris LQ et al (2021) COVID-19 vaccine second-dose completion and interval between first and second doses among vaccinated persons—United States, December 14, 2020–February 14, 2021. MMWR Morb Mortal Wkly Rep 70:389–395. https://doi.org/10.15585/mmwr.mm7011e2
La Regina M, Woods L, Klender P, Gaertner DJ, Paturzo FX (1992) Transmission of sialodacryoadenitis virus (SDAV) from infected rats to rats and mice through handling, close contact, and soiled bedding. Lab Anim Sci 42:344–346
Lamprinou M, Sachinidis A, Stamoula E, Vavilis T, Papazisis G (2023) COVID-19 vaccines adverse events: potential molecular mechanisms. Immunol Res 71:356–372. https://doi.org/10.1007/s12026-023-09357-5
Lee C (2015) Porcine epidemic diarrhea virus: an emerging and re-emerging epizootic swine virus. Virol J 12:193. https://doi.org/10.1186/s12985-015-0421-2
Leung NHL (2021) Transmissibility and transmission of respiratory viruses. Nat Rev Microbiol 19:528–545. https://doi.org/10.1038/s41579-021-00535-6
Levin EG, Lustig Y, Cohen C, Fluss R, Indenbaum V, Amit S et al (2021) Waning immune humoral response to BNT162b2 Covid-19 vaccine over 6 months. N Engl J Med 385:e84. https://doi.org/10.1056/NEJMoa2114583
Li SM, Zhu FC (2024) The era of SARS-CoV-2 variants calls for an urgent strategy for COVID-19 vaccination in children. Lancet Infect Dis 24:666–667. https://doi.org/10.1016/S1473-3099(24)00157-9
Li R, Pei S, Chen B, Song Y, Zhang T, Yang W et al (2020) Substantial undocumented infection facilitates the rapid dissemination of novel coronavirus (SARS-CoV-2). Science 368:489–493. https://doi.org/10.1126/science.abb3221
Lin SY, Chen HW (2017) infectious bronchitis virus variants: molecular analysis and pathogenicity investigation. Int J Mol Sci 18. https://doi.org/10.3390/ijms18102030
Lundberg AL, Lorenzo-Redondo R, Ozer EA, Hawkins CA, Hultquist JF, Welch SB et al (2022) Has omicron changed the evolution of the pandemic? JMIR Public Health Surveill 8:e35763. https://doi.org/10.2196/35763
Ma C, Chen X, Mei F, Xiong Q, Liu Q, Dong L et al (2022) Drastic decline in sera neutralization against SARS-CoV-2 omicron variant in Wuhan COVID-19 convalescents. Emerg Microbes Infect 11:567–572. https://doi.org/10.1080/22221751.2022.2031311
Macchia A, Ferrante D, Angeleri P, Biscayart C, Mariani J, Esteban S et al (2021) Evaluation of a COVID-19 vaccine campaign and SARS-CoV-2 infection and mortality among adults aged 60 years and older in a middle-income country. JAMA Netw Open 4:e2130800. https://doi.org/10.1001/jamanetworkopen.2021.30800
Masiá M, Telenti G, Fernández M, García JA, Agulló V, Padilla S et al (2021) SARS-CoV-2 seroconversion and

viral clearance in patients hospitalized With COVID-19: viral load predicts antibody response. Open Forum Infect Dis 8:ofab005. https://doi.org/10.1093/ofid/ofab005

Massonis G, Banga JR, Villaverde AF (2021) Structural identifiability and observability of compartmental models of the COVID-19 pandemic. Annu Rev Control 51:441–459. https://doi.org/10.1016/j.arcontrol.2020.12.001

Mathieu E, Ritchie H, Ortiz-Ospina E, Roser M, Hasell J, Appel C et al (2021) A global database of COVID-19 vaccinations. Nat Hum Behav 5:947–953. https://doi.org/10.1038/s41562-021-01122-8

Meir R, Rosenblut E, Perl S, Kass N, Ayali G, Perk S et al (2004) Identification of a novel nephropathogenic infectious bronchitis virus in Israel. Avian Dis 48:635–641. https://doi.org/10.1637/7107

Mistry D, Litvinova M, Pastore YPA, Chinazzi M, Fumanelli L, Gomes MFC et al (2021) Inferring high-resolution human mixing patterns for disease modeling. Nat Commun 12:323. https://doi.org/10.1038/s41467-020-20544-y

Moghadas SM, Vilches TN, Zhang K, Wells CR, Shoukat A, Singer BH et al (2021) The impact of vaccination on COVID-19 outbreaks in the United States. Clin Infect Dis 73:2257. https://doi.org/10.1093/cid/ciab079

Mullin S, Wyk BV, Asher JL, Compton SR, Allore HG, Zeiss CJ (2022) Modeling pandemic to endemic patterns of SARS-CoV-2 transmission using parameters estimated from animal model data. PNAS Nexus 1:pgac096. https://doi.org/10.1093/pnasnexus/pgac096

Nerpel A, Yang L, Sorger J, Käsbohrer A, Walzer C, Desvars-Larrive A (2022) SARS-ANI: a global open access dataset of reported SARS-CoV-2 events in animals. Sci Data 9:438. https://doi.org/10.1038/s41597-022-01543-8

Nohynek H, Wilder-Smith A (2022) Does the world still need new Covid-19 vaccines? N Engl J Med 386:2140–2142. https://doi.org/10.1056/NEJMe2204695

Park S, Marcus GM, Olgin JE, Carton T, Hamad R, Pletcher MJ (2023) Unreported SARS-CoV-2 home testing and test positivity. JAMA Netw Open 6:e2252684. https://doi.org/10.1001/jamanetworkopen.2022.52684

Pensaert M, Cox E, van Deun K, Callebaut P (1993) A sero-epizootiological study of porcine respiratory coronavirus in Belgian swine. Vet Q 15:16–20. https://doi.org/10.1080/01652176.1993.9694361

Percy DH, Hanna PE, Paturzo F, Bhatt PN (1984) Comparison of strain susceptibility to experimental sialodacryoadenitis in rats. Lab Anim Sci 34:255–260

Percy DH, Bond SJ, Paturzo FX, Bhatt PN (1990) Duration of protection from reinfection following exposure to sialodacryoadenitis virus in Wistar rats. Lab Anim Sci 40:144–149

Pušnik J, Monzon-Posadas WO, Zorn J, Peters K, Baum M, Proksch H et al (2023) SARS-CoV-2 humoral and cellular immunity following different combinations of vaccination and breakthrough infection. Nat Commun 14:572. https://doi.org/10.1038/s41467-023-36250-4

Rai B, Shukla A, Dwivedi LK (2021) Incubation period for COVID-19: a systematic review and meta-analysis. Z Gesundh Wiss 1–8. https://doi.org/10.1007/s10389-021-01478-1

Rani S, Khatoon M, Hiremath J, Suresh KP, Anandakumar J, Barman NN et al (2024) Decoding bovine coronavirus immune targets: an epitope informatics approach. Animal Diseases 4:15

Read AF, Baigent SJ, Powers C, Kgosana LB, Blackwell L, Smith LP et al (2015) Imperfect vaccination can enhance the transmission of highly virulent pathogens. PLoS Biol 13:e1002198. https://doi.org/10.1371/journal.pbio.1002198

Spicer KB, Glick C, Cavanaugh AM, Thoroughman D (2022) Protective immunity after natural infection with severe acute respiratory syndrome coronavirus-2 (SARS-CoV-2)—Kentucky, USA, 2020. Int J Infect Dis 114:21–28. https://doi.org/10.1016/j.ijid.2021.10.010

Stein C et al (2023) Past SARS-CoV-2 infection protection against re-infection: a systematic review and meta-analysis. Lancet 401:833–842. https://doi.org/10.1016/S0140-6736(22)02465-5

Turlewicz-Podbielska H, Pomorska-Mól M (2021) Porcine coronaviruses: overview of the state of the art. Virol Sin 36:833–851. https://doi.org/10.1007/s12250-021-00364-0

Usherwood T, LaJoie Z, Srivastava V (2021) A model and predictions for COVID-19 considering population behavior and vaccination. Sci Rep 11:12051. https://doi.org/10.1038/s41598-021-91514-7

van Doremalen N, Bushmaker T, Morris DH, Holbrook MG, Gamble A, Williamson BN et al (2020) Aerosol and surface stability of SARS-CoV-2 as compared with SARS-CoV-1. N Engl J Med 382:1564–1567. https://doi.org/10.1056/NEJMc2004973

Vijgen L, Keyaerts E, Moës E, Thoelen I, Wollants E, Lemey P et al (2005) Complete genomic sequence of human coronavirus OC43: molecular clock analysis suggests a relatively recent zoonotic coronavirus transmission event. J Virol 79:1595–1604. https://doi.org/10.1128/JVI.79.3.1595-1604.2005

Wang Q, Guo Y, Bowen A, Mellis IA, Valdez R, Gherasim C et al (2024) XBB.1.5 monovalent mRNA vaccine booster elicits robust neutralizing antibodies against XBB subvariants and JN.1. Cell Host Microbe 32:315–321. https://doi.org/10.1016/j.chom.2024.01.014

Wassie GT, Azene AG, Bantie GM, Dessie G, Aragaw AM (2020) Incubation period of severe acute respiratory syndrome novel Coronavirus 2 that causes coronavirus disease 2019: a systematic review and meta-analysis. Curr Ther Res Clin Exp 93:100607. https://doi.org/10.1016/j.curtheres.2020.100607

Weir EC, Jacoby RO, Paturzo FX, Johnson EA (1990) Infection of SDAV immune rats with SDAV and rat coronavirus. Lab Anim Sci 40:363–366

Wesley RD, Woods RD, Cheung AK (1991) Genetic analysis of porcine respiratory coronavirus, an attenuated variant of transmissible gastroenteritis virus. J Virol

65:3369–3373. https://doi.org/10.1128/JVI.65.6.3369-3373.1991
Wiemken TL, Khan F, Puzniak L, Yang W, Simmering J, Polgreen P et al (2023) Seasonal trends in COVID-19 cases, hospitalizations, and mortality in the United States and Europe. Sci Rep 13:3886. https://doi.org/10.1126/science.ads6028
Wise J (2024) Covid-19: two rare vaccine side effects detected in large global study. BMJ 384:q488. https://doi.org/10.1136/bmj.q488
Wojcinski ZW, Percy DH (1986) Sialodacryoadenitis virus-associated lesions in the lower respiratory tract of rats. Vet Pathol 23:278–286. https://doi.org/10.1177/030098588602300308
World Health Organization. https://data.who.int/dashboards/covid19/variants. Accessed 22 June 22
Xiong T, Xie H, Li L, Liang S, Huang M, Yu C et al (2024) Prevalence, genotype diversity, and distinct pathogenicity of 205 gammacoronavirus infectious bronchitis virus isolates in China during 2019–2023. Viruses 16. https://doi.org/10.3390/v16060930
Yang YL, Qin P, Wang B, Liu Y, Xu GH, Peng L et al (2019) Broad cross-species infection of cultured cells by Bat HKU2-related swine acute diarrhea syndrome coronavirus and identification of its replication in murine dendritic cells in vivo highlight its potential for diverse interspecies transmission. J Virol 93. https://doi.org/10.1128/JVI.01448-19
Yoo D, Pei Y, Christie N, Cooper M (2000) Primary structure of the sialodacryoadenitis virus genome: sequence of the structural-protein region and its application for differential diagnosis. Clin Diagn Lab Immunol 7:568–573. https://doi.org/10.1128/CDLI.7.4.568-573.2000
Zeiss C (2021) An in vivo driven SEIRS (susceptible-exposed-infectious-recovered susceptible) model of coronaviral infection. Columbia Academic Commons, Academic Unit: Northeast Big Data Innovation Hub, Series: COVID-19 Research Lightning Talks. https://doi.org/10.7916/ed19-af61
Zeiss CJ, Asher JL, Vander Wyk B, Allore HG, Compton SR (2021) Modeling SARS-CoV-2 propagation using rat coronavirus-associated shedding and transmission. PLoS One 16:e0260038. https://doi.org/10.1371/journal.pone.0260038.
Zhang R, Li Y, Zhang AL, Wang Y, Molina MJ (2020a) Identifying airborne transmission as the dominant route for the spread of COVID-19. Proc Natl Acad Sci U S A 117:14857–14863. https://doi.org/10.1073/pnas.2009637117
Zhang J, Litvinova M, Liang Y, Wang Y, Wang W, Zhao S et al (2020b) Changes in contact patterns shape the dynamics of the COVID-19 outbreak in China. Science 368:1481–1486. https://doi.org/10.1126/science.abb8001
Zhou P, Fan H, Lan T, Yang XL, Shi WF, Zhang W et al (2018) Fatal swine acute diarrhoea syndrome caused by an HKU2-related coronavirus of bat origin. Nature 556:255–258. https://doi.org/10.1038/s41586-018-0010-9

Sarah Mullin is an Assistant Professor in the Department of Biostatistics and Bioinformatics at Roswell Park Comprehensive Cancer Center and Director of the Biomedical Research Informatics Shared Resource. She is passionate about modeling information contained in various population health, epidemiologic, basic research, and clinical data sources. Mullin specializes in epidemiologic transmission modeling, including compartmental models, that synthesize public population health data sources with animal-derived estimates. She has used rat experiment-derived estimates to predict the translational direct and indirect transmission of SARS-CoV-2 in humans and is currently working on modeling SARS-CoV-2 transmission within and across multiple species. Additionally, Mullin has focused on natural language processing within the oncology domain. Her research has spanned using machine learning and deep learning in conjunction with terminological and ontological information and multiple sources of data in heterogeneous formats. Her training includes a postdoctoral fellowship at the Yale Center for Medical Informatics, a PhD in Biomedical Informatics from the SUNY University at Buffalo, and a MS in Statistics from The Ohio State University.

Susan R. Compton is a Senior Research Scientist in Comparative Medicine at the Yale School of Medicine. Her research is focused on the pathogenesis, prevention, and diagnosis of viral infections in cultured cells and in mice and rats. She has been studying mouse and rat coronavirus infections for over 40 years. Her recent research has focused on mouse and rat coronaviruses as models for COVID. She has also researched infections with rodent parvoviruses, noroviruses, astroviruses, and Seoul virus. Susan is the Chief of Molecular and Serological Diagnostics Unit in the Department of Comparative Medicine which provides testing for rodent viruses, bacteria, fungi, and parasites. Susan received her PhD in Microbiology from the Uniformed Service University of the Health Sciences where she studied murine coronavirus replication and the species specificity of coronavirus receptors. Her postdoctoral research at the University of Colorado focused on poliovirus uncoating mechanisms. She served as an editor for the American Association for Laboratory Animal Science journals for over 20 years.

Sylvester Ibemgbo is a Postdoctoral Associate at Caroline Zeiss Lab, Department of Comparative Medicine, Yale University School of Medicine. He completed his PhD in life sciences at Jawaharlal Nehru University (ICGEB-JNU), New Delhi, where he characterized arbovirus (chikungunya and dengue) infections and co-infections causing febrile illness. His work involved studying the growth patterns of different clinical isolates of these viruses, including in vitro coinfections of dengue and chikungunya. He also analyzed the neutralization capacity of antibodies elicited by chikungunya virus. Surveillance and characterization of the circulating isolates revealed different mutations in the viral genes, which were associated with disease severity. Sylvester is currently involved in the coronaviral surveillance project at the Zeiss Lab, where he tests farmed, free-ranging, wild, and domestic animals in the Northeast and generates data for SEIR models. He is also involved in BSL-3 experiments to isolate coronaviruses from animal samples and to assess the capacity of human adjacent rodents to maintain an enzoonotic SARS-CoV-2 cycle.

Caroline Zeiss is a Professor of Comparative Medicine and Ophthalmology and Visual Science at the Yale School of Medicine. She is Chief of Pathology in the Department of Comparative Medicine, in which she founded and led a Research Pathology Core for 20 years. She co-directs the Neuropathology Core of Yale's Alzheimer's Disease Research Center. Her research has centered on developing a wide range of translational animal models, with emphasis on neurologic and retinal disease models. The animal to human translational gap is an active area of research using biomedical informatics to identify large-scale patterns of animal model use. Most recently, her translational work has broadened to include epidemiologic models of pandemic to endemic transition for COVID-19, One-Health driven coronaviral surveillance in the Northeast, and mechanisms of viral interference. Her clinical expertise is directed almost exclusively toward non-human primate care, with emphasis on anesthesia, pain management, imaging, internal medicine, and pathology of macaques and marmosets. Zeiss received a PhD from Cornell University and a DVM veterinary degree from the University of Pretoria, South Africa and is dual-boarded in laboratory animal medicine and veterinary anatomic pathology.

Contributions from Structural Biology to the Pandemic Response During the SARS-CoV-2 Pandemic: Lessons Learned for the Future

3

Miranda L. Lynch and Sarah E. J. Bowman

Abstract

Structural biology—the scientific area devoted to understanding the three-dimensional architecture of biological molecules—became a lynchpin in the effort to combat the SARS-CoV-2 virus from the very inception of the pandemic. Determining the structures of key components of the virus, and how those components work together to execute the viral life cycle, began literally as soon as we had the initial gene sequences from patient samples in the first weeks after the virus was identified; the first sequence was made available online on January 10, 2020 (Virological-Website, https://virological.org/t/novel-2019-coronavirus-genome/319; Wu et al. Nature 579:265–269, 2020a). The structures fueled the development of both vaccines and therapeutics, and helped unravel details of how the virus infected hosts and spread. This chapter highlights two NSF-funded RAPID projects, one experimental and one computational, focused on structural biology. The first project set up resources to accelerate structure discovery by enabling rapid screening for crystallization. The second project used computational tools to model the interaction of proteins from the virus with those from the host. Both projects are placed within the larger context of the contributions of the field of structural biology to the

Research funded by

NSF Award #2029885—RAPID (Rapid Response Research Award) IIBR (Infrastructure Innovation Biological Research): Informatics Computational methods for utilizing SARS-CoV2 sequence and structure data in predicting host–pathogen protein–protein interactions;
NSF Award #2029943—RAPID (Rapid Response Research Award) Enhanced SARS-CoV-2 High-Throughput Crystallization for Structural Studies;
NIH National Institute of General Medical Sciences (NIGMS) R24GM141256.

COVID Information Commons Video Presentations

The COVID Information Commons (CIC) Research Lightning Talk video presentations and individual multilingual transcripts for this project can be found in the Columbia Academic Commons (AC) (Bowman 2020, 2022).

M. L. Lynch · S. E. J. Bowman (✉)
University at Buffalo Hauptman-Woodward Institute, Buffalo, NY, USA
e-mail: sebowman@buffalo.edu

M. L. Lynch
e-mail: mllynch2@buffalo.edu

M. L. Lynch
Department of Structural Biology, Jacobs School of Medicine and Biomedical Sciences, State University of New York at Buffalo, Buffalo, NY, USA

S. E. J. Bowman
Department of Biochemistry, Jacobs School of Medicine and Biomedical Sciences, State University of New York at Buffalo, Buffalo, NY, USA

F. D. Hudson (ed.), *The COVID Information Commons*, Sustainable Development Goals Series,
https://doi.org/10.1007/978-3-032-06125-6_3

fights against COVID-19, and underscore the role that structural biology plays in our ability to address emerging pandemic threats.

Keywords

Structural biology · Virus · X-ray crystallography · Computational biology · Cryo-electron microscopy · Cryo-EM · Crystallization · SARS-CoV-2 · COVID-19

3.1 Introduction

Structural biology provides information about the three-dimensional architecture of macromolecules, which played an essential role along two different fronts in the worldwide response to the global COVID-19 pandemic.

First, knowledge gained from structures of the component parts of the severe acute respiratory syndrome coronavirus 2 (SARS-CoV-2) was essential to answering basic biological questions about the virus (Baker 2020; Subramaniam 2020). *How does the virus connect and interact with human host cells during infection? What are the mechanisms by which the virus carries out its life cycle, including replication and packaging the genome and creating new virions?* Answers to these essential questions were made possible by knowing the structural details of the parts of the virus that were involved.

Second, structural details were critical for generating therapeutic responses, including treatments for SARS-CoV-2 infections and development of vaccines to prevent COVID-19 (Yang and Rao 2021; Lubin et al. 2022; Lynch et al. 2021). Early therapies included structure-based design of antibodies and antibody cocktails that could neutralize the virus (Yuan et al. 2020; Huo et al. 2020; Barnes et al. 2020). The spike protein is embedded in the envelope surrounding the virus and is the interaction point at which a host cell contacts the virus. Vaccines were developed to train the immune system to recognize the spike protein, establishing an immune response (Duan et al. 2020; Bangaru et al. 2020). Structure-based drug design enabled identification of inhibitors capable of reducing the negative health impacts of a COVID-19 infection (Owen et al. 2021).

3.2 Goals of the Research

In the very early days of what would become the COVID-19 pandemic, many research communities heard and heeded the call to bring immediate results and expertise to bear on this novel, nebulous, and growing threat. The structural biology community was one of the research communities that immediately pivoted toward study of the novel SARS-CoV-2 virus. The World Health Organization (WHO) declared a pandemic on March 11, 2020. By then, the genomic information on the SARS-CoV-2 virus had already been released to the research community (in early January) from strains isolated in Wuhan, China (Wu et al. 2020b). The combination of the SARS-CoV-2 genetic details and the previous research on other coronaviruses (SARS-CoV-1 and MERS) were used to guide rapid paths to key structures of the new virus, notably the main viral protease (Mpro) and the spike protein. By the time the WHO declared a pandemic, repositories for global exchange of structural data were already being set up. The very first structures were publicly released to the worldwide Protein Data Bank (wwPDB), the worldwide repository for biological structural models, in February 2020 (Jin et al. 2020a; Wrapp et al. 2020; Berman et al. 2003). Experimental structural approaches included macromolecular X-ray crystallography (X-ray) and cryo-electron microscopy (Cryo-EM), as well as the techniques Nuclear Magnetic Resonance (NMR) spectroscopy and other diffraction methods (neutron and electron). The three-dimensional models generated with structural methods helped us to visualize the component parts of the virus (Fig. 3.1). Each of the structures shown in Fig. 3.1 is the first structural model deposited to the wwPDB for that particular protein or protein–protein complex (PDB IDs 7MSW (QCRG_Structural_Biology_Consortium 2021), 6VXS (Michalska et al. 2020), 7KAG (Stogios et al. 2020), 6WIQ (Wilamowski et al. 2021),

6M71 (Gao et al. 2020), 6XEZ (Chen et al. 2020), 6W4H (Rosas-Lemus et al. 2020), 6XDC (Kern et al. 2021), 7JTL (Flower et al. 2021), 6W37 (Nelson et al. 2020), 6Z4U (Weeks et al. 2020), 7K3N (Semper et al. 2021), 6W9C (Osipiuk et al. 2020), 6LU7 (Jin et al. 2020b), 6W4B (Chen et al. 2020), 6ZCT (Rogstam et al. 2020), 6ZSL (Newman et al. 2021), 6VWW (Kim et al. 2020), 6VSB (Wrapp et al. 2020), 7K3G (Mandala et al. 2020), 8CTK (Dolan et al. 2022), and 6VYO (Chang et al. 2020)). Cleavage sites of the proteases are depicted by a green triangle (papain-like protease) and a black arrow (main protease). Figure 3.1 updated from Lynch et al. (2021); the new structures included in Fig. 3.1 include 2nsp and M protein.

The open data repositories already established and in wide use in the structural biology community, notably the wwPDB, made these worldwide research efforts possible (Lubin et al. 2022). The research efforts from structural biologists were challenged during the pandemic in the following months, as labs were shut down to contain spread or forced to work on staggered schedules. Further, supply chain issues for laboratory consumables dramatically impacted the ability even for those labs that remained open. Even with these obstacles, the response from the structural science community was breathtakingly rapid. In this chapter, we highlight two National Science Foundation (NSF)-supported Rapid Response Research (RAPID) projects that were part of this global push to understand the structure of the virus and its component parts. One project contributed directly to experimental efforts; the other project focused on computational structural approaches. Both focused on using structure as a way of understanding the virus, and are placed in the context of related work emerging from the structure research community.

3.2.1 Experimental Goals for NSF RAPID Award #2029943—Enhanced SARS-CoV-2 High-Throughput Crystallization for Structural Studies

When the SARS-CoV-2 virus emerged as a global threat in December 2019, the entire scientific community pivoted to focus attention on solving the crisis. As the world went into lockdown to prevent the spread of the virus, many academic laboratories were also shut down and unable to operate as normal. New York State (NYS), where the National Crystallization Center is located, mandated that all nonessential businesses were to be closed statewide in March 2020. In the same month, the Crystallization Center at University at Buffalo Hauptman-Woodward Research Institute (UB HWI) began receiving requests from all over the United States (U.S.) for access to our crystallization screening for proteins from the virus. We were declared "essential" for the work we were doing crystallizing proteins from the virus, enabling us to remain in operation, and received our first sample of the Mpro protein in March 2020. We determined conditions for crystallization for Mpro within one week using our novel crystallization screening approach (Fig. 3.2). The resulting structural model from the initial crystals (PDB 6WQF) (Kneller et al. 2020a) and imaging from the Crystallization Center, including enhanced state-of-the-art crystal imaging capabilities (UV-TPEF and SHG), are shown in Fig. 3.2a, b. In early April 2020, we applied for NSF RAPID funding to facilitate access for those working on SARS-CoV-2 protein samples to the unique resources in the Crystallization Center.

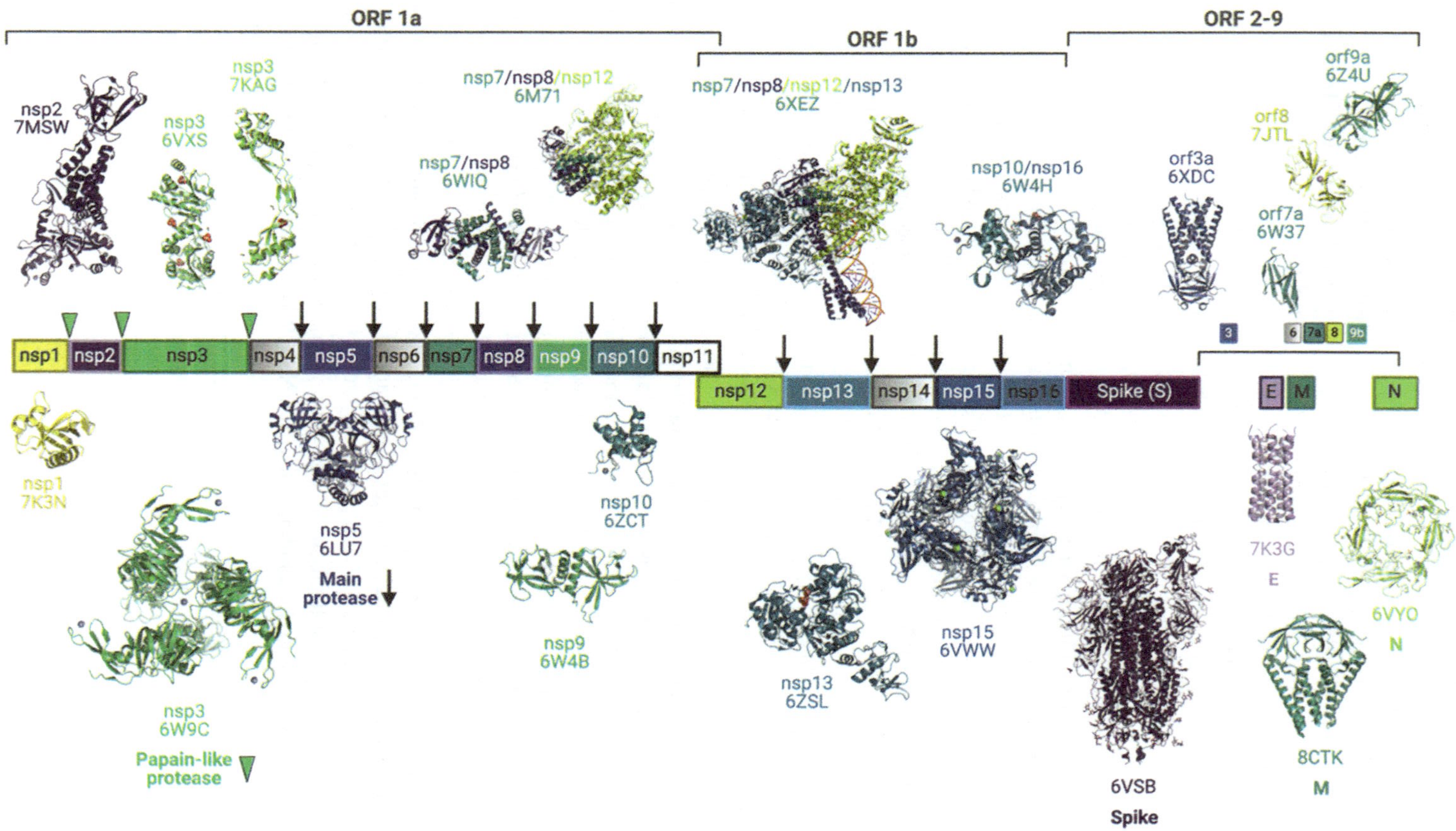

Fig. 3.1 Map of protein structures across the open reading frames of the SARS-CoV-2 viral genome (updated from Lynch et al. (2021))

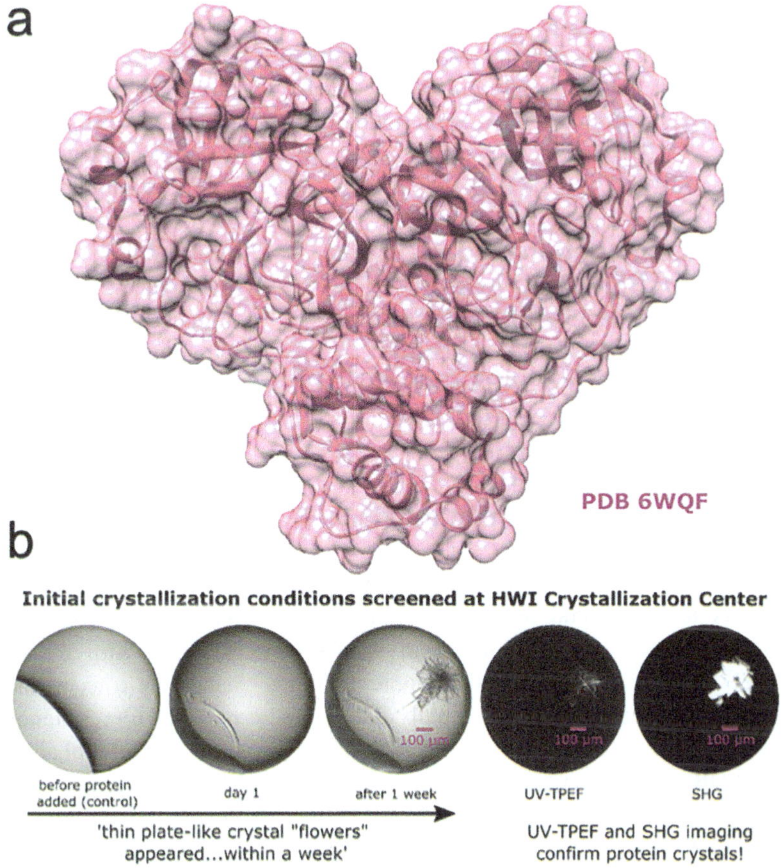

Fig. 3.2 The first sample from SARS-CoV-2 Mpro protein was set up in the Crystallization Center in March 2020: **a** resulting structural model from the initial crystals (PD8 6WQF), **b** initial crystallization conditions observed with three imaging modalities

At the time, the application for NSF RAPID award #2029943 was submitted to the NSF, just over 100 structures had been deposited to the wwPDB. Our goals with the funding were to provide an efficient, centralized resource for crystal-based structural approaches to study proteins from the SARS-CoV-2 virus.

The major goals of the project included:

1. To provide coordinated and centralized resources for experimental screening for crystallization of SARS-CoV-2-related samples.
2. To provide enhanced imaging for rapid detection of crystal growth formation for SARS-CoV-2-related samples.
3. To develop a Graphical User Interface (GUI) for easy implementation of the MAchine Recognition of Crystallization Outcomes (MARCO) autoscoring algorithm (Bruno et al. 2018) for analyzing crystallization screen images.
4. To engage in outreach and educational opportunities for broader impacts.

Approximately 84% of protein structural information comes from structural techniques using biomolecular crystals (X-ray, neutron diffraction, electron diffraction, etc.) (Budziszewski et al. 2023). One of the major bottlenecks in crystal-based experiments is finding the conditions in which a protein will crystallize. The Crystallization Center at UB HWI is a high-throughput crystallization facility that specializes in rapid, efficient crystallization of biomolecular targets. We are the only resource in the world available to external users that employs a unique high-throughput microassay plate that enables screening 1536 conditions in one experimental plate and has state-of-the-art imaging methods capable of detecting protein crystals very rapidly. As such, we were uniquely poised to contribute in rapid response to the threats presented by the SARS-CoV-2 virus.

3.2.2 Computational Goals for NSF RAPID Award #2029885—IIBR Informatics Computational methods for utilizing SARS-CoV-2 Sequence and structure Data in Predicting Host–Pathogen Protein–Protein Interactions

Computational work made key contributions to efforts to understand and combat the SARS-CoV-2 virus. The goals of the in silicostructural biology work included using computing to improve and enhance the rapidly determined structures that were being deposited into the wwPDB, developing computational models for predicting and modeling interactions between host and viral proteins, predicting the effects of mutations in the viral genome, and much more. An excellent example of the first activity was the emergence of the Coronavirus Structural Task Force, spearheaded by Dr. Andrea Thorn and with participation from a global team of structural biologists (Croll et al. 2021a; Coronavirus Structural Task Force, https://insidecorona.net/). Starting in February 2020, this group undertook the task of automating the computational evaluation, error correction, revision, and improvement of the experimental structural data appearing in the wwPDB, and created a dedicated open web site for making the revised and updated structures available, along with curated information resources for each of the viral proteins. These types of resources provided invaluable tools for the research community and did a lot to ensure that downstream computational work on the viral protein structures were working with the most accurate models possible (Croll et al. 2021a, b). The Research Collaboratory for Structural Bioinformatics Protein Data Bank, the U.S. data center for the wwPDB, also established a dedicated page for the deposited SARS-CoV-2 structures, to help researchers locate and access the most recent structural work (RCSB COVID-19/SARS-CoV-2 Resources, www.rcsb.org/covid19).

The computational community also set itself to using data mining, text mining, literature curation, and machine learning to amass and organize important repositories of information, including interactome browsers based on the comprehensive protein interaction screening study of Gordon, et al. (Gordon et al. 2020). A variety of computational resources and databases emerged for other structural aspects of the SARS-CoV-2 virus, including the 3D SARS-CoV-2—Human Interactome Browser (Wierbowski et al. 2021, 3D SARS-CoV-2-Human Interactome Browser, https://3d-sars2.yulab.org/), the STRING Coronavirus interaction (COVID-19 STRING, https://string-db.org/cgi/covid.pl), and databases to access the variant mutations and their potential for resistance, such as the Stanford Coronavirus Antiviral and Resistance Database (Coronavirus antiviral and resistance database, https://covdb.stanford.edu/). Many of these computational tools were very large scale, assembled by teams of researchers and making use of enormous computing resources. Accompanying these large-scale efforts was a flurry of activity from groups in the computational chemistry community to bring the computational tools of the structural biology trade to bear on the problems of understanding the viral proteins. These projects ranged from modeling glycosylation of the spike protein (Woo et al. 2020; Coutinho et al. 2022; Huang et al. 2022; Ghorbani et al. 2021; Casalino et al. 2021), to computational prediction of viral protein interactions (Karunakaran et al. 2022; Horlacher et al. 2023; Chakraborty et al. 2023), to analysis of in silico viral mutagenesis (Zou et al. 2020; Sobitan et al. 2021; Guzzi et al. 2023). This was a titanic effort on the part of the computational community, and only a handful of representative studies are cited here, as the total number of such studies is too numerous to list; an assessment of the important contributions coming out of the computational simulation work is given in Abduljalil et al. (2023).

Our own work as part of our NSF RAPID project employed molecular dynamics (MD) simulations to probe the functional binding aspects of interaction of the viral spike protein with the binding domain of the human-interacting partner, the ACE2 (angiotensin converting enzyme 2) receptor protein (Rynkiewicz et al. 2022). The study provided a computationally based comparative analysis of the dynamics of the host–pathogen binding interaction at the SARS-CoV-2 spoke and the ACE2 partner (Fig. 3.3), by using MD trajectories for the human receptor protein coupled with multiple versions of the spike protein from both the emergent Betacoronavirus forms (SARS-CoV-2, along with SARS-CoV and MERS) and the endemic forms (hCoV-OC43 and hCoV-HKU1). Included as well were models of a hypothetical interaction of the spike protein from a bat progenitor strain with a human receptor protein. The goal of this modeling was to uncover whether there were potential conserved dynamics of the targeted receptors and different viral spike proteins that were functionally relevant to their interactions, to provide understanding of the role dynamics of protein interactions play in zoonotic spillover events.

3.3 Results

3.3.1 Results: Experimental Outcomes for NSF RAPID Award #2029943

Over the funding period, the Crystallization Center worked with multiple SARS-CoV-2 protein samples from laboratories across the USA, providing efficient sample set up and enhanced imaging for all of these samples and empowering rapid identification of crystals. One of the key targets for developing antiviral inhibitors as treatments for COVID-19 infections is the main protease (Mpro). The functional role of Mpro is to cleave the polyproteins encoded by the SARS-CoV-2 genome in ORF 1a and ORF 1b (marked by black arrows on Fig. 3.1). Mpro plays a central role in enabling viral replication and transcription; therefore, it has been an attractive target for identifying potential inhibitors to the virus. In the Crystallization Center, we successfully identified conditions to crystallize Mpro both with and without inhibitors. The first sample that we screened in March 2020 resulted in the first room temperature structure of Mpro (Fig. 3.2), which revealed conformational flexibility of the active site of the enzyme at physiologically relevant temperatures (Kneller et al. 2020a). Additional Mpro structures that were supported by crystallization efforts in the Crystallization Center included additional room temperature Mpro structures (Kneller et al. 2020b, 2021a), Mpro structures solved using neutron crystallography (Fig. 3.4a, (Kneller et al. 2020c; Kneller et al. 2021b)), and a series of structures with small molecule inhibitors and proto-drugs bound in the Mpro active site (Fig. 3.4b, (Kneller et al. 2021c; Kneller et al. 2022; Iketani et al. 2021; Liu et al. 2022)). Further, some of these structures provided a basis for computational virtual screening and iterative drug design for more potent inhibitors (Clyde et al. 2021; Glaser et al. 2022). While many of the SARS-CoV-2 samples handled in the Crystallization Center during the two-year NSF RAPID award were the Mpro protein, we also assisted with crystallization of a nanobody drug candidate that targets the receptor binding domain (RBD) of the spike protein (Fig. 3.4c (Ye et al. 2020)). Our efforts to provide a centralized resource for crystallization, including enhanced imaging for fast and efficient detection of crystals for SARS-CoV-2 related samples, were extremely successful.

As part of the NSF RAPID award, we also aimed to develop a GUI for the recently published MARCO autoscoring algorithm, which is an algorithm designed to score crystallization images (Bruno et al. 2018). A postbaccalaureate student worked with us remotely through the first summer of the pandemic in 2020, building a GUI called MARCO-Polo that is integrated into the Crystallization Center pipeline and now available to all users (Holleman et al. 2021). All coding for MARCO-Polo is open-source and available on GitHub.

Finally, a tremendous amount of outreach and educational efforts were accomplished. We

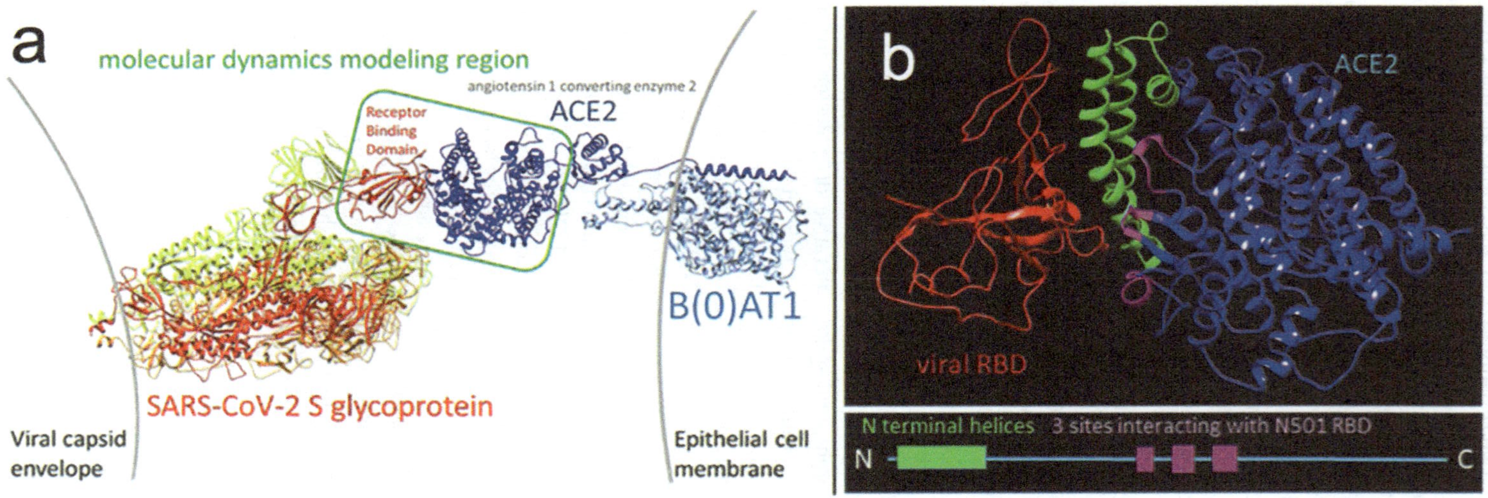

Fig. 3.3 The specific region of the SARS-CoV-2 spike and its ACE2 partner modeled in our MD simulations: **a** Full spike trimer (left, warm colors) and full ACE2 receptor (right, cool colors), with modeled contact area shown in box, **b** Highlighted regions of contact revealed in simulations (reproduced with permission from Taylor & Francis Ltd. (www.tandfonline.com) (Rynkiewicz et al. 2022))

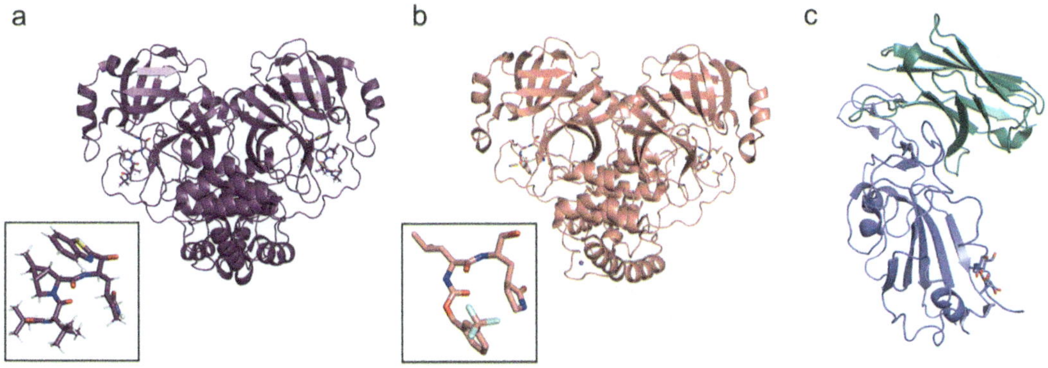

Fig. 3.4 Structures of SARS-CoV-2 proteins from screening: **a** Mpro with a potential inhibitor solved using X-ray and neutron crystallography, **b** Mpro with a potential inhibitor solved using X-ray crystallography, **c** SARS-CoV-2 RBD interaction solved using X-ray crystallography

participated in podcasts on the coronavirus (Bell Ringer Podcast (Crystal Clear: Buffalo-Niagara Invest Bell Ringer Podcast, https://investbuffaloniagara.podbean.com/e/crystal-clear-1597858264/), Point of Learning Podcast (Crystallizing Coronavirus with Sarah Bowman: Point of Learning Podcast, https://hornedconsulting.org/pointoflearningpodcast/crystals), on panels, and in a pilot episode of "Nerdy Jobs" with Dr. Raven "the Science Maven" Baxter (Dr. Raven the Science Maven Baxter, https://www.youtube.com/watch?v=siFfzoFv61I). Results have been distributed through outreach efforts to both the structural biology community, as well as to non-scientific audiences to describe the role of structural biology in the fight against COVID-19. Finally, we presented our project and results at the COVID Information Commons Webinar Series.

3.3.2 Results: Computational Outcomes for NSF RAPID Award #2029885

The comparative evolutionary perspective and use of a multi-agent machine learning approach resulted in some novel insights from our computational study of Betacoronavirus host–pathogen protein–protein interactions (Fig. 3.3a). First, we were able to discern conserved dynamic regions on the ACE2 receptor, one a highly stable area on the N-terminal helix that underwent dampened motion upon binding with the receptor binding domain (RBD) of the spike protein, and a second, more transient site that appears to contact the N501 residue of the spike RBD (Fig. 3.3b). Our simulations indicate that, compared to the endemic strains of Betacoronaviruses, the highly transmissible SARS-CoV-2 variants (B.1.1.7, B.1.351 and P.1) have evolved more stable viral binding via more focused interactions between the viral N501 and ACE2 K353 alone at the second contact region of the host–pathogen protein interaction. The results from our simulation studies involving multiple endemic and emergent Betacoronavirus spike proteins led us to propose a model of protein interaction for coronavirus RBD evolution and specificity of contact with host cell receptors that indicate a role for two distinct touch points between the viral RBD and the human target proteins. This two-touch model underscores the contact regions of the viral entry protein with its cellular receptor, which include the N-terminal helices of ACE2 that is conserved in all known strains of potentially human-interacting Betacoronavirus that we studied, both endemic and emergent. The model provides strong insights into a common path for the functional evolution of these pathogenic Betacoronaviruses, including different variants of the SARS-CoV-2 virus, that occurs at the protein interaction interface of the entry spike protein and its targeted human receptor protein. Such insight underscores the importance of molecular dynamics modeling to uncover evolutionary conserved motions that could provide key targets for therapeutic development in emerging zoonotic spillover events (Pierri 2020).

3.3.3 Results: Structural Biology in the Time of COVID-19

We have discussed in this chapter how our two NSF RAPID awards contributed to basic science efforts to comprehend the novel viral threat. We have placed our work in the larger context of the important role played by the global structural biology community during and since the declared COVID-19 pandemic. Indeed, that role has been so central to the effort to combat the virus, that we undertook a similar collaborative assessment of how structural biology research on SARS-CoV-2 viral proteins was carried out in the early stages of the pandemic in 2021 and published in the International Union of Crystallography Journal (IUCrJ) (Lynch et al. 2021). The COVID-19 pandemic provided a unique opportunity to investigate structural methods; we were able to do a real-time snapshot of the contributions of the different structural methods to understand the structures of the proteins from the SARS-CoV-2 virus. Some of the results in this chapter serve to update the assessments from the IUCrJ paper with the most up-to-date data and after the primary wave of the pandemic has passed, with the goal of gaining insight into what was (and could be) accomplished under the strenuous setting of a novel viral outbreak and subsequent global pandemic. Figure 3.1 is an update from the IUCrJ paper that now includes viral protein structures for some of the proteins for which the community had not yet solved structures in 2021, and which have since yielded to structure determination. Similarly, Figs. 3.5 and 3.6 appeared in similar form in Lynch et al. (2021), but we have updated them to include the most up-to-date set of experimental structural models. Figure 3.5.a shows that of the 4443 total structures (as of July 31, 2024), ~ 67% were solved using X-ray and ~ 32% using Cryo-EM. When considering the subset of structures with inhibitors or fragments (proto-drugs), the ratio is ~ 93% X-ray and ~ 6% Cryo-EM (Fig. 3.5.b). When considering the subset of structures with antibodies bound, the ratio is ~ 65% X-ray and ~ 35% Cryo-EM (Fig. 3.5c). NMR contributes some structures, but the overall number is few (21 of the 4443 structures) relative to X-ray and Cryo-EM.

There are now 4443 total structures of SARS-CoV-2 proteins in the wwPDB, more than four times as many as when our 2021 study was performed. This entailed including the contribution of structures from NMR methods, and adds a further assessment of the structural methods that contributed to determining viral proteins coupled to antibodies, nanobodies, and Fabs. Figure 3.6 shows a scatter plot displaying resolution versus molecular weight (on the log2 scale) for the 4422 deposited structures solved using X-ray (orange triangles) and Cryo-EM (blue circles). Results from our original IUCrJ investigation of structural biology contributions are underscored dramatically with these new data, highlighting the fact that diffraction-based methods like X-ray crystallography are essential to rapid structure determination of smaller-to-medium sized proteins, and especially to those bound to ligands and inhibitors. Notably, to cover the entire molecular weight range of the proteins that are essential for SARS-CoV-2 function requires both X-ray and Cryo-EM experimental techniques. Again, the results demonstrate how necessary Cryo-EM has been to probing the very large spike protein trimer. The spike structures have given us invaluable information about the spike protein dynamics, including how the protein shifts from closed to open states in ways that are key to the ability of the virus to make contact with the host receptor. We take the opportunity of putting together this chapter to not only update the portrait of structural biology work in the COVID-19 pandemic, but to learn afresh the important lesson that no single method will suffice in rapidly evaluating the proteins of a new virus, but all the tools of structural biology must be leveraged in order to gain the knowledge needed to mount a response to a novel agent.

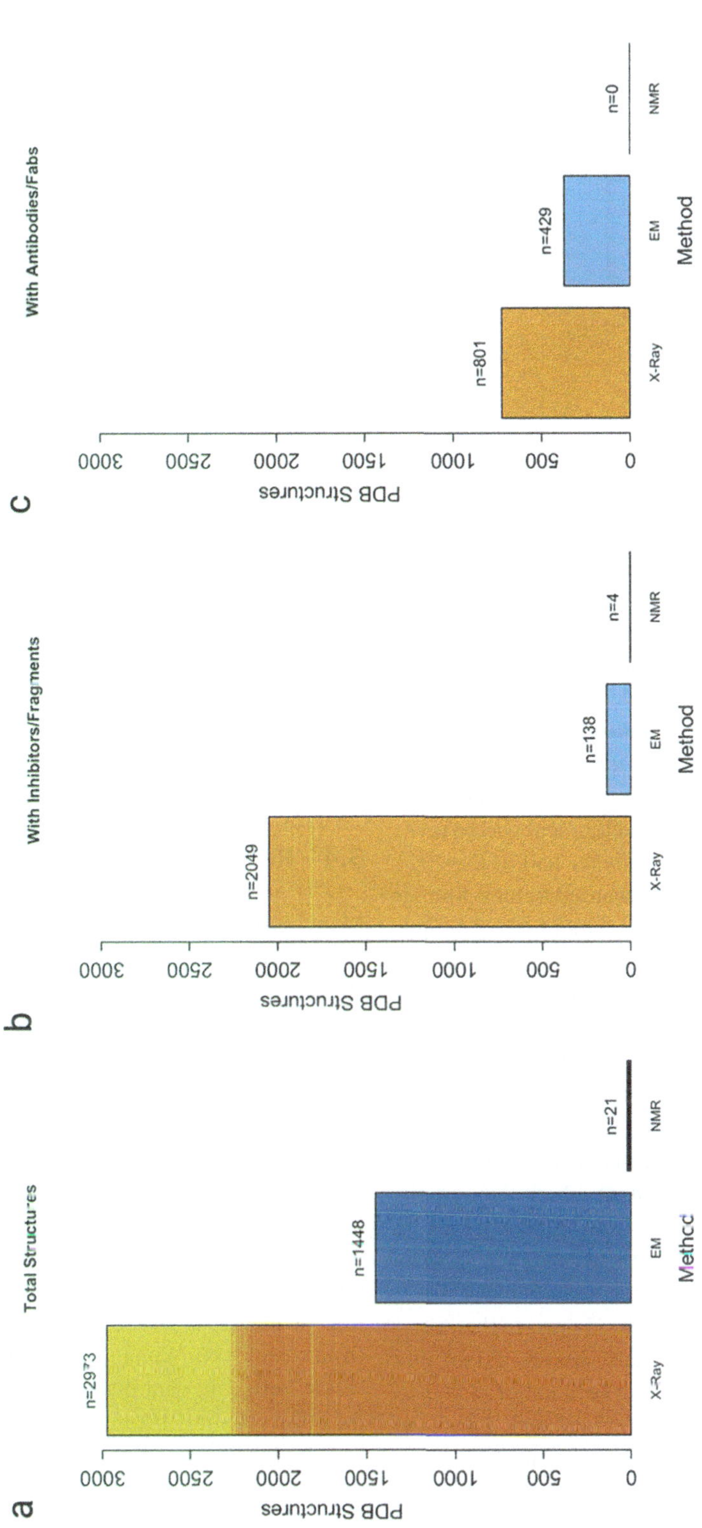

Fig. 3.5 Bar plots of released structures of SARS-CoV-2 samples from the wwPDB as of July 31, 2024: **a** total structures, **b** with inhibitors/fragments, **c** with antibodies/fabs

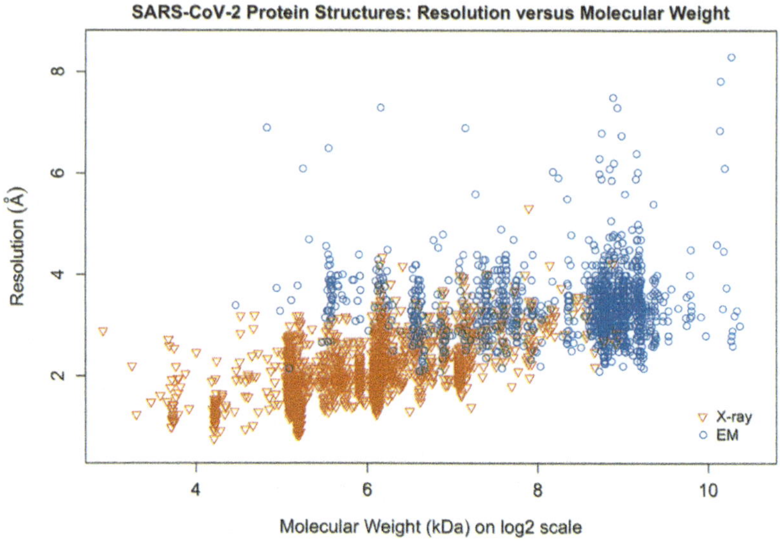

Fig. 3.6 SARS-CoV-2 structures deposited in the wwPDB as of July 31, 2024

This chapter would be remiss if we did not mention a major change that swept over the scientific community approximately midway through the major part of the pandemic: the appearance of Alphafold as a computational algorithmic tool capable of providing high-quality predictions of protein structures (Jumper et al. 2021; Abramson et al. 2024). Indeed, this development was so revolutionary as to form the basis for half of the 2024 Nobel Prize in Chemistry awarded to John Jumper and Demis Hassabis. The Alphafold computational tools underscore the very themes of this chapter—that experimentally derived structural data of biomolecules are essential to moving the field forward, and that the wwPDB as a curated repository of structural models is a crucial centralized resource. Indeed, the Alphafold algorithms were trained on the entire body of extant structures in the wwPDB, without which the computational tools would not have been able to be developed. It is difficult to assess exactly what impact Alphafold could have had on the COVID-19 pandemic if it had appeared prior to the initial outbreak. We can be certain, however, that this computational resource will be a major player in structural biology in any pandemics that develop in the future. We are now, several years after the first iteration of Alphafold appeared, learning how to use the predicted structural models. Each subsequent version of Alphafold uses more and more data to address the gaps in its predictive abilities. The combination of experimental structural biology and computational predictions will be essential to address challenges that we face as a society that impact our global health, from future pandemics to climate change.

3.4 Recommendations

Many important lessons emerged from our experiences using computational and experimental structural biology approaches to combat COVID-19. First, there was the critical role of "pre-information" in the form of structural studies of proteins from related betacoronaviruses, notably the SARS-CoV and MERS viruses. This information was used to accelerate the initial structure determination by providing adjacent models. Second, the high level of data sharing between scientists globally did a tremendous amount to remove barriers to progress. This was especially true of the rapid sharing of genomic sequences from scientists in Wuhan, China, at the very outset of the outbreak, data that would prove crucial to the efforts to study the virus as it emerged, spread, and mutated. The global network of collaborating scientists working with focus on a unifying problem is essential for pandemic preparedness. In the structural biology efforts, the combination of X-ray crystallography, Cryo-EM, and computational biology work was crucial to developing rapid therapeutic responses. Third, the availability of centralized resources in the way of national facilities was essential—from our Crystallization Center to the synchrotrons, microscopy centers, and computational centers that enabled fast and safe data collection, processing, and analyses. Fourth, the history of data availability for structural data from the wwPDB, which serves as a public good according to the FAIR (Findable, Accessible, Interoperable, Reusable) data principles, was vital. Finally, the funding that was made available worldwide, including from NSF RAPID awards, filled a crucial role in enabling the research on the SARS-CoV-2 virus to occur.

Acknowledgements The following funding is acknowledged: NSF RAPID Awards #2029885 and #2029943 for SARS-CoV-2 specific work, and NIH NIGMS R24GM141256 for support of the National Crystallization Center high-throughput crystallization capabilities.

References

3D SARS-CoV-2-Human Interactome Browser, Available from: https://3d-sars2.yulab.org/

Abduljalil JM, Elghareib AM, Samir A et al (2023) How helpful were molecular dynamics simulations in shaping our understanding of SARS-CoV-2 spike protein dynamics? Int J Biol Macromol 242:125153. https://doi.org/10.1016/j.ijbiomac.2023.125153

Abramson J, Adler J, Dunger J et al (2024) Accurate structure prediction of biomolecular interactions with AlphaFold 3. Nature 630:493–500. https://doi.org/10.1038/s41586-024-07487-w

Baker EN (2020) Visualizing an unseen enemy; mobilizing structural biology to counter COVID-19. IUCrJ 7:366–367. https://doi.org/10.1107/S2052252520004571

Bangaru S, Ozorowski G, Turner HL et al (2020) Structural analysis of full-length SARS-CoV-2 spike protein from an advanced vaccine candidate. Science 370:1089–1094. https://doi.org/10.1126/science.abe1502

Barnes CO, Jette CA, Abernathy ME et al (2020) SARS-CoV-2 neutralizing antibody structures inform therapeutic strategies. Nature 588:682–687. https://doi.org/10.1038/s41586-020-2852-1

Berman H, Henrick K, Nakamura H (2003) Announcing the worldwide Protein Data Bank. Nat Struct Biol 10:980. https://doi.org/10.1038/nsb1203-980

Bowman S (2020) RAPID enhanced SARS-CoV-2 high-throughput crystallization for structural studies. Columbia Academic Commons, Academic Unit: Northeast Big Data Innovation Hub, Series: COVID-19 Research Lightning Talks. https://doi.org/10.7916/q3x9-k309

Bowman S (2022) Enhanced SARS-CoV-2 high-throughput crystallization for structural studies. Columbia Academic Commons, Academic Unit: Northeast Big Data Innovation Hub, Series: COVID-19 Research Lightning Talks. https://doi.org/10.7916/5jvt-3n94

Bruno AE, Charbonneau P, Newman J et al (2018) Classification of crystallization outcomes using deep convolutional neural networks. PLoS ONE 13:e0198883. https://doi.org/10.1371/journal.pone.0198883

Budziszewski GR, Snell ME, Wright TR et al (2023) High-throughput screening to obtain crystal hits for protein crystallography. J Vis Exp e65211.https://doi.org/10.3791/65211

Casalino L, Dommer AC, Gaieb Z et al (2021) AI-driven multiscale simulations illuminate mechanisms of SARS-CoV-2 spike dynamics. Int J High Perform Comput Appl 35:432–451. https://doi.org/10.1177/10943420211006452

Chakraborty A, Mitra S, Bhattacharjee M et al (2023) Determining human-coronavirus protein-protein interaction using machine intelligence. Med Nov Technol Devices 18:100228. https://doi.org/10.1016/j.medntd.2023.100228

Chang C, Michalska K, Jedrzejczak R et al (2020) Crystal structure of RNA binding domain of nucleocapsid phosphoprotein from SARS coronavirus 2. https://doi.org/10.2210/pdb6VYO/pdb

Chen J, Malone B, Llewellyn E et al (2020) Structural basis for helicase-polymerase coupling in the SARS-CoV-2 replication-transcription complex. Cell 182(1560–1573):e13

Clyde A, Galanie S, Kneller DW et al (2021) High-throughput virtual screening and validation of a SARS-CoV-2 main protease noncovalent inhibitor. J Chem Inf Model 62:116–128

Coronavirus Antiviral and Resistance Database, Available from: https://covdb.stanford.edu/

Coronavirus Structural Task Force, Available from: https://insidecorona.net/

Coutinho JVP, Macedo-da-Silva J, Mule SN et al (2022) Glycoprotein molecular dynamics analysis: SARS-CoV-2 spike glycoprotein case study. In: Donev R (ed) Advances in protein chemistry and structural biology. Academic, pp 277–309

COVID-19 STRING, Available from: https://string-db.org/cgi/covid.pl

Croll TI, Diederichs K, Fischer F et al (2021a) Making the invisible enemy visible. Nat Struct Mol Biol 28:404–408. https://doi.org/10.1038/s41594-021-00593-7

Croll TI, Williams CJ, Chen VB et al (2021b) Improving SARS-CoV-2 structures: peer review by early coordinate release. Biophys J 120:1085–1096. https://doi.org/10.1016/j.bpj.2020.12.029

Crystal Clear: Buffalo-Niagara Invest Bell Ringer Podcast, Available from: https://investbuffaloniagara.podbean.com/e/crystal-clear-1597858264/

Crystallizing Coronavirus with Sarah Bowman: Point of Learning Podcast, Available from: https://hornedconsulting.org/pointoflearningpodcast/crystals
Dolan KA, Dutta M, Kern DM et al (2022) Structure of SARS-CoV-2 M protein in lipid nanodiscs. Elife 11:e81702. https://doi.org/10.7554/eLife.81702
Dr. Raven the Science Maven Baxter, *Nerdy Jobs S1E1*. https://www.youtube.com/watch?v=siFfzoFv61I
Duan L, Zheng Q, Zhang H et al (2020) The SARS-CoV-2 Spike glycoprotein biosynthesis, structure, function, and antigenicity: implications for the design of spike-based vaccine immunogens. Front Immunol 11:576622. https://doi.org/10.3389/fimmu.2020.576622
Flower TG, Buffalo CZ, Hooy RM et al (2021) Structure of SARS-CoV-2 ORF8, a rapidly evolving immune evasion protein. Proc Natl Acad Sci 118:e2021785118. https://doi.org/10.1073/pnas.2021785118
Gao Y, Yan L, Huang Y et al (2020) Structure of the RNA-dependent RNA polymerase from COVID-19 virus. Science 368:779–782. https://doi.org/10.1126/science.abb7498
Ghorbani M, Brooks BR, Klauda JB (2021) Exploring dynamics and network analysis of spike glycoprotein of SARS-COV-2. Biophys J 120:2902–2913. https://doi.org/10.1016/j.bpj.2021.02.047
Glaser J, Sedova A, Galanie S et al (2022) Hit expansion of a noncovalent SARS-CoV-2 main protease inhibitor. ACS Pharmacol Transl Sci 5:255–265. https://doi.org/10.1021/acsptsci.2c00026
Gordon DE, Jang GM, Bouhaddou M et al (2020) A SARS-CoV-2 protein interaction map reveals targets for drug repurposing. Nature 583:459–468. https://doi.org/10.1038/s41586-020-2286-9
Guzzi PH, di Paola L, Puccio B et al (2023) Computational analysis of the sequence-structure relation in SARS-CoV-2 spike protein using protein contact networks. Sci Rep 13:2837. https://doi.org/10.1038/s41598-023-30052-w
Holleman ET, Duguid E, Keefe LJ et al (2021) Polo: an open-source graphical user interface for crystallization screening. J Appl Crystallogr 54:673–679. https://doi.org/10.1107/S1600576721000108
Horlacher M, Oleshko S, Hu Y et al (2023) A computational map of the human-SARS-CoV-2 protein-RNA interactome predicted at single-nucleotide resolution. NAR Genom Bioinform 5:lqad010. https://doi.org/10.1093/nargab/lqad010
Huang Y, Harris BS, Minami SA et al (2022) SARS-CoV-2 spike binding to ACE2 is stronger and longer ranged due to glycan interaction. Biophys J 121:79–90. https://doi.org/10.1016/j.bpj.2021.12.002
Huo J, Zhao Y, Ren J et al (2020) Neutralization of SARS-CoV-2 by destruction of the prefusion spike. Cell Host Microbe 28(445–454):e6. https://doi.org/10.1016/j.chom.2020.06.010
Iketani S, Forouhar F, Liu H et al (2021) Lead compounds for the development of SARS-CoV-2 3CL protease inhibitors. Nat Commun 12:2016. https://doi.org/10.1038/s41467-021-22362-2
Jin Z, Du X, Xu Y et al (2020a) Structure of M(pro) from SARS-CoV-2 and discovery of its inhibitors. Nature 582:289–293. https://doi.org/10.1038/s41586-020-2223-y
Jin Z, Du X, Xu Y et al (2020b) Structure of Mpro from SARS-CoV-2 and discovery of its inhibitors. Nature 582:289–293. https://doi.org/10.1038/s41586-020-2223-y
Jumper J, Evans R, Pritzel A et al (2021) Highly accurate protein structure prediction with AlphaFold. Nature 596:583–589. https://doi.org/10.1038/s41586-021-03819-2
Karunakaran KB, Balakrishnan N, Ganapathiraju MK (2022) Interactome of SARS-CoV-2 modulated host proteins with computationally predicted PPIs: insights from translational systems biology studies. Front Syst Biol 2.https://doi.org/10.3389/fsysb.2022.815237
Kern DM, Sorum B, Mali SS et al (2021) Cryo-EM structure of the SARS-CoV-2 3a ion channel in lipid nanodiscs. bioRxiv. https://doi.org/10.1101/2020.06.17.156554
Kim Y, Jedrzejczak R, Maltseva NI et al (2020) Crystal structure of Nsp15 endoribonuclease NendoU from SARS-CoV-2. Protein Sci 29:1596–1605
Kneller DW, Phillips G, O'Neill HM et al (2020a) Structural plasticity of SARS-CoV-2 3CL M(pro) active site cavity revealed by room temperature X-ray crystallography. Nat Commun 11:3202. https://doi.org/10.1038/s41467-020-16954-7
Kneller DW, Phillips G, Kovalevsky A et al (2020b) Room-temperature neutron and X-ray data collection of 3CL M(pro) from SARS-CoV-2. Acta Crystallogr F Struct Biol Commun 76:483–487. https://doi.org/10.1107/S2053230X20011814
Kneller DW, Phillips G, Weiss KL et al (2020c) Unusual zwitterionic catalytic site of SARS-CoV-2 main protease revealed by neutron crystallography. J Biol Chem 295:17365–17373. https://doi.org/10.1074/jbc.AC120.016154
Kneller DW, Zhang Q, Coates L et al (2021a) Michaelis-like complex of SARS-CoV-2 main protease visualized by room-temperature X-ray crystallography. IUCrJ 8:973–979. https://doi.org/10.1107/S2052252521010113
Kneller DW, Phillips G, Weiss KL et al (2021b) Direct observation of protonation state modulation in SARS-CoV-2 main protease upon inhibitor binding with neutron crystallography. J Med Chem 64:4991–5000. https://doi.org/10.1021/acs.jmedchem.1c00058
Kneller DW, Li H, Galanie S et al (2021c) Structural, electronic, and electrostatic determinants for inhibitor binding to subsites S1 and S2 in SARS-CoV-2 main protease. J Med Chem 64:17366–17383. https://doi.org/10.1021/acs.jmedchem.1c01475
Kneller DW, Li H, Phillips G et al (2022) Covalent narlaprevir- and boceprevir-derived hybrid inhibitors of SARS-CoV-2 main protease. Nat Commun 13:2268. https://doi.org/10.1038/s41467-022-29915-z
Liu H, Iketani S, Zask A et al (2022) Development of optimized drug-like small molecule inhibitors of the

SARS-CoV-2 3CL protease for treatment of COVID-19. Nat Commun 13:1891. https://doi.org/10.1038/s41467-022-29413-2

Lubin JH, Zardecki C, Dolan EM et al (2022) Evolution of the SARS-CoV-2 proteome in three dimensions (3D) during the first 6 months of the COVID-19 pandemic. Proteins 90:1054–1080. https://doi.org/10.1002/prot.26250

Lynch ML, Snell EH, Bowman SEJ (2021) Structural biology in the time of COVID-19: perspectives on methods and milestones. IUCrJ 8:335–341. https://doi.org/10.1107/S2052252521003948

Mandala VS, McKay MJ, Shcherbakov AA et al (2020) Structure and drug binding of the SARS-CoV-2 envelope protein transmembrane domain in lipid bilayers. Nat Struct Mol Biol 27:1202–1208. https://doi.org/10.1038/s41594-020-00536-8

Michalska K, Kim Y, Jedrzejczak R et al (2020) Crystal structures of SARS-CoV-2 ADP-ribose phosphatase: from the apo form to ligand complexes. IUCrJ 7

Nelson C, Minasov G, Shuvalova L et al (2020) Structure of the SARS-CoV-2 orf7a encoded accessory protein. https://doi.org/10.2210/pdb6W37/pdb

Newman JA, Douangamath A, Yadzani S et al (2021) Structure, mechanism and crystallographic fragment screening of the SARS-CoV-2 NSP13 helicase. Nat Commun 12:4848. https://doi.org/10.1038/s41467-021-25166-6

Osipiuk J, Jedrzejczak R, Tesar C et al (2020) The crystal structure of papain-like protease of SARS CoV-2. https://doi.org/10.2210/pdb6W9C/pdb

Owen DR, Allerton CMN, Anderson AS et al (2021) An oral SARS-CoV-2 M(pro) inhibitor clinical candidate for the treatment of COVID-19. Science 374:1586–1593. https://doi.org/10.1126/science.abl4784

Pierri CL (2020) SARS-CoV-2 spike protein: flexibility as a new target for fighting infection. Signal Transduct Target Ther 5:254. https://doi.org/10.1038/s41392-020-00369-3

QCRG_Structural_Biology_Consortium (2021) CryoEM and AI reveal a structure of SARS-CoV-2 Nsp2, a multifunctional protein involved in key host processes. https://doi.org/10.2210/pdb7MSW/pdb

RCSB COVID-19/SARS-CoV-2 Resources, Available from: www.rcsb.org/covid19

Rogstam A, Nyblom M, Christensen S et al (2020) Crystal structure of non-structural protein 10 from severe acute respiratory syndrome coronavirus-2. Int J Mol Sci 21:7375

Rosas-Lemus M, Minasov G, Shuvalova L et al. (2020) High-resolution structures of the SARS-CoV-2 2′-O-methyltransferase reveal strategies for structure-based inhibitor design. Science Signaling 13:eabe1202. https://doi.org/10.1126/scisignal.abe1202

Rynkiewicz P, Lynch ML, Cui F et al (2022) Functional binding dynamics relevant to the evolution of zoonotic spillovers in endemic and emergent Betacoronavirus strains. J Biomol Struct Dyn 40:10978–10996. https://doi.org/10.1080/07391102.2021.1953604

Semper C, Watanabe N, Savchenko A (2021) Structural characterization of Nonstructural protein 1 from SARS-CoV-2. iScience 24:101903

Sobitan A, Mahase V, Rhoades R et al (2021) Computational saturation mutagenesis of SARS-CoV-1 spike glycoprotein: stability, binding affinity, and comparison with SARS-CoV-2. Front Mol Biosci 8:784303. https://doi.org/10.3389/fmolb.2021.784303

Stogios P, Skarina T, Chang C et al (2020) Crystal structure of the ubiquitin-like domain 1 (Ubl1) of Nsp3 from SARS-CoV-2. https://doi.org/10.2210/pdb7KAG/pdb

Subramaniam S (2020) COVID-19 and cryo-EM. IUCrJ 7:575–576. https://doi.org/10.1107/S2052252520008799

VirologicalWebsite, Available from: https://virological.org/t/novel-2019-coronavirus-genome/319

Weeks S, De Graef S, Munawar A (2020) X-ray Crystallographic Structure of Orf9b from SARS-CoV-2. https://doi.org/10.2210/pdb6Z4U/pdb

Wierbowski SD, Liang S, Liu Y et al (2021) A 3D structural SARS-CoV-2-human interactome to explore genetic and drug perturbations. Nat Methods 18:1477–1488. https://doi.org/10.1038/s41592-021-01318-w

Wilamowski M, Hammel M, Leite W et al (2021) Transient and stabilized complexes of Nsp7, Nsp8, and Nsp12 in SARS-CoV-2 replication. Biophys J 120:3152–3165. https://doi.org/10.1016/j.bpj.2021.06.006

Woo H, Park SJ, Choi YK et al (2020) Developing a fully glycosylated full-length SARS-CoV-2 spike protein model in a viral membrane. J Phys Chem B 124:7128–7137. https://doi.org/10.1021/acs.jpcb.0c04553

Wrapp D, Wang N, Corbett KS et al (2020) Cryo-EM structure of the 2019-nCoV spike in the prefusion conformation. Science 367:1260–1263. https://doi.org/10.1126/science.abb2507

Wu F, Zhao S, Yu B et al (2020a) A new coronavirus associated with human respiratory disease in China. Nature 579:265–269. https://doi.org/10.1038/s41586-020-2008-3

Wu F, Zhao S, Yu B et al (2020b) A new coronavirus associated with human respiratory disease in China. Nature 579:265–269

Yang H, Rao Z (2021) Structural biology of SARS-CoV-2 and implications for therapeutic development. Nat Rev Microbiol 19:685–700. https://doi.org/10.1038/s41579-021-00630-8

Ye G, Gallant JP, Massey C et al (2020) The development of a novel nanobody therapeutic for SARS-CoV-2. bioRxiv https://doi.org/10.1101/2020.11.17.386532

Yuan M, Wu NC, Zhu X et al (2020) A highly conserved cryptic epitope in the receptor binding domains of SARS-CoV-2 and SARS-CoV. Science 368:630–633. https://doi.org/10.1126/science.abb7269

Zou J, Yin J, Fang L et al (2020) Computational prediction of mutational effects on SARS-CoV-2 binding by relative free energy calculations. J Chem Inf Model 60:5794–5802. https://doi.org/10.1021/acs.jcim.0c00679

Dr. Miranda L. Lynch is a Research Scientist at the University at Buffalo Hauptman-Woodward Institute. She also has an affiliate appointment in the Department of Structural Biology in the Jacobs School of Medicine and Biomedical Sciences at the State University of New York at Buffalo. Her research as a quantitative scientist involves mathematical and Bayesian modeling, as well as functional analysis and image analysis. Her application areas include problems in macromolecular structural biology, in cancer research, and in infectious disease research, using mathematical and computational tools to advance understanding of human health and disease.

Dr. Sarah E. J. Bowman is an Associate Professor in the Department of Biochemistry, Jacobs School of Medicine and Biomedical Sciences at the State University of New York at Buffalo. She is the Director of the National Crystallization Center at the University at Buffalo Hauptman-Woodward Institute, a major resource for the structural biology community worldwide. Her research expertise is in using structural and biophysical methods to probe biomolecular structure, function, and dynamics, with specific focus on metalloproteins. Her research laboratory develops new techniques for crystallization of biological macromolecules, for detection of small crystals, and for building new pipelines for cutting-edge structural approaches.

The Human Milk Antibody Response to SARS-CoV-2 Infection

4

Rebecca L. R. Powell

Abstract

There is an enormous body of evidence demonstrating the benefits of human milk antibodies (Abs) in protecting infants against a wide range of infections. We have collected milk longitudinally from hundreds of previously infected participants, in studies designed originally as no-contact at the onset of COVID-19 cases in New York City in March 2020. We have found that ~ 90% of samples are positive for Spike-specific Immunoglobulin A (IgA). Levels of IgA and secretory Abs were shown to be strongly correlated. The secretory Immunoglobulin A (sIgA) response was dominant compared to the Immunoglobulin G (IgG) response. Seventy five percent of samples tested exhibited significant neutralizing activity strongly correlated with IgA titer. Nine to twelve months after infection, 94% of samples continued to exhibit positive anti-Spike titers, with 50% of participants exhibiting < two-fold reduction of specific IgA, indicating a highly durable Ab response. Milk IgA was extracted from high-titer donors, pooled, and tested for its capacity to reduce viral lung burden in SARS-CoV-2 susceptible mice. Extracted IgA significantly reduced SARS-CoV-2 titers compared to control IgA, with a maximum TCID50 reduction of 4.9 logs. Our work highlights the importance of continued chest/breastfeeding after SARS-CoV-2 infection and supports further development of milk-derived, polyclonal sIgA therapeutics.

Keywords

COVID-19 · SARS-CoV-2 · Lactation · Milk · Immunity · Infants · Secretory immunoglobulin A · sIgA · Chest/breastfeeding

Research funded by

NIH/NIAID (National Institute of Allergy and Infectious Diseases) Award R01 AI158214-01—Comprehensive assessment of SARS-CoV-2-reactive antibodies in human milk to determine their potential as a COVID-19 therapeutic and as a means to prevent infection of breastfed babies.

COVID Information Commons Video Presentation

The COVID Information Commons (CIC) Research Lightning Talk video presentation and individual multilingual transcripts for this project can be found in the Columbia Academic Commons (AC) (Powell 2023).

R. L. R. Powell (✉)
Department of Medicine, Division of Infectious Diseases, Icahn School of Medicine at Mount Sinai, New York, NY, USA
e-mail: rebecca.powell@mssm.edu

4.1 Introduction

There is an enormous body of evidence demonstrating the benefits of human milk in protecting against a wide range of infections for the first

F. D. Hudson (ed.), *The COVID Information Commons*, Sustainable Development Goals Series,
https://doi.org/10.1007/978-3-032-06125-6_4

6–12 months and even through the first five years of life (Frank et al. 2019; Ardic and Yavuz 2018; Ladomenou et al. 2010; Duijts et al. 2010). Though milk antibodies (Abs) are thought to play a major role, their contribution has not been well-elucidated, especially in terms of pathogen-specific Abs. Human milk contains ~ 0.7 g/L IgA, comprising ~ 90% of the total antibody (Ab), nearly all in secretory form (Weaver et al. 1998; Hurley and Theil 2011). Secretory Immunoglobulin A (sIgA) is critical in protecting all mucosae from infection, creating an essential barrier from the outside world (Li et al. 2020). sIgA is most often comprised of dimeric IgA linked by joining (j) chain. This dimer is secreted by the B cell, with j-chain marking it for ultimate transport by the polymeric Ig receptor (pIgR) on epithelial cells into the milk (Cerutti and Rescigno 2008). pIgR is cleaved as it transports, leaving one of its domains, secretory component (SC), attached, thus creating sIgA (Low et al. 2010; Brandtzaeg 2010). SC protects these secretory antibodies (sAbs), rendering them highly stable and resistant to enzymatic degradation in relatively harsh mucosal environments such as human milk, the oral/nasal cavity, the stomach, and the gastrointestinal (GI), genitourinary, and respiratory tracts (Hurley and Theil 2011; Demers-Mathieu et al. 2018). Human milk Immunoglobulin G (IgG) comprises ~ 2% of total milk Ab but both the total amount and particularly pathogen-specific IgG can increase after infection or vaccination. Non-secretory Abs may contribute to infant protection, particularly if titers are high. Milk-derived Abs are likely most active in the oral/nasal cavity and upper GI tract of infants, though ~ 30–50% of these Abs have been shown to resist degradation in the stomach for as long as two hours suggesting that they are functional throughout the GI tract and possibly beyond (Demers-Mathieu et al. 2018). B cells in the mammary gland that ultimately produce IgA that becomes sIgA predominately originate from the gut-associated lymphoid tissue (GALT), exemplifying a critical *entero-mammary pathway,* wherein the sAbs found in human milk echo the immunogens in the lactating parent's GI tract (Brandtzaeg 2003, 2010). sIgA can mediate protective functions by various mechanisms, all of which might contribute to pathogen clearance. sIgA can block pathogen entry (neutralization), which has been demonstrated for various bacterial and viral pathogens and in our own work (Corthésy 2013; Fox et al. 2022). The neutralization effect in many of these studies has been shown to be significantly more potent using sIgA compared to monomeric IgA, likely due in part to its enhanced durability and increased avidity as a polymer (Corthésy 2013; Perrier et al. 2006). It is also important to keep in mind that many Abs that are unable to prevent pathogen entry facilitate a variety of "non-neutralizing", protective activities (Holmes et al. 1986; Chenoweth et al. 2020). These functions are mediated by the crystallizable fragment of the Ig (Fc), via interactions with Fc receptors (FcRs) found on most leukocytes (Ackerman and Nimmerjahn 2014). Notably, like IgG, sIgA has been shown to elicit FcR-mediated functions (Corthésy 2013; Perrier et al. 2006; Monteiro and Winkel 2003; Franca et al. 2011; Musich et al. 2015; Hiemstra et al. 1987).

When we initiated our COVID-19 human milk immune studies in April 2020, virtually nothing was known about the antibody (Ab) response in human milk to SARS-CoV-2. Additional research was urgently needed to test human milk for SARS-CoV-2-specific antibodies (Abs) and their functions. Knowing the typology and degree of COVID-19-specific Abs in human milk will help inform smart policy and treatment decisions for the many pregnant and breastfeeding mothers who are or will become infected by SARS-CoV-2.

4.2 Methodology and Results

4.2.1 Milk Ab Elicited Shortly Following Infection

Initially, we sought to characterize the types and magnitude of targeted Abs in human milk against SARS-CoV-2 in 15 milk samples obtained from donors previously-infected with COVID-19, three to four weeks after symptoms had abated. Milk samples were initially evaluated for IgA binding reactivity by human IgA-specific ELISA

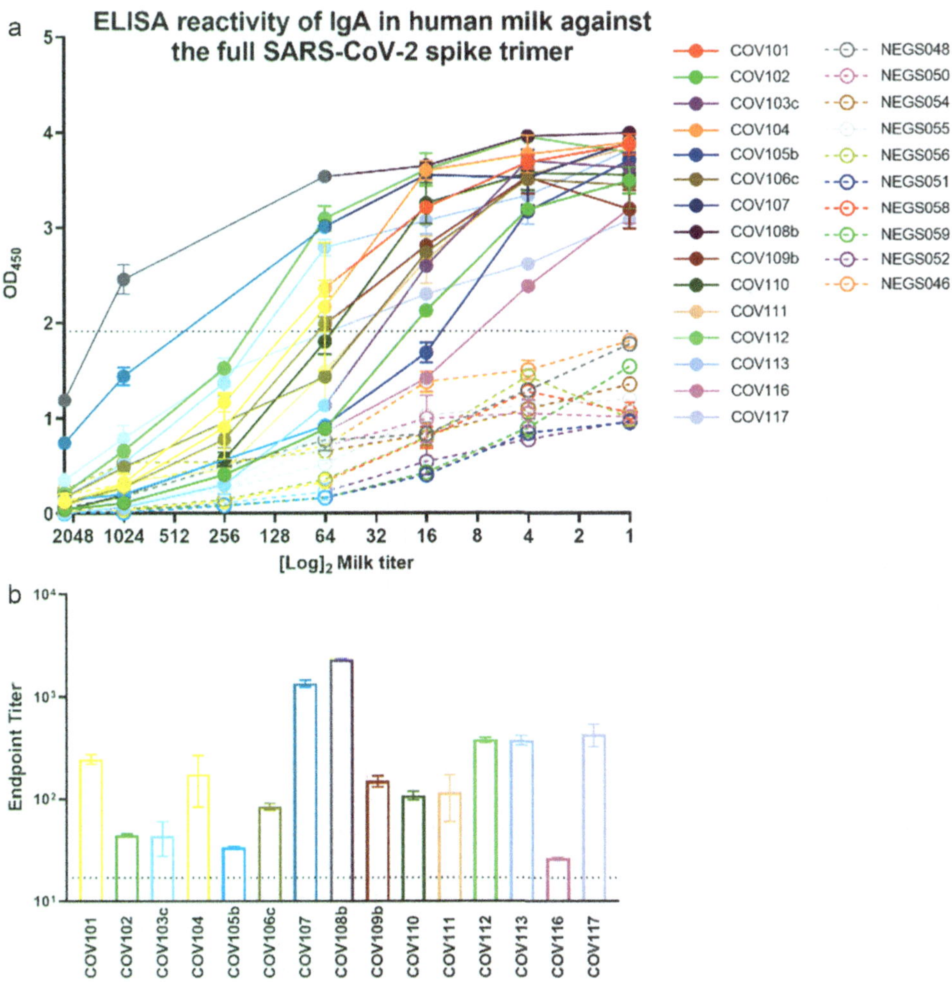

Fig. 4.1 Human milk IgA reactivity to SARS-CoV-2 Spike following infection. Top panel **a** ELISA titration curve. Bottom Panel **b** Endpoint titers. Dotted lines represent pre-pandemic control samples. Each color represents a unique participant

(enzyme-linked immunosorbent assay) to the full trimeric SARS-CoV-2 Spike (Fig. 4.1). It was evident that all samples obtained from COVID 19-recovered donors (100%), in undiluted form, exhibited binding activity significantly above that of the pre-pandemic control milk samples, which did exhibit some low-level non-specific or cross-reactive activity. Milk samples were titrated and endpoint titer values were determined. Twelve of 15 samples (80%) obtained from previously-COVID-19-infected donors exhibited significant IgA binding activity to receptor binding domain (RBD), as determined by human IgA-specific ELISA, compared to controls in undiluted form (Fig. 4.2a). Milk samples were titrated and endpoint titer values determined. It was found that nine of the 12 samples with reactive IgA to RBD when undiluted (75%) exhibited significant endpoint titers (Fig. 4.2c). Notably, all 12 of the milk samples with significant IgA reactivity in undiluted form to RBD (100%) were also positive for RBD-specific secretory Ab reactivity as determined by human SC-specific ELISA (Fig. 4.2b). All 12 milk samples also exhibited positive secretory Ab endpoint titers upon dilution compared to controls (Fig. 4.2d).

In our larger follow-up study, 66 of 74 samples (89%) were positive for Spike-specific

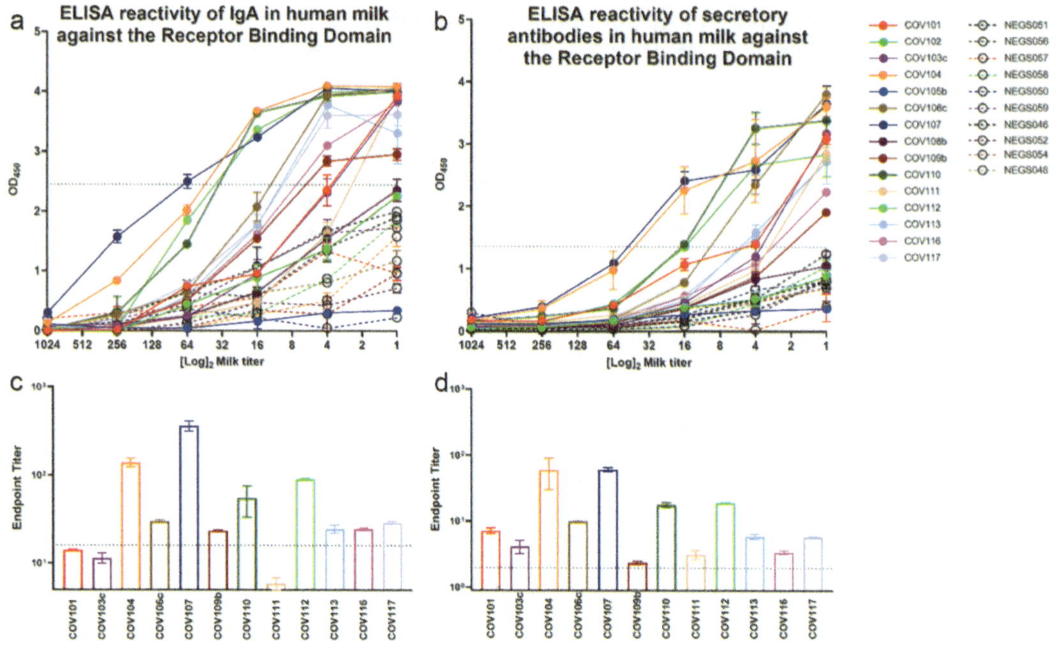

Fig. 4.2 Human milk Ab reactivity to Spike RBD following infection

IgA, with the COVID-19 samples exhibiting significantly higher Spike-specific IgA binding compared to controls (Fig. 4.3a; $p < 0.0001$). Following this initial screening, 40 of the Spike-positive samples were further titrated to determine binding endpoint titers as an assessment of Ab affinity and/or quantity (Fig. 4.3b). Thirty-eight of 40 (95%) Spike-reactive samples exhibited positive IgA endpoint titers, and 19 of these samples (50%) were $\geq$ five times higher than the endpoint titer of the positive cutoff value and were therefore designated as "high-titer" (Fig. 4.3c).

Additionally, 20 samples assayed for Spike-specific IgA were also assessed for Spike-specific secretory Ab (by detecting for SC) and IgG. Nineteen of these undiluted milk specimens (95%) from convalescent COVID-19 donors were positive for Spike-specific secretory Abs compared to pre-pandemic control milk (Fig. 4.4a). One sample (COV125b) was negative for specific IgA but positive for specific secretory Ab, while another sample (COV123b) was positive for specific IgA but negative for specific secretory Ab. Eighteen undiluted milk samples (95%) exhibiting Spike-specific secretory Ab activity also exhibited positive endpoint titers (Fig. 4.4c). Of the samples found to be high-titer for Spike-specific IgA, seven were also high-titer for specific secretory Ab (70%). Mean optical density (OD) values for undiluted milk and endpoint titers were used in separate Spearman correlation tests to compare IgA and secretory Ab reactivity (Fig. 4.4e). It was found that IgA and secretory Ab levels were positively correlated (using ODs: $r = 0.77, p < 0.0001$; using endpoint titers: $r = 0.86$, $p < 0.0001$). Additionally, 15/20 undiluted milk samples from COVID-19-recovered donors were positive for Spike-specific IgG compared to pre-pandemic controls (75%; Fig. 4.4b), with 13/15 of these samples exhibiting a positive endpoint titer (87%; Fig. 4.4d), and 2/15 designated as high titer with values $\geq$ five times cutoff (13%). No correlation was found between IgG and IgA titers or between IgG and SC titers.

Total IgA was extracted from eight COVID-19 samples and eight control milk samples. All eight COVID-19 samples had been shown to exhibit positive Spike-specific IgA and secretory Ab titers. Neutralization capacity was tested using

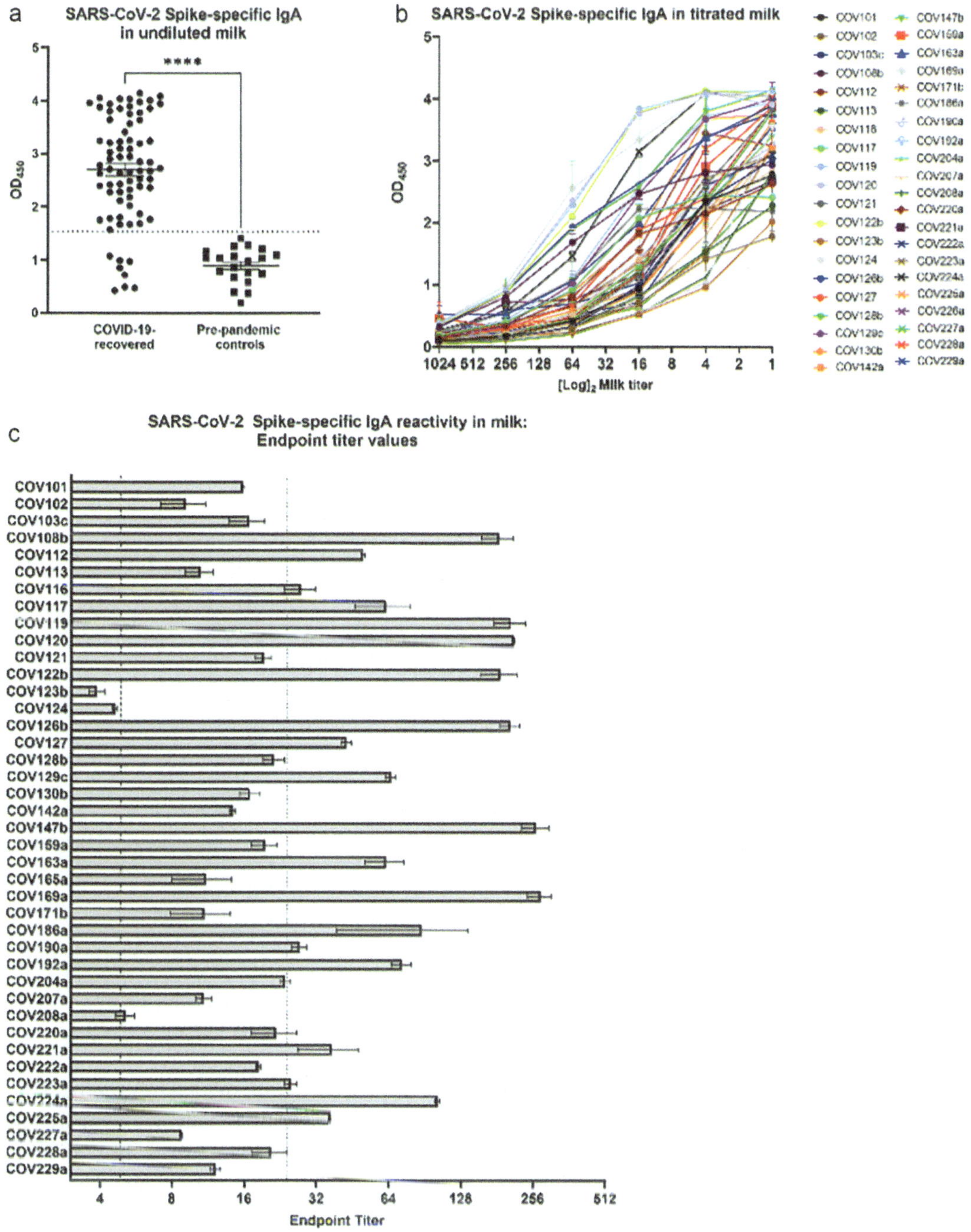

Fig. 4.3 Expanded follow-up analysis of anti-Spike IgA in milk

a Vesicular Stomatitis Virus (VSV)-based pseudovirus assay, wherein the native VSV surface protein G is replaced by the SARS-CoV-2 Spike, as described previously (Fig. 4.5) (Oguntuyo et al. 2021). At the maximum concentration tested (200 μg/mL total purified milk IgA), 6/8 (75%) COVID-19 samples exhibited > 50% neutralization (mean, 87% neutralization; range, 70–100%), while only 1/8 control samples (13%) achieved this benchmark (94% neutralization;

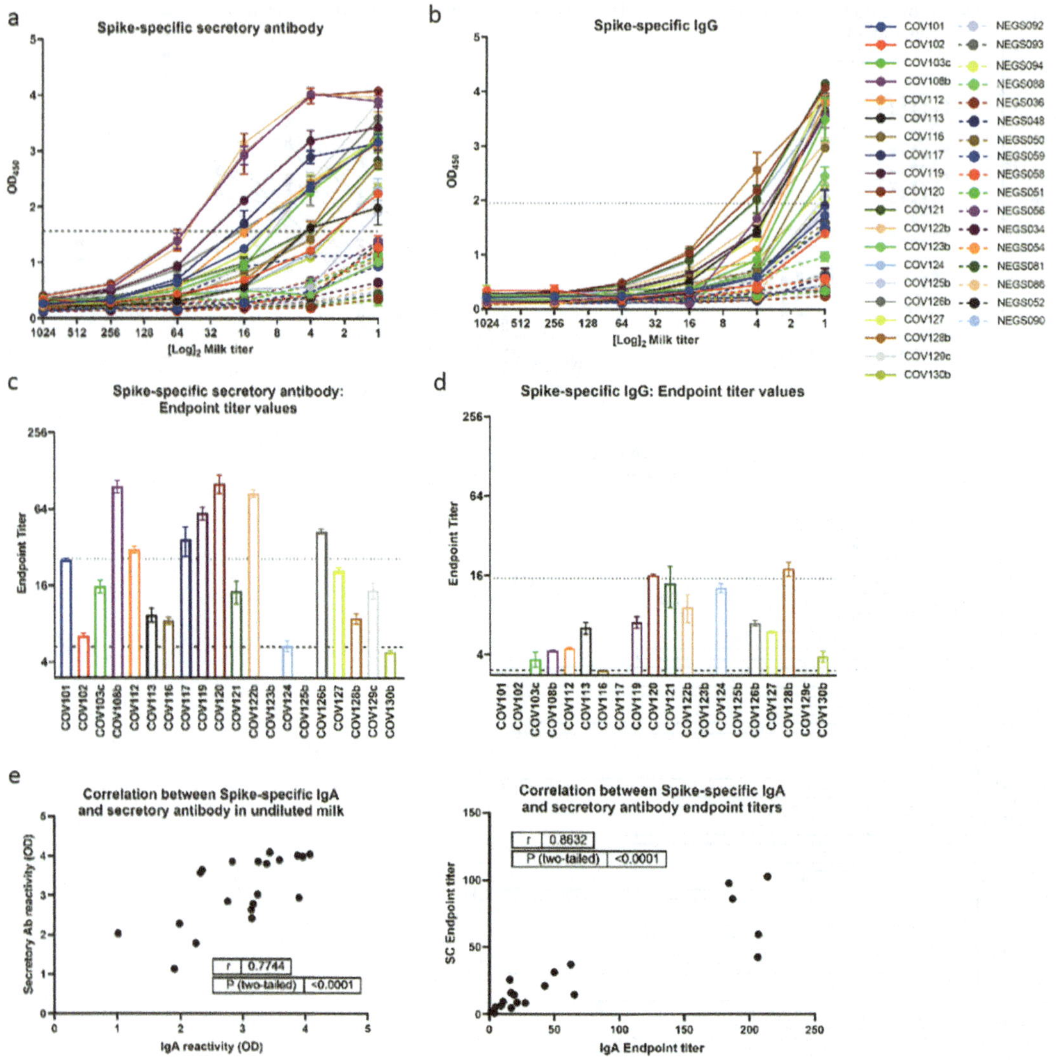

Fig. 4.4 Milk IgG and sAb anti-Spike reactivities in milk following infection

Fig. 4.5a). Mean percent neutralization values at 50 ug/mL extracted IgA were grouped and compared among COVID-19 and pre-pandemic control samples. COVID-19 samples exhibited significantly greater neutralization compared to controls ($p = 0.0064$; Fig. 4.5b). As well, when the concentration of IgA required to achieve 50% neutralization (IC50) was determined, 7/8 pre-pandemic controls did not achieve 50% neutralization (IC50 > 200 ug/mL) while, for the COVID-19 samples, 2/8 did not achieve 50% neutralization, and the mean IC50 for the six COVID-19 specimens that displayed neutralizing activity was 33.6 ug/mL of total IgA (range, 2.39–89.4 ug/mL; Fig. 4.5c). Finally, we compared the neutralization IC50 titers to the IgA endpoint titers measured for these samples. There was a significant positive correlation between IgA binding and neutralization capacities ($r = 0.83$, $p = 0.0154$; Fig. 4.5d). Notably, the two non-neutralizing COVID-19 IgA samples also exhibited the lowest IgA endpoint titers (COV121, COV130; mean IgA endpoint titers of 19 and 17, respectively), while the six neutralizing samples exhibited high Spike-specific IgA binding titers (Fig. 4.5c).

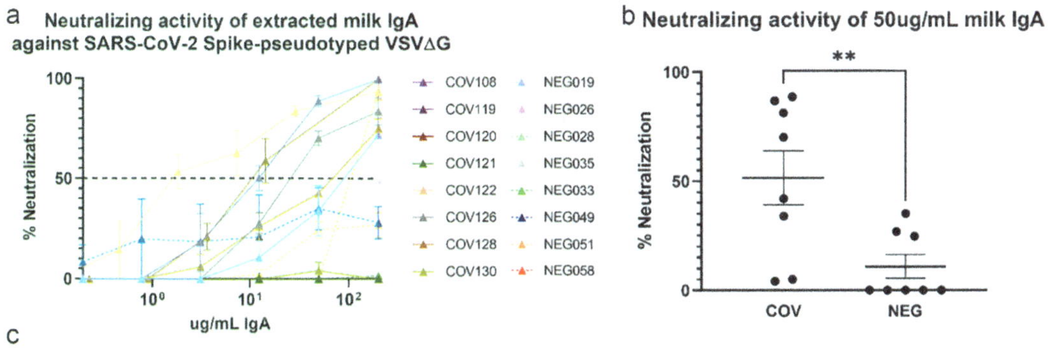

c

Neutralization IC50 values

COV108	COV119	COV120	COV121	COV122	COV126	COV128	COV130
89.42313	11.81476	10.39056	>200	2.391491	23.84901	64.03523	>200
NEG019	**NEG026**	**NEG028**	**NEG033**	**NEG035**	**NEG038**	**NEG049**	**NEG051**
>200	>200	>200	>200	>200	178.6628	>200	>200

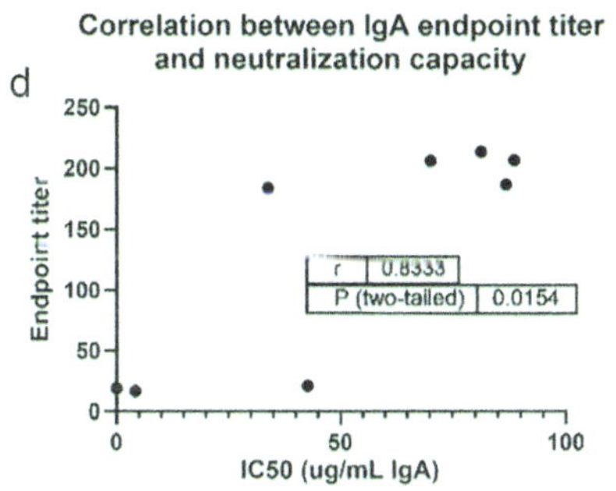

Fig. 4.5 SARS-CoV-2 neutralization by extracted milk IgA

4.2.2 Durability of the Milk Ab Response Over Time

As a further longitudinal study, 16 sets of milk samples obtained at sequential time points from COVID-19-recovered lactating individuals were examined. Samples were analyzed for IgA and sAb reactivities against SARS-CoV-2 Spike using a Luminex-based immunoassay as described in methods. All milk samples obtained from COVID-19-recovered participants exhibited positive median fluorescence intensity (MFI) indicating Spike-specific IgA and sAb 3–8 weeks post-infection (Fig. 4.6a, b). Longitudinal milk samples were similarly assessed, and endpoint titers determined for each assay. It was found that at the last time point available (9–12 months) 14/16 participants (88%) continued to exhibit positive IgA endpoint titers against Spike (Fig. 4.6c), and 15/16 participants (94%) continued to exhibit positive sAb endpoint titers against Spike (Fig. 4.6d).

Endpoint titers were grouped by time point in order to determine if significant changes had occurred over the study period (Fig. 4.6e, f). Values were compared by paired ANOVA test. At the initial three to eight weeks post-infection time point, samples exhibited a mean IgA endpoint titer against Spike of 1354 (range 43–3465). At the following time point examined, 5–6 months post-infection, samples exhibited a mean IgA endpoint titer against Spike of 960 (range 20–4323), which was not significantly different from the initial time point (Fig. 4.6e). At nine months samples exhibited a mean IgA endpoint titer of Spike of 740 (range 33–3437), which was also not significantly different compared to either the initial time point or the 5–6 month time point. Following

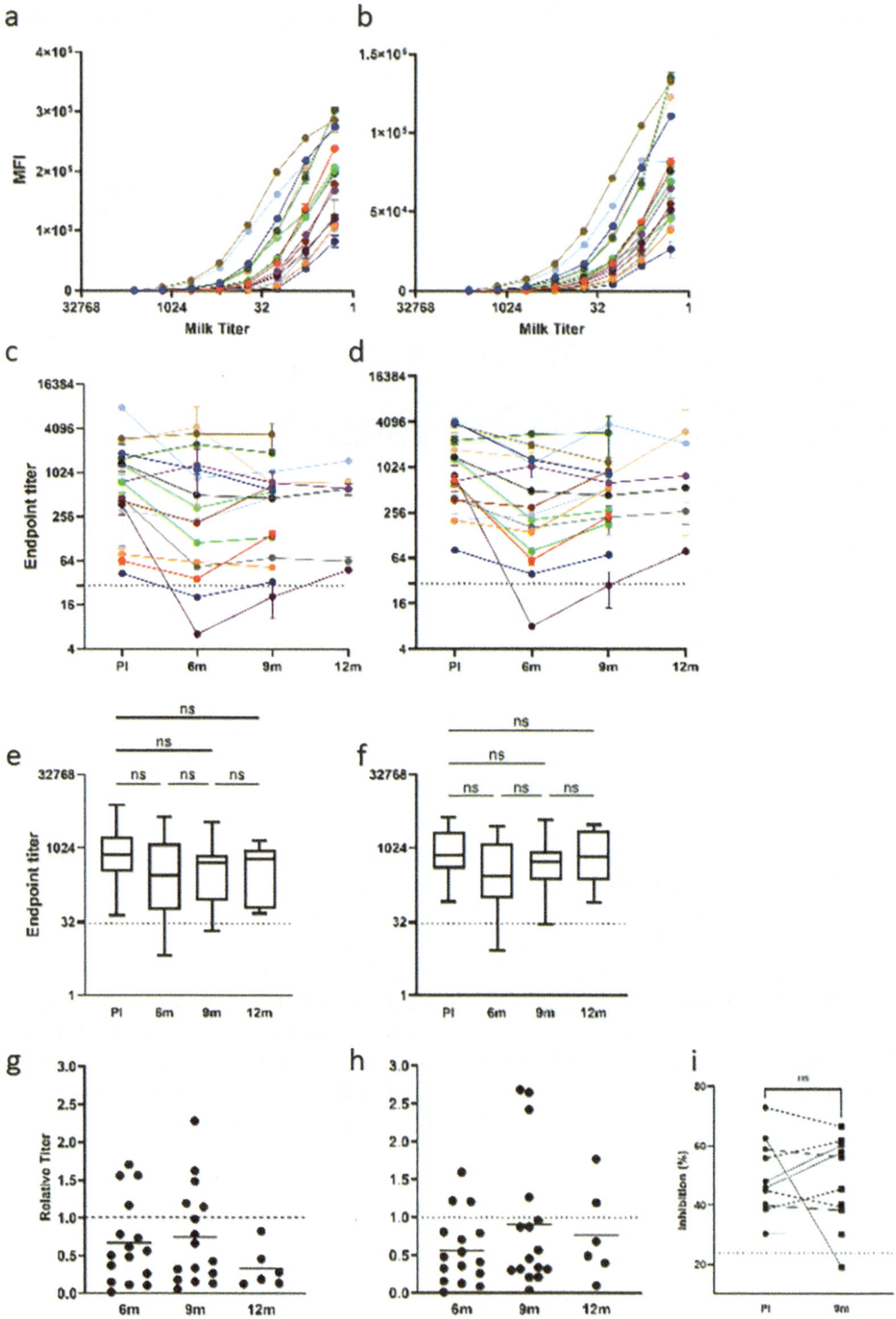

Fig. 4.6 Durability of milk anti-Spike IgA and sAb over time. IgA: **a**, **c**, **e**, **g**. sAb: **b**, **d**, **f**, **h**. **i**: neutralization activity

a highly similar trend, mean sAb titers against Spike at 3–8 weeks (mean 1436, range 82–4000), 5–6 months (mean 721, range 8–2835), and 9–12 months (mean 1007, range 28–3816) were not significantly different (Fig. 4.6f).

To evaluate the relative change of IgA and sAb levels over time, endpoint titers for each longitudinal milk sample were normalized by the individual's endpoint titer of their initial milk sample obtained 3–8 weeks post-infection (Fig. 4.6g, h). Notably, 94% of the participants exhibited < tenfold reduction in IgA endpoint titer 5–6 months post-infection, and 56% exhibited < twofold reduction. By nine months, 50% of participants exhibited < two-fold decrease in IgA titer, having decreased by only 25% on average compared to those at 3–8 weeks (range: 73% decrease to 127% increase). By the end of the study period, IgA had decreased by 64% (17% to 87%); however, this was not significantly different from the nine month time point (Fig. 4.6g). Examining the durability of Spike-specific sAb, COV296 and COV157 exhibited > tenfold decrease in titer at 5–6 months post-infection, with the remaining participants exhibiting < tenfold decreases (titers were 57% of those compared to those observed at 3–8 weeks). At nine months post-infection, sAb titers were on average maintained at 90% of those measured at 3–8 weeks post-infection, with only COV295 exhibiting > tenfold decrease, though this titer was also similarly increased compared to that measured at 5–6 months. Fifty percent of participants exhibited < twofold reduction by the end of the study period (Fig. 4.6h). A proxy neutralization assay that measures Abs specific for the Spike receptor binding domain (RBD) that blocks the RBD-ACE2 interaction was performed using a premade kit (GenScript). Eleven pairs of samples analyzed above were examined for changes in neutralization capacity from 3–8 weeks post-infection to nine months post-infection. At 3–8 weeks post-infection, the mean % neutralization of milk at neat was 49% (range 30–73%), with all samples exhibiting neutralization activity that was significantly greater than that exhibited by naïve milk ($p = 0.0029$). At nine months, the mean % neutralization was 47% (range 30–67%) (Fig. 4.6i). Although one sample's neutralization capacity dropped below that of the naïve controls at 9 months (COV364, from 63 to 19%). Notably, COV364 dropped 67% in IgA titer and 89% in sAb titer at nine months post-infection. It was found that no significant change in neutralization was detected between the early and late groups.

4.2.3 Safety of Human Milk Feeding Following SARS-CoV-2 Infection

Undoubtedly, milk from a COVID-19-recovered person is highly likely to contain a high amount of specific sIgA that may have a protective effect on the breastfed infant. One major, important question from the outset of the pandemic has been whether a SARS-CoV-2-infected person's milk might be a vehicle for SARS-CoV-2 transmission. This question stirred considerable panic and caused mothers and babies to be separated, particularly at birth, often with adverse consequences to establishment of the breastfeeding relationship, in some cases, irrevocably (Spatz et al. 2020). Now, more than 18 months into the pandemic, this question has been explored by several groups.

To date, there has been no evidence that SARS-CoV-2 transmits via *human milk.* Numerous studies of colostrum and mature milk from women with acute SARS-CoV-2 infection have failed to find any viral RNA in milk samples. This includes several early studies conducted in Wuhan, China, that examined milk from infected mothers of newborns by RT-PCR (Liu et al. 2020; Chen et al. 2020c, 2020a, 2020b; Dong et al. 2020; Cui et al. 2020; Fan et al. 2021; Kam et al. 2020; Wang et al. 2020), reviewed in Rose et al. 2020; Mirbeyk et al. 2021; Lackey et al. 2020). One early study from Italy and several of the Chinese reports documented both symptomatic and asymptomatic SARS-CoV-2-infected infants with probable mother-to-infant transmission (likely via respiratory droplets) or unknown transmission routes, still with no evidence of viral RNA in the mothers' milk samples (Kam et al. 2020; Wang et al. 2020; Salvatori et al. 2020). Notably, as cases of infected mothers continued to be analyzed, viral RNA was ultimately identified in a small minority of milk samples studied (Gross et al. 2020; Chambers et al. 2020; Zhu et al. 2020). In a single case, SARS-CoV-2 RNA was detected in milk at days four, five, and seven after symptom onset, with subsequent samples found to be negative (Gross et al. 2020). In another study that included 18 women providing 64 milk samples, one milk sample had detectable SARS-CoV-2 RNA, on the day of symptom onset (Chambers et al. 2020). Subsequent samples were negative for viral RNA. Notably, collection methods in these reports do not always include masking, cleaning of the breast, or even handwashing to avoid contamination of viral RNA from the donor's respiratory droplets. Indeed, viral RNA in such studies has been found on the breast skin (Pace et al. 2021).

It is also critical to note that even where viral RNA was detected in milk, infectious SARS-CoV-2 particles were not. Several meta-analyses to date have failed to find evidence of worsened health outcomes for infants of any age breastfed by infected mothers, or increased rates of infection for these infants compared to those who are formula-fed (Centeno-Tablante et al. 2021; Perez-Bermejo et al. 2021). In one prospective study of 19 women in New York City testing positive for SARS-CoV-2 infection at delivery, RT-PCR of colostrum samples found all but one to be negative for viral RNA, with no evidence of infection in the baby of the mother with the positive milk sample, nor in any other baby in the study (Kunjumon et al. 2021). A larger New York City-based study of mothers with confirmed or suspected SARS-CoV-2 infection at delivery included 101 newborns monitored and tested while the mothers were still in hospital, with 55 followed for 10–25 days after birth (Dumitriu et al. 2021). This study found only two infants to be SARS-CoV-2-positive (very low viral load) just after birth, with none of the infants in the study exhibiting any COVID-19 pathology, despite most newborns breastfeeding directly from the mother (with appropriate mask and hand hygiene encouraged) and rooming-in Dumitriu et al. (2021).

Given the clear safety profile of milk from SARS-CoV-2-infected mothers, the significant amount of immunological data demonstrating a robust and rapid specific Ab response, and the general immense benefits of breastfeeding, particularly in low-income countries, the World Health Organization (WHO), Centers for Disease Control and Prevention (CDC), and all major relevant associations recommend that infants not be separated from SARS-CoV-2-infected mothers and that breastfeeding should be established and not disrupted (depending on the mothers' desire to do so), in combination with masking and other hygiene measures (Team W 2020; CDC 2020).

4.2.4 Therapeutic Applications of Extracted Milk IgA

sIgA is challenging to produce recombinantly, but is naturally found in human milk, which could be considered a global resource for this biologic, justifying its development as a mucosal therapeutic. In this vein, in collaboration with Lactiga Inc. (Lactiga™ 2024), we utilized SARS-CoV-2 as a model mucosal pathogen, and methods were developed to efficiently extract human milk sIgA from donors who were naïve to SARS-CoV-2 or had recovered from infection that elicited high-titer anti-SARS-CoV-2 Spike sIgA in their milk (pooled to make LCTG-002). IgA was extracted from pooled human milk samples from either pre-pandemic, COVID-naive lactating people as a negative control (control IgA) or COVID-recovered lactating people (LCTG-002) based on methods disclosed in patent U.S. 11,124,560, *Stable pooled breastmilk antibodies for oral delivery*. Samples were run on SDS-PAGE, stained with Coomassie blue, and imaged. A pattern of three protein bands was consistently visible in both batches corresponding to the known molecular weights of SC, heavy alpha chain, and light chain. Two additional batches were prepared from pooled human milk and imaged and demonstrated identical protein bands (data not shown). Protein purity was assessed across all four batches by SDS-PAGE imaging and ranged from 93.2 to 99.2% across batches (data not shown). Samples from the control IgA and LCTG-002 batches were also evaluated by western blot to compare sIgA content. The blot was immunostained for human IgA, and the signal was identical in both batches.

A separate blot was immunostained for human SC, and the signal was also identical in both batches, confirming that both batches had similar human IgA content which was predominantly in the secretory (sIgA) form. To quantitatively assess relative protein abundance, mass spectrometry was performed on the control IgA and LCTG-002 batches, and data were analyzed using the search engine MaxQuant with label-free quantification to generate intensity-based absolute quantification (iBAQ) values. Proteins with a relative abundance of 1.0% or greater were plotted in descending order. These proteins were all associated with sIgA, specifically: immunoglobulins, SC (which is a cleavage product of polymeric immunoglobulin receptor (PIGR)), and J chain (JCHAIN) (data not shown). None of the proteins exhibited statistically significant differences between batches, establishing that protein composition was comparable.

The binding specificity of LCTG-002 against SARS-CoV-2 Spike was assessed by ELISA and binding curves were used to determine endpoint titers. Using a secondary Ab specific for human IgA, Spike-specific IgA binding endpoint was found to be significantly greater for LCTG-002 compared to control IgA (mean endpoint of 0.87 vs. 5.87; $p = 0.0133$; data not shown). Using a secondary Ab specific for human SC to confirm that the Spike-specific human IgA were largely in secretory form (sIgA), it was similarly found that LCTG-002 exhibited significantly higher Spike-specific binding compared to control IgA (mean endpoint of 1.82 vs. 6.90; $p = 0.0032$). To further probe the biological characteristics of LCTG-002, an ELISA-based proxy neutralization assay was used to measure LCTG-002's ability to compete with the interaction between ACE2 receptor and the Spike RBD. IC50 was calculated from competition binding curves, and it was shown that inhibition activity of RBD-ACE2 binding by LCTG-002 was significantly greater compared to control (mean IC50 of 154 ug/mL for LCTG-002; 50% inhibition not achieved for control; $p = 0.0002$; Fig. 4.7a, b). Additionally, a SARS-CoV-2 pseudovirus cell-based neutralization assay was also performed using VERO-ACE2-TMPRSS2 cells and a Spike-pseudotyped VSV as described previously (Ikegame et al. 2021). Titrated LCTG-002 and control IgA were used to generate neutralization curves, and IC50 was calculated. It was found that LCTG-002 exhibited significantly greater

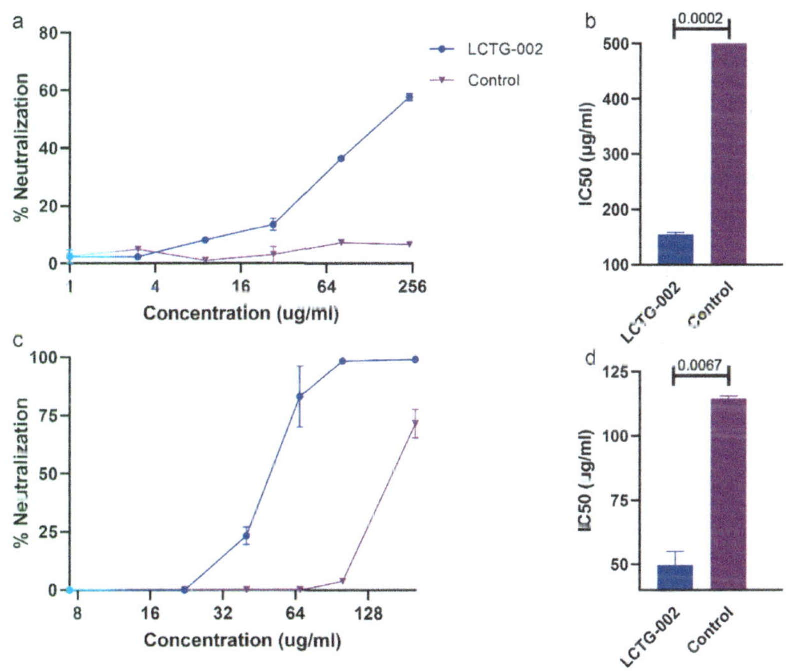

Fig. 4.7 SARS-CoV-2 Neutralization by pooled extracted milk IgA from previously-infected donors (LCTG-002)

neutralization activity compared to control (mean IC50 of 49.8 ug/mL vs. 114.5 ug/mL; $p = 0.0067$; Fig. 4.7c, d).

In order to test the potential antiviral efficacy of LCTG-002 against SARS-CoV-2 infection, K18-hACE2 transgenic mice were selected for use, based on literature indicating that SARS-CoV-2 productively replicates in this mouse model that expresses human ACE2 and models clinical manifestations of COVID-19 including severe lung disease, thrombosis, and vasculitis (Zheng et al. 2021). All mice were bright, alert, and responsive to manual restraint throughout the study. Mean percent weight loss was not more than 10% for any study group (data not shown). PBS, control IgA, or LCTG-002 were administered both intranasally and intratracheally to model IgA deposition across the nasal cavity, airways, and lungs, which are targets of SARS-CoV-2 infection in humans. Administration occurred one day before, the day of, and each day after intranasal inoculation with an infectious dose of 4.6×10^4 TCID50 SARS-CoV-2, as outlined in Table 4.1. A positive control group of mice received Nirmatrelvir, the SARS-CoV-2 protease inhibitor component of Paxlovid, by the oral route, as this compound has been demonstrated previously to inhibit SARS-CoV-2 replication in the K18 + hACE2 transgenic mouse model (Owen et al. 2021; Jeong et al. 2022).

Viral titers were measured at the end of the study in the lungs 72 h post inoculation via TCID50, plaque forming units (PFU) and qRT-PCR assay. Mice receiving 250 ug/day of 2.5 mg/mL LCTG-002 exhibited statistically significant reductions in viral titer as measured by all 3 assays compared to mice receiving 250 ug/day of control IgA ($p = 0.02$–$p < 0.0001$; Fig. 4.8a). Mice receiving 1 mg/day of 10 mg/mL LCTG-002 exhibited even greater reductions in viral titer across all three assays, ($p < 0.0001$), notably exhibiting a 4.9-log reduction in TCID50. Compared to control, this group (1 mg/day of LCTG-002) exhibited significantly less weight loss (6.6% vs. 0% mean loss by day 3; $p = 0.019$, data not shown), as did the Nirmatrelvir control treatment (6.6% vs. 0.6% mean loss by day 3; $p = 0.033$) (data not shown). Nirmatrelvir treatment caused significant reductions in viral titer by TCID50 and PFU assays but not qRT-PCR, as similarly observed in prior unpublished studies using the same dose and route.

An additional mouse study was performed measuring viral titers in nasal washes and lung homogenates 24 h after SARS-CoV-2 inoculation to determine whether prophylactic administration of LCTG-002 inhibited nasal and/or lung replication of SARS-CoV-2 in an earlier stage of infection. Compared to mice receiving PBS, mice receiving 250 ug/day LCTG-002 exhibited statistically significant reductions in viral titers in

Table 4.1 Experimental design for K18 + hACE2 mouse studies

Group	Animals	Test article	Dose/route	Frequency	Duration of infection
1	3 males + 3 females	0.25 mg control IgA	0.125 mg in 50uL (25uL per naris) IN (am); 0.125 mg in 50uL IT (pm)	24 h prior to inoculation[a], then daily to day 2 (4 doses total)	3 days
2		0.25 mg LCTG-002			
3		1 mg LCTG-002	0.5 mg in 50uL (25uL per naris) IN (am); 0.5 mg in 50uL IT (pm)		
4		Nirmatrelvir	250 mg/kg in 10 mL/kg volume PO		
5		PBS	50uL (25uL per naris) IN (am); 50uL IT (pm)	24 h prior to, and on the day of inoculation* (2 doses total)	1 day
6		0.25 mg LCTG-002	0.125 mg in 50uL (25uL per naris) IN (am); 0.125 mg in 50uL IT (pm)		

[a] Mice were inoculated with 4.6×10^4 TCID50 SARS-CoV-2. *IN* intranasal; *IT* intratracheal; *PO* oral administration

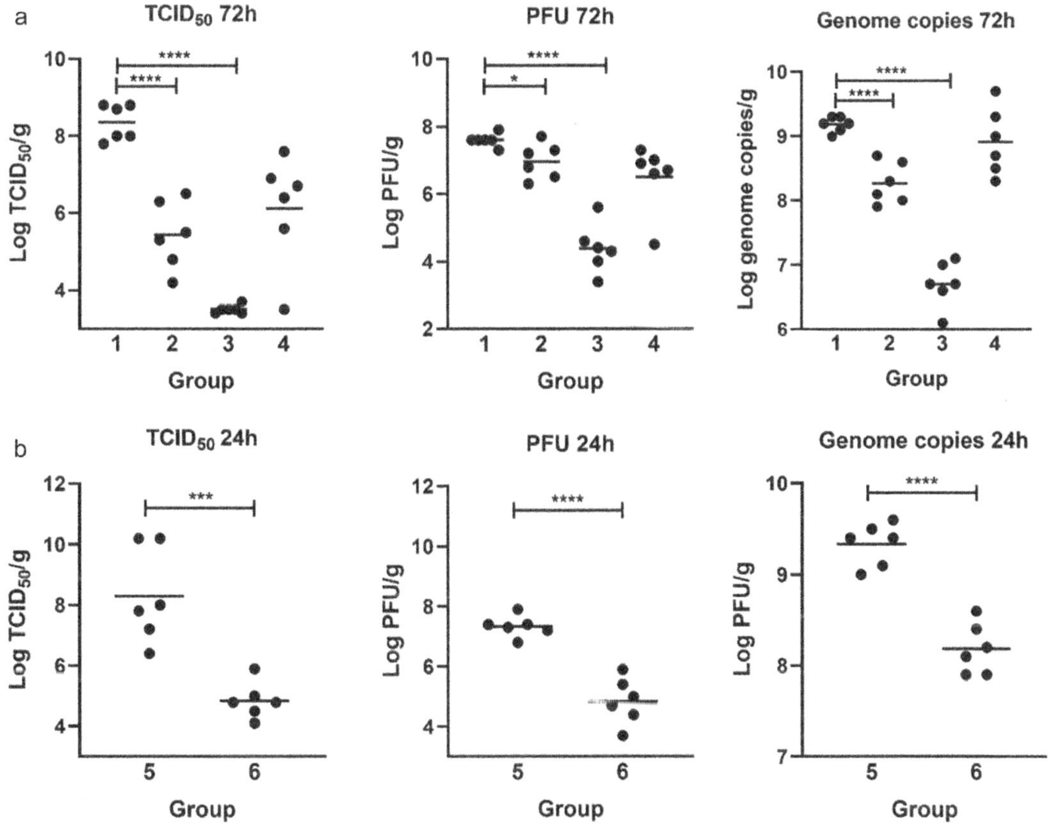

Fig. 4.8 In-vivo antiviral efficacy of extracted pooled milk IgA from COVID-19 recovered donors (LCTG-002) against SARS-CoV-2

lung across TCID50, PFU, and qRT-PCR assays ($p = 0.005$–$p < 0.0001$; Fig. 4.8b). However, no significant reductions were observed in the nasal washes of mice receiving 250 ug/day LCTG-002 compared to PBS treated mice (data not shown).

4.3 Conclusions

Nearly all sAb derives from the GALT, via the *entero-mammary* pathway—homing of Ab-secreting B cells from the gut to the mammary gland. Various animal studies have demonstrated this migration and homing during late pregnancy and lactation. Homing appears to be controlled hormonally, as well as by various adhesion factors on the B cells and the maternal vasculature including MadCAM-1, integrin α4β7, CCL28, and CCR10 (Rodriguez et al. 2021). This link is an evolutionarily critical mechanism facilitating specific protection to a vulnerable infant against the pathogens in the maternal/infant environment sampled by the maternal GALT and provides key immunological training for the infant (Rodriguez et al. 2021). Infants benefit greatly from the sAb provided in human milk, as the neonate mucosal immune system is relatively deficient in sAb production as well as other key immune factors. Even past the neonatal period, these Abs can supplement the infant's own immunity to provide protection against pathogens against which the infant does not yet have immunological protection. sAbs are critical due to their relative durability in harsh environments including the milk itself and all mucosal fluids including the oral/nasal cavity and GI tract (Hurley and Theil 2011; Demers-Mathieu et al. 2018).

SARS-CoV-2 does not transmit via *human milk*. Though viral RNA has been identified in a small minority of milk samples studied (Gross et al. 2020; Chambers et al. 2020; Zhu et al. 2020), infectious SARS-CoV-2 particles have not. Several meta-analyses to date have failed to find evidence of worsened health outcomes for infants of any age chest/breastfed by a SARS-CoV-2-infected person or increased rates of infection for these infants compared to those who are formula-fed (Centeno-Tablante et al. 2021; Perez-Bermejo et al. 2021; Dumitriu et al. 2021). Unlike the observed decay over time of the serum response that has been widely cited in other reports (Seow et al. 2020; Bruel et al. 2022; Muecksch et al. 2022; Pan et al. 2021), we observed little change in Spike-specific milk titers over time. Notably, 50% of participants exhibited < two-fold reduction of Spike-specific IgA, and 94% of participants sustained positive sAb titers against Spike through 12 months. IgA and sAb titers maintained a strong significant positive correlation throughout the study period. The serological response against SARS-CoV-2 varies by variant, disease severity, and pre-existing conditions (Röltgen et al. 2020). Generally, the decay rate of Spike/RBD-specific serum IgA is high. The half-life of RBD-specific IgA has been found to be ~ 71 days, with certain reports finding serum IgA levels to be undetectable after 60 days (Seow et al. 2020; Iyer et al. 2020; Sterlin et al. 2021). Serum IgG half-life has been reported in multiple studies to be > three months, with 57% of individuals who were asymptomatically infected to be IgG seropositive after 8 months, with 71% of those who experienced mild symptoms maintaining positive IgG titers (Seow et al. 2020; Iyer et al. 2020; Sterlin et al. 2021). In the present study, Spike-specific sIgA titers in milk were shown to be significantly more durable over time. The highly durable titers observed in the present study may be reflective of long-lived plasma cells in the GALT and/or mammary gland, as well as continued antigen stimulation in these compartments, possibly by other human coronaviruses, or repeated exposures to SARS-CoV-2. Consistent with the present data, it has been shown in general that although an initial IgA plasmablast response quickly declines, IgA-producing plasma cells can persist for decades in the human GALT, even when not measurable in the periphery (Landsverk et al. 2017). Given the established entero-mammary link, the highly durable Spike-specific sIgA in milk observed herein strongly suggests similarly long-lived plasma cells in the mammary gland.

The present data suggest that lactating individuals who recover from COVID-19 are likely to continue producing specific sIgA in their milk for one year or more. Based on studies of other mucosal pathogens, milk sIgA likely provides critical passive immunity to infants against SARS-CoV-2 infection and COVID-19 pathology. The significant durability of the Spike-specific sIgA response in milk highlights the importance of continued chest/breastfeeding after SARS-CoV-2 infection. Based on the in vivo efficacy results presented herein, future studies will assess nasal formulation strategies for LCTG-002 which will inform nonclinical toxicology studies to be performed in support of clinical trials. The clinical trial design may contemplate pre- and post-exposure prophylaxis against COVID-19 in adults and adolescents with moderate to severe immune compromise due to a medical condition or immunosuppressive medications. sIgA delivered to the nasal cavity and/or lungs would not be expected to bypass the respiratory epithelium into the systemic circulation, instead favoring local deposition (Chow et al. 2023). In support of this, a SARS-CoV-2-neutralizing mAb delivered intranasally into mice resulted in a 30-fold lower concentration of Ab in serum and higher lung lavage levels as compared to the same dose delivered intravenously (Cao et al. 2022). In the context of respiratory mucosal infections like COVID-19, deposition and retention of a therapeutic such as LCTG-002 in the local mucosa is highly desirable.

As Spike-specific sIgA titers are detectable in human milk for 12 months or more after SARS-CoV-2 infection (Fox et al. 2022; Yang et al. 2023), with COVID-19 vaccination prior to or following infection boosting this response (Powell 2023), human milk represents a sustainable and durable source of protective Ab against COVID-19 with an enormous potential donor pool globally. Unlike monomeric Ab, sIgA is highly stable and resistant to degradation in all relatively harsh mucosal compartments and fluids (Hurley and Theil 2011; Fouda et al. 2017). This study establishes a novel proof of concept for airway administration of human milk-derived, polyclonal, pooled sIgA as a well-tolerated and repeatable prophylactic antiviral treatment, with broader applicability beyond SARS-CoV-2.

References

Ackerman M, Nimmerjahn F (2014) Antibody Fc, 1st edn. Elsevier, San Diego

Ardic C, Yavuz E (2018) Effect of breastfeeding on common pediatric infections: a 5-year prospective cohort study. Arch Argent Pediatr 116(2):126–132. https://doi.org/10.5546/aap.2018.eng.126

Brandtzaeg P (2003) Mucosal immunity: integration between mother and the breast-fed infant. Vaccine 21(24):3382–3388

Brandtzaeg P (2010) The mucosal immune system and its integration with the mammary glands. J Pediatr 156(2):S8-15. https://doi.org/10.1016/j.jpeds.2009.11.014

Bruel T, Stéfic K, Nguyen Y et al (2022) Longitudinal analysis of serum neutralization of SARS-CoV-2 Omicron BA. 2, BA. 4, and BA. 5 in patients receiving monoclonal antibodies. Cell Rep Med 3(12):100850

Cao X, Maruyama J, Zhou H et al (2022) Unbiased approach to identify and assess efficacy of human SARS-CoV-2 neutralizing antibodies. Sci Rep 12(1):15517. https://doi.org/10.1038/s41598-022-19780-7

CDC (2020) Coronavirus Disease (COVID-19) and Breastfeeding, 4 June 2020

Centeno-Tablante E, Medina-Rivera M, Finkelstein JL et al (2021) Transmission of SARS-CoV-2 through breast milk and breastfeeding: a living systematic review. Ann N Y Acad Sci 1484(1):32–54. https://doi.org/10.1111/nyas.14477

Cerutti A, Rescigno M (2008) The biology of intestinal immunoglobulin A responses. Immunity 28(6):740–750. https://doi.org/10.1016/j.immuni.2008.05.001

Chambers C, Krogstad P, Bertrand K et al (2020) Evaluation for SARS-CoV-2 in breast milk from 18 infected women. JAMA 324(13):1347–1348. https://doi.org/10.1001/jama.2020.15580

Chen H, Guo J, Wang C et al (2020a) Clinical characteristics and intrauterine vertical transmission potential of COVID-19 infection in nine pregnant women: a retrospective review of medical records. Lancet 395(10226):809–815. https://doi.org/10.1016/S0140-6736(20)30360-3

Chen Y, Peng H, Wang L et al (2020b) Infants born to mothers with a new coronavirus (COVID-19). Front Pediatr 8:104. https://doi.org/10.3389/fped.2020.00104

Chen L, Li Q, Zheng D et al (2020) Clinical Characteristics of Pregnant Women with Covid-19 in Wuhan, China. N Engl J Med 382(25):e100. https://doi.org/10.1056/NEJMc2009226

Chenoweth AM, Wines BD, Anania JC et al (2020) Harnessing the immune system via FcγR function in immune therapy: a pathway to next-gen mAbs. Immunol Cell Biol. https://doi.org/10.1111/imcb.12326

Chow MYT, Pan HW, Seow HC et al (2023) Inhalable neutralizing antibodies—promising approach to combating respiratory viral infections. Trends Pharmacol Sci 44(2):85–97. https://doi.org/10.1016/j.tips.2022.11.006

Corthésy B (2013) Multi-faceted functions of secretory IgA at mucosal surfaces. Front Immunol 4:185. https://doi.org/10.3389/fimmu.2013.00185

Cui Y, Tian M, Huang D et al (2020) A 55-day-old female infant infected with 2019 novel coronavirus disease: presenting with pneumonia, liver injury, and heart damage. J Infect Dis 221(11):1775–1781. https://doi.org/10.1093/infdis/jiaa113

De Rose DU, Piersigilli F, Ronchetti MP et al (2020) Novel coronavirus disease (COVID-19) in newborns and infants: what we know so far. Ital J Pediatr 46(1):56. https://doi.org/10.1186/s13052-020-0820-x

Demers-Mathieu V, Underwood MA, Beverly RL, et al (2018) Comparison of human milk immunoglobulin survival during gastric digestion between preterm and term infants. Nutrients 10(5). https://doi.org/10.3390/nu10050631

Dong L, Tian J, He S et al (2020) Possible vertical transmission of SARS-CoV-2 from an infected mother to her newborn. JAMA 323(18):1846–1848. https://doi.org/10.1001/jama.2020.4621

Duijts L, Jaddoe VW, Hofman A et al (2010) Prolonged and exclusive breastfeeding reduces the risk of infectious diseases in infancy. Pediatrics 126(1):e18-25. https://doi.org/10.1542/peds.2008-3256

Dumitriu D, Emeruwa UN, Hanft E et al (2021) Outcomes of neonates born to mothers with severe acute respiratory syndrome coronavirus 2 infection at a large medical center in New York City. JAMA Pediatr 175(2):157–167. https://doi.org/10.1001/jamapediatrics.2020.4298

Fan C, Lei D, Fang C et al (2021) Perinatal transmission of 2019 coronavirus disease-associated severe acute respiratory syndrome coronavirus 2: should we worry? Clin Infect Dis 72(5):862–864. https://doi.org/10.1093/cid/ciaa226

Fouda GG, Eudailey J, Kunz EL et al (2017) Systemic administration of an HIV-1 broadly neutralizing dimeric IgA yields mucosal secretory IgA and

virus neutralization. Mucosal Immunol 10(1):228–237. https://doi.org/10.1038/mi.2016.32
Fox A, Marino J, Amanat F et al (2022) The IgA in milk induced by SARS-CoV-2 infection is comprised of mainly secretory antibody that is neutralizing and highly durable over time. PLoS ONE 17(3):e0249723. https://doi.org/10.1371/journal.pone.0249723
Franca EL, Morceli G, Fagundes DL et al (2011) Secretory IgA-Fcalpha receptor interaction modulating phagocytosis and microbicidal activity by phagocytes in human colostrum of diabetics. APMIS 119(10):710–719. https://doi.org/10.1111/j.1600-0463.2011.02789.x
Frank NM, Lynch KF, Uusitalo U et al (2019) The relationship between breastfeeding and reported respiratory and gastrointestinal infection rates in young children. BMC Pediatr 19(1):339. https://doi.org/10.1186/s12887-019-1693-2
Gross R, Conzelmann C, Muller JA et al (2020) Detection of SARS-CoV-2 in human breastmilk. Lancet 395(10239):1757–1758. https://doi.org/10.1016/S0140-6736(20)31181-8
Hiemstra PS, Gorter A, Stuurman ME et al (1987) Activation of the alternative pathway of complement by human serum IgA. Eur J Immunol 17(3):321–326. https://doi.org/10.1002/eji.1830170304
Holmes MJ, Callow KA, Childs RA et al (1986) Antibody dependent cellular cytotoxicity against coronavirus 229E-infected cells. Br J Exp Pathol 67(4):581–586
Hurley WL, Theil PK (2011) Perspectives on immunoglobulins in colostrum and milk. Nutrients 3(4):442–474. https://doi.org/10.3390/nu3040442
Ikegame S, Siddiquey MNA, Hung CT et al (2021) Neutralizing activity of Sputnik V vaccine sera against SARS-CoV-2 variants. Res Sq. https://doi.org/10.21203/rs.3.rs-400230/v1
Iyer AS, Jones FK, Nodoushani A et al (2020) Persistence and decay of human antibody responses to the receptor binding domain of SARS-CoV-2 spike protein in COVID-19 patients. Sci Immunol 5(52):eabe0367
Jeong JH, Chokkakula S, Min SC et al (2022) Combination therapy with nirmatrelvir and molnupiravir improves the survival of SARS-CoV-2 infected mice. Antiviral Res 208:105430. https://doi.org/10.1016/j.antiviral.2022.105430
Kam KQ, Yung CF, Cui L et al (2020) A well infant with coronavirus disease 2019 with high viral load. Clin Infect Dis 71(15):847–849. https://doi.org/10.1093/cid/ciaa201
Kunjumon B, Wachtel EV, Lumba R et al (2021) Breast Milk and breastfeeding of infants born to SARS-CoV-2 positive mothers: a prospective observational cohort study. Am J Perinatol 38(11):1209–1216. https://doi.org/10.1055/s-0041-1731451
Lackey KA, Pace RM, Williams JE et al (2020) SARS-CoV-2 and human milk: what is the evidence? Matern Child Nutr 16(4):e13032. https://doi.org/10.1111/mcn.13032
Lactiga™ is an award-winning, venture-backed biotherapeutics company developing novel biologics to treat and prevent infections with a focus on improving the quality of life in patients with Primary Immunodeficiency Diseases. https://lactiga.com/ (2024)
Ladomenou F, Moschandreas J, Kafatos A et al (2010) Protective effect of exclusive breastfeeding against infections during infancy: a prospective study. Arch Dis Child 95(12):1004–1008. https://doi.org/10.1136/adc.2009.169912
Landsverk OJ, Snir O, Casado RB et al (2017) Antibody-secreting plasma cells persist for decades in human intestine. J Exp Med 214(2):309–317. https://doi.org/10.1084/jem.20161590
Li Y, Jin L, Chen T (2020) The effects of secretory IgA in the mucosal immune system. Biomed Res Int 2020:2032057. https://doi.org/10.1155/2020/2032057
Liu W, Wang J, Li W et al (2020) Clinical characteristics of 19 neonates born to mothers with COVID-19. Front Med 14(2):193–198. https://doi.org/10.1007/s11684-020-0772-y
Low EN, Zagieboylo L, Martino B et al (2010) IgA ASC accumulation to the lactating mammary gland is dependent on VCAM-1 and alpha4 integrins. Mol Immunol 47(7–8):1608–1612. https://doi.org/10.1016/j.molimm.2010.01.015
Mirbeyk M, Saghazadeh A, Rezaei N (2021) A systematic review of pregnant women with COVID-19 and their neonates. Arch Gynecol Obstet 304(1):5–38. https://doi.org/10.1007/s00404-021-06049-z
Monteiro RC, Van De Winkel JG (2003) IgA Fc receptors. Annu Rev Immunol 21:177–204. https://doi.org/10.1146/annurev.immunol.21.120601.141011
Muecksch F, Wise H, Templeton K et al (2022) Longitudinal variation in SARS-CoV-2 antibody levels and emergence of viral variants: a serological analysis. The Lancet Microbe. 3(7):e493–e502
Musich T, Demberg T, Morgan IL et al (2015) Purification and functional characterization of mucosal IgA from vaccinated and SIV-infected rhesus macaques. Clin Immunol 158(2):127–139. https://doi.org/10.1016/j.clim.2015.03.020
Oguntuyo KY, Stevens CS, Hung CT et al (2021) Quantifying absolute neutralization titers against SARS-CoV-2 by a standardized virus neutralization assay allows for cross-cohort comparisons of COVID-19 sera. mBio 12(1). https://doi.org/10.1128/mBio.02492-20
Owen DR, Allerton CMN, Anderson AS et al (2021) An oral SARS-CoV-2 M(pro) inhibitor clinical candidate for the treatment of COVID-19. Science 374(6575):1586–1593. https://doi.org/10.1126/science.abl4784
Pace RM, Williams JE, Jarvinen KM et al (2021) Characterization of SARS-CoV-2 RNA, antibodies, and neutralizing capacity in milk produced by women with COVID-19. mBio 12(1). https://doi.org/10.1128/mBio.03192-20
Pan Y, Jiang X, Yang L et al (2021) SARS-CoV-2-specific immune response in COVID-19 convalescent individuals. Signal Transduct Target Ther 6(1):256
Perez-Bermejo M, Peris-Ochando B, Murillo-Llorente MT (2021) COVID-19: relationship and impact on

breastfeeding-a systematic review. Nutrients 13(9). https://doi.org/10.3390/nu13092972
Perrier C, Sprenger N, Corthésy B (2006) Glycans on secretory component participate in innate protection against mucosal pathogens. J Biol Chem 281(20):14280–14287. https://doi.org/10.1074/jbc.M512958200
Powell R (2023) Unpublished
Powell R (2023) Comprehensive assessment of SARS-CoV-2-reactive antibodies in human milk to determine their potential as a COVID-19 therapeutic and as a means to prevent infection of breastfed babies. Columbia Academic Commons, Academic Unit: Northeast Big Data Innovation Hub, Series: COVID-19 Research Lightning Talks. https://doi.org/10.7916/53bq-d932
Rodriguez JM, Fernandez L, Verhasselt V (2021) The GutBreast Axis: programming health for life. Nutrients 13(2). https://doi.org/10.3390/nu13020606
Röltgen K, Powell AE, Wirz OF et al (2020) Defining the features and duration of antibody responses to SARS-CoV-2 infection associated with disease severity and outcome. Sci Immunol 5(54):eabe0240
Salvatori G, De Rose DU, Concato C et al (2020) Managing COVID-19-positive maternal-infant dyads: an Italian experience. Breastfeed Med 15(5):347–348. https://doi.org/10.1089/bfm.2020.0095
Seow J, Graham C, Merrick B et al (2020) Longitudinal observation and decline of neutralizing antibody responses in the three months following SARS-CoV-2 infection in humans. Nat Microbiol 5(12):1598–1607
Spatz DL, Davanzo R, Muller JA et al (2020) Promoting and protecting human milk and breastfeeding in a COVID-19 world. Front Pediatr 8:633700. https://doi.org/10.3389/fped.2020.633700
Sterlin D, Mathian A, Miyara M et al (2021) IgA dominates the early neutralizing antibody response to SARS-CoV-2. Sci Transl Med 13(577). https://doi.org/10.1126/scitranslmed.abd2223
Team W. Breastfeeding and COVID-19. WHO, Geneva, 23 June 2020. WHO/2019-nCoV/Sci_Brief/Breastfeeding/2020.1
Wang S, Guo L, Chen L et al (2020) A case report of neonatal 2019 coronavirus disease in China. Clin Infect Dis 71(15):853–857. https://doi.org/10.1093/cid/ciaa225
Weaver LT, Arthur HM, Bunn JE et al (1998) Human milk IgA concentrations during the first year of lactation. Arch Dis Child 78(3):235–239
Yang X, Fox A, DeCarlo C et al (2023) The secretory IgA (sIgA) response in human milk against the SARS-CoV-2 Spike is highly durable and neutralizing for at least 1 year of lactation post-infection. medRxiv. https://doi.org/10.1101/2023.05.19.23290192
Zheng J, Wong LR, Li K et al (2021) COVID-19 treatments and pathogenesis including anosmia in K18-hACE2 mice. Nature 589(7843):603–607. https://doi.org/10.1038/s41586-020-2943-z
Zhu C, Liu W, Su H et al (2020) Breastfeeding risk from detectable severe acute respiratory syndrome coronavirus 2 in breastmilk. J Infect 81(3):452–482. https://doi.org/10.1016/j.jinf.2020.06.001

Rebecca L.R. Powell, PhD, CLC is Assistant Professor in the Departments of Medicine and Immunology and Immunotherapy at the Icahn School of Medicine at Mount Sinai Hospital in New York City (NYC). Her early post-doctoral research was performed at the International AIDS Vaccine Initiative Design and Development Lab, where she was involved in pre-clinical testing of replicating viral vector vaccines against HIV. Dr. Powell's background in HIV research brought her to Mount Sinai, where she first found her niche studying the impact of the cellular immune response in milk on mitigating mother-to-child transmission of HIV via breastfeeding. Dr. Powell has continued that work and currently has several other projects aimed to fill knowledge gaps about human milk immunology, exploring: the antibody response to COVID-19, influenza, and RSV; therapeutic use of extracted human secretory IgA; and the pre-clinical modeling and design of maternal vaccines aimed to elicit robust and protective milk antibody responses.

5 Agent-in-Cell Modeling of Pandemics

Amir Mohammad Esmaieeli Sikaroudi, Alon Efrat, and Michael Chertkov

Abstract

This study introduces an intermediate-level modeling approach that bridges the gap between complex agent-based models (ABMs) and traditional compartmental models for infectious diseases. This novel, geospatial information-aware ABM, termed agent-in-cell (AIC), leverages key concepts such as high-risk regions and points-of-interest (POIs). Its main contribution lies in the development of a robust and flexible methodology for constructing practicable ABMs, enhancing disease modeling in urban environments, and supporting public health strategies that require geographic specificity. AIC is designed to integrate a diverse set of data, including high-resolution mobility data from SafeGraph and coarser medical datasets. We demonstrate that AIC successfully replicates historical trends observed during the COVID-19 pandemic and provides reliable short-term spatiotemporal predictions of how pandemics may evolve from their current state. This chapter is complemented by Chap. 6 in this book by Sikaroudi, Efrat, and Chertkov, *Agent-In-Cell Modeling of a Pandemic: Harnessing Super-Agents for Predictive Modeling.*

Keywords

Agent based modeling · Pandemics · Mobility · COVID-19

5.1 Agent-Based Simulation: Literature Review and Our Contributions

In the field of epidemiology, conventional models such as the susceptible-infectious-recovered (SIR) compartmental framework, originating from the papers of Ross (Ross 1910), Kermack (Kermack et al. 1927), Anderson and May (1991), and many others, offer a highly aggregated view of disease spread within populations. While these models are valuable for certain applications, they

Research funded by

NSF Award #2027072—RAPID (Rapid Response Research Award): Infer and Control Global Spread of Corona-Virus with Graphical Models.

COVID Information Commons Video Presentation

The COVID Information Commons (CIC) Research Lightning Talk video presentation and individual multilingual transcripts for this project can be found in the Columbia Academic Commons (AC) (Chertkov 2021).

A. M. Esmaieeli Sikaroudi · A. Efrat · M. Chertkov (✉)
University of Arizona, Tucson, AZ, USA
e-mail: chertkov@arizona.edu

A. M. Esmaieeli Sikaroudi
e-mail: amesmaieeli@arizona.edu

A. Efrat
e-mail: alon@cs.arizona.edu

F. D. Hudson (ed.), *The COVID Information Commons*, Sustainable Development Goals Series,
https://doi.org/10.1007/978-3-032-06125-6_5

often simplify interactions and behaviors to the point where individual-level details are lost. In contrast, our approach serves as an intermediary, striking a balance between the complexity of agent-based models (ABMs) (Perez and Dragicevic 2009; Auchincloss and Diez Roux 2008; Epstein et al. 2008; Epstein 2009; Nianogo and Arah 2015) and the extreme aggregation of compartmental models (Ross 1910; Kermack et al. 1927; Anderson and May 1991).

In this chapter, we focus on designing new ABM methodology aimed specifically at improving modeling in epidemiology. However, it is also worth mentioning that the applicability approach of ABM is much broader. ABM simulations find applications in numerous fields, including sociology, economics, biology, ecology, computer science, and urban planning, among others. It serves as a virtual laboratory for exploring complex scenarios and conducting "what-if" analyses, making it an indispensable tool for decision-making, policy formulation, and understanding the dynamics of complex systems.

The core concept of ABM simulation is in the creation of virtual entities known as "agents," each equipped with individual characteristics, behaviors, and decision-making abilities. These agents interact with each other and their environment, giving rise to emergent behavior at the system level. An agent can represent a wide range of entities, such as individuals, animals, organizations, or even abstract entities like ideas or market forces. The primary advantage of ABM simulation lies in its ability to capture the intricacies of interactions and behaviors within a system, making it particularly suitable for studying systems with nonlinear, dynamic, and uncertain characteristics. While an ABM can benefit from introducing different agents, we work in this chapter only with agents representing individuals.

To facilitate the simulation process, agents are programmed with predefined rules or algorithms that dictate their actions and responses to various stimuli and events. As the simulation progresses, agents continuously adapt their behaviors based on internal states, external inputs, and interactions with other agents. Consequently, the aggregate behavior of the system emerges from the cumulative actions of individual agents, allowing us to gain insights into the collective behavior and outcomes of the modeled system.

Since the onset of the COVID-19 pandemic, a plethora of new open-source ABM tools have emerged. Among these, Covasim stands out, offering pandemic transmission modeling, intervention scenarios, and a versatile modeling library (Kerr et al. 2021). Shamil and colleagues (Shamil et al. 2021) have developed an agent-based simulation incorporating non-pharmaceutical interventions like lockdowns to control the reproduction number of COVID-19. However, their approach does not account for geographic heterogeneity, although they do differentiate agents' activities based on their occupations. Similarly, Silva and team (Silva et al. 2020) considered various lockdown scenarios and assessed their economic implications, but omitted geographical information. DESSABNeT is another notable ABM designed to simulate individual behavior and infection dynamics, though it lacks the geospatial trajectories of agents' activities (Stapelberg et al. 2021). Gupta and colleagues (Gupta et al. 2020) delved into the impact of COVID-19 contact tracing on virus spread and analyzed the economic consequences of interventions in Montreal, relying on aggregated geographical data.

OpenABM-COVID-19, as outlined by Hinch et al. (2021), has concentrated on medical aspects like different strains and vaccination patterns. Although geospatial information is not a focal point, it emphasizes demographic data relevant to geo-aggregated medical modeling. A distinctive approach was presented by Mahmood et al. (Mahmood et al. 2020), who implemented Flu And Coronavirus Simulator (FACS), an ABM accounting for the geospatial inhomogeneity of agent mobility in London. They evaluated their model by comparing the number of hospitalizations between historical data and their model's output. FACS-CHARM (dynamiC Hospital wARd Management) builds upon the FACS simulator by integrating hospital simulations, including aspects like ward management and intensive care unit operations. This extension offers a more comprehensive view on the utilization of medical services and hospital bed occupancy (Anagnostou et al. 2022).

In response to valuable insights from recent systematic reviews, notably the comprehensive analysis of ABMs in COVID-19 research highlighted by Lorig et al. (Lorig et al. 2021), we would like to acknowledge the diverse and heterogeneous application of ABMs and geographic information systems (GIS) data in pandemic modeling. Recognizing the identified gaps in data representation and the scarcity of open-source implementations, our approach uniquely integrates detailed geographic tessellation and provides an open-source framework. This methodology not only enhances the granularity and accuracy of pandemic simulations but also promotes transparency and reproducibility in the field.

Juxtaposing ABMs to traditional compartmental models, we notice that the ABMs excel at capturing the detailed behaviors and interactions of individuals within a population, offering a high-resolution lens through which complex systems can be examined. This granularity enables the exploration of scenario-specific dynamics and the effects of individual-level interventions, which are often overlooked in more aggregated modeling approaches. Conversely, compartmental models provide a streamlined, macroscopic view of disease dynamics, allowing for the efficient analysis of population-wide trends and outcomes. Such models are valued for their mathematical simplicity and computational efficiency, facilitating the examination of long-term epidemic trends and the impact of broad-scale public health measures.

We focus on ABM modeling for monitoring and predicting epidemiological situations because, despite the extensive body of work on using ABM in epidemiology, especially during the COVID-19 pandemic, we believe that the fundamental approach for setting up ABM models, calibrating them with data, and making them practical is still in a relatively early stage of development. With this chapter, we contribute to the development of this methodology by focusing specifically on impactful features of ABM modeling, such as geographical representation and disease transmission.

In the second chapter of this series, which is Chap. 6 in this book, *The COVID Information Commons - Research Insights from the Coronavirus* (Esmaieeli Sikaroudi et al. 2025), we extend the agent-in-cell (AIC) model by developing a more principled approach to coarse-graining, introducing the concept of "super-agents" to represent collective agent behaviors. That chapter places special emphasis on model calibration and validation, ensuring that the models we build are both computationally efficient and sufficiently accurate. We also compare these coarse-grained models to higher-resolution ABMs, highlighting the balance between efficiency and precision.

5.2 Agent-in-Cell Methodology

While numerous ABM implementations for pandemics exist, each varying in levels of detail, we opted to develop our own model—termed agent-in-cell (AIC)—to ensure direct access to and control over the functionalities of individual agents, rather than relying on a "black box" approach. The flexibility and transparency of the AIC model are key advantages, allowing us to fine-tune its components.

Since AIC uniquely assigns each agent to a tessellation cell, we have employed established ABM methodologies, especially those relevant to epidemiology, in crafting the AIC, as discussed in the literature. Our AIC draws inspiration from the framework introduced by Shamil et al. (2021) for basic activities and interactions at a point-of-interest (POI). However, there are also several elements in AIC that, while fundamentally imported from other ABM approaches, required our custom adaptations. These are discussed in the following subsections. We discuss standard agents—those associated with individuals—and our choices for modeling daily trip activities, activities at home and at work in Sect. 5.2.2, disease transmission modeling which occurs at POIs, and our use of mobility patterns to generate paths for the agents.

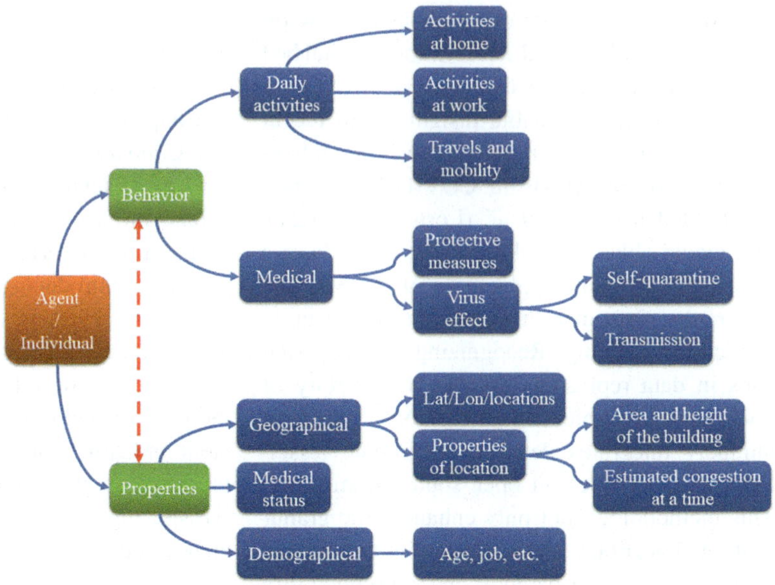

Fig. 5.1 Fundamental components of our customized ABM model of a pandemic—the AIC model

Figure 5.1 offers an overview of the interplay between behavioral patterns and related properties within the AIC framework. It is crucial to highlight that all these factors are linked to geographical locations. The spread of infection is treated as a stochastic process, occurring at each time step with a Poisson rate tailored to the agent's situation and interactions with peers. Agents can be either actively engaged at specific POIs or within distinct city areas, referred to as tessellation cells. Among the available tessellation options, the Census Block Group (CBG) is highlighted. An illustration of the CBG-based tessellation is presented in Fig. 5.2 for the city of Seattle. Rooted in United States (USA) federal geographical identifiers, CBGs correspond to census areas, which inherently implies smaller CBG areas in densely populated regions and larger ones in sparser locales. Each CBG typically hosts a population ranging from 600 to 3000 (US Census Bureau 2021).

5.2.1 Daily Trip Activities

Apart from considering agents' activities at their residences and workplaces, we also incorporate their daily trip-related engagements, encompassing activities such as shopping and visits to various POIs. These daily excursions are significant as they introduce a geographical dimension to the model. An essential modeling hurdle lies in accurately representing the mobility patterns of each individual agent. Understanding their destinations and interactions is crucial for effectively tracking pandemic transmission. However, acquiring such detailed information poses privacy concerns. To address this challenge, we turn to the SafeGraph dataset (SafeGraph COVID-19 Data Consortium 2022a), offering a partial solution to our data needs.

The SafeGraph dataset is a valuable resource for tracking mobility patterns while addressing privacy concerns. It provides (a) aggregated counts of visits to specific POIs and (b) aggregated CBGs as the origins of these trips. This dataset offers insights into the movement from CBG i to location j with an hourly time granularity and includes the classification of destination types, such as accommodation, food service, schools, and more. The data is collected by monitoring cellphone users, acknowledging that it might not capture the entire population and could potentially introduce biases. Despite these limitations, it remains a widely employed open dataset for related research purposes.

Fig. 5.2 Map of Seattle depicts census tracts with thicker boundaries and Census Block Groups (CBGs) with narrow boundaries. *Source* Office of Planning & Community Development, Seattle (2010)

5.2.2 Activities at Home and at Work

We adhere to the standard ABM procedure by initiating the simulation with the creation of random household instances. This is accomplished based on the population density of the corresponding CBGs and the household size statistics retrieved from the U.S. Census Bureau (US Census Bureau 2019). Each employed agent is then assigned a workplace location, selected randomly according to the relevant CBG-specific statistical input. Conversely, non-working agents are assumed to remain at home. Grouping an agent's work-related activities together facilitates the evaluation of new infection transmission between colleagues who share a common work location. This transmission occurs probabilistically and is influenced by the time spent at the same workplace. Job-related trips are discerned based on the information furnished by SafeGraph (SafeGraph COVID-19 Data Consortium 2022b), particularly the daytime CBG. If an agent's daytime CBG matches their home CBG, it is inferred that the agent is working from home.

5.2.3 Disease Transmission Modeling

The efficacy of contact between two agents hinges on both their physical proximity and the chosen proximity metric. To account for uncertainty and stochastic effects, it is logical to introduce a proximity measure influenced by an agent's activity group. A fundamental contribution of this chapter lies in the development of a fitting proximity measure for travel scenarios. Infection transmission occurs as agents interact with infected individuals or come into contact with contaminated surfaces. We address these mechanisms through parameterized models, extracting pertinent parameters from the previous research. For instance, parameters such as the typical duration of pathogen viability on surfaces and the likelihood of infection through surface contact are adopted from Mwalili et al. (2020). Transmission through pathogens on surfaces depends on the agents' actions and duration of stay at a POI. An action defined for the agents is to touch the surfaces, while cleaning hands is another action that serves as a countermeasure. Agent interactions are characterized with granular detail, including considerations of the geometric attributes of the agent's dwelling or visited locations. We draw inspiration from Agrawal and Bhardwaj (2021) and Mittal et al. (2020), leveraging their reported estimates on transmission probabilities. Furthermore, we incorporate insights from Agrawal and Bhardwaj (2021) regarding various aspects of disease transmission, including aerosol dynamics, underlying fluid mechanics, and probability estimations based on the distance between infected and healthy individuals. The various inputs integrated into our infection model are illustrated in Fig. 5.3, while more specific information regarding the physical modeling of infection spread within a POI is provided in Fig. 5.4.

We assume a uniform distribution of people within a building and thus introduce the density of people within building (POI) i at time t as $\delta(t, i)$. This density is influenced by N_{t_i}, the estimated number of people in a POI at time t,, and by the building geometry derived from the OpenStreetMap (OSM) dataset (OpenStreetMap contributors 2023). While many POIs like groceries, schools, and offices are likely present on OSM, for unidentified POIs we estimate them based on the average size of buildings in the area. The building geometry is assumed to be dependent on the land area of the building, A_i, and the number of floors, F_i. N_{t_i} is extracted from the SafeGraph database (SafeGraph COVID19 Data Consortium 2022b), and then, the expected

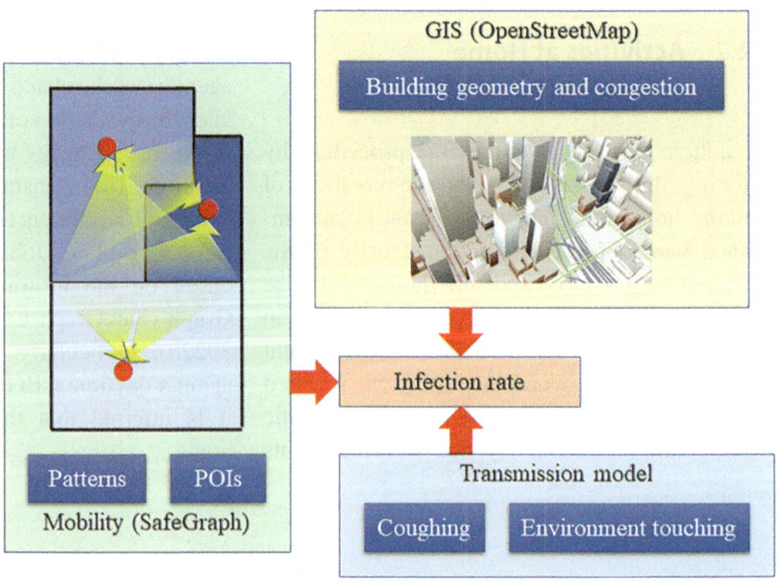

Fig. 5.3 Schematic view of the infection model

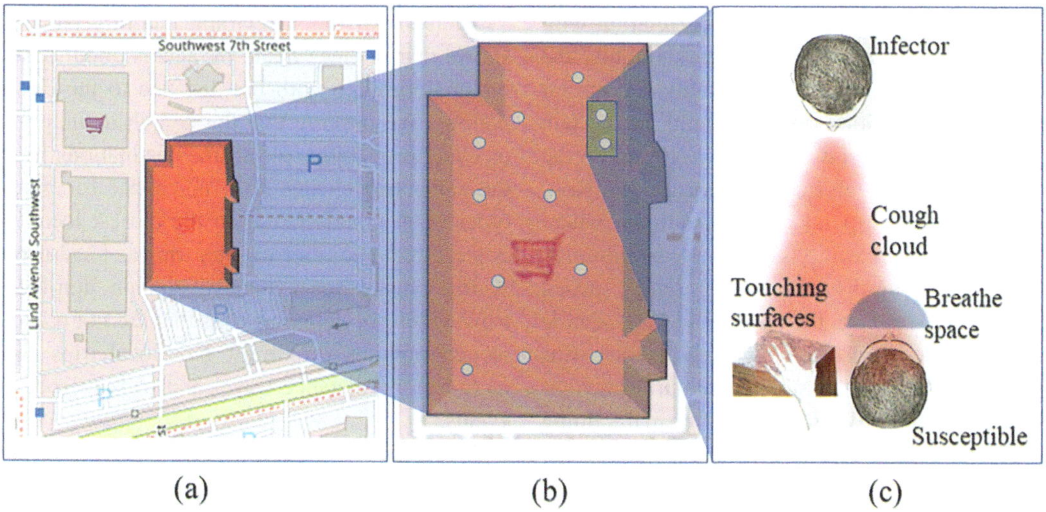

Fig. 5.4 Infection model for trips. **a** Building location and geometry, **b** estimated congestion, **c** physical transmission models (based on Agrawal and Bhardwaj 2021)

distance between a pair of agents at time t, co-located in building/POI i, is estimated as shown in Eq. 5.1:

$$\delta(t, i) = \frac{N_{t_i}}{A_i \cdot F_i}, \quad \text{for every POI } i \text{ in every time } t \tag{5.1}$$

Focusing on a specific POI, each agent is assigned a dwell time (w) when visiting the POI, which is sampled from the SafeGraph database (SafeGraph COVID-19 Data Consortium 2022b). Recognizing that not all contacts involve an infected agent, we incorporate the fraction of infected agents (I_f) out of the total number of agents to adjust individual contact rates. We assume that agents diligently practice social distancing within buildings and are uniformly distributed within them. To calculate the distance between a pair of agents within a building, we utilize the inverse of the density of people within that building ($1/\delta(t, i)$) as the basis for each agent's contact area (A_c). Each agent's contact area is represented as a square with a side length denoted as r. The value of r is simply calculated by $\sqrt{A_c}$. Consequently, the distance d between a pair of agents is approximated by r. The probability of transmission between two agents within the building is denoted as $P_c(d, m)$, where d signifies the distance, and m serves as a binary index indicating whether the agents are wearing masks or not. This information is sourced from Agrawal and Bhardwaj (2021) where the dynamics of airflow and droplets are discussed. C represents the contact rate based on the age groups that are present at the POI based on the research on the contacts per day for different age groups by Feehan and Mahmud (2021). We assumed that the daily contacts occurred during the active hours of the day, excluding nighttime. It should be noted that P_{tr} is updated by the arrival and departure of an agent to the POI. Lastly, we calculate the probability of infection transmission for a randomly selected pair of agents sharing the same building using the formula in Eq. 5.2:

$$P_{tr} = 1 - P_c(d, m)^{C \cdot w \cdot I_f} \tag{5.2}$$

5.2.4 Mobility Patterns

This subsection focuses on reconstructing mobility patterns, which describe destination types and geographic information extracted from historical data. Agent mobility plays a pivotal role in pandemic dynamics. Analyzing SafeGraph mobility patterns (SafeGraph COVID-19 Data Consortium 2022b) reveals that agent choices regarding destinations exhibit a correlation with the distance to those destinations. Notably, the assumption of equal visit probability for similar types of destinations does not hold for certain types, where preference is often given to the closest location—referred to as local trips. Non-local trips correspond to destinations where geographical proximity's impact is minimal, describing long-distance travel and special events. Notice that the COVID-19 pandemic notably led to a substitution of long-distance trips with local trips, resulting in structural changes within the mobility network while maintaining overall mobility, as demonstrated by Schlosser et al. (2020).

Our examination of the SafeGraph mobility data underscores the significance of proximity for three distinct types of POIs: 1) Groceries and department stores, 2) Education services, and 3) Religious organizations. Within the total mobility data analyzed for the year 2021 in the USA, these three POI types contribute 25, 6.8, and 1.8 of trips, respectively. Impressively, over 75% of the trips are directed toward the five closest destinations. These mobility insights not only aid in the development of a realistic simulation but also hold potential for refining the geographical abstraction of the simulation scope, which will be elaborated upon in subsequent discussions in this chapter.

5.3 Analyzing Mobility Patterns

This section is dedicated to analyzing mobility patterns, e.g., associated with type of destinations and geospatial information extracted from historical data. This investigation provides many valuable elements, including destination types, journey distances, and geospatial travel information, which might enhance pandemic modeling methodologies.

We categorize the trips by the type and functionality of the destinations. The trips with similar destinations are grouped together in our analysis. We refer to a group as a trip cluster; for example, trips to different general purpose grocery stores are considered as the same trip cluster.

Analysis of the SafeGraph mobility patterns shows that choices made by agents about destinations correlate with the distance to the destinations. We observe that the default assumption of equal probability for all the destinations from the same trip cluster does not apply for many types of destinations. Often the closest location is preferred. We refer to such trips as local trips. A trip cluster set may also contain non-local trips which we define as trips that are not related to geographical proximity. Non-local trips also include long-distance travel and trips to special events.

Schlosser et al. (2020) presented that during the COVID-19 pandemic the long-distance trips were often substituted by local trips, and even though the mobility network showed structural changes, the overall mobility remained unchanged. Moreover, our analysis of the SafeGraph data suggests that a sizeable portion of all the trips fall under the category of local, i.e., those decided according to their proximity to the agent's household. For example, if several supermarkets are available, and each offers comparable selections and services, individuals are most likely to choose the nearest supermarket. Similar arguments could be made about schools, places of worship, etc. Distribution of all the SafeGraph trips in the USA from September 2020 to February 2021 is shown in Fig. 5.5 which is classified according to the destination's North American Industry Classification System (NAICS) code.

The observations made on the selection of the trip cluster sets based on the SafeGraph data are summarized below:

- Many trips are to destinations where distance is a secondary consideration. Typically, individuals do not seek employment at the closest location to their residence. On the other hand, we also observe that during the early period of the COVID-19 pandemic (3/2020 to 7/2021) over 30% of all the trips are to the nearest locations specified in Table 5.1, namely to groceries and food stores, schools, places of worship, and to destinations providing healthcare and social assistance. The mobility type

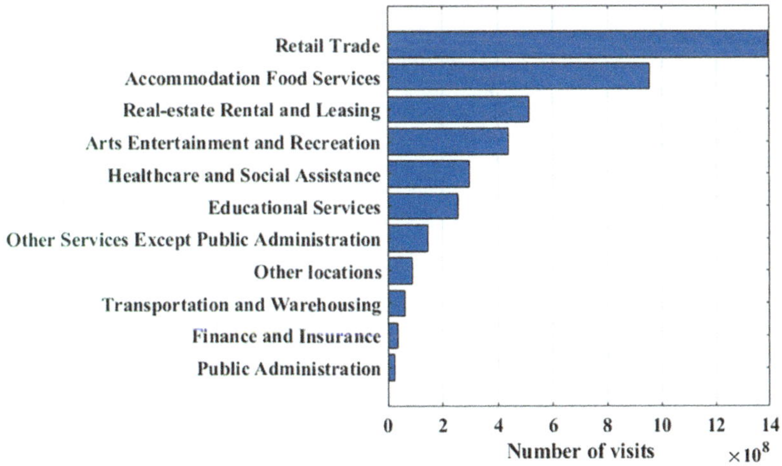

Fig. 5.5 Travels based on SafeGraph and NAICS code of destinations from September 1, 2020, to February 1, 2021 (insignificant types of destinations are removed from the graph)

Table 5.1 Destination categories and the percentage of travels to the nearest location (for instance, from all trips to groceries and retailing POIs, 27.8 percent were to the nearest location)

Groceries and retailing	27.8%
Health care and social assistance (clinics and hospitals)	7.2%
Education services	8.2%
Worship places	1.8%

(of trips) related to the destinations of the aforementioned types is referred to as local trips.

- The term "nearest" refers to the minimum commuting time which is not necessarily the closest in terms of the geographic proximity. As discussed earlier, the commuting time may include walking, using public transportation and/or private vehicles.
- We observed from SafeGraph that the longer the distances are, the more people tend to focus on nearest locations. For instance, if the distance difference between consecutive orders of closeness is less than 20%, people almost uniformly attend the preferred POIs. If such relative distance exceeds 20%, people will highly likely visit the nearest POI. For instance, if the second nearest POI is twice the distance to the first nearest, and the same between the second and third nearest and so on, then 70.1% of the times, people visit the nearest POI.

5.4 Simulation Experiments

We investigate the performance of AIC by simulating the city of Seattle under two starting scenarios. The first scenario is the one in which the simulation begins on March 1, 2020, and takes place in the early stages of the COVID-19 pandemic. In the second scenario, AIC is started on March 1, 2021, while the epidemic is prevalent. An advantage of the city of Seattle is that it has a relatively distinguishable urban delineation. We initiate the infected agents based on the historical COVID-19 cases sourced from the Johns Hopkins University dataset (Dong et al. 2020). Additionally, we compare the simulation output with historical pandemic data. We use infection-to-population ratio (IPR) as a measure of comparison. Figure 5.6 shows the simulation for both scenarios, and it can be seen that AIC shows superior performance for following the historical pandemic trend for the early stage of the pandemic. For the second scenario, AIC shows overshooting and large oscillatory behavior.

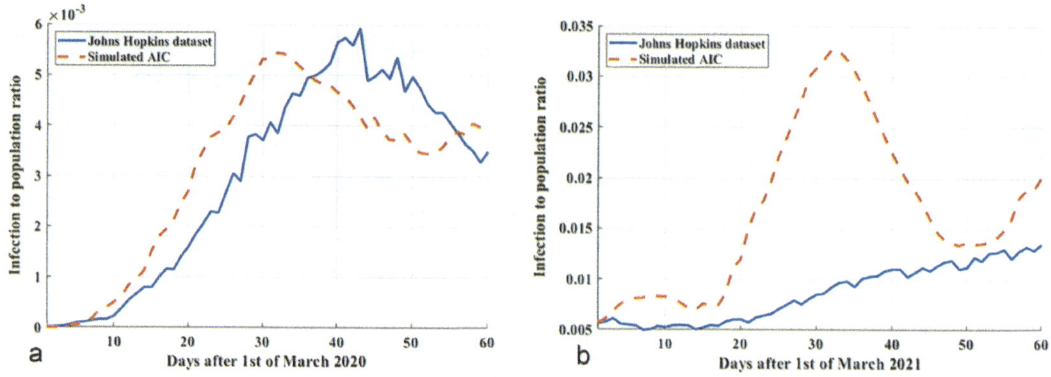

Fig. 5.6 AIC simulation of the city of Seattle under two initial conditions: **a** shows the simulation initiated in the early pandemic, **b** shows the simulation during the peak of the pandemic

5.5 Conclusion and Path Forward

This chapter introduces the agent-in-cell model, an innovative new ABM. The AIC model uniquely integrates realistic mobility with all actions represented within specific geographical regions, thereby enhancing the simulation's fidelity to real-world scenarios. We depict infection transmission through interaction of agents at public places—POIs.

In the future, our overarching objective is to position the AIC model as a foundational component within a hierarchy of interconnected and cross-validated models. These models will be characterized by their robustness, data-driven nature, and the ability to predict the spread of infectious diseases across diverse spatiotemporal scales. Our ultimate aim is to contribute significantly to effective pandemic prevention strategies.

The methodology underlying the AIC model, as developed in this chapter, primarily focuses on representing the finest level of granularity, i.e., individual agents or people. We showed that AIC is capable of following the realistic pandemic specifically in early stages. Since AIC has intensive geographical features, it is possible to point out high-risk regions of the urban environment and high-risk public buildings such as groceries. This can provide valuable information for policymakers.

Moving forward, our objective is to enhance the performance of AIC by increasing its efficiency and achieving comparable predictive accuracy with fewer computational resources. There is a clear need for reliable, versatile, and flexible methods to manage open-source and transparent ABMs, such as AIC, and to build or adjust them in a principled manner. This involves automating adjustments to agent behavior, interaction, and representability based on a variety of datasets and inputs. Notably, the issue of representability is addressed in the second chapter of this series in this book, where we focus on equipping AIC with built-in spatiotemporal coarse-graining capabilities (Esmaieeli Sikaroudi et al. 2025).

We conclude this chapter by emphasizing the importance of further testing and extending AIC's capabilities to integrate larger and more diverse datasets. While we plan to pursue this development ourselves, we also welcome collaboration with others to refine and train richer versions of AIC on larger, more complex datasets. We believe that initiating this development with synthetic datasets generated by higher-resolution models is a critical step before transitioning to real-world data, which is often incomplete and inconsistent.

References

Agrawal A, Bhardwaj R (2021) Probability of COVID-19 infection by cough of a normal person and a super-spreader. Phys Fluids 33:031704

Anagnostou A, Groen D, Taylor SJ, Suleimenova D, Abubakar N, Saha A, Mintram K, Ghorbani M, Daroge H, Islam T et al (2022) FACS-CHARM: a hybrid agent-based and discrete-event simulation approach for Covid-19 management at regional level. In: 2022 Winter Simulation Conference (WSC). IEEE, pp 1223–1234

Anderson RM, May RM (1991) Infectious disease of humans: dynamics and control. Oxford University Press, Oxford

Auchincloss AH, Diez Roux AV (2008) A new tool for epidemiology: the usefulness of dynamic-agent models in understanding place effects on health. Am J Epidemiol 168:1–8. https://doi.org/10.1093/aje/kwn118

Chertkov M (2021) Graphical (and agent based) models of pandemic. Columbia Academic Commons, Academic Unit: Northeast Big Data Innovation Hub, Series: COVID-19 Research Lightning Talks. https://doi.org/10.7916/46jp-3f96

Dong E, Du H, Gardner L (2020) An interactive web-based dashboard to track COVID-19 in real time. Lancet Infect Dis 20:533–534

Epstein JM (2009) Modelling to contain pandemics. Nature 460:687–687. https://doi.org/10.1038/460687a

Epstein JM, Parker J, Cummings D, Hammond RA (2008) Coupled contagion dynamics of fear and disease: mathematical and computational explorations. PLoS ONE 3:e3955. https://doi.org/10.1371/journal.pone.0003955

Esmaieeli Sikaroudi AM, Efrat A, Chertkov M (2025) Agent-in-cell modeling of pandemics. In: Hudson FD (ed) The COVID information commons—research insights from the coronavirus. Sustainable Development Goals Series. Springer

Feehan DM, Mahmud AS (2021) Quantifying population contact patterns in the United States during the COVID-19 pandemic. Nat Commun 12:893

Gupta P, Maharaj T, Weiss M, Rahaman N, Alsdurf H, Sharma A, Minoyan N, Harnois-Leblanc S, Schmidt V, Charles P-LS et al (2020) COVI-AgentSim: an agent-based model for evaluating methods of digital contact tracing. arXiv preprint arXiv:201016004

Hinch R, Probert WJ, Nurtay A, Kendall M, Wymant C, Hall M, Lythgoe K, Bulas Cruz A, Zhao L, Stewart A et al (2021) OpenABM-Covid19—an agent-based model for non-pharmaceutical interventions against COVID-19 including contact tracing. PLoS Comput Biol 17:e1009146

Kermack WO, McKendrick AG, Walker GT (1927) A contribution to the mathematical theory of epidemics. Proc R Soc Lond Ser a, Contain Papers Math Phys Char 115:700–721. https://doi.org/10.1098/rspa.1927.0118

Kerr CC, Stuart RM, Mistry D, Abeysuriya RG, Rosenfeld K, Hart GR, Núñez RC, Cohen JA, Selvaraj P, Hagedorn B et al (2021) Covasim: an agent-based model of COVID-19 dynamics and interventions. PLoS Comput Biol 17:e1009149

Lorig F, Johansson E, Davidsson P (2021) Agent-based social simulation of the COVID-19 pandemic: a systematic review. J Artif Soc Soc Simul 24(3):1–5

Mahmood I, Arabnejad H, Suleimenova D, Sassoon I, Marshan A, Serrano-Rico A, Louvieris P, Anagnostou A, Taylor SJ, Bell D et al (2020) FACS: a geospatial agent-based simulator for analysing COVID-19 spread and public health measures on local regions. J Simul 1–19

Mittal R, Meneveau C, Wu W (2020) A mathematical framework for estimating risk of airborne transmission of COVID-19 with application to face mask use and social distancing. Phys Fluids 32:101903

Mwalili S, Kimathi M, Ojiambo V, Gathungu D, Mbogo R (2020) SEIR model for COVID-19 dynamics incorporating the environment and social distancing. BMC Res Notes 13:1–5

Nianogo RA, Arah OA (2015) Agent-based modeling of noncommunicable diseases: a systematic review. Am J Public Health 105:e20–e31. https://doi.org/10.2105/AJPH.2014.302426

Office of Planning & Community Development, Seattle (2010) Census tract map of Seattle

OpenStreetMap contributors (2023) Planet dump. Retrieved from https://planet.osm.org

Perez L, Dragicevic S (2009) An agent-based approach for modeling dynamics of contagious disease spread. Int J Health Geogr 8:1–17. https://doi.org/10.1186/1476-072X-8-50

Ross R (1910) The prevention of malaria. John Murray, London

SafeGraph COVID-19 Data Consortium (2022a) San Francisco, CA. https://www.safegraph.com/covid-19-data-consortium

SafeGraph COVID-19 Data Consortium (2022b). SafeGraph Inc., San Francisco, CA

Schlosser F, Maier BF, Jack O, Hinrichs D, Zachariae A, Brockmann D (2020) COVID-19 lockdown induces disease-mitigating structural changes in mobility networks. Proc Natl Acad Sci 117:32883–32890

Shamil MS, Farheen F, Ibtehaz N, Khan IM, Rahman MS (2021) An agent-based modeling of COVID-19: validation, analysis, and recommendations. Cognitive Computation 1–12

Silva PC, Batista PV, Lima HS, Alves MA, Guimarães FG, Silva RC (2020) COVID-ABS: an agent-based model of COVID-19 epidemic to simulate health and economic effects of social distancing interventions. Chaos Solitons Fractals 139:110088

Stapelberg NJ, Smoll NR, Randall M, Palipana D, Bui B, Macartney K, Khandaker G, Wattiaux A (2021) A discrete-event, simulated social agent-based network transmission (DESSABNeT) model for communicable

diseases: method and validation using SARS-CoV-2 data in three large Australian cities. PLoS ONE 16:e0251737
US Census Bureau (2019) American Community Survey 5-Year Data (2009–2019)
US Census Bureau (2021) Glossary: Block Groups

Amir Mohammad Esmaieeli Sikaroudi is currently a Ph.D. Student in computer science at the University of Arizona. He graduated from the University of Arizona in 2023 with a master's degree in systems and industrial engineering. Prior to enrolling in the computer science degree, he had prior experience in mathematical modeling and data mining. Graphical models, computational geometry, and agent-based models are some of the disciplines that he is interested in researching. With a particular emphasis on pandemics, he is interested in conducting research on the mobility and interaction of individuals. With the goal of accurately and efficiently modeling pandemics, he is working on a variety of techniques from computational geometry and machine learning. The control of pandemics, facility layout, employee turnover prediction, hospital scheduling, and image processing are some of the topics that he has published. For both academic and professional goals, as well as for his own hobby, he is enthusiastic about high-performance computing and the creation of visualization tools.

Dr. Alon Efrat is an Associate Professor at the Computer Science Department at the University of Arizona. He was trained as a Computational Geometer. His research includes optimization problems in geometric settings and in particular their instances in geographic domains, networking and sensors networks, and mobility. Dr. Efrat received his Ph.D. degree in computer science from the Tel-Aviv University in 1998. He was also a Postdoctoral Research Assistant with Stanford University, Stanford, California, and with the IBM Almaden Research Center, San Jose, California. Dr. Efrat has contributed more than 150 publications. Dr. Efrat has served many National Science Foundation (NSF) panels and technical program committees in different areas. He served on the Editorial Board of various venues such as International Journal of Computational Geometry and Applications (IJCGA), Journal of Discrete Algorithms (JDA), and conferences such as Foundations of Computer Science (FOCS), International Symposium on Computational Geometry (SoCG), INFOCOM, ACM SIGSPATIAL International Conference on Advances in Geographic Information Systems (SIGSPATIAL), International Conference on the Design of Reliable Communication Networks (DRCN), IEEE Military Communications Conference (MILCOM), and many others. Dr. Efrat was Co-chair of the sensor network algorithms track of the International Symposium on Algorithms and Experiments for Sensor Systems, Wireless Networks and Distributed Robotics (ALGOSENSORS14), and a co-organizer of the workshop on Geometric Optimization in Wireless Communication and Sensing, in conjunction with the Thirtieth Annual Symposium on Computational Geometry (SoCG'14).

Dr. Michael Chertkov is a Professor of Mathematics and Chair of the Graduate Interdisciplinary Program in Applied Mathematics at the University of Arizona. His research addresses foundational challenges in mathematics, statistics, machine learning, and artificial intelligence, particularly as they apply to and are inspired by physical systems like fluid mechanics. He also works on applications in the control of engineered systems, such as energy grids, and bio-social systems. Dr. Chertkov received his Ph.D. in physics from the Weizmann Institute of Science in 1996. After obtaining his Ph.D., he spent three years as R.H. Dicke Fellow in the Department of Physics at Princeton University. In 1999, he joined the Los Alamos National Laboratory, first as a J.R. Oppenheimer Fellow, later becoming a Technical Staff Member in the Theory Division. He transitioned to the University of Arizona in 2019. Throughout his career, Dr. Chertkov has contributed about 300 research papers. He holds the title of Fellow in both the American Association for the Advancement of Science (AAAS) and the American Physical Society and is a Senior Member of the Institute of Electrical and Electronics Engineers (IEEE).

6 Agent-In-Cell Modeling of Pandemics: Harnessing Super-Agents for Predictive Modeling

Amir Mohammad Esmaieeli Sikaroudi, Alon Efrat, and Michael Chertkov

Abstract

In this chapter, we propose incorporating the concept of "super-agents" as efficient and versatile techniques to enhance the agent-in-cell (AIC) model for pandemics, the agent-based model (ABM) introduced in the *Agent-in-Cell Modeling of Pandemics* in Chap. 5 in *The COVID Information Commons—Research Insights from the Coronavirus*. These super-agents enable the efficient simulation of more complex infection dynamics in urban environments, while reducing computational complexity, all while maintaining the granularity of individual-level interactions. We extend the AIC model by employing geospatial tessellation, specifically through the construction of Voronoi diagrams based on street network locations. Our approach demonstrates that flexibility in tessellations—particularly the introduction of different tessellations for various types of points-of-interest (POI) visited by agents—achieves a better balance between accuracy and computational efficiency compared to standard tessellations based on Census Block Groups (CBGs). Special attention is given to calibrating AIC using super-agents, validating coarser tessellations against their higher-resolution counterparts (considered as ground truth), and benchmarking the model against an existing open-source ABM. This flexible, hybrid approach optimizes both precision and performance in modeling infection spread in urban settings.

Keywords

Agent based modeling · Tessellations · Super-agent · Pandemics · Mobility · COVID-19

Research funded by

NSF Award #2027072—RAPID*: Infer and Control Global Spread of Corona-Virus with Graphical Models.
*Rapid Response Research Award.

COVID Information Commons Video Presentation

The COVID Information Commons (CIC) Research Lightning Talk video presentation and individual multilingual transcripts for this project can be found in the Columbia Academic Commons (AC) (Chertkov 2021).

A. M. Esmaieeli Sikaroudi · A. Efrat · M. Chertkov (✉)
University of Arizona, Tucson, AZ, USA
e-mail: chertkov@arizona.edu

A. M. Esmaieeli Sikaroudi
e-mail: amesmaieeli@arizona.edu

A. Efrat
e-mail: alon@cs.arizona.edu

6.1 Introduction

Agent-based simulation is a powerful tool for modeling complex systems where individual entities, or "agents," interact with each other and their environment. However, due to its inherently

F. D. Hudson (ed.), *The COVID Information Commons*, Sustainable Development Goals Series,
https://doi.org/10.1007/978-3-032-06125-6_6

granular nature, it is highly resource-intensive, often requiring significant computational power and memory (Chertkov 2021). As the number of agents and interactions grow, the demand for processing resources can quickly escalate, leading to longer run times and inefficiencies. These may include the use of advanced coarsening techniques, parallel processing, and efficient data structures. By optimizing agent-based simulations, researchers and engineers can explore more complex systems with greater accuracy while conserving resources, enabling faster analysis and more cost-effective studies.

The topic of ABM abstraction and simulation optimization, e.g., by exploring reduced representation of the population, remains relatively unexplored. Rhodes et al. (2016) delved into the potential of modeling complex systems with a reduced number of agents, examining interaction rates, iteration durations, and messaging overhead. Similarly, Tregubov and Blythe (2021) investigated sub-sampling agents to represent groups and teams in a social media model, concluding that the representative agent approach effectively reduces runtime while maintaining reasonable performance levels.

In this chapter, we extend the development of the agent-in-cell modeling of pandemics introduced in Chap. 5 *Agent-in-Cell Modeling of Pandemics* in this book, *The COVID Information Commons—Research Insights from the Coronavirus* (Esmaieeli Sikaroudi et al. 2025). We focus on spatial coarse graining, implemented through super-agents and geospatial tessellations. These advancements are made possible by the consistent and transparent "white-box" approach used in the construction of the AIC model. Further, we explore various approaches to enhance the performance of AIC, which are summarized below:

Agent Representability: ABM with as many agents as the population would be ideal, but it is computationally and prohibitively expensive for monitoring epidemiological situations over large geographical areas, such as the United States (USA), a large state, or a major city. How may we reduce computing limits while maintaining city-scale accuracy? We overcome the computational challenge by using super-agents (SAs), which represent a group of original agents/individuals. As shown below, SAs will move around the city during the day and visit points-of-interest (POIs) like normal agents, but they will also represent statistical interactions of agents at POIs where epidemiological transmissions occur. SAs aim to balance typical ABM models, which resolve all individuals, with compartmental models, which aggregate everyone. The SA technique maintains individual-level interactions and behaviors, allowing us to examine urban disease spread with high precision and processing efficiency.

Geographical Resolution: When addressing an urban region, we like to account for the specifics and details of the geography, such as an agent's (or SA's) residence's closeness to POIs. We introduce tessellation within the ABM to improve geographic resolution. By representing a more abstract sample space during simulations, we can better simulate agent and SA travel. Creating excellent tessellation requires considerable study, as we show later.

6.2 Tessellation Types

Tessellation plays a pivotal role in facilitating agent-based simulations by dividing the simulation space into discrete cells or polygons, effectively organizing and governing agents' interactions. This spatial partitioning not only streamlines agent tracking and interaction management but also enhances the simulation's scalability to encompass larger systems. Each tessellation cell is furnished with a mobility schedule, outlining a roster of POIs and corresponding visit probabilities for distinct demographic groups. The connection between mobility from SafeGraph and demographic groups is not explicitly available. We used the needs of agents from Mahmood et al. (2020). We aggregated their needs matrix into three age categories, and we connected the needs to the North American Industry Classification System (NAICS) codes.

The mobility schedule of a tessellation cell is the list of POIs historically visited by the tessellation cell based on SafeGraph.

Tessellations need to be understood in the context of large-scale agent-based simulations while examining them. Trajectories connecting tessellation cells are the main reason tessellation is naturally needed for large-scale general-purpose mobility. While certain mobility datasets pertain to specific point-to-point journeys, such as public transportation data, the preservation of individual privacy is inherent due to use of similar routes by a large number of people. This is not the case for a general-purpose mobility dataset with which point-to-point information can violate privacy of people due to sparsity of the trajectories or the association of trajectories with a narrowed geographical region. Another reason to embrace tessellation is the complexity of interactions. When mobility patterns in a tessellation cell are more uniform, abstract definitions might be used. Depending on the modeling approach, agent activity may be limited to cells representing communities or neighborhoods, reducing the computation cost of simulation. Thirdly, tessellation cells serve as a valuable framework for implementing control policies, and a reduced number of cells is favorable while ensuring that the requirements of each cell are met. When a tessellation cell adequately caters to most of an individual's needs within their residing area, it leads to diminished mobility between distinct cells.

The geographical delineation provided by tessellation cells facilitates a more nuanced understanding of disease spread within distinct regions of a city. This, in turn, could aid policymakers and public health officials in devising targeted interventions based on the spatial dynamics of an outbreak. By defining regions within a city with precision, our model lays the groundwork for future studies to explore how control policies can be optimally applied in a real-world context, thereby contributing to the broader goal of improving disease modeling and public health response strategies.

6.3 Voronoi Diagram on Network

Using a Voronoi diagram (VD) to tessellate a city is useful for urban planning and analysis. Geometrically, the VD divides space into areas based on POI proximity. Public amenities, transportation hubs, and landmarks are key urban division points. Some VD regions are closer to a seed point than others. Urban designers and politicians can better arrange infrastructure and services using Voronoi tessellation. This method helps define service coverage zones for different facilities. Planting seed points in appropriate locations ensures equitable access to these vital services across each region, improving urban infrastructure efficiency and effectiveness. We use the term "local trips" to describe trips to the nearest destination, allowing us to construct VDs based on commuting times along the street network, taking into account route types and average traffic conditions.

To create VDs using commuting time, a method akin to Dijkstra's search algorithm is employed. Rather than searching for destinations, this method calculates travel times from multiple sources. Each source leaves traces on the map while calculating travel times, and the node in the graph is won by the source with minimum commuting time. The source code for the implemented ABM is accessible online.[1]

6.4 Euclidean Clustering

To underscore the significance of network-based commuting time, we turn to the concept of Euclidean clustering, where each street node functions as a data point, and a tessellation cell emerges from clusters of street nodes sharing proximity. In this approach, conventional clustering algorithms, like the K-means algorithm, can be applied.

K-means clustering, a popular unsupervised machine learning technique, serves the purposes

[1] https://bitbucket.org/pandemic-ames/aicsimulation.

of data clustering and pattern recognition. The algorithm's objective is to partition a given dataset into k distinct clusters, where each data point is allocated to the cluster with the nearest mean (centroid). K-means is sensitive to the initial positioning of centroids, which can result in suboptimal clustering outcomes or convergence to disparate local optima. To counteract this concern, the algorithm is often executed multiple times with varying initializations, and the most favorable clustering result is selected based on a pre-established evaluation metric (Arthur and Vassilvitskii 2007).

When using K-means clustering in our experiments, we fix the number of clusters, selecting it based on the considerations of fair comparison with benchmarks. For instance, in the case of the VD tessellation, the number of VD cells is kept equal to the number of POIs, and the same number of clusters is used in the K-means clustering.

6.5 Super-Agents—Reducing the Number of Agents

The pandemic dynamic simulation mechanism relies on agents. An ABM simulation's realism, dynamics, and efficacy depend on its agent count. Since agents directly affect computing time, reducing their numbers can yield substantial performance gains. Simulating a large number of agents allows analysis of rare or unexpected events that may only occur at certain agent densities, revealing deeper system dynamics and potentially identifying pivotal tipping points. Analyzing the influence of agent numbers on simulated pandemic patterns offers valuable insights. Evidently, sparse agent representation can lead to pronounced deviations from authentic mobility trajectories.

To ensure robustness, pandemic simulations should minimize the sensitivity to the agent count. For instance, an agent's at-home activities should be contingent on family size, remaining independent of the total agent count within the study scope. This rationale should be thoughtfully extended to all aspects of the ABM, promoting a design that fosters stability and meaningful results regardless of the specific agent population.

Super-agents can be applied to any ABM to reduce complexity. Any ABM has a flow of functions to simulate one iteration; for instance, in AIC, mobility of the agents is the first item, then activities are processed, and then transmission is processed. Applying SA to any step of the simulation requires the information needed downstream to be preserved as much as possible. For AIC, this means that the required information for activities and transmission needs to be preserved to the maximal possible extent when SA is applied to mobility.

A SA represents a group of agents for mobility activities. A SA move from a CBG to a POI suggests a group of agents traveled. However, inside a POI and during the activities, the agents represented by SAs act similarly to a group of real agents. The reason that SAs can act similarly to a group of real agents is that the activities and the disease transmission model only require the number of agents and the number of infected agents. Thus, agents inside a SA may transmit disease at a POI with just one SA. SA preserves infection fraction accuracy, I_f, according to the disease transmission model. The construction of SA is a post-processing step when initializing the ABM. We first initialize the ABM according to the population of the scope, and then, SA is applied to reduce the number of agents. We find agents with the highest demographic and home location similarities to build a SA. We search for a Pareto-efficient solution by simply searching for the nearest agent and checking for demographic similarity. Iteratively, agents with similar demographics from the same tessellation cell are combined. If a tessellation is utilized, a SA employs the mobility schedule generated for the cell it resides in; otherwise, it covers the union of the agent schedules it represents. Each tessellation cell must have at least one SA to make sure that the information of that geographical area is preserved.

6.6 Experiments: Setting

Assuming population is a key factor, the cities of Santa Fe, New Mexico, Seattle, Washington, and Chicago, Illinois, represent low, medium, and high populations, respectively. We focused on the cities that have less complications in terms of being contained within a county and have better data availability. We provide three measures based on the most visited point of interest (MVPOI) to examine the effects of tessellation type and number of agents over a specific time period: (a) The Number of Visits (NOV) to the MVPOI; (b) The Average Visit Duration (AVD) at the MVPOI; (c) The Probability of Co-Visiting (PCV) the MVPOI by two agents. The focus of the measures on MVPOI is motivated by the fact that less frequently visited POIs might exhibit a significantly lower count of visits, thereby rendering the collection of reliable statistics unfeasible. Therefore, we concentrate on MVPOIs, as they provide a robust basis for analysis and interpretation. We select one MVPOI for each POI category, such as grocery and medical services.

We select CBG_1 and CBG_2 by the highest source of visits to MVPOI. Assuming that an agent from CBG_1 (a_1) visits MVPOI, if an agent from CBG_2 (a_2) is not present at MVPOI, a counter related to the trip without a co-visit is incremented. If a_1 and a_2 are present at MVPOI and they stay for more than five minutes together, the counter for co-visit is incremented. PCV is the fraction of the co-visit counter and the total visits from CBG_1 and CBG_2.

We introduce a fundamental ground truth point of reference for our experiments, denoted as the No Tessellation (NT) approach. In the NT method, we leverage statistical data extracted from SafeGraph (2022) and commute times to POIs. However, unlike considering mobility from CBG to POI, NT considers mobility originating from precise locations to the respective POIs. Subsequently, we develop tessellations based on a map featuring an array of potential trajectories. Naturally, we anticipate that tessellations exhibiting more uniform mobility patterns will offer distinct advantages.

In each experiment, employing a specific tessellation variant alongside a reduced number of agents, we yield metrics including NOV, AVD, and PCV, all comparable with the NT baseline. We explore three distinct sets of tessellations categorized by the number of generated cells. The first group maintains an equivalent (the same) number of cells (s) as the original CBGs, while the second group produces fewer (reduced number of) cells (r), and the third group generates an increased quantity of cells (i). For the tessellations with fewer cells, the count is determined by the VDs of the POIs. In contrast, tessellations generating more cells than the base case result from an overlay of VDs and CBGs.

To further emphasize the influence of tessellation type, we also introduce a tessellation created by merging neighboring CBGs, referred to as randomly merged CBG (RMCBG) tessellation. The tessellations examined in our experiments are summarized below:

- VD_r: VD constructed based on commuting time to the POIs related to retail services, educational services, and religious organizations. Here, the number of generated cells is less than the number of CBGs.
- K-$Means_r$: Involves clustering streets, treated as nodes, using Euclidean distance. The parameter k is set to the number of cells generated within VD_r.
- RMCBG: Constructed by (randomly) merging adjacent CBG cells to match the cell count generated in VD_r.
- CBG: This tessellation is based on census data.
- VD_s: Based on the VD of all POIs sorted by visit frequency. In this case, the number of cells matches the number of CBG cells.
- K-$Means_s$: Like K-$Means_r$, but with k set to the number of CBG cells.
- CBGVD: A hybrid tessellation overlaying VD tessellation and CBG tessellation results in breaking the original cells into a larger number of smaller cells.
- VD_i: Like VD_s, but with the number of cells determined by the CBGVD tessellation.
- K-$Means_i$: Like K-$Means_r$, but with k equal to the number of cells in the CBGVD tessellation.

The agents are placed on the map according to their home locations, determined by a combination of population density and street types (treated as nodes in a geo-graph) extracted from census data and OpenStreetMap (OSM). We designed this placement to prioritize residential streets, making it highly likely for agents to be positioned there while virtually eliminating the probability of agents being placed on highways. After the agents are positioned on the map, the SafeGraph dataset is used to determine the CBG that each agent is assigned as the work region during the day.

To ensure comprehensive insights into the variations across experiments, we conduct each experimental setup several times, thereby providing a robust depiction of the spectrum of changes compared to other trials. We observed that the average of the performance measures and runtime converges after 72 replications. We used change point detection on mean and standard deviation, and we used the largest number of replicates that show convergence to ensure that all measures are converged (Killick et al. 2012).

Table 6.1 Extracted information from analyzing the case studies

	Chicago	Seattle	Santa Fe
Reduced cells	493	242	52
Cells in CBG	2179	485	62
Increased cells	5051	1032	373
Retailing services	74	56	13
Educational services	161	81	21
Religious organizations	49	32	14
Population (2020 census)	2,746,388	737,015	87,505

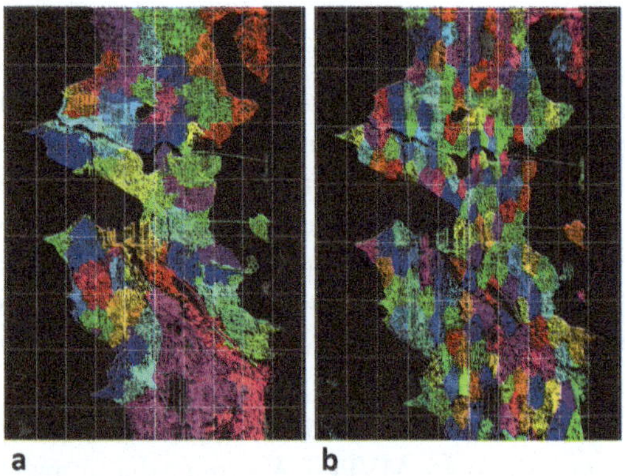

Fig. 6.1 Two tessellation types for the city of Seattle: **a** VD tessellation, **b** K-means clustering

6.7 Experiments: Results

The counts of cells resulting from the reduced and increased tessellations are presented in Table 6.1 for the test cities. In smaller cities, these tessellations match CBG cell counts closely. For the three cities, the cell counts generated through the POIs are fewer than the CBG cell numbers. We observe that VD cells can exhibit intricate boundaries since the VDs are constructed upon the street network. Figure 6.1 shows two tessellation types VD_r and K-Means$_r$ for the city of Seattle, Washington.

We studied the performance of the AIC model over a week, initializing agent placement on the map based on the historical COVID-19 cases sourced from the Johns Hopkins University dataset (Dong et al. 2020). The simulation begins on March 1, 2021, and concludes on March 8, 2021. Aggregating performance outcomes across various agent reduction scenarios, Fig. 6.2 summarizes the effectiveness of all tessellation strategies. Aggregate absolute error is calculated by averaging the absolute error between no-tessellation and different tessellation types over different agent reduction scenarios (i.e., 75, 50, 25, and 10%). Notably, tessellations relying on K-means Euclidean clustering exhibit suboptimal performance. VD and CBGVD tessellations show superior performance, specifically CBGVD in PCV and VD in NOV.

To gain further insight, we closely examine the three most successful tessellations (showing in Fig. 6.2 least aggregate absolute error) in Fig. 6.3—cbg, VD_r, and CBGVD—scrutinizing their behavior across three key measures and their response to reductions in the number of agents. Figure 6.3 illustrates the AVD, NOV, and PCV metrics, with all the recorded data pertaining to

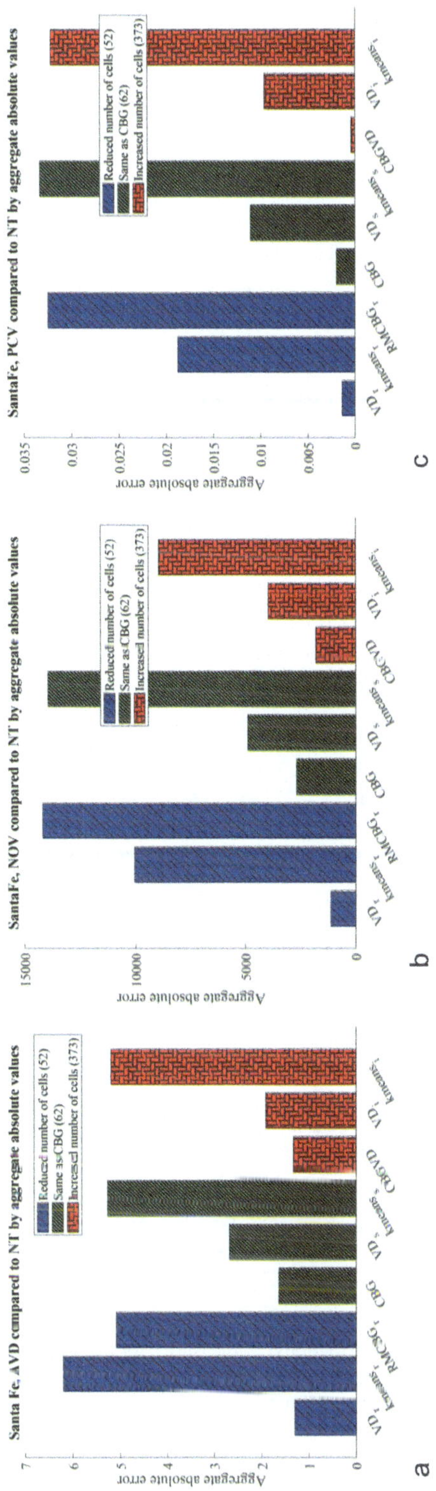

Fig. 6.2 Tessellations for the Santa Fe case study showing the aggregate absolute error for **a** AVD, **b** NOV, **c** PCV measures

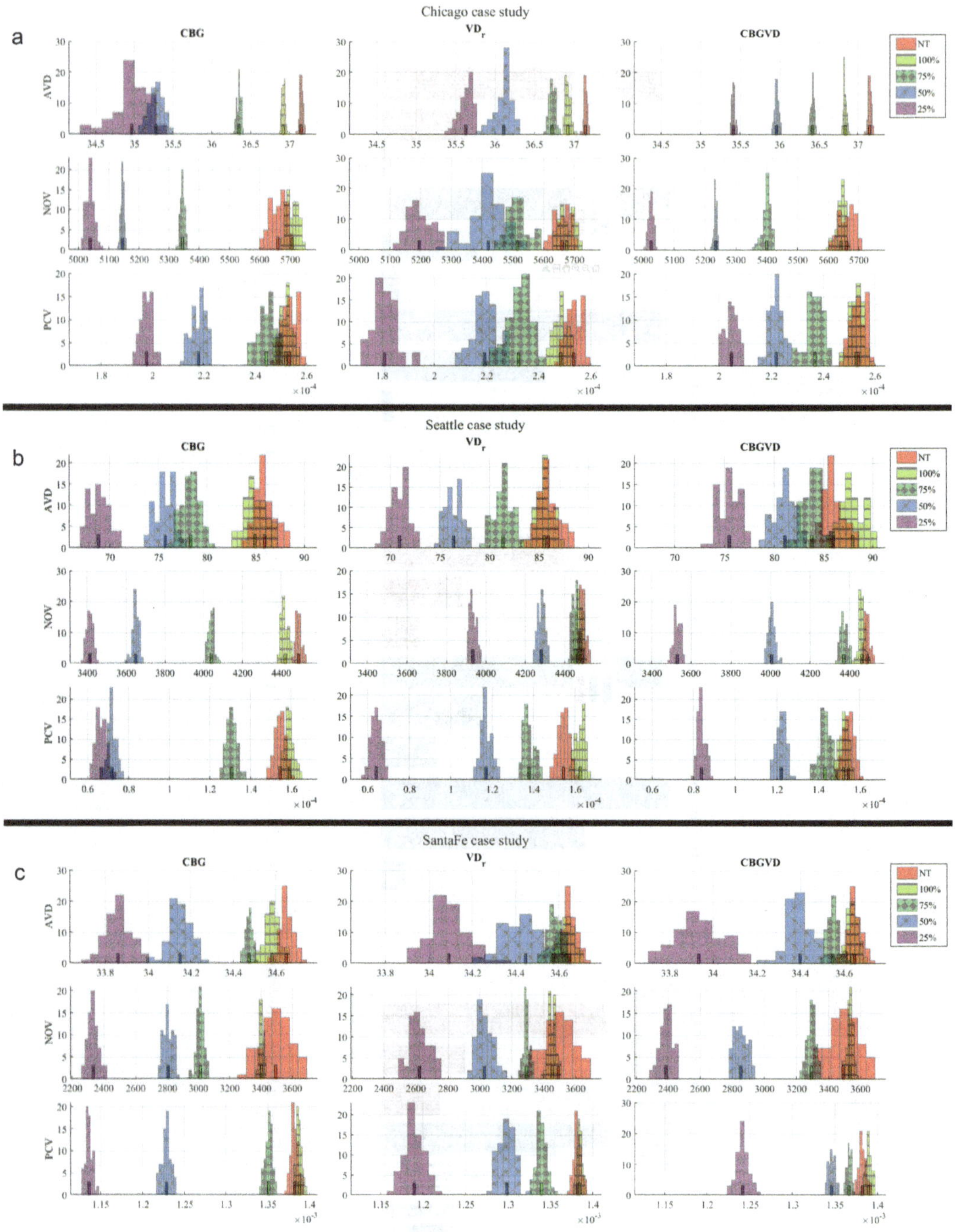

Fig. 6.3 Histograms (binned probability distributions) illustrating the impact of three primary tessellations and a reduced number of agents, quantified through three distinct measures, for the **a** Chicago, **b** Seattle, and **c** Santa Fe case study

the MVPOI and originating from the SafeGraph dataset, presenting histograms (binned probability distributions) where varying color codes correspond to distinct levels of coarse graining, representing different numbers of SAs. The reference point for comparison is the "red" No Tessellation (NT), serving as the ground truth. Therefore, a more pronounced deviation from the NT histogram signifies a greater divergence from the precise representation. The following insights are derived from a comprehensive analysis of Fig. 6.3:

- Impact of Agent Count: A discernible decline is noted with a reduction in the number of agents. This underscores the significance of our methodology in determining an optimal level of coarsening, essential for striking a balance between desired quality and result assurance.
- Depletion in PCV: The most substantial degradation is observed in the PCV metric. This observation is logical given the inherent complexity of this feature among the three under consideration.
- Comparable Performance: The three winning strategies (VD, CBG, and CBGVD) demonstrate approximately equivalent performance. Notably, CBGVD outperforms the rest in modeling the most intricate feature (PCV) and fares the best in the case of AVD.
- VD for Visit Count: The VD strategy emerges as the most effective in capturing the modeling of visit counts (NOV).
- City-Specific Intricacies: The characterization of certain cities exhibits nuances that transcend mere population size. Notably, the AVD in Chicago (the largest city) and Santa Fe (the smallest) reveals striking similarity, while Seattle (a medium-sized city) stands out with a nearly twofold longer average duration.
- Variation in Simulations: With an increase in the number of cells in a tessellation, the variation from repeated simulations diminishes. Furthermore, the overlap extent across different levels of agent reduction notably diminishes as the number of cells rises.

These observations collectively provide valuable insights into the behavior and efficiency of different tessellation strategies across various dimensions of our analysis.

To distinguish the impact of SAs from tessellations, we conducted simulations for the city of Seattle without employing SAs, thereby reducing the number of agents without the explicit use of SAs. The contribution of SAs to the simulation metrics is quantified by comparing outcomes with and without the use of SAs. For instance, if the AVD without tessellation and including the SA is $\mathrm{AVD}_{\mathrm{nt}_s}$, the AVD with VD tessellation using SAs is $\mathrm{AVD}_{\mathrm{VD}_s}$, and the AVD for VD tessellation without SAs is $\mathrm{AVD}_{\mathrm{VD}_{ns}}$, then the contribution of the SA is calculated by Eq. 6.1:

$$\tau = \frac{\mathrm{AVD}_{\mathrm{VD}_s} - \mathrm{AVD}_{\mathrm{VD}_{ns}}}{\mathrm{AVD}_{nt_s} - \mathrm{AVD}_{\mathrm{VD}_{ns}}} \tag{6.1}$$

The intuition of τ is that if τ is zero, this means that SA has no contribution to the measure. If $\tau < 0$, it means that SA deteriorates the measure; for instance, if there is an underestimation of AVD_{nt_s} by AVD_{nt_s}, SA caused a larger underestimation. Similarly, if $\tau > 2$, it means that SA had a detrimental effect; for instance, it changed an overestimation into a large underestimation. If $\tau > 1$, then it means that the SA effect is large enough to change an underestimation into an overestimation and vice versa.

This approach is also applied to quantify the effects of SAs on the NOV and PCV metrics. Figure 6.4 presents the contribution of SAs to these simulation outcomes under various tessellation strategies for the Seattle case study, revealing that SAs significantly improve the PCV metric, whereas their impact on AVD and NOV is more marginal. It is important to note that SAs enhance the accuracy of disease transmission dynamics; an infected agent's need for hospitalization or self-quarantine indirectly influences mobility patterns. From Fig. 6.4, it is evident that SAs substantially affect the PCV measure, with a marginal effect on AVD and NOV.

AIC's use of SAs helps maintain accurate pandemic dynamics, and it is capable of parallelizing certain aspects of the simulation. While

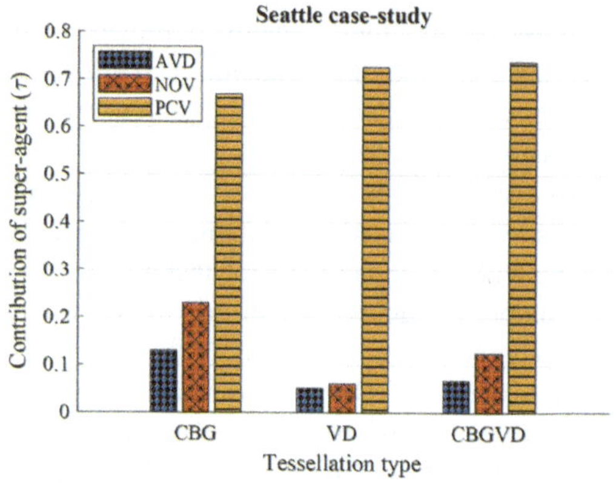

Fig. 6.4 Evaluating the impact of SAs across strategies on simulation outcomes in the Seattle case study

agent behaviors are assessed in parallel, the simulator undergoes a regrouping and synchronization process every hour. It is essential for all processing threads to align during each synchronization, limiting the extent of feasible parallelization. Figure 6.5 demonstrates the impact of adding more processing threads to the AIC simulation. Beyond the addition of 14 threads, there is an unexpected increase in runtime, likely due to the heightened synchronization demands among threads, which hampers performance gains. To assess the computational efficiency of our simulations, we meticulously measured memory usage across various tessellation types while simulating different numbers of agents within the

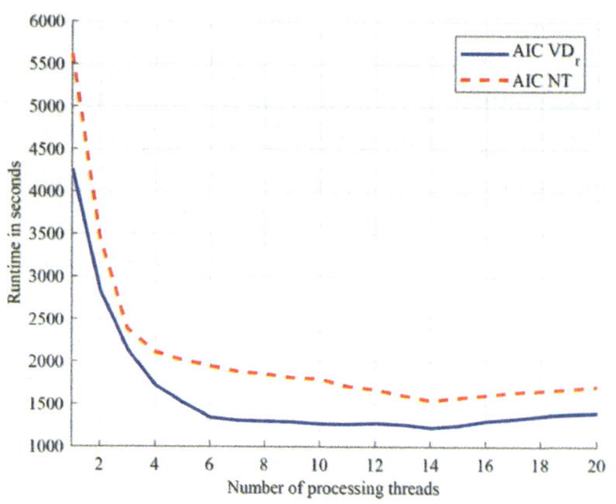

Fig. 6.5 AIC runtime as a function of various Tessellation processing thread count for NT and VD_r tessellation methods

AIC framework, which operates atop the Java Virtual Machine. We used the VisualVM tool to accurately measure memory consumption on the virtual machine and enhanced our measurement precision by initiating a garbage collection process during periods of peak agent activity. Reducing the number of agents while using SAs shows a significant effect on memory saving. For instance, for the city of Seattle, 100% number of agents requires 20.08 GB memory while using only 50% of agents with SAs only requires 11.15 GB memory. Moreover, the choice of tessellation type can marginally optimize memory resources in the simulation of large-scale urban environments. For instance, using VD tessellation needs 20.27 GB memory while CBG tessellation needs 20.68 GB.

6.7.1 Comparison of AIC with FACS-CHARM

We compare our model—the AIC—with Flu And Coronavirus Simulator–dynamiC Hospital wARd Management (FACS-CHARM) (Anagnostou et al. 2022), a state-of-the-art and open-source ABM that aligns sufficiently closely with our approach. FACS-CHARM builds upon the FACS pandemic simulator (Mahmood et al. 2020), offering enhanced details within hospital simulations, including discrete simulations of wards and intensive care units. As one of the most recent advancements in ABMs, it incorporates the geospatial features of its predecessor, the FACS simulator. The FACS simulation integrates building data extracted from OSM, and agent mobility is determined by individual needs and distances, positioning FACS as one of the most geographically precise simulators available. This level of detail and alignment with our model's aims explains why we chose to compare AIC with FACS-CHARM.

However, the level of detail in AIC and FACS-CHARM differs in several ways: (a) FACS reschedules and regroups agents once per day, while AIC does this hourly to align with their tasks and locations, enhancing the accuracy of agent contacts; (b) FACS operates solely on a daily timescale without simulating actions, whereas AIC conducts minute-by-minute checks for hourly-specific actions, thereby simulating the pandemic's reproduction number more accurately; (c) AIC simulates building congestion and pathogen persistence every minute, providing a more detailed transmission simulation, especially in scenarios involving brief visits by infected individuals.

The mobility of agents in FACS-CHARM reflects their individual needs, whereas in AIC, mobility is determined based on probabilities extracted from commuting times and SafeGraph data. The FACS simulator requires data on the types and sizes of POIs, as well as the age distribution within the simulation scope. The CHARM simulator requires hospital capacities.

Therefore, to conduct a sensible comparison between AIC and FACS-CHARM, we utilize SafeGraph data to extract relevant POIs and use OSM to retrieve building geometries, enabling both simulations to be run on the same city. We used the hospital capacities for the Seattle city area, necessary for simulating hospitalizations (Reno 2014). Our focus is on assessing the infection trends over a ten-day period for both simulators in Seattle. The simulation commences on March 1, 2021, and concludes on March 10, 2021.

The comparison results between the two models are shown in Fig. 6.6. Due to the differences in the settings of AIC and FACS-CHARM, the results differ even when no reduction in the number of agents is applied (100% in Fig. 6.6). However, we observe that the AIC model fares better at replicating pandemic dynamics with fewer (super-) agents compared to the FACS-CHARM model (Mahmood et al. 2020). The AIC model, while using only 75% of the agent count, remains sufficiently close to the comprehensive simulation (100% of the agent count). This better representation of reality with fewer (super-) agents in the AIC is attributed to the use of SAs and VD tessellation.

6.8 Conclusion and Path Forward

This chapter introduces the agent-in-cell model, an innovative new ABM. The AIC model uniquely integrates realistic mobility with all actions represented within specific geographical regions, thereby enhancing the simulation's fidelity to real-world scenarios. We depict infection transmission through interaction of agents at public places (POIs). Our proposed model introduces various computational strategies to optimize these detailed simulations, notably aggregation of agents into super-agents and optimal geographical tessellations, striking a balance between accuracy and efficiency.

The AIC framework retains the dynamism of a comprehensive simulation while employing a reduced number of agents via the introduction of super-agents. We have demonstrated that meticulous urban tessellation offers dual benefits: it not only minimizes the count of tessellation cells for memory efficiency but also leads to a reduction in agent numbers, enhancing simulation efficiency. Employing the SA concept, we aggregate similar agents within tessellation cells while minimizing perturbations to the dynamics. Our investigation showcases the feasibility of implementing tessellations of varying intricacy through VDs while upholding satisfactory quality. Our analysis encompasses diverse city sizes in the USA, spanning small, medium, and large urban areas. The analysis also yields valuable insights: agent count reduction affects results; the PCV metric is the most sensitive; VD, CBG, and CBGVD exhibit comparable performance with CBGVD excelling in complex features; and city-specific intricacies impact visit duration.

While applied to varied case studies, our proposed model consistently mirrors the disease spread patterns observed in the FACS-CHARM model—a state-of-the-art ABM we use as a benchmark. AIC's disease transmission model is intricate, warranting simulations at minute

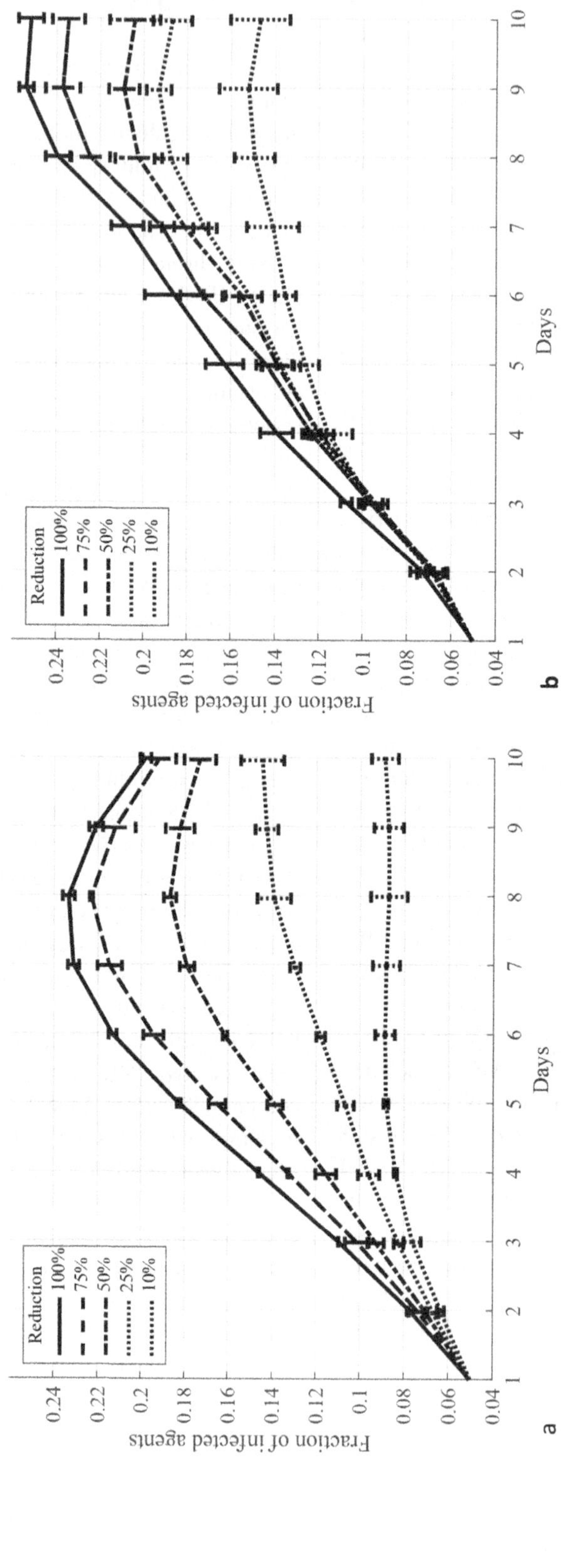

Fig. 6.6 Fraction of infected agents over a ten-day simulation between **a** AIC and **b** FACS-CHARM with standard deviation error bars for the city of Seattle

intervals, complemented by hourly synchronization. This simulation encapsulates building occupancy dynamics and myriad activities transpiring within each hour. Furthermore, AIC adeptly manages complex mobility patterns by directly sampling from historical datasets like SafeGraph. Runtime optimization in AIC is anchored in three pivotal strategies: leveraging reduced agent counts through SAs; tessellating the area of study; and harnessing parallelization techniques.

In the future, our overarching objective is to position the AIC model as a foundational component within a hierarchy of interconnected and cross-validated models. These models will be characterized by their robustness, data-driven nature, and the ability to predict the spread of infectious diseases across diverse spatiotemporal scales. Our aim is to contribute significantly to effective pandemic prevention strategies.

The methodology underlying the AIC model, as developed in this chapter, primarily focuses on representing the finest level of granularity (i.e., individual agents or people). Moving forward, we intend to integrate this fine-grained modeling framework with a coarser modeling approach known as geo-graphical models. These models draw upon insights from fields such as artificial intelligence (AI), data science, optimization, and applied mathematics and theoretical engineering (Chertkov et al. 2021; Krechetov et al. 2022). Geo-graphical models serve as tools for aggregating individual agents into geo-compartments, which can be thought of as cells within the AIC framework. These compartments could represent various levels of granularity, ranging from households and census tracts to entire cities or states. The resulting aggregated models strike a balance between computational efficiency and accuracy. They will undergo rigorous calibration and validation processes, benchmarked against higher-resolution ABMs (including AIC). Importantly, these models will leverage data-driven approaches, taking full advantage of recent breakthroughs in AI, machine learning, and data science.

Looking ahead, our vision also encompasses collaborative efforts to create more comprehensive datasets. Specifically, we aim to incorporate synthetic data generated by high-resolution models like AIC to train and improve coarser models. This approach not only enhances the modeling pipeline but also bridges gaps in existing data resources. Our commitment is to continuously advance the field of infectious disease modeling and contribute to more effective pandemic preparedness and response strategies.

References

Anagnostou A, Groen D, Taylor SJ, Suleimenova D, Abubakar N, Saha A, Mintram K, Ghorbani M, Daroge H, Islam T et al (2022) FACS-CHARM: a hybrid agent-based and discrete-event simulation approach for COVID-19 management at regional level. In: 2022 Winter simulation conference (WSC). IEEE, pp 1223–1234

Arthur D, Vassilvitskii S (2007) K-means++ the advantages of careful seeding. In: Proceedings of the eighteenth annual ACM-SIAM symposium on discrete algorithms, pp 1027–1035

Chertkov M (2021) Michael Chertkov: graphical (and agent based) models of pandemic. Columbia Academic Commons, Academic Unit: Northeast Big Data Innovation Hub, Series: COVID-19 Research Lightning Talks. https://doi.org/10.7916/46jp-3f96

Chertkov M, Abrams R, Sikaroudi AME, Krechetov M, Slagle CN, Efrat A, Fulek R, Oren E (2021) Graphical models of pandemic. medRxiv

Dong E, Du H, Gardner L (2020) An interactive web-based dashboard to track COVID-19 in real time. Lancet Infect Dis 20:533–534

Esmaieeli Sikaroudi AM, Efrat A, Chertkov M (2025) Agent-in-cell modeling of pandemics. In: Hudson FD (ed) The COVID information commons—research insights from the coronavirus. Sustainable development goals series. Springer

Killick R, Fearnhead P, Eckley IA (2012) Optimal detection of changepoints with a linear computational cost. J Am Stat Assoc 107:1590–1598

Krechetov M, Esmaieeli Sikaroudi AM, Efrat A, Polishchuk V, Chertkov M (2022) Prediction and prevention of pandemics via graphical model inference and convex programming. Sci Rep 12:7599. https://doi.org/10.1038/s41598-022-11705-8

Mahmood I, Arabnejad H, Suleimenova D, Sassoon I, Marshan A, Serrano-Rico A, Louvieris P, Anagnostou A, Taylor SJ, Bell D et al (2020) FACS: a geospatial agent-based simulator for analysing COVID-19 spread and public health measures on local regions. J Simul 1–19

Reno Gazette Journal (2014) COVID-19 hospital capacity in King County and surrounding area

Rhodes DM, Holcombe M, Qwarnstrom EE (2016) Reducing complexity in an agent based reaction model—benefits and limitations of simplifications in relation to run time and system level output. Biosystems 147:21–27

SafeGraph COVID-19 Data Consortium (2022) SafeGraph Inc, San Francisco, CA. https://www.safegraph.com/covid-19-data-consortium

Tregubov A, Blythe J (2021) Optimization of large-scale agent-based simulations through automated abstraction and simplification. In: Multi-agent-based simulation XXI: 21st International workshop, MABS 2020. Auckland, New Zealand, May 10, 2020, Revised Selected Papers 21. Springer, pp 81–93

Amir Mohammad Esmaieeli Sikaroudi is currently a Ph.D. Student in computer science at the University of Arizona. He graduated from the University of Arizona in 2023 with a master's degree in systems and industrial engineering. Prior to enrolling in the computer science degree, he had prior experience in mathematical modeling and data mining. Graphical models, computational geometry, and agent-based models are some of the disciplines that he is interested in researching. With a particular emphasis on pandemics, he is interested in conducting research on the mobility and interaction of individuals. With the goal of accurately and efficiently modeling pandemics, he is working on a variety of techniques from computational geometry and machine learning. The control of pandemics, facility layout, employee turnover prediction, hospital scheduling, and image processing are some of the topics that he has published. For both academic and professional goals, as well as for his own hobby, he is enthusiastic about high-performance computing and the creation of visualization tools.

Alon Efrat is an Associate Professor at the Computer Science Department at the University of Arizona. He was trained as Computational Geometer. His research includes optimization problems in geometric settings and in particular their instances in geographic domains, networking and sensors networks, and mobility. Dr. Efrat received his Ph.D. degree in computer science from the Tel-Aviv University in 1998. He was also Postdoctoral Research Assistant with Stanford University, Stanford, California, and with the IBM Almaden Research Center, San Jose, California. Dr. Efrat has contributed more than 150 publications. Dr. Efrat has served many National Science Foundation (NSF) panels and technical program committees in different areas. He served on the Editorial Board of various venues such as International Journal of Computational Geometry and Applications (IJCGA), Journal of Discrete Algorithms (JDA), and conferences such as Foundations of Computer Science (FOCS), International Symposium on Computational Geometry (SoCG), INFOCOM, ACM SIGSPATIAL International Conference on Advances in Geographic Information Systems (SIGSPATIAL), International Conference on the Design of Reliable Communication Networks (DRCN), IEEE Military Communications Conference (MILCOM), and many others. Dr. Efrat was Co-chair of the sensor network algorithms track of the International Symposium on Algorithms and Experiments for Sensor Systems, Wireless Networks and Distributed Robotics (ALGOSENSORS14), and co-organizer of the workshop on Geometric Optimization in Wireless Communication and Sensing, in conjunction with the Thirtieth Annual Symposium on Computational Geometry (SoCG'14).

Michael Chertkov is a Professor of Mathematics and Chair of the Graduate Interdisciplinary Program in Applied Mathematics at the University of Arizona. His research addresses foundational challenges in mathematics, statistics, machine learning, and artificial intelligence, particularly as they apply to and are inspired by physical systems like fluid mechanics. He also works on applications in the control of engineered systems, such as energy grids, and bio-social systems. Dr. Chertkov received his Ph.D. in physics from the Weizmann Institute of Science in 1996. After obtaining his Ph.D., he spent three years as R.H. Dicke Fellow in the Department of Physics at Princeton University. In 1999, he joined the Los Alamos National Laboratory, first as a J.R. Oppenheimer Fellow, later becoming a Technical Staff Member in the Theory Division. He transitioned to the University of Arizona in 2019. Throughout his career, Dr. Chertkov has contributed about 300 research papers. He holds the title of Fellow in both the American Association for the Advancement of Science (AAAS) and the American Physical Society and is a Senior Member of the Institute of Electrical and Electronics Engineers (IEEE).

7 Preventing the Next Pandemic: Ethical Management of Persistent Viruses

John Yin

Abstract

Prevention of future pandemics will depend on the development of robust antiviral therapeutics and strategies. Here we focus on defective interfering particles (DIPs) and their engineered counterparts, therapeutic interfering particles (TIPs). Such particles can mitigate viral replication, resist virus escape, and provide broad-spectrum protection against diverse viral infections, including COVID-19. Drawing from historical and recent studies, we briefly review DIPs and TIPs in both natural and laboratory settings, including their potential as transmissible therapeutics, their co-evolution with standard viruses, and their ability to modulate host immune responses. We also delve into the ethical considerations linked to their deployment, including long-term risks to individual patients and wider communities. This work highlights the importance of interdisciplinary research, integrating virology, immunology, bioinformatics, and ethics to address the challenges posed by evolving pathogens. Finally, we underscore the importance of a balanced approach to the development of TIPs, evaluating the benefits and risks, with a focus on ensuring their safe and ethical use as a strategy for pandemic prevention.

Keywords

COVID-19 · Pandemic · Ethical management · Virus · Defective interfering particles (DI particles, DIPs) · Therapeutic interfering particles (TIPs)

Research Funded by
NSF/MCB* Award #2029281—RAPID**: Ecological Dynamics of Human Coronavirus;

NSF/CBET*** Award #2030750—EAGER****: Rapid and Sensitive Drug Testing for COVID-19.

*Division of Molecular and Cellular Biosciences.
**Rapid Response Research Award.
***Division of Chemical, Bioengineering, Environmental, and Transport Systems.
****EArly-Concept Grants for Exploratory Research.

COVID Information Commons Video Presentations
The COVID Information Commons (CIC) Research Lightning Talk video presentations and individual multilingual transcripts for this project can be found in the Columbia Academic Commons (AC) (Yin 2020, 2024).

J. Yin (✉)
University of Wisconsin-Madison, Madison, WI, USA
e-mail: john.yin@wisc.edu

7.1 Introduction and Goal

Despite the remarkable success of mRNA vaccines to protect against COVID-19, the virus responsible for the pandemic, SARS-CoV-2, continues to cause disease globally (Yin 2020, 2024). The mRNA vaccines developed by Pfizer/BioNTech and Moderna were highly effective

F. D. Hudson (ed.), *The COVID Information Commons*, Sustainable Development Goals Series,
https://doi.org/10.1007/978-3-032-06125-6_7

in protecting against SARS-CoV-2 infection, preventing an estimated two million deaths, 17 million hospitalizations, and $900 billion in healthcare costs (Schneider et al. 2021). However, more than 190,000 new cases of COVID-19 were reported by the World Health Organization over the 28-day period ending July 28, 2024 (COVID-19 Cases 2024). New and continuing cases of COVID-19 can be attributed in part to the facile ability of SARS-CoV-2 to quickly adapt to new environments, contributing to its transmissibility, drug and immune escape, and reinfections (Markov et al. 2023).

A long-term goal of our research has been to explore and develop antiviral strategies that have potential to resist escape (Endy and Yin 2000; Kim and Yin 2004). Specifically, we have focused on strategies based on defective interfering particles (Stauffer Thompson et al. 2009; Thompson and Yin 2010; Baltes et al. 2017; Bhat et al. 2022).

7.2 Insights and Results

7.2.1 DIPs and TIPs

Defective interfering particles (or DIPs) were first found in the 1950s and 1960s during studies of viral replication in cell cultures, where they were observed as truncated or mutated viral particles that interfered with the replication of their parent viruses (von Magnus 1985, 1954). DIPs lack essential genetic information required for replication but can hijack the replication machinery of co-infecting standard viruses, leading to a reduction in the production of infectious virus particles (Thompson and Yin 2010; Huang and Baltimore 1970). Owing to their ability to interfere or "prey on" normal virus infections, DIPs were proposed and explored as therapeutics to treat virus infections. DIPs of Rift Valley fever virus, Semliki Forest virus, influenza virus, and flaviviruses (e.g., West Nile, Yellow Fever, and Zika) derived from laboratory cultures were shown to have protective effects against infections in mice (Mims 1956; Barrett and Dimmock 1984; Dimmock et al. 2012; Rezelj et al. 2021). More recently, DIPs based on deletion variants of poliovirus and SARS-CoV-2 have been engineered and demonstrated protection in rodents (Chaturvedi et al. 2021; Xiao et al. 2021); to differentiate DIPs that arise spontaneously from lab culture from interfering particles that have been engineered, these have been named therapeutic interfering particles (or TIPs). It is noteworthy that the TIP derived from poliovirus was able to protect against infections by SARS-CoV-2, influenza, rhino, and enteroviruses, indicating stimulation of a potentially broadly protective host immune response (Xiao et al. 2021). Moreover, the TIP derived from SARS-CoV-2 was shown to compete for the virus replication machinery and inhibit the virus over multiple cycles of replication, indicating potential to resist virus escape (Chaturvedi et al. 2021).

Although DIPs were initially thought to be a byproduct of culturing viruses in the laboratory, the isolation of defective genomes of hepatitis A virus from clinical samples (Nüesch et al. 1989) and further studies have demonstrated their prevalence in natural and clinical infections by dengue virus (Aaskov et al. 2006), West Nile virus (Pesko et al. 2012), influenza virus (Saira et al. 2013), and HIV-1 (Imamichi et al. 2016). With the arrival of COVID-19, we wondered whether SARS-CoV-2 might form DIPs and, if so, how they might contribute to disease severity, where some COVID-positive individuals are asymptomatic, while others experienced life-threatening symptoms (Frater et al. 2020; García 2020; Rahi et al. 2021). It is plausible that the interfering effects of DIPs could inhibit virus growth and spread, resulting in less severe disease, as observed for TIPs. But initial sequence analyses by others have found DIPs or their associated defective genomes are more prevalent in patients who *are* symptomatic (Wong et al. 2021; Zhou et al. 2023). Our preliminary analyses of SARS-CoV-2 genome sequences from COVID patient swabs have highlighted a need for improved bioinformatic methods; specifically, one needs to distinguish actual deletion variants from false positives associated with multiplex-PCR (polymerase chain reaction) sequencing, the standard method of swab sample analyses. Ongoing work in our lab aims to identify natural DIPs and their genome sequences from patients with known disease severity. Moreover, we seek the design rules for natural TIPs associated with less severe or asymptomatic disease.

7.2.2 How Might TIPs Resist Escape?

Escape is the ability of viruses to evade drugs or human immune responses to viral infection. Antiviral drugs are developed to bind and inhibit or shut down specific enzymes or functions that are essential for virus growth. Random mutations that occur during replication of the virus genome can result in subtle changes to virus enzymes, enabling the enzymes to escape inhibition by the drug; so, virus variants that possess such mutations enrich in the presence of the drug compared to variants that lack the mutations. In short, viruses escape drugs by generating mutations that enable their enzymes or other functions to evade interactions with the drugs. Similarly, human immune responses are mediated by antibodies that target proteins on the surface of virus particles, which the virus uses to dock with host cells. By binding to such proteins, antibodies can prevent the virus-cell encounters that are needed to initiate infection. As with drug escape, random mutations caused during virus replication can cause subtle changes in the virus surface proteins, so they escape surveillance by the host antibodies; hence, the infections proceed. For both drug and immune escape, the generation of mutations during error-prone replication of the virus genome enables variant viruses to arise and escape variants to enrich as dominant variants. TIPs typically interfere with virus growth by diverting replication and virus packaging resources away from the virus and toward their own growth (Kim and Yin 2004; Chaturvedi et al. 2021); in so doing, TIPs use the same error-prone replication machinery of the virus and generate TIP variants. The generation of mutations during virus and TIP or DIP replication opens the possibility that they may exhibit co-evolutionary dynamics, behavior that has been observed in lab evolution studies of vesicular stomatitis virus (DePolo et al. 1987), a model system for studies of virus evolution. It is an open frontier to understand what role co-evolutionary processes among TIPs, DIPs, and viruses may contribute to the persistence and pathogenesis of virus infections in natural and clinical settings.

7.2.3 How About TIP Transmission Between Humans?

DIPs as well as TIPs are virus-like particles that may transmit between humans or other hosts (Bhat et al. 2022; Rezelj et al. 2021; Aaskov et al. 2006). To anticipate how TIP evolution and transmission might play out, consider the case of polio, where the Sabin oral polio vaccine (OPV), a live attenuated virus strain, can be transmitted by fecal–oral routes. Risks associated with OPV are vaccine-associated paralytic poliomyelitis (VAPP) cases and the emergence of vaccine-derived polio-virus strains (VDPV) (Baicus 2012). VAPP arises when the vaccine strain mutates back to the neurovirulent strain-causing paralysis, while VDPVs implicated in most polio outbreaks are associated with regions of low vaccination coverage, prolonged circulation, and virus mutation (Diop et al. 2015). OPV contains DIPs (McLaren and Holland 1974), but it is not known what role they may play in VAPP, VDPV, or stimulating immune protection. DIPs are also present in attenuated vaccines of other highly transmissible viruses, including measles (Bellocq et al. 1990) and influenza (Gould et al. 2017), but their precise roles in activating protective immunity are not known.

7.3 Ethics of Adaptable Transmissible Therapeutics

The development of adaptable transmissible therapeutics, based on natural or engineered DIPs, should be guided by ethical considerations of their risks. Current risks include incomplete understanding of their molecular mechanisms, specifically how they stimulate host immune responses, as well as their potential for transmission and evolution.

DIPs and TIPs have shown efficacy in rodent studies, consistent with molecular mechanisms that divert growth and packaging resources away from the production of a standard virus. They have also shown potential to counter diverse virus infections, granting protection by activation of host innate immune responses. Despite these advances, DIPs can activate inflammatory immune responses (Genoyer and López 2019). When dysregulated, inflammatory responses have been associated with cytokine storms and high disease severity (Hu et al. 2021), and higher DIP levels have been associated with symptomatic rather than asymptomatic COVID (Wong et al. 2021; Zhou et al. 2023). It is currently unknown to what extent DIPs or TIPs can trigger host immune responses that are either non-protective or adversely affect disease severity. These unknowns define direct risks to the patient, highlighting the need for informed consent by the patient to employ such treatments.

Further, unlike traditional antiviral treatments, TIPs can be transmitted between individuals, which introduces risks akin to those associated with live attenuated vaccines. Although they are not "live" in the sense of attenuated vaccines that can autonomously cause infection, trigger immune responses and spread, DIPs and TIPs can autonomously activate host immune responses; and during co-infection with a standard virus, they can replicate, spread, and activate or modulate host immune responses. It is currently unknown whether naturally occurring DIPs evolve or co-evolve with standard virus; our bioinformatic analyses of virus and virus-like sequences from immunocompromised COVID patients provides preliminary evidence that viruses and their DIPs can co-evolve. The possibility of TIPs co-evolving with viruses in ways that could lead to unforeseen disease outcomes demands careful consideration of their safety over the long term. The unknowns define risks not only to patients, but also their social networks and beyond, so a broader societal engagement will be needed to weigh the risks against the benefits of transmissible therapeutics.

7.4 Summary and Recommendations

Therapeutic interfering particles (TIPs) offer hope for preventing future pandemics. They have potential as transmissible broad-spectrum drugs that resist virus escape. However, the development of TIPs should aim for a comprehensive assessment of their risks at both the individual and societal level. Striking a balance between their benefits and risks will be key to their safe and ethical deployment.

Acknowledgements We are grateful to the National Science Foundation (NSF), the National Institutes of Health (NIH), the Wisconsin Alumni Research Foundation, the Wisconsin Institute for Discovery, and the Office of the Vice Chancellor for Research at the University of Wisconsin-Madison for their support. Collaborators Colin Dewey and Nan Jiang have contributed to our initial study of DIPs in COVID. We thank James Conway, Norman Fost, Christine Grady, Jo Handelsman, Paul Kelleher, and H. Clifford Lane for thoughtful discussions.

References

Aaskov J, Buzacott K, Thu HM, Lowry K, Holmes EC (2006) Long-term transmission of defective RNA viruses in humans and Aedes mosquitoes. Science 311:236–238. https://doi.org/10.1126/science.1115030

Baicus A (2012) History of polio vaccination. World J Virol 1:108–114. https://doi.org/10.5501/wjv.v1.i4.108

Baltes A, Akpinar F, Inankur B, Yin J (2017) Inhibition of infection spread by co-transmitted defective interfering particles. PLoS ONE 12:e0184029. https://doi.org/10.1371/journal.pone.0184029

Barrett AD, Dimmock NJ (1984) Modulation of Semliki forest virus-induced infection of mice by defective-interfering virus. J Infect Dis 150:98–104. https://doi.org/10.1093/infdis/150.1.98

Bellocq C, Mottet G, Roux L (1990) Wide occurrence of measles virus subgenomic RNAs in attenuated live-virus vaccines. Biologicals 18:337–343. https://doi.org/10.1016/1045-1056(90)90039-3

Bhat T, Cao A, Yin J (2022) Virus-like particles: measures and biological functions. Viruses 14. https://doi.org/10.3390/v14020383

Chaturvedi S, Vasen G, Pablo M, Chen X, Beutler N, Kumar A et al (2021) Identification of a therapeutic interfering particle—a single-dose SARS-CoV-2 antiviral intervention with a high barrier to resistance. Cell 184:6022–6036.e18. https://doi.org/10.1016/j.cell.2021.11.004

COVID-19 Cases. In: datadot [Internet]. Available: https://data.who.int/dashboards/covid19/cases. Cited 10 Aug 2024

DePolo NJ, Giachetti C, Holland JJ (1987) Continuing coevolution of virus and defective interfering particles and of viral genome sequences during undiluted passages: virus mutants exhibiting nearly complete resistance to formerly dominant defective interfering particles. J Virol 61:454–464. https://doi.org/10.1128/JVI.61.2.454-464.1987

Dimmock NJ, Dove BK, Meng B, Scott PD, Taylor I, Cheung L et al (2012) Comparison of the protection of ferrets against pandemic 2009 influenza A virus (H1N1) by 244 DI influenza virus and oseltamivir. Antiviral Res 96:376–385. https://doi.org/10.1016/j.antiviral.2012.09.017

Diop OM, Burns CC, Sutter RW, Wassilak SG, Kew OM, Centers for Disease Control and Prevention (CDC) (2015) Update on vaccine-derived polioviruses—worldwide, January 2014–March 2015. MMWR Morb Mortal Wkly Rep. 64:640–646

Endy D, Yin J (2000) Toward antiviral strategies that resist viral escape. Antimicrob Agents Chemother 44:1097–1099. https://doi.org/10.1128/AAC.44.4.1097-1099.2000

Frater JL, Zini G, d'Onofrio G, Rogers HJ (2020) COVID-19 and the clinical hematology laboratory. Int J Lab Hematol 42(Suppl 1):11–18. https://doi.org/10.1111/ijlh.13229

García LF (2020) Immune response, inflammation, and the clinical spectrum of COVID-19. Front Immunol 11:1441. https://doi.org/10.3389/fimmu.2020.01441

Genoyer E, López CB (2019) The impact of defective viruses on infection and immunity. Annu Rev Virol 6:547–566. https://doi.org/10.1146/annurev-virology-092818-015652

Gould PS, Easton AJ, Dimmock NJ (2017) Live attenuated influenza vaccine contains substantial and unexpected amounts of defective viral genomic RNA. Viruses 9. https://doi.org/10.3390/v9100269

Hu B, Huang S, Yin L (2021) The cytokine storm and COVID-19. J Med Virol 93:250–256. https://doi.org/10.1002/jmv.26232

Huang AS, Baltimore D (1970) Defective viral particles and viral disease processes. Nature 226:325–327. https://doi.org/10.1038/226325a0

Imamichi H, Dewar RL, Adelsberger JW, Rehm CA, O'Doherty U, Paxinos EE et al (2016) Defective HIV-1 proviruses produce novel protein-coding RNA species in HIV-infected patients on combination antiretroviral therapy. Proc Natl Acad Sci U S A 113:8783–8788. https://doi.org/10.1073/pnas.1609057113

Kim H, Yin J (2004) Quantitative analysis of a parasitic antiviral strategy. Antimicrob Agents Chemother 48:1017–1020. https://doi.org/10.1128/aac.48.3.1017-1020.2004

Markov PV, Ghafari M, Beer M, Lythgoe K, Simmonds P, Stilianakis NI et al (2023) The evolution of SARS-CoV-2. Nat Rev Microbiol 21:361–379. https://doi.org/10.1038/s41579-023-00878-2

McLaren LC, Holland JJ (1974) Defective interfering particles from poliovirus vaccine and vaccine reference strains. Virology 60:579–583. https://doi.org/10.1016/0042-6822(74)90352-3

Mims CA (1956) Rift valley fever virus in mice. IV. Incomplete virus; its production and properties. Br J Exp Pathol 37:129–143

Nüesch JP, de Chastonay J, Siegl G (1989) Detection of defective genomes in hepatitis A virus particles present in clinical specimens. J Gen Virol 70(Pt 12):3475–3480. https://doi.org/10.1099/0022-1317-70-12-3475

Pesko KN, Fitzpatrick KA, Ryan EM, Shi PY, Zhang B, Lennon NJ et al (2012) Internally deleted WNV genomes isolated from exotic birds in New Mexico: function in cells, mosquitoes, and mice. Virology 427:10–17. https://doi.org/10.1016/j.virol.2012.01.028

Rahi MS, Jindal V, Reyes S-P, Gunasekaran K, Gupta R, Jaiyesimi I (2021) Hematologic disorders associated with COVID-19: a review. Ann Hematol 100:309–320. https://doi.org/10.1007/s00277-020-04366-y

Rezelj VV, Carrau L, Merwaiss F, Levi LI, Erazo D, Tran QD et al (2021) Defective viral genomes as therapeutic interfering particles against flavivirus infection in mammalian and mosquito hosts. Nat Commun 12:1–14. https://doi.org/10.1038/s41467-021-22341-7

Saira K, Lin X, DePasse JV, Halpin R, Twaddle A, Stockwell T et al (2013) Sequence analysis of in vivo defective interfering-like RNA of influenza A H1N1 pandemic virus. J Virol 87:8064–8074. https://doi.org/10.1128/JVI.00240-13

Schneider EC, Shah A, Sah P, Moghadas SM, Vilches T, Galvani AP (2021) The U.S. COVID-19 vaccination program at one year: how many deaths and hospitalizations were averted? Commonwealth Fund. https://doi.org/10.26099/3542-5N54

Stauffer Thompson KA, Rempala GA, Yin J (2009) Multiple-hit inhibition of infection by defective interfering particles. J Gen Virol 90:888–899. https://doi.org/10.1099/vir.0.005249-0

Thompson KAS, Yin J (2010) Population dynamics of an RNA virus and its defective interfering particles in passage cultures. Virol J. 7:257. https://doi.org/10.1186/1743-422X-7-257

von Magnus P (1954) Incomplete forms of influenza virus. Adv Virus Res 2:59–79. https://doi.org/10.1016/s0065-3527(08)60529-1

von Magnus P (1951) Propagation of the PR8 strain of influenza A virus in chick embryos. III. properties of the incomplete virus produced in serial passages of undiluted virus. Acta Pathol Microbiol Scand 29:157–181. https://doi.org/10.1111/j.1699-0463.1951.tb00114.x

Wong CH, Ngan CY, Goldfeder RL, Idol J, Kuhlberg C, Maurya R et al (2021) Reduced subgenomic RNA expression is a molecular indicator of asymptomatic SARS-CoV-2 infection. Commun Med 1:1–12. https://doi.org/10.1038/s43856-021-00034-y

Xiao Y, Lidsky PV, Shirogane Y, Aviner R, Wu C-T, Li W et al (2021) A defective viral genome strategy elicits broad protective immunity against respiratory viruses. Cell. 184:6037–6051.e14. https://doi.org/10.1016/j.cell.2021.11.023

Yin J (2020) John Yin: ecological dynamics of human coronavirus, EAGER: rapid and sensitive drug testing for COVID-19. Columbia Academic Commons, Academic Unit: Northeast Big Data Innovation Hub, Series: COVID-19 Research Lightning Talks. https://doi.org/10.7916/wax8-bb43

Yin J (2024) John Yin: evolutionary dynamics of human coronaviruses: implications for strategies to prevent future pandemics. Columbia Academic Commons, Academic Unit: Northeast Big Data Innovation Hub, Series: COVID-19 Research Lightning Talks. https://doi.org/10.7916/89kv-pz57

Zhou T, Gilliam NJ, Li S, Spandau S, Osborn RM, Connor S et al (2023) Generation and functional analysis of defective viral genomes during SARS-CoV-2 infection. Mbio 14:e0025023. https://doi.org/10.1128/mbio.00250-23

John Yin blends science, engineering, and other creative endeavors to tackle pressing challenges in biology. Born in Boston and raised across the USA, Yin's early curiosity spanned both the sciences and the arts, leading him to pursue degrees in chemical engineering and liberal arts at Columbia University. Alongside his studies, he trained in molecular pharmacology at the National Cancer Institute and performed the Schumann Piano Concerto with the university orchestra. After earning his Ph.D. at the University of California, Berkeley, where he patented a bio-mimetic material for environmental cleanup, Yin's fascination with evolution took him to Germany. There, under Nobel laureate Manfred Eigen, he discovered a way to track viral growth and evolution—insights that continue to shape his research today. Now a professor of chemical and biological engineering at the University of Wisconsin-Madison and founding-five faculty member of the Wisconsin Institute for Discovery, Yin investigates the emergence and persistence of molecular information in nature. His work on therapeutic interfering particles offers new hope for preventing future pandemics by leveraging natural molecular parasites of virus growth and spread. Beyond the lab, Yin's interests range from chamber music to table tennis to culinary experiments—each a reflection of his passion for exploration and innovation.

Part III

COVID and Public Health

8 Physician Perspectives on Using Telehealth with Older Adults

Natalie M. Davoodi, Megan E. Bounds, and Elizabeth M. Goldberg

Abstract

Since the onset of the COVID-19 pandemic in March 2020, hospitals and medical offices had to quickly pivot from traditional in-person care to telehealth services. However, these medical institutions and private practices sometimes faced challenges digitally connecting with older adults because of age-related differences that can make technology use difficult, such as changes in vision, hearing, cognition, and dexterity. Physicians have shared different suggestions for best practices when using telehealth with the older adult population, as well as lessons learned surrounding adopting new technology workflows in health care that can be applied during future pandemics. This chapter will discuss the findings from a qualitative research study with 48 physicians who used telehealth with older adults during the COVID-19 pandemic. First, we will broadly explain the importance of focusing on telehealth use within the older adult population. Next, we will explore how physician perspectives in the qualitative study differed depending on their specialty. Facilitators and barriers to telehealth use in older adults varied depending on whether the specialty was procedural, and whether physicians specialized in treating acute complaints or mostly managed minor or chronic health conditions. Finally, we will describe the key lessons learned that can be applied to future pandemics as needed

Keywords

COVID-19 · Pandemic · Telehealth · Telemedicine · Older adults · Physician · Primary care · Geriatric · Emergency medicine

Research Funded by

NIH Award K76AG059983—GAPcare II: The Geriatric Acute & Post-acute Care Coordination Program for Fall Prevention in the Emergency Department.

COVID Information Commons Video Presentation

The COVID Information Commons (CIC) Research Lightning Talk video presentation and individual multilingual transcripts for this project can be found in the Columbia Academic Commons (AC) (Goldberg 2021).

N. M. Davoodi
Johns Hopkins Bloomberg School of Public Health, Baltimore, MD, USA
e-mail: ndavood1@jh.edu

M. E. Bounds · E. M. Goldberg (✉)
University of Colorado Anschutz, Aurora, CO, USA
e-mail: elizabeth.goldberg@cuanschutz.edu

M. E. Bounds
e-mail: Megan.Bounds@cuanschutz.edu

F. D. Hudson (ed.), *The COVID Information Commons*, Sustainable Development Goals Series,
https://doi.org/10.1007/978-3-032-06125-6_8

8.1 Older Adults and Telehealth

Older adults ages 65 years and older have faced disproportionately high levels of COVID-induced morbidity and mortality in the United States (U.S.) (Goldberg 2021). Given that the older adult population consumes the most health care in the U.S. and often has complex care needs, telehealth technologies should be designed with older adult users in mind (Ahn et al. 2020) (Fig. 8.1). Before the pandemic, Medicare beneficiaries did not have the flexibility to use telehealth services, and, more broadly, many physicians had never offered telehealth (Roberts and Mehrotra 2020). Coupled with the lack of a preexisting telehealth "blueprint," hospitals and medical offices had to quickly shift from offering traditional in-person care to telehealth services, thus lacking the time to incorporate age-friendly modifications to their technology.

As a result of the lack of existing experience and urgency with which physicians needed to adopt telehealth, telehealth was not designed with older adults in mind. Importantly, members of this population likely have one or more impairments that can make telehealth technology use difficult (Chen et al. 2022; Lam et al. 2020). For instance, impairments in hearing, vision, cognition, and dexterity can make accessing telehealth services challenging. Prior to the pandemic, older adults in the U.S. had already reported that they were less comfortable using digital health services than younger populations (Lam et al. 2020). There is a "digital divide" whereby older adults did not use technology as young adults, compared to younger generations who are "digital natives". It is also necessary to highlight the need for Internet connectivity in using telehealth. Unfortunately, many older adults in the U.S. do not have Internet in their homes or devices that are Internet accessible. This lack of access is particularly problematic among older adults who are low income, have low educational attainment, or are Black or Hispanic (Chen et al. 2022; Hunsaker and Hargittai 2018).

8.2 The Qualitative Study

To explore how telehealth usage did or did not work with older adults in the U.S., we conducted a qualitative study with 48 U.S. physicians who cared for older adults during the pandemic from various specialties (15 primary care, 18 geriatrics, and 15 emergency medicine physicians) (Goldberg et al. 2021). Eligible physicians must have cared for patients 65 years of age and older (with or without COVID-19) in-person and/or via telehealth during the pandemic. Physicians were recruited using Twitter/X, Facebook, and two society listservs (Academy for Geriatric Emergency Medicine with 158 members, and the American Geriatric Society Member Forum with 7600 members in 2020). The research team

Fig. 8.1 Older adults using technology (Image taken from Canva)

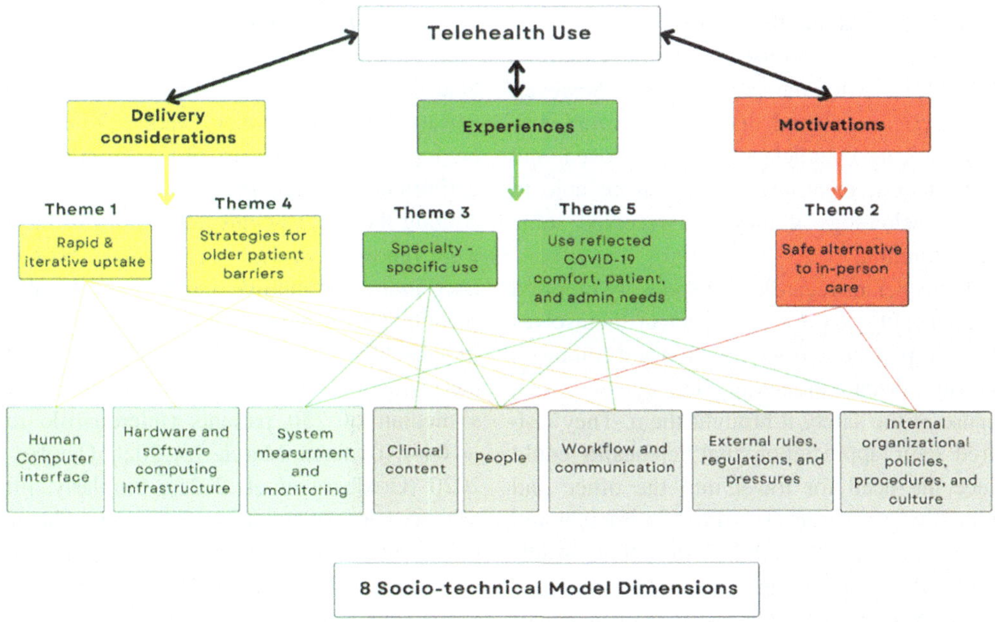

Fig. 8.2 Qualitative study themes using Sittig's sociotechnical model (Adapted from Goldberg, 2021)

created a semi-structured interview guide and pilot-tested the guide amongst themselves.

Physicians were asked about (1) telehealth delivery considerations (ability to use telehealth during the pandemic, adaptations made to clinical practice to incorporate telehealth, and timing of telehealth adoption), (2) motivations for telehealth use (positives and negatives regarding telehealth use, and perceived challenges with initiating or continuing telehealth care), and (3) experiences with telehealth during the pandemic (specialty-specific or other factors influencing use or non-use and trajectories or changes in use). The study period for interviews was March 13, 2020–October 16, 2020 (Goldberg et al. 2021).

We identified five key themes from the qualitative study in the context of Sittig's sociotechnical model for studying health information technology in complex adaptive healthcare systems (Sittig and Singh 2010): (1) telehealth uptake was rapid and iterative, (2) telehealth was seen as a safe alternative to in-person care, (3) specific uses for telehealth varied based on physician specialty, (4) physicians had several strategies for using telehealth with older patients, and (5) telehealth use was specific to patient and administrative needs, as well as severity of the COVID-19 pandemic (Fig. 8.2).

More information on the qualitative study procedures and Sittig's sociotechnical model can be found in our published manuscript (Goldberg et al. 2021; Sittig and Singh 2010). In the next sections, we will break down the findings from our qualitative study by physician specialty.

8.3 Physician Views on Using Telehealth with Older Adults

Broadly, all participating physicians shared the opinion that the uptake of audiovisual telehealth platforms in their practice was rapid and often disorganized (Goldberg et al. 2021). After the initial onset, physicians continued to make changes as needed to their telehealth delivery, to improve equipment, platforms, and scheduling mechanisms, thus emphasizing the need for practices to be iterative. Even though 42% of participating physicians ($N = 20$) had not used any telehealth prior to the pandemic, many physicians shared that they received very little or no training on how to use telehealth (Goldberg et al. 2021). After the initial period, however, most physicians shared that their comfort with delivering care via telehealth had improved, and all physicians, regardless of age, were able to use the technology if they had received some type of training.

All physicians regardless of specialty reported that telehealth was a way to protect themselves during the pandemic from infection and embraced telehealth, despite the initial challenges it posed, because of the safety it brought them. They also shared their appreciation that telehealth could reduce the need for travel into the office and potentially improve access to care for homebound patients or those with mobility limitations. Additionally, all physicians agreed that telehealth use could be beneficial for chronic disease management and lower-acuity complaints, whereas higher acuity concerns necessitated an in-person visit. All physicians shared that they were concerned about missing diagnoses via telehealth, although they were unaware of any adverse events related to telehealth during the study period (Goldberg et al. 2021).

In a web-based cross-sectional national survey of U.S. geriatricians, primary care physicians (PCPs), and emergency medicine physicians, three key findings were identified: (1) 50% of emergency physicians, 33% of geriatricians, and 18% of primary care physicians considered telehealth to be a poor substitute for providing end-of-life care ($p = 0.68$); (2) for treating patients with hearing, vision, and cognitive impairments, 61%, 58%, and 54%, respectively, saw telehealth as a good or fair substitute for providing care ($p = 0.14$); and (3) 98% indicated that telehealth was a good or fair substitute for in-person care for those with mobility impairment ($p < 0.001$) (Serina et al. 2023) (Fig. 8.3). Preferences and comfort using telehealth with older adults varied by clinical context, patient population, and physician specialty, requiring tailored adaptations (Serina et al. 2023).

8.3.1 Geriatricians and Primary Care Physicians

Due to similar findings and practice structures, geriatricians and primary care physicians will be reported together in this section. Eighteen geriatricians participated in the qualitative study. Among these physicians, all regions of the U.S. were represented, 44% were female, 50% practiced in the community, and 50% practiced in an academic institution. Among participating geriatricians, 50% had not used any telehealth prior to the pandemic. These physicians reported seeing a median of 250 patients (interquartile range = 64–640) from March 13, 2020–October 16, 2020 (Goldberg et al. 2021). Similarly, fifteen primary care physicians participated in the qualitative study. Among these physicians, all regions of the U.S. (Northeast, Midwest, South, and West) were represented, all physicians resided in either metro or suburban areas, 80% were female, 67% practiced in the community, and 33% practiced in an academic institution (Goldberg et al. 2021). Among participating primary care physicians, 33% had not used any telehealth prior to the pandemic. These physicians reported seeing a median of 500 patients (interquartile range = 200–960) from March 13, 2020–October 16, 2020 (Goldberg et al. 2021).

Geriatricians and primary care physicians were surprised with how much care they could deliver remotely to their older patients (Goldberg et al. 2021). These physicians were often successful when taking patient history and examining patients. During telehealth appointments, it was common for these physicians to ask caregivers to help with exams, to teach patients how to examine themselves, and to ask patients to use remote devices, such as blood pressure cuffs, to share their vital signs. Although geriatricians and primary care physicians shared concerns about missing diagnoses because of the lack of an in-person physical exam, they were unable to recount an instance in which they missed a diagnosis via telehealth (Chen et al. 2022; Goldberg et al. 2021).

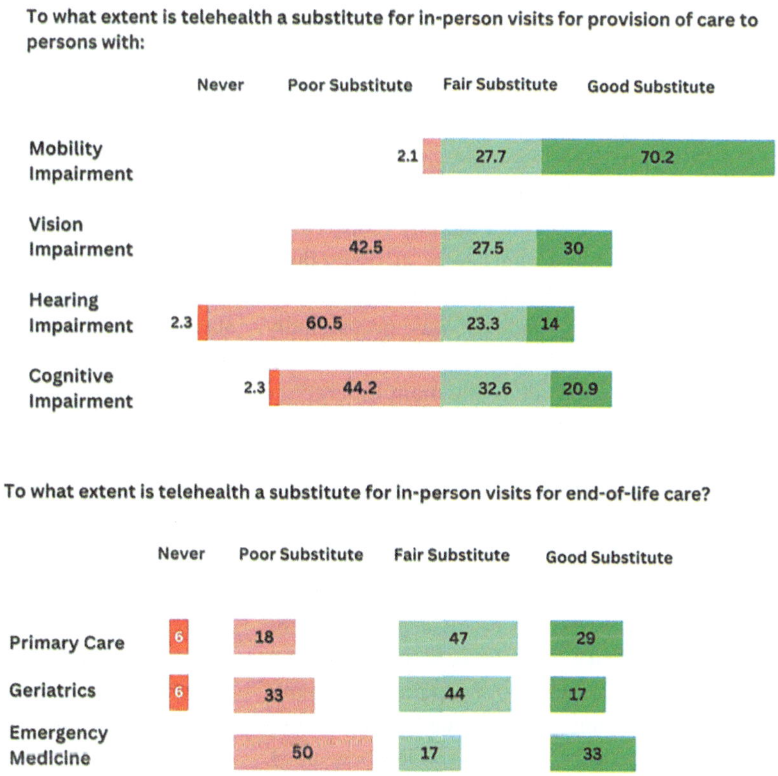

Fig. 8.3 Findings from web-based national survey (Adapted from Serina, 2023)

Primary care physicians reported that they preferred to use telehealth whenever possible so that they could keep their offices safe during the pandemic. They especially appreciated using telehealth for mental health visits, medication refills, as well as for certifying both home health and durable medical equipment needs (Goldberg et al. 2021). Geriatricians, on the other hand, were more willing to have in-person appointments. For instance, they preferred in-person visits for older adults who had cognitive impairment, lacked broadband connectivity or necessary technology, or did not have strong familial or social support. However, geriatricians preferred to use telehealth for Medicare Annual Wellness Visits, for patients living in facilities (such as assisted living), for those who primarily had chronic illnesses, for follow-up appointments, and for medication reconciliation (Chen et al. 2022; Goldberg et al. 2021).

8.3.2 Emergency Medicine Physicians

Fifteen emergency medicine physicians participated in the qualitative study. Among these physicians, all regions of the U.S. were represented, 47% were female, 33% practiced in the community, and 67% practiced in an academic institution. Among participating emergency medicine physicians, 40% had not used any telehealth prior to the pandemic. These physicians reported seeing a median of 100 patients (interquartile range = 35–400) from March 13, 2020–October 16, 2020 (Goldberg et al. 2021).

As opposed to primary care physicians and geriatricians, emergency medicine physicians did not use telehealth as frequently during the pandemic (Goldberg et al. 2021). They primarily used telehealth to connect with patients' families during the pandemic-induced hospital visitation restrictions. These physicians also used video-enabled devices and phone calls to obtain patient histories and reduce their exposure time (Goldberg et al. 2021; Davoodi et al. 2021). For instance, some emergency physicians would call into the patient room to obtain the necessary history and could thereby reduce their time at the bedside. Exams were at times done several feet away with observation only. Additionally, triaging of patients was occasionally done by telehealth, whereby patients would sit in a dedicated room outside of the main Emergency Department to get an initial assessment by telehealth only and determine if further in-person care was needed. However, critical patients often required in-person care. Interestingly, some emergency physicians communicated with older adults using chat-based platforms during the pandemic and were able to fill a gap in public health and primary care services, especially for medication refills and helping patients navigate to COVID testing sites or understand when in-person visits were necessary (Goldberg et al. 2021; Davoodi et al. 2021).

Emergency physicians commonly shared that telehealth should continue to have a place in emergency care as a supplement, rather than a replacement, for in-person care because most critical patients required a physical exam and procedures (Davoodi et al. 2021). For instance, one emergency physician shared that, "At least in our telemedicine program, [...] you can only do simple diagnoses [like rashes], but you're not supposed to do abdominal pain and you're not supposed to do things that obviously are not amenable to telemedicine and are higher acuity, because ...it's just a waste of time for both the patient and me, if they're going to come in with, whatever, something more complicated that obviously ... needs to be evaluated on an emergent basis." (Participant 14, Academic, Metro) (Goldberg et al. 2021).

8.4 Physician Strategies for Using Telehealth with Older Adults

Most physicians, regardless of specialty, reported difficulties with using video visits with older adults, not only due to patient age, but also because of cognitive or sensory impairments, lack of Internet connectivity, and low socioeconomic status (Chen et al. 2022; Goldberg et al. 2021) (Fig. 8.4). One primary care physician importantly described this phenomenon, "the digital divide, it's far less age and far more technical literacy for one reason or another" (Participant 22, Western U.S., Community) (Goldberg et al. 2021).

To best engage older adults in telehealth visits, participating physicians shared various strategies that worked for them (Chen et al. 2022; Goldberg et al. 2021). First, many physicians recommended calling patients before visits to see whether they had the necessary equipment to join a telehealth visit. If older patients were unable to join a video visit, physicians preferred reverting to a phone call visit. Another common strategy was asking caregivers of the older adult patient to be present during the visit to help

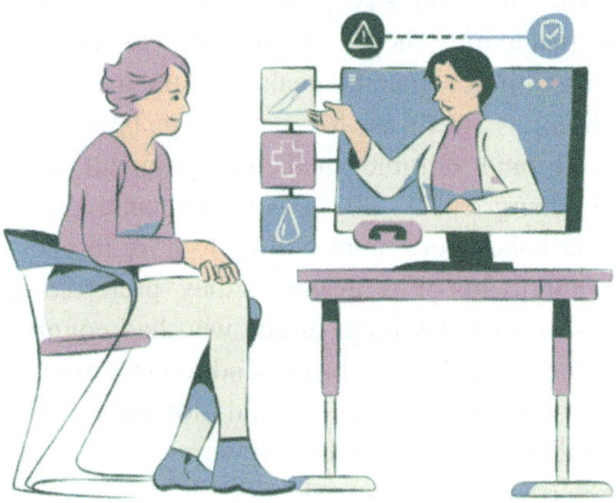

Fig. 8.4 Physician using telehealth with an older patient (Image taken from Canva)

as needed, especially in instances when the patient was lacking in technological literacy or necessary equipment. Some medical offices even distributed tablets to their older patients who did not have a computer or smartphone. Regardless of specialty, physicians reported that to improve the telehealth experience for their older patients, they were intentional to speak slower and more clearly and to use chats when audio features were not connecting properly (Goldberg et al. 2021). Some physicians also reported that they connected their older patients who had hearing impairments with Bluetooth-enabled headphones or pocket talkers so they could better engage in their visits (Chen et al. 2022; Goldberg et al. 2021; Davoodi et al. 2021).

8.5 Ultimate Decline in Telehealth Use

Despite the positive uses for telehealth with older patients and strategies used to improve engagement, participating physicians shared that as the perceived risk of the pandemic decreased, so did their use of telehealth with their older patients (Goldberg et al. 2021). Once more was learned about the pandemic and protective equipment was in better supply, there was less incentive to continue using telehealth as the primary mode of care delivery. Physicians also reported that they began to miss interacting with patients in-person, that the technology had become inefficient, or that their older patients preferred to come in-person for visits. Physicians also shared that they had some medicolegal concerns with telehealth, specific to inaccurate or missed diagnoses, and that changing regulations were impacting their reimbursement levels for telehealth, thus lowering their incentive to deliver care remotely. Regardless of whether an individual physician wanted to continue using telehealth with older patients, they needed to have support from their administrators to continue using telehealth in their office or hospital. At times administrators mandated that healthcare visits now be in-person only.

8.6 Lessons Learned for Future Pandemics

There are several lessons that can be learned for telehealth care delivery moving forward (Chen et al. 2022; Goldberg et al. 2021, 2022; Davoodi et al. 2021). Ideally, telehealth technology and platforms should be designed with the needs of older adults and those with disabilities in mind. For instance, telehealth systems should use one-click mechanisms or provider-initiated visits to reduce confusion when initiating a visit. However, if this is not possible, it is important that hospitals and medical offices adjust their telehealth delivery systems to accommodate any challenges with vision, hearing, dexterity, or cognition. Engaging with caregivers, family members, or others in the patient's social network can improve the visit. Physicians should be willing to change the cadence and tone when speaking with older patients and have their staff reach out to older patients prior to their visit to ensure that they can join their visit.

To incentivize administrators and physicians to continue using telehealth currently and in future pandemics, there are several important concerns brought up during our qualitative study that must be addressed (Goldberg et al. 2021, 2022). First, there should be legal protections in place for medicolegal concerns surrounding misdiagnosis. Telehealth offers convenience and efficiency but limits physical exam capabilities. Next, it is imperative that integration of electronic health records and telehealth platforms is enhanced to improve efficiency when using telehealth. Many physicians reported using separate platforms for documenting, billing, and initiating telehealth visits. Finally, policymakers must ensure cross-state licensure and competitive reimbursement. Specifically, pay parity with in-person visits and improved reimbursement for phone calls may address administrator-level concerns with using telehealth.

As telehealth will likely continue to play an important role in the care of many older adults, further research on patient-centered design and clinician and patient training in telehealth is warranted (Goldberg et al. 2021). Many of our physicians emphasized the importance of tailoring their telehealth practice to the needs of older adults, yet they also reported receiving limited to no training on how to optimally use telehealth with older patients. Our findings can be incorporated into future training of physicians as strategies and tips for improving care before, during, and after patient appointments.

There is a need for continued investment in and improvement of telehealth to prepare for future pandemics and for physician shortages, unrelated to pandemics, that make on-site care challenging (Goldberg et al. 2022). For instance, while the population of older adults in the United States is projected to increase exponentially in the upcoming decades, the workforce will likely be short-staffed to take care of this population. With the proper investments and design modifications, telehealth can play an important role in filling these care gaps for older adults moving forward.

References

Ahn S, Hussein M, Mahmood A, Smith ML (2020) Emergency department and inpatient utilization among U.S. older adults with multiple chronic conditions: a post-reform update. BMC Health Serv Res 20(1):77. https://doi.org/10.1186/s12913-020-4902-7

Chen K, Davoodi NM, Strauss DH, Li M, Jiménez FN, Guthrie KM, Goldberg EM (2022) Strategies to ensure continuity of care using telemedicine with older adults during COVID-19: a qualitative study of physicians in primary care and geriatrics. J Appl Gerontol Official J Southern Gerontol Soc 41(11):2282–2295. https://doi.org/10.1177/07334648221109728

Davoodi NM, Chen K, Zou M, Li M, Jiménez F, Wetle TF, Goldberg EM (2021) Emergency physician perspectives on using telehealth with older adults during COVID-19: a qualitative study. J Am Coll Emerg Physicians Open 2(5):e12577. https://doi.org/10.1002/emp2.12577

Goldberg EM, Jiménez FN, Chen K, Davoodi NM, Li M, Strauss DH, Zou M, Guthrie K, Merchant RC (2021) Telehealth was beneficial during COVID-19 for older Americans: a qualitative study with physicians. J Am Geriatr Soc 69(11):3034–3043. https://doi.org/10.1111/jgs.17370

Goldberg EM, Lin MP, Burke LG, Jiménez FN, Davoodi NM, Merchant RC (2022) Perspectives on telehealth for older adults during the COVID-19 pandemic using the quadruple aim: interviews with 48 physicians. BMC Geriatr 22(1):188. https://doi.org/10.1186/s12877-022-02860-8

Goldberg L (2021) Liz Goldberg: meeting the health needs of older Americans with telehealth during COVID-19, A supplement to GAPcare II: the geriatric acute & post-acute care coordination program for fall prevention in the emergency department. Columbia Academic Commons, Academic Unit: Northeast Big Data Innovation Hub, Series: COVID-19 Research Lightning Talks. https://doi.org/10.7916/22da-c235

Hunsaker A, Hargittai E (2018) A review of Internet use among older adults. New Media Soc 20(10):3937–3954. https://doi.org/10.1177/1461444818787348

Lam K, Lu AD, Shi Y, Covinsky KE (2020) Assessing telemedicine unreadiness among older adults in the United States during the COVID-19 pandemic. JAMA Intern Med 180(10):1389–1391. https://doi.org/10.1001/jamainternmed.2020.2671

Roberts ET, Mehrotra A (2020) Assessment of disparities in digital access among medicare beneficiaries and implications for telemedicine. JAMA Intern Med 180(10):1386–1389. https://doi.org/10.1001/jamainternmed.2020.2666

Serina PT, Davoodi NM, Guthrie KM, Merchant RC, Goldberg EM (2023) Physician views of telehealth for special populations of older adults: preliminary findings. J Patient Exp 10:23743735231160416. https://doi.org/10.1177/23743735231160418

Sittig DF, Singh H (2010) A new sociotechnical model for studying health information technology in complex adaptive healthcare systems. Qual Saf Health Care 19(Suppl 3):i68–i74. https://doi.org/10.1136/qshc.2010.042085

Natalie M. Davoodi is a Ph.D. Student in the Department of Health Policy and Management at the Johns Hopkins Bloomberg School of Public Health. Her research interests include examining policies and programs pertaining to the health care of older adults in the United States, specifically in the areas of financing, delivery, spending, organization, and health outcomes. Natalie has a Master of Public Health from the Brown University School of Public Health and a Bachelor of Science in Physiology and Neuroscience from the University of California, San Diego. Prior to entering her Ph.D., Natalie was a Clinical Research Coordinator at Rhode Island Hospital and a Senior Data Analyst with the Center for Gerontology and Healthcare Research at the Brown University School of Public Health. She also worked as an Emergency Medical Technician throughout the COVID-19 pandemic.

Megan E. Bounds earned her Master's in Public Health with an emphasis on global health from the University of Arizona in 2023. She was a Global Health Initiative Scholar and completed a research internship with the University of Limerick, Ireland, gaining qualitative research experience in an international setting focusing on older adults' experience in the emergency department during COVID-19. Ms. Bounds currently works as a Research Coordinator for Dr. Elizabeth Goldberg's lab focusing on older adult care and emergency medicine. She has served as Project Lead on a fall prevention trial as well as multiple projects aiming to increase older adults' access to care using different digital tools and interventions. Ms. Bounds also serves as a Coordinator for an interdisciplinary lab team, GeriTech, which advances collaborative efforts in technology and older adult care.

Elizabeth M. Goldberg is a practicing Emergency Physician and Associate Professor of Emergency Medicine at the University of Colorado School of Medicine. After her residency, she completed a post-doctoral research fellowship in gerontology and epidemiology at Brown University. Dr. Goldberg is Recipient of the prestigious Paul B. Beeson Career Development Award from the National Institute on Aging which funds a fall prevention trial using the Apple Watch to track cognition and fitness. Her research focus is creating effective interventions to improve emergency care for geriatric patients and using digital tools to help older adults age in place. Dr. Goldberg is Past President of the Academy for Geriatric Emergency Medicine for the Society of Academic Emergency Medicine. In this capacity, she led the research and education efforts of over 300 academic clinicians to improve geriatric emergency care in the USA.

Measuring the Effect of the COVID-19 Pandemic on a Longitudinal Study of Patient Reported Symptoms in Patients with Breast Cancer

9

Susan W. Wesmiller, Katrina Carr, Yvette Conley, and Susan M. Sereika

Abstract

COVID-19 introduced unique challenges for individuals with breast cancer, a population already vulnerable to significant symptom burden. While increased psychological distress during the pandemic has been well-documented, the intersection of symptom burden, socioeconomic factors, and resilience remains less explored. Understanding how financial insecurity, social support, and resilience influenced the symptom experience can provide valuable insights for future healthcare strategies. Women with breast cancer faced disruptions in employment and social interactions, contributing to worsening symptom burden. Higher resilience was associated with lower symptom severity, highlighting its protective role. However, individuals who experienced job or income loss reported greater depression, sleep disturbances, and difficulties engaging in social roles. Those living alone reported increased heightened symptom burden, emphasizing the importance of social support in managing health. The findings of this study call for more research on an integrated precision health care approach to care that considers treatment-related factors alongside environmental, social, biological, and economic influences. Recognizing these factors, healthcare providers can develop effective support systems to mitigate symptom burden and enhance well-being, particularly during crises.

Keywords

Breast cancer · COVID-19 · Symptom burden · Resilience · Employment loss · Social support · Psychological distress

Research funded by

NIH/NINR* Award 3R01NR016695-03S1—Genomic Underpinnings for Breast Cancer Treatments Induced Nausea and Vomiting.
*National Institute of Nursing Research

COVID Information Commons Video Presentation

The COVID Information Commons (CIC) Research Lightning Talk video presentation and individual multilingual transcripts for this project can be found in the Columbia Academic Commons (AC) (Wesmiller 2021).

S. W. Wesmiller (✉) · K. Carr · Y. Conley · S. M. Sereika
University of Pittsburgh, Pittsburgh, PA, USA
e-mail: swe100@pitt.edu

K. Carr
e-mail: klc190@pitt.edu

Y. Conley
e-mail: yconley@pitt.edu

S. M. Sereika
e-mail: ssereika@pitt.edu

F. D. Hudson (ed.), *The COVID Information Commons*, Sustainable Development Goals Series,
https://doi.org/10.1007/978-3-032-06125-6_9

9.1 Introduction

The COVID-19 pandemic, which surfaced in the United States in early 2020, introduced an unexpected layer of complexity for individuals with cancer—a population already at high risk for significant symptom burden. At the same time, our team was actively recruiting participants for a 12-month longitudinal National Institute of Nursing Research (NINR) funded study (R01NR016695) (Wesmiller 2021). Our parent study focused on precisely phenotyping symptoms women experienced for one year following surgery and during adjuvant therapy for breast cancer and collection of biospecimens to address the biological underpinnings of variation in symptom burden. As data collection continued during the beginning of the pandemic, we began noticing shifts in patient-reported outcomes (Table 9.1) as measured by the Patient Reported Outcomes Measurement System® 29 (PROMIS®-29) v2.0 profile (Hartmann et al. 2023; Hays et al. 2018) and qualitative responses to open-ended study questions. Participants expressed concerns about job security, especially as many had already experienced job loss within their families. One study participant shared, "*I am trying to hold on to my job through chemotherapy treatments. I just have to because both my son and my husband were laid off this week. I do not know what we will do.*" A major concern for participants was the loss of employer-based insurance, if job loss occurred. Others described losing access to resources like community pools, which helped alleviate joint pain, and the distress of attending chemotherapy or radiation treatments alone. Recognizing the impact of these changes, we applied for an administrative supplement (NOT-OD-20-097) to investigate these evolving challenges further and were awarded the administrative supplement (NIH award 3R01NR016695-03S1) to the parent study.

These qualitative concerns align with other evidence reported globally at the beginning of the pandemic. During the early stages of the COVID-19 pandemic, many studies focused on changes that women with breast cancer experienced including heightened levels of depression, anxiety, loneliness, and sleep deprivation (Wesmiller and Harpel 2021; Miaskowski et al. 2020; Clark-Snow and Rittenberg 2021; Seven et al. 2021; Zhao et al. 2022). In France, 30% of patients with cancer reported anxiety, and 15% met the criteria for Post-traumatic Stress Disorder (PTSD), with women at higher risk for both (Stanizzo et al. 2022). A study in Italy found that patients with breast cancer in 2019 had lower levels of anxiety and PTSD compared to those experiencing Italy's stay-at-home orders one year later (Marino et al. 2022). These findings highlight the need for further investigation into the pandemic's impact on symptom burden among women with breast cancer.

Thus, based on these findings, including our own report (Wesmiller and Harpel 2021), the purpose of this study, which was perfectly aligned with the parent study, was to investigate the effect of the COVID-19 pandemic on the symptoms experienced by women with breast cancer and to identify those women at highest risk for increased symptom burden.

9.2 Methods

9.2.1 Research Design, Setting, and Sample

This supplemental study recruited study participants from the parent longitudinal study who were scheduled for breast cancer surgery (without reconstruction) at a large, National Cancer Institute (NCI) designated cancer center hospital. Patients were eligible for study participation if they met the following criteria: (1) biological female aged 18–90 years; (2) diagnosed with early stage breast cancer (Stage I, II, IIIA); (3) had a negative history for previous neurological conditions that could also be a cause of nausea and vomiting; (4) were assigned an American Society of Anesthesiologists (ASA) physical status of I, II, or III; (5) had access to a computer or a telephone to report symptoms experienced; and (6) were able to read and write in English. Study participants in the supplemental study were three months post-operative for their breast cancer surgery before or after April 1, 2020. The supplemental study was approved by the University of Pittsburgh Institutional Review Board (IRB). All study participants engaged in active informed written addendum consent prior to data collection.

9.2.2 Procedures and Study Variables

Socio-demographic data were collected at the time of recruitment and included age and race/ethnicity. Given the known associations between socioeconomic factors and symptom burden in cancer patients, we included additional variables such as resilience, changes in employment status, living arrangements, and neighborhood deprivation to capture the multifaceted impact of COVID-19 on patient outcomes. We used the time frame of April 1, 2020, to add the presence of COVID-19 as a variable in the model to determine changes across symptoms, coinciding with our observed heightened response to symptoms measured by the PROMIS®-29, as reported by study participants in addition to their qualitative comments (Table 9.1).

PROMIS® tools are an NIH initiative to develop standardized items that measure feelings, function, and perceptions applicable to a range of conditions effectively, flexibly, and precisely (Hays et al. 2018). The PROMIS®-29 includes brief but valid assessments of quality of life including anxiety, depressive symptoms, fatigue, pain interference, sleep disturbance, functional status and ability to participate in social activities. An 11-point pain rating scale is also included. Norm based scores have been developed for the PROMIS® tools, a score of 50 represents the mean of the general population. A score above 50 indicates a worse outcome, as compared to the general population.

9.2.3 Resiliency

The American Psychological Association (APA) defines resilience as the process and outcome of successfully adapting to difficult or challenging life experiences, especially through mental, emotional, and behavioral flexibility and adjustment to external and internal demands (APA 2023). Resiliency was measured by the Connor-Davidson RISC 10 survey (CD-RISC 10) (Davison 2020), a well validated 10-item tool used previously in women following breast cancer surgery. The CD-RISC 10 measures the ability to adapt when changes occur and the ability to bounce back after illness, injury or other hardship. It has also been found to be a predictor of health status (Chen 2020). COVID-19 was a public health emergency of international concern and posed a challenge to psychological resilience (Wang et al. 2020). The previous research has shown a resilient personality in older adults can moderate the effect of ill-health (Windle et al. 2020). COVID-19 posed a significant challenge to psychological resilience, and research at the time highlighted a lack of data on resilience-based interventions that could reduce symptom burden (Marroquín et al. 2016; Langford et al. 2020).

Table 9.1 Selected items from PROMIS®-29 in January 2020 and April 2020 (Preliminary data)

Item from PROMIS®-29	January 2020 ($n = 102$)	April 2020 ($n = 91$)
I felt fearful (Never or rarely)	84 (89%)	73 (79%)
I felt helpless (Never or rarely)	99 (89%)	85 (84%)
My sleep was refreshing (quite a bit or very much)	51 (50%)	38 (42%)
I have trouble doing all my regular leisure activities (Never or rarely)	76 (75%)	53 (57%)

9.2.4 Living Arrangements

Research prior to the emergence of COVID-19 found women treated for breast cancer were at risk to experience loneliness significant enough to cause depression (Marroquín et al. 2016). Other researchers have identified one of the most stressful impacts on patient outcomes is living alone. Langford and colleagues further identified the factors that place patients with cancer in a stressed subgroup, as living alone, most life stress and low resilience (Langford et al. 2020). Finally, Kroenke et al. (2020) found that living alone could lead to significantly poorer health outcomes, including increased mortality from colorectal cancer. Living arrangements was measured as a categorical variable that included partnered, living with family, living alone with close family support or living alone without support.

9.2.5 Change in Employment Status

In the United States alone, more than 30 million adults filed for unemployment benefits in the first year of the pandemic, suggesting an unemployment rate comparable only to that of the Great Depression (Rugaber 2020). Loss of employment, and potentially the employer-based insurance especially for patients undergoing cancer, is a cause of increased distress and anxiety (Mimoun et al. 2020). These economic and employment impacts are likely to endure past the peak of the pandemic, and it may take several years for these jobs to become available once again (Mimoun et al. 2020). Employment loss or change of income was measured as a categorical variable addressing employment loss, decreased income due to change of employment status, or no change in employment.

9.2.6 The Area Deprivation Index

The Area Deprivation Index (ADI) (Kind and Buckingham 2018) measures deprivation rankings of neighborhoods by socioeconomic disadvantage in a region of interest (e.g., at the state or national level). Factors include income, education, employment, and housing quality. The ADI is measured on both a statewide and national level. The national measure is based on a 100-point scale, with low deprivation receiving a lower score, and increased deprivation receiving a higher score (Index 2018).

9.2.7 Statistical Analysis

To be included in the analysis for the supplemental study, study participants had to have both the CD-RISC 10 score and PROMIS®-29 *T*-scores at three months. That said, the 150 participants included in our analysis had both the CD-RISC 10 score and PROMIS®-29 *T*-scores at three months.

SAS version 9.4 (SAS Institute, Inc., Cary, NC) was used for all analyses. Data were first explored yielding descriptive statistics to characterize the study sample as well the results from data screening to check for outliers, the amount and pattern of missing data, and underlying statistical assumptions for the primary analysis strategy, linear regression analysis. When performing simple linear regression analyses, we treated the *T*-scores from the PROMIS®-29 at the three-month assessment for the symptoms of anxiety, depression, fatigue, sleep disturbance, and participation with social activities as the dependent variables to investigate the association between the symptom and predictor variables, and considered each targeted predictor variable in the linear regression model one at a time where the level of statistical significance was set at

0.05 for non-directional hypothesis testing. We performed model assessment for each regression model, including residual analysis and application of influence diagnostics. For each combination of symptom outcome and predictor variable, we report the unstandardized regression coefficient along with the lower and upper limits for its 95% confidence interval and the p-value from the model F-statistic.

9.3 Results

9.3.1 Study Participants

Participants ($n = 150$) were mostly non-Hispanic White (90.0%, $n = 135$) with a mean age of 60.6 years (SD = 11.6) and body mass index of 30.1 (SD = 24.8) kg/m^2. For more than a quarter of the sample (28.7%, $n = 43$) someone in their household experienced job or income loss due to the COVID-19 pandemic. Most (71.3%, $n = 107$) lived with immediate family members and on average lived in a neighborhood with an area deprivation index based on the ADI national percentile of 58.8 (SD = 23.2), with 68.0% ($n = 102$) living in more disadvantaged neighborhoods (ADI national percentile $\geq$ 50). Despite the challenges of the pandemic, study participant retention remained at 100%, and data collection was successfully maintained via telephone interviews, facilitated by the fact that most participants were homebound.

All participants were diagnosed with early stage breast cancer and were mostly Estrogen Receptor (ER) positive (81.3%, $n = 122$), Progesterone Receptor (PR) positive (74.0%, $n = 111$), and Human Epidermal Growth Factor Receptor 2 (HER-2) negative (71.3%, $n = 107$). At three months post-surgery, 70.7% ($n = 106$) were currently on one or more treatments for breast cancer, including radiation (28.0%, $n = 42$), chemotherapy (14.0%, $n = 21$), and aromatase inhibitor therapy (21.3%, $n = 32$).

For resilience, participants had CD-RISC 10 scores on average of 32.2 (SD = 6.4) with scores ranging from 12 to 40. Regarding symptoms ratings based on the PROMIS®-29 T-scores at three months post-surgery, T-scores on average fell within the normal range for anxiety (mean = 45.9, SD = 8.1), depression (mean = 44.8, SD = 6.4), fatigue (mean = 49.6, SD = 10.4), sleep disturbance (mean = 48.7, SD = 8.8), and ability to participate in social roles and activities (mean = 55.1, SD = 10.8). While most participants had T-scores falling within the normal range (82.0% for anxiety, 88.0% for depression, 65.3% for fatigue, 80.0% for sleep disturbance, 74.7% for ability to participate in social roles and activities), significant percentages had T-scores falling in the mild to moderate levels for anxiety (mild: 10.0%, moderate: 8.0%) and depression (mild: 8.0%, moderate: 4.0%) and mild to severe levels for fatigue (mild: 18.7%, moderate: 13.3%, severe: 2.7%), sleep disturbance (mild: 8.0%, moderate: 11.3%, severe: 0.7%), and ability to participate in social roles and activities (mild: 11.3%, moderate: 13.3%, severe: 0.7%).

9.3.2 Associations Between PROMIS®-29 Symptom Ratings and Targeted Predictors

Table 9.2 summarizes the results from applying simple linear regression for each combination of a PROMIS®-29 T-score and a targeted predictor variable. Higher resilience scores were fairly strongly and negatively associated with more severe PROMIS®-29 T-scores for domains of interest ($p < 0.010$), except for fatigue ($p = 0.265$). Significant negative associations were also observed between age with fatigue ($p = 0.041$) and sleep disturbance ($p = 0.040$). Job or income loss for any member of the participant's household (including the participant) during the COVID-19 pandemic was associated with significantly poorer T-scores for depression, sleep disturbance and ability to engage in social roles and activities. Specifically, higher T-scores for depression ($p = 0.029$) and sleep disturbance ($p = 0.035$) and lower scores for one's ability to engage in social roles and activities ($p = 0.008$) were associated with any job or income loss during the COVID-19 pandemic.

Table 9.2 Results of simple linear regression of PROMIS®-29 *T*-scores at 3-months post-surgery

Predictor	Anxiety	Depression	Fatigue	Sleep disturbance	Participation in social roles and activities
COVID-19 After 4/1/2020	− 0.571 (− 3.335, 2.193) 0.858	− 0.741 (− 2.923, 1.441) 0.503	− 0.372 (− 3.932, 3,188) 0.837	1.047 (− 1.973, 4.067) 0.494	− 3.029 (− 6.707, 0.649) 0.106
Connor-Davidson RISC 10 score	− 0.668 (− 0.843, − 0.493) **< 0.001**	− 0.567 (− 0.701, − 0.433) **< 0.001**	− 0.149 (− 0.413, 0.114) 0.265	− 0.389 (− 0.605, − 0.173) **< 0.001**	0.414 (0.147, 0.682) **0.003**
High neighborhood disadvantage (ADI National Percentile ≥ 50)	− 1.456 (− 4.333, 1.402) 0.314	− 1.964 (− 4.178, 0.251) 0.082	0.018 (− 3.661, 3.698) 0.992	− 1.470 (− 4.586, 1.645) 0.352	2.125 (− 1.690, 5.941) 0.273
Age (years)	− 0.102 (− 0.213, 0.009) 0.072	− 0.052 (− 0.140, 0.036) 0.247	− 0.149 (− 0.292, − 0.007) **0.041**	− 0.127 (− 0.248, − 0.006) **0.040**	0.059 (− 0.092, 0.209) 0.443
Lives alone	1.568 (− 1.672, 4.807) 0.341	1.231 (− 1.329, 3.790) 0.344	− 0.712 (− 4.807, 3.383) 0.732	− 0.879 (− 4.421, 2.663) 0.625	3.238 (− 0.985, 7.461) 0.132
Any job or income loss during the COVID-19 pandemic	1.954 (− 0.911, 4.820) 0.180	2.509 (0.268, 4.750) **0.029**	2.453 (− 1.238, 6.143) 0.191	3.350 (0.244, 6.456) 0.**035**	− 5.132 (− 8.910, − 1.355) **0.008**

Note The values reported in cells are the unadjusted unstandardized regression coefficient and its corresponding 95% confidence interval and the *p*-value for the regression model *F*-test

9.4 Discussion

This study examined the impact of the COVID-19 pandemic on symptom burden among women undergoing treatment for early stage breast cancer. Our qualitative findings suggest that the pandemic intensified psychosocial distress, particularly in areas of anxiety, depression, sleep disturbance, and social participation. These results align with previous research indicating that the COVID-19 pandemic significantly increased psychological distress among cancer patients, particularly women (Wesmiller and Harpel 2021; Miaskowski et al. 2020; Clark-Snow and Rittenberg 2021; Seven et al. 2021; Zhao et al. 2022; Stanizzo et al. 2022; Marino et al. 2022). Colomer-Lahigua discusses the increase in symptoms experienced by patients with cancer during the pandemic as psychological morbidity (Colomer-Lahiguera et al. 2024). One of the more striking findings was the association between job or income loss and increased symptom burden. This is consistent with Han's work, which found increased morbidity and mortality during COVID-19 in individuals with lower median income and greater age (Han et al. 2024).

It is also consistent with research that has focused on the effect of a higher area deprivation index score (ADI) and symptoms experienced (Balogun et al. 2024). Women who experienced employment or income disruptions reported significantly higher levels of depression ($p = 0.029$) and sleep disturbance ($p = 0.035$), alongside a reduced ability to engage in social roles and activities ($p = 0.008$). This finding is consistent with previous studies showing that financial insecurity and loss of employer-based health insurance during COVID-19 heightened psychological distress in cancer patients (Kind and Buckingham 2018; Index 2018). Economic hardship is well-documented as a major contributor to health disparities, and our results underscore the need for continued policy-level interventions to support financially vulnerable populations, particularly those undergoing active cancer treatment.

Another key finding was the role of resilience in mitigating symptom burden. Higher resilience scores were significantly associated with lower symptom severity across multiple domains, except for fatigue ($p = 0.265$). This finding supports prior research suggesting that resilience serves as a protective factor against stress-related health outcomes, including anxiety and depression, in cancer patients (Davison 2020; Chen 2020; Wang et al. 2020; Windle et al. 2020). Barthelemy describes resilience during the pandemic, as "how to bend and not break" (Barthélemy et al. 2021). Living alone was also a notable predictor of worsened symptom burden. Research has long suggested that social isolation exacerbates distress in cancer patients, and our findings reinforce these concerns. However, Liu found that resilience mitigated the adverse health outcomes associated with loneliness (Liu et al. 2023). Women who lived alone during the pandemic may have been particularly vulnerable to heightened anxiety and depression due to reduced social interactions and the inability to receive in-person support during treatment. These findings emphasize the importance of identifying and supporting at-risk individuals who may lack strong social networks, particularly during public health crises.

An important finding from this study was the feasibility of remote biospecimen collection during public health emergencies. Despite the COVID-19 pandemic and associated restrictions on in-person contact, participants consistently returned mailed saliva collection kits, suggesting that mail-based protocols are a reliable alternative to in-person collection. This has important implications for the design of future longitudinal studies requiring biospecimens, particularly under circumstances that limit face-to-face interactions.

9.5 Limitations

A limitation of this study was the single measurement of resilience. However, there are some researchers who feel that resilience is a trait, not a state (Hannoodee and Dhamoon 2025; Calo et al. 2024; Mondolin et al. 2024). Those who recognize the concept of hardiness as a personal, and usually a protective, trait would also agree that in this study, one measure of resilience is appropriate.

Nonetheless, the results of this study contribute to a growing body of literature highlighting the psychosocial impact of a pandemic on cancer patients and the protective nature of resilience.

9.6 Conclusion

The COVID-19 pandemic had a profound impact on the symptom burden experienced by women with breast cancer, with job loss, social isolation, and neighborhood deprivation identified in this study as key risk factors. Resilience, however, emerges as a critical protective factor, underscoring the importance of examining the impact of low resilience levels. These findings call for an integrated approach to cancer care that considers not only biological and treatment-related factors but also social and economic determinants of health; what we are moving toward calling the exposome. Moving forward, ensuring robust support systems for cancer patients—particularly during global crises—will be critical in improving health outcomes and quality of life.

References

APA (2023) APA dictionary of psychology

Balogun Z, Gardiner LA, Li J, Moroni EA, Rosenzweig M, Nilsen ML (2024) Neighborhood deprivation and symptoms, psychological distress, and quality of life among head and neck cancer survivors. JAMA Otolaryngol Head Neck Surg 150:295–302. https://doi.org/10.1001/jamaoto.2023.4672

Barthélemy EJ, Thango NS, Höhne J, Lippa L, Kolias A, Germano IM (2021) Resilience in the face of the COVID-19 pandemic: how to bend and not break. World Neurosurg 146:280–284. https://doi.org/10.1016/j.wneu.2020.11.105

Calo M, Judd B, Peiris C (2024) Grit, resilience and growth-mindset interventions in health professional students: a systematic review and meta-analysis. Med Educ 58:902–919. https://doi.org/10.1111/medu.15391

Chen LK (2020) Older adults and COVID-19 pandemic: resilience matters. Arch Gerontol Geriatrics 89

Clark-Snow RA, Rittenberg C (2021) Oncology nursing supportive care during the COVID-19 pandemic: reality and challenges. Support Care Cancer 29:2259–2262. https://doi.org/10.1007/s00520-021-06005-2

Colomer-Lahiguera S, Pozzar RA, Cooper BA, Paul SM, Snowberg K, Kenfield SA, Chang SM, Abbott M, Van Blarigan EL, Levine JD et al (2024) Worse psychological profiles are associated with higher levels of stress and symptom burden in patients with cancer during the COVID-19 pandemic. Oncol Nurs Forum 51:529–546. https://doi.org/10.1188/24.Onf.529-546

Davison JR (2020) Connor-Davidson resilience scale (CDRISC) manual

Han Y, Li N, Zhang Y, Sun J, Jiang Z (2024) Epidemiology and risk factors for adult gliomas died by respiratory diseases during the COVID-19 pandemic: a population-based study. Neurosurg Rev 47:700. https://doi.org/10.1007/s10143-024-02943-7

Hannoodee S, Dhamoon AS (2025) Nursing Neuman systems model. In: StatPearls. StatPearls Publishing Copyright © 2025, StatPearls Publishing LLC

Hartmann C, Fischer F, Klapproth CP, Röhle R, Rose M, Karsten MM (2023) PROMIS-29 and EORTC QLQ-C30: an empirical investigation towards a common conception of health. Qual Life Res 32:749–758. https://doi.org/10.1007/s11136-022-03324-7

Hays RD, Spritzer KL, Schalet BD, Cella D (2018) PROMIS(®)-29 v2.0 profile physical and mental health summary scores. Qual Life Res 27:1885–1891. https://doi.org/10.1007/s11136-018-1842-3

Index UoWSoMaPHAd (2018) Area deprivation index

Kind A, Buckingham W (2018) Making neighborhood disadvantage metrics accessible: the neighborhood atlas. New Engl J Med 378:2456–2458. PMCID: PMC6051533. https://doi.org/10.1056/NEJMp1802313

Kroenke CH, Paskett ED, Cené CW, Caan BJ, Luo J, Shadyab AH, Robinson JRM, Nassir R, Lane DS, Anderson GL (2020) Prediagnosis social support, social integration, living status, and colorectal cancer mortality in postmenopausal women from the women's health initiative. Cancer 126:1766–1775. https://doi.org/10.1002/cncr.32710

Langford DJ, Cooper B, Paul S, Humphreys J, Hammer MJ, Levine J, Conley YP, Wright F, Dunn LB, Miaskowski C (2020) Distinct stress profiles among oncology patients undergoing chemotherapy. J Pain Symptom Manage 59:646–657. https://doi.org/10.1016/j.jpainsymman.2019.10.025

Liu J, Gou RY, Jones RN, Schmitt EM, Metzger E, Tabloski PA, Arias F, Hshieh TT, Travison TG, Marcantonio ER et al (2023) Association of loneliness with change in physical and emotional health of older adults during the COVID-19 shutdown. Am J Geriatr Psychiatry 31:1102–1113. https://doi.org/10.1016/j.jagp.2023.07.015

Marino P, Touzani R, Pakradouni J, Ben Soussan P, Gravis G (2022) The psychological distress of cancer patients following the COVID-19 pandemic first lockdown: results from a large French survey. Cancers (Basel) 14. https://doi.org/10.3390/cancers14071794

Marroquín B, Czamanski-Cohen J, Weihs KL, Stanton AL (2016) Implicit loneliness, emotion regulation, and depressive symptoms in breast cancer survivors. J Behav Med 39:832–844. https://doi.org/10.1007/s10865-016-9751-9

Miaskowski C, Paul SM, Snowberg K, Abbott M, Borno H, Chang S, Chen LM, Cohen B, Cooper BA, Hammer MJ et al (2020) Oncology patients' perceptions of and experiences with COVID-19. Support Care Cancer 1-10. https://doi.org/10.1007/s00520-020-05684-7

Mimoun E, Ben Ari A, Margalit D (2020) Psychological aspects of employment instability during the COVID-19 pandemic. Psychol Trauma 12:S183-s185. https://doi.org/10.1037/tra0000769

Mondolin V, Karlsson H, Perasto L, Tuulari JJ, Karlsson L, Kataja EL (2024) Understanding resilience in parents: longitudinal examination of trait resilience, stressful life events, and psychological distress symptoms-insights from the FinnBrain study. Stress Health 40:e3516. https://doi.org/10.1002/smi.3516

Rugaber C (2020) Another 3.8 million US workers file for unemployment; coronavirus has put 30 million out of work. Chicao Tribune, April 30

Seven M, Bagcivan G, Pasalak SI, Oz G, Aydin Y, Selcukbiricik F (2021) Experiences of breast cancer survivors during the COVID-19 pandemic: a qualitative study. Support Care Cancer 29:6481–6493. https://doi.org/10.1007/s00520-021-06243-4

Stanizzo MR, Castelli L, Di Nardo C, Brunetti M, De Sanctis C, Ghiggia A (2022) Psychological distress in breast cancer patients during the Italian COVID-19 pandemic. Int J Environ Res Public Health 19. https://doi.org/10.3390/ijerph191811433

Wang C, Pan R, Wan X, Tan Y, Xu L, Ho CS, Ho RC (2020) Immediate psychological responses and associated factors during the initial stage of the 2019 coronavirus disease (COVID-19) epidemic among the general population in China. Int J Environ Res Public Health 17. https://doi.org/10.3390/ijerph17051729

Wesmiller S (2021) Susan Wesmiller: measuring the effect of Sars-CoV2 on a longitudinal study (A supplement to genomic underpinnings for breast cancer treatments induced nausea and vomiting). Columbia Academic Commons, Academic Unit: Northeast Big Data Innovation Hub, Series: COVID-19 Research Lightning Talks. https://doi.org/10.7916/2tsq-ny10

Wesmiller S, Harpel C (2021) COVID-19 influence on self-reported symptoms in women with breast cancer. Support Care Cancer 29:S11

Windle G, Woods R, Markland D (2020) Living with ill-health in older age: the role of a resilient personality.

J Happiness 11:763–777. https://doi.org/10.1007/s10902-009-9172-3
Zhao J, Mazanec SR, Rosenzweig M (2022) Anxiety and PTSD symptoms during the COVID-19 pandemic in women with breast cancer. Oncol Nurs Forum 49:201–205. https://doi.org/10.1188/22.Onf.201-205

Susan W. Wesmiller is an Associate Professor and Department Chair for Health, Promotion and Development at the University of Pittsburgh School of Nursing and a member of the Biobehavioral Cancer Control Program at the University of Pittsburgh Medical Center (UPMC) Hillman Cancer Center. Her research focuses on the genomic and epigenomic underpinnings of symptoms experienced by women treated for breast cancer. Dr. Wesmiller is a fellow in the American Academy of Nursing, and the past president of the International Society of Nurses in Genetics.

Katrina Carr is currently a predoctoral fellow at the University of Pittsburgh School of Nursing. She originally majored in math and computer science and worked as a computer scientist before transitioning to nursing through an accelerated Bachelor of Science in Nursing (BSN) program. Prior to beginning her doctoral work, she was the Project Director for a National Institute of Nursing Research (NINR) funded study focused on symptoms experienced by women following breast cancer surgery. Ms. Carr is also a seasoned Trauma Intensive Care Unit (ICU) nurse, where she developed her passion for working with patients who experience the psychoneurological symptoms of depression, anxiety, and sleep disturbance.

Yvette Conley is a Distinguished Service Professor of Nursing and Senior Associate Dean for Research and Scholarship at the University of Pittsburgh School of Nursing, with a secondary appointment as Professor of Human Genetics at the University of Pittsburgh School of Public Health. Her research focuses on using omics approaches to better understand variability in patient outcomes following cancer, treatment for cancer, traumatic brain injury, and stroke. Dr. Conley's research also investigates the role of omics in variability in response to a variety of interventions including nonpharmacological interventions.

Susan M. Sereika is a Professor in the Department of Health and Community Systems and Associate Dean for Computing and Information Technology in the University of Pittsburgh School of Nursing with a secondary appointment in the Departments of Biostatistics and Epidemiology in the School of Public Health. She is also an affiliated faculty statistician in the Biostatistical Faculty at the University of Pittsburgh Medical Center (UPMC) Hillman Cancer Center. Dr. Sereika has 35 years of experience designing research studies and managing cross-sectional and longitudinal data. Her expertise includes multivariate data analysis from single- and multi-site clinical trials evaluating behavioral, educational, and pharmaceutical interventions.

10 Approaches to Dementia Family Caregiving During the COVID-19 Pandemic: Utilization of Health Care and Community-Based Services

Amanda N. Leggett, Laura N. Gitlin, Cathleen M. Connell, Natasha Nemmers, Wenhua Lai, and Helen C. Kales

Abstract

The COVID-19 pandemic dramatically changed health and community-based service-seeking patterns for individuals with dementia and their care partners, together referred to as the care dyad. This chapter describes a mixed-methods study of a convenience sample of primary family caregivers of community-dwelling individuals with dementia during the pandemic, mostly living in a private home ($n = 96$), with a small subset in a long-term care facility ($n = 4$). We explore: (1) caregiver readiness and dementia management strategies as they relate to service utilization of individuals living with dementia, and (2) family caregivers' perspectives on pandemic service utilization and its impacts on their care management practice. In particular, we found that more caregiver readiness was associated with lower odds of urgent care service utilization and less difficulty in accessing medical care. Additionally, caregivers described challenges with accessing care, changes in care quality, benefits, and challenges of virtual technologies, and pandemic-specific changes to their roles and responsibilities as caregivers. Strategies for intervention and mitigation of future pandemic challenges for family caregivers are detailed, as well as general lessons for caregivers adapting to challenging circumstances.

Research funded by

NIH Award 3K01AG056557-04S1—Style and Substance: Characterizing Dementia Caregiving Styles and Associated Biopsychosocial and Health Services Utilization Outcomes.

COVID Information Commons Video Presentation

The COVID Information Commons (CIC) Research Lightning Talk video presentation and individual multilingual transcripts for this project can be found in the Columbia Academic Commons (AC) (Leggett 2021).

A. N. Leggett (✉) · N. Nemmers · W. Lai
Wayne State University and University of Michigan, Detroit, MI, USA
e-mail: leggetta@wayne.edu

N. Nemmers
e-mail: natasha.nemmers@wayne.edu

W. Lai
e-mail: laiwenhua@wayne.edu

L. N. Gitlin
College of Nursing & Health Professions, AgeWell Collaboratory, Drexel University, and Johns Hopkins School of Nursing, Philadelphia, PA, USA
e-mail: lng45@drexel.edu

C. M. Connell
University of Michigan, Ann Arbor, MI, USA
e-mail: cathleen@umich.edu

H. C. Kales
University of California, Davis, Sacramento, CA, USA
e-mail: hckales@ucdavis.edu

F. D. Hudson (ed.), *The COVID Information Commons*, Sustainable Development Goals Series,
https://doi.org/10.1007/978-3-032-06125-6_10

Keywords

Alzheimer's disease and related dementias · Family caregiving · Health services utilization · Hospitalization · Emergency department · Care management · Caregiver readiness · COVID-19

10.1 Goals and Approach of the Research

Family caregivers are key medical decision-makers for persons living with dementia and play a critical role in interfacing with community-based services and the healthcare system. The COVID-19 pandemic, however, had a dramatic impact on service utilization, with many services and appointments suspended, delayed, or moved to telehealth, and with fewer resources attributed to non-COVID-19 medical conditions like Alzheimer's disease and related dementias (Couteur et al. 2020; Rimmer 2020). Additionally, as individuals with dementia were at increased risk for both complications from COVID-19 and COVID-19 mortality, many care partners aimed to keep individuals with dementia physically distanced to reduce exposure risk, which may have been elevated in healthcare settings. These changes impacted caregivers' care management practices in a plethora of ways. For example, while rates of residential care admission and paid care utilization declined during the first year of the pandemic, national data on older adults suggests that family members compensated for these declines in paid care by increasing the hours of unpaid family caregivers (even among those already co-residing with their care partner), and with increased rates of co-residence with the care partner (Truskinovsky and Wiemers 2023; Vislapuu et al. 2021).

In terms of urgent care, emergency room visits and hospitalizations, while utilization trended down during the early waves of the pandemic (Giannouchos et al. 2021; Bronskill et al. 2022), individuals with dementia had greater risk for mortality and were more likely to present with delirium in such settings (Bronskill et al. 2022; Harb et al. 2021; Wang et al. 2021). Additionally, visitation restrictions to ensure physical distancing and reduce exposure to the virus often altered whether family caregivers were permitted in healthcare appointments and how they were able to advocate for the individual with dementia in these settings. In fact, using a nationally representative sample of emergency departments and hospitals during the first wave of the pandemic, Lo and colleagues (Lo et al. 2022) found that while these institutions almost universally had visitor restrictions, only 39% of hospitals and 29% of emergency departments had exceptions to these policies for their patients with cognitive impairment. More broadly, two systematic reviews found that the switch to healthcare services offered via telemedicine was viewed by caregivers as beneficial for maintaining continuity of care for individuals with dementia. Yet telehealth had upfront challenges for patients, providers and caregivers in learning how to navigate the new technology, and in familiarizing and engaging the individual living with dementia, an additional point of focus for family caregivers (Elbaz et al. 2021; Liang and Aranda 2023).

Similar declines in terms of home care services, such as personal care and therapy, were reported during the first pandemic wave in 2020 (Jones et al. 2021). Though many services started to rebound over the course of the pandemic, this was at a slower process for some demographic groups such as rural populations of community-dwelling individuals with dementia (Jones et al. 2021). In general, healthcare and community-based service disruptions were found to be greater for individuals with dementia than for older adults without dementia (Tannous et al. 2022). The difficulties in seeking services, accessing care, and family care management were of key importance, as data suggests the pandemic challenges relating to service use increased stress for family caregivers. In a national sample of caregivers for adults with a chronic or disabling condition, 13% reported difficulty in getting needed in-home and out-of-home services and 19% reported difficulty in accessing medical care for one's care recipient

(Leggett et al. 2021). Such disruptions in caregiving patterns, including taking on new caregiving roles and increasing one's caregiving intensity during the pandemic, were associated with increased stress and negative mental health symptoms for the family caregivers (Truskinovsky and Wiemers 2023; Leggett et al. 2021, 2022). Cessation of adult day respite programs, e.g., was also related to increased family caregiver stress (Wong et al. 2021). Yet, how service utilization difficulties impacted and were associated with dementia family caregivers' care management and practice is unknown.

This is a critical area of public health focus as medical care utilization was significantly higher for individuals living with dementia than their same-aged peers without dementia, both within and outside pandemic contexts (Phelan et al. 2012; Bynum et al. 2004; Feng et al. 2014). Additionally, urgent care such as emergency department visits and subsequent hospitalizations can alter health trajectories and increase the risk for institutionalization of persons living with dementia (Yaffe et al. 2002; Stern et al. 1997; Smith et al. 2000, 2001). While the decision to hospitalize a patient is a physician-led decision, family caregivers are key medical decision-makers and often act as gatekeepers to whether an individual with dementia enters an urgent care setting.

Yet, while both pandemic-related disruptions in care and increased strain on family caregivers are well-documented, less is understood about family caregivers' care management practices and levels of readiness and how this influenced outcomes for the dementia care dyad (i.e., both the caregiver and individual living with dementia) during COVID-19. This chapter focused on two facets of care management: (1) caregiver readiness and dementia management strategies as they relate to service utilization of individuals living with dementia, and (2) family caregivers' perspectives on pandemic service utilization and its impacts on their care management practice. In contrast with formal care providers, most family caregivers have not received training in nursing or medical tasks and do not have time to prepare for their care role (Alzheimer's Association 2020; McClendon and Smyth 2013). This lack of readiness may lead to an array of adaptive and maladaptive caregiving management strategies as family members seek to provide care and meet the needs of an individual with dementia (Alzheimer's Association 2020; McClendon and Smyth 2013; Gitlin and Rose 2016a). Caregiver readiness assesses a caregiver's willingness and ability to utilize different care techniques. Greater readiness has been associated with caregiver confidence, better mood, greater therapeutic engagement in a caregiver intervention, and reduced frequency of problem behaviors in the individual living with dementia (Gitlin and Rose 2014, 2016a). Dementia management strategies, on the other hand, include active strategies caregivers employ when managing care (e.g., critical approaches like yelling, active management approaches such as making environmental modifications, and encouraging approaches such as praising the individual for appropriate behaviors) (Hinrichsen and Niederehe 1994). Specific dementia management approaches of criticism, anger, lack of acceptance, and lower encouragement are also associated with caregiver outcomes along with increased caregiver burden, a desire to institutionalize the person living with dementia, and less experienced caregiving gains (i.e., positive reported aspects of caregiving such as fulfillment and growth in the care role) (Hinrichsen and Niederehe 1994; Lim et al. 2011; Bakker et al. 2013; Hong et al. 2013; Davis et al. 2014; Vugt et al. 2004). Care management can impact outcomes for the care dyad, however COVID-19 may have altered caregiver routines. This chapter seeks to share research insights to better understand how care management styles relate to difficulties in the seeking and utilization of care services during the pandemic with an eye toward response strategies and interventions in future pandemic contexts.

10.1.1 The Current Study Methods

Data for this study came from a COVID-19 administrative supplement to a National Institutes of Health (NIH) K01 career development award (K01AG056557) focused on dementia care management styles (the STYLE study). The COVID-19 supplement (3K01AG056557-04S1) extended the work on care management styles to consider care management during a global pandemic utilizing mixed-methods interviews with 100 primary, unpaid, family, and friend caregivers for a person with a diagnosis of an age-related dementia. Participants from the core STYLE study could be included if they were still actively providing care for the same relative or friend during the pandemic ($n = 34$; *note.* We included persons living with dementia residing in a long-term care facility to maintain as much of the core STYLE study sample as possible, $n = 4$). To attain a sample of 100 caregivers, new participants were also recruited ($n = 66$) who were caring for a community-dwelling person living with dementia who was not at the terminal stage of disease.

The qualitative portion of the interview (45 min on average) covered care management during the pandemic, changes in routines of care, access to home and community-based services and healthcare, changes in social supports, and any barriers or facilitators of care specific to the pandemic. The quantitative interview (45 min on average) covered demographics, the dementia care context (e.g., behavioral symptoms, functional impairment), caregiver stress, care management, and the COVID-19 context. All interviews were conducted via Zoom between February and October 2021, and this study was approved by the Michigan Medicine Institutional Review Board. Participants received a small stipend for their time.

10.1.2 Study Aims

We specifically aimed to detail how care management practices and readiness for care were related to health and community-based services utilization during the pandemic. Mixed-methods analysis was run to consider: (1) how caregivers' level of readiness for care and dementia management strategies (quantitative) were associated with difficulties accessing services for the person living with dementia and the person with dementia's urgent services utilization; and (2) caregivers' perceptions of service utilization during the pandemic and how it impacted their caregiving practice (qualitative).

10.1.3 Measures

We examined three urgent care needs: whether, since March of 2020, the individual with dementia had an emergency medical services call made on his/her behalf by the caregiver, had an emergency room visit, and/or had a fall (yes/no). We also examined three types of difficulty the caregiver reported in accessing care for the individual with dementia: in-home and out-of-home services, medical care, and friends and family as respite support (yes/no). We sought to examine whether caregivers' care management during the pandemic impacted urgent service utilization and the caregivers' difficulty in navigating services. Caregiver readiness was assessed with a 17-item Likert sum scale (1 not at all true to 4 very much true), where higher scores indicate greater readiness (e.g., "I am willing to change the way I communicate with my care recipient") (Gitlin and Rose 2014, 2016a). The three subscales of the Dementia Management Strategies scale (Hinrichsen and Niederehe 1994) were utilized: criticism (11 items; e.g., "I blamed my older relative for having created the difficulties"), active management (9 items; e.g., "I tried to arrange situations I hoped would be stimulating to my older relative (mentally or emotionally)"), and encouragement (8 items; e.g., "I tried to help my older relative look on the bright side of things"), all assessed on a Likert scale of 1 (never) to 5 (most of the time), where a higher score indicated great use of each strategy.

10.1.4 Analysis

First, logistic regression analyses considered caregivers' level of readiness and management strategies as key predictors of urgent services utilization and difficulty accessing services adjusting for caregivers' age, gender, educational attainment (college degree or higher = 1), race (non-white = 1), co-residence with the person living with dementia (co-residence = 1), and the stage of dementia (assessed using the Functional Assessment Staging (FAST) of Alzheimer's Disease scored from 1 = normal aging to 7 = severe dementia) (Reisberg 1984).

Next, Watkins' (Watkins 2017) rigorous and accelerated data reduction technique was used to analyze qualitative data and identify themes specific to family caregivers' access to and use of community-based and healthcare services. Interviews were recorded, transcribed, and redacted of identifiable information for analysis. Thematic qualitative analysis was conducted through the iterative process of note taking, coding (open and focused), and consensus meetings undertaken by a team of six coders including the lead author. Finally, conventional content analysis was employed to see which codes were most frequent across the themes.

10.2 Results

10.2.1 Sample Description

Caregivers were 63.3 years of age on average, 75% female, 17% non-white, 61% spouses of the person living with dementia, and highly educated (67% had a college degree). Urgent healthcare needs were common during the pandemic, with 20% of caregivers making a call to emergency medical services on behalf of the person living with dementia, 28% of individuals with dementia having an emergency room visit, and 39% of individuals living with dementia experiencing a fall from March 2020 through date of interview (2021). Difficulty of access to supports during the pandemic was also high, with 31% of caregivers reporting difficulty accessing in home and out-of-home services, 24% reporting difficulty accessing medical care, and 47% reporting difficulty accessing friends and family for respite support on behalf of the person living with dementia.

10.2.2 Quantitative Associations: Care Management and COVID-19 Service Utilization and Difficulties

Care management and readiness to deliver care were significantly related to the degree of urgent service use and difficulties in seeking services. A more critical care management style appeared to relate to relational care challenges during the pandemic. For example, greater use of criticism in care was significantly associated with greater odds of reporting difficulty in accessing respite support from friends and family (Odds ratio, OR = 1.10, 95% Confidence interval, CI = 1.003–1.21, $p < 0.05$), accounting for controls. A more critical management style in caregiving may carry through to other social relationships or relate to a personality style that is more socially combative, hence access to support from other family and friends may be harder to find if relationships are strained. On the other hand, the use of encouragement and active management strategies in caregiving was not significantly associated with service utilization during the pandemic.

Greater caregiver readiness was associated with reduced odds of making an emergency medical service call on behalf of the person living with dementia (OR = 0.85, 95% CI = 0.74–0.97, $p < 0.05$), reduced odds of the person living with dementia having an emergency department visit (OR = 0.89, 95% CI = 0.80–0.99, $p < 0.05$), and reduced odds of reporting difficulty accessing medical services for the person living with dementia (OR = 0.85, 95% CI = 0.75–0.97, $p < 0.05$), adjusting for covariates. Readiness encompasses an overall understanding of dementia and its symptoms, as well as how adaptable and willing one is to make changes to one's caregiving approach. While an emergency call or visit may be medically necessary, family caregivers serve as gatekeepers for such utilization and existing research suggests caregiver factors may impact likelihood of utilization.

For example, prior work has found that caregivers' health and strain impacts risk for an individual with dementia's services utilization (Maust et al. 2017; Ankuda et al. 2017). However, the current study uniquely shows that a caregiver's level of readiness may be a modifiable trait and readiness may empower caregivers to reduce unnecessary service utilization and better navigate medical care, particularly in a challenging service environment like a global pandemic. A caregiver with greater readiness may have more efficacy in managing care challenges at home or have been able to better navigate the changes to pandemic healthcare in such a way that helped to prevent medical emergencies. Behavioral care management and readiness, on the other hand, were not significantly associated with whether a person living with dementia had fallen or whether the caregiver had difficulty accessing needed in-home and out-of-home services.

10.2.3 Qualitative Analysis: Service Utilization and Caregiving Management

Qualitative analysis also revealed barriers and facilitators which caregivers experienced around service utilization for the person living with dementia that impacted their caregiving practice. We identified five themes of health and community-based service use during the pandemic including access to services, the quality of care experienced, caregiving roles and responsibilities, technology, and instrumental support.

Access Caregivers described barriers and facilitators around accessing both medical and community-based services during the pandemic (e.g., transportation challenges, a shortage of paid caregivers, waitlists for care services), yet many also recognized continuity of services despite the pandemic. For example, one caregiver described an accessibility challenge around seeing a counselor for mental health support surrounding their caregiving activities, *"[Counseling] was hard to schedule… I craved seeing [someone] more than every five weeks."* Caregivers also noted having to navigate delays and cancelations of scheduled services. On the other hand, when available, caregivers described how beneficial services were, and also noted that in many instances there was continuity in their normal care providers, *"He sees a heart doctor, a neurologist, a psychiatrist… and he sees his regular family care doctor. So of those doctors, he was able to see all of them."*

Quality of Care Caregivers described ways in which their caregiving experience shifted during the pandemic due to variations in quality and quantity of care received and accommodations made or not made to support the caregiver. Caregivers noted how supported they felt by many community-based and medical services, *"I was one of the lucky ones that we had the support from the Alzheimer's [Association]. I could call Dr [Name] with any of my questions. Zoom, honestly saved my life in being something to look forward to."* Caregivers also found that services often made accommodations or relaxed restrictions for them to be included as a care partner of the individual with dementia. Yet equally, caregivers reported struggling with poor communication with providers or visit restrictions where a caregiver could not attend an appointment with the individual with dementia. For example, one caregiver described, *"[He had a] grand mal seizure… I went to the hospital, followed the ambulance, and they wouldn't let me in…And I said, 'I've been married 56 years and I'm not going to let the man die alone'. And of course they had just come down with the closures. So the doctor came out and… she talked to me about what he had said…so I thought okay, he's all right. I, you know I won't make a spectacle and say I have to go to see him."* Caregivers also expressed concerns over the insufficiency and quality of care received such as rushed visits with emergency services or feeling that the number of appointments prescribed was less than what was needed.

Caregiving Roles and Responsibilities Caregivers described how their priorities and care management changed by necessity.

For example, significant time was devoted to taking precautions to reduce COVID-risk for the caregiver and person living with dementia such as reprioritizing travel to care appointments, canceling unnecessary medical appointments, and finding ways for the person living with dementia to stay home. A caregiver described, *"prioritizing what needs to be done today. Like what doctor's appointment absolutely needs to be done today, and what can wait until things get better."* This necessitated increased focus on care coordination for caregivers around making plans for medical care and pharmacy refills. In general, caregivers found this increased their responsibility and perceptions of burden around care. As access to formal care providers declined, one caregiver described, *"The real kicker was because of the pandemic they don't even have caregivers that could even provide three half days a week... it left the family kind of filling in a lot more. Instead of me just stopping over and visiting for a bit and feeling like he had care, it was me staying with him and making a breakfast and making sure we ate lunch and calling every evening at six o'clock to make sure that he remembered how to heat up his dinner and so it just seemed like it was more because of the pandemic."* Further, relating to our theme of challenges around access to services, caregivers often took on a new role of patient advocate and having to be assertive to ensure the individual with dementia got needed care. For example, one caregiver detailed, *"It was a battle for me to convince them that I need to be with him at all times. He cannot go into an appointment alone."*

Technology Caregivers detailed both benefits and challenges regarding the transition of many services to telehealth or virtual modalities. For example, some caregivers found it more difficult to engage or communicate via technology, were disappointed in not receiving a physical exam for the individual with dementia and had upfront difficulties in learning how to adopt the technology. One caregiver described, *"When you're seeing the doctor for her cognitive difficulties, I think it's just kind of hard to do that on a Zoom call. And one of the things is that she doesn't understand technology. So I really have to do most of the talking for her, which is fine, but I don't think it's optimal in terms of the doctor being able to get a good sense of how she's doing."* In contrast, caregivers valued the convenience and added safety of not having to leave home for appointments, appreciated engaging with other dementia caregivers who may have had similar challenges (e.g., a Lewy Body Dementia specific support group) but were not in close geographic proximity, and valued access to cultural activities that became virtual (e.g., virtual museum tours or exercise classes). A caregiver explains, *"I joined those caregiving workshops and really that's given me a lot of support... I never could go to any of the caregiver workshops or caregiver events that were here in the area, because I couldn't leave him alone... the fact that all of these events have gone virtual has [been] amazing to me."*

Instrumental support. One service caregivers repeatedly described as helpful to their ability to physically distance and provide care from home was pickup or delivery services. These were generally services caregivers did not have prior familiarity with, yet became valuable during the pandemic. One caregiver described, *"You know the delivery services have become available. Like Shipt- that's a free service for seniors so that was probably the biggest assistant... So it was a wonderful service and very helpful."*

10.3 Recommendations Regarding Mitigation, Response, and Recovery from Future Pandemics

Caregivers showed great resilience during the pandemic, in filling in the care gaps when services were interrupted or advocating for the continuation or quality of services. There were a variety of positive outcomes for caregivers during the pandemic including supportive staff and programs, telehealth and meal delivery services offering support while allowing care dyads to stay home, and increased access to services through virtual modalities. On the other hand, services often needed better plans for incorporating caregivers in visits while maintaining health and safety (e.g., physical distancing). Caregivers also had added burden when services canceled or reduced availability, requiring families to compensate for formal care.

Utilizing our mixed-methods data, we also found that caregivers' management styles were correlated to utilization of services, with caregivers who had greater readiness levels reporting that the individual with dementia utilized less urgent care and that the caregiver had less difficulty in accessing medical care on behalf of the individual with dementia. While prior research has identified caregiver stress and fatigue as being associated with increased utilization of healthcare services and medical care costs (Maust et al. 2017; Ankuda et al. 2017), we found that readiness levels may reduce the need for urgent care services. While the findings are cross-sectional and causality cannot be implied, existing nonpharmacological intervention work with dementia family caregivers found that individuals at lower levels of caregiver readiness were less likely to engage in intervention strategies to minimize the occurrence of behavioral and psychological symptoms of dementia (Rose and Gitlin 2017; Gitlin and Rose 2016b), whereas greater readiness was associated with confidence and a reduction in behavioral and psychological symptoms of dementia. Hence, interventions that help build readiness in family caregivers may have great benefit in how caregivers seek resources and manage care both in and out of pandemic contexts. Additionally, as mechanisms of action for interventions supporting caregivers are not always clear, readiness may be a mechanism on which focus is needed prior to other benefits occurring from interventions.

Our work also highlights other important strategies to support individuals with dementia and their family caregivers in the ongoing COVID-19 landscape and future pandemics. First, continuity of care was critical for individuals living with dementia (Amjad et al. 2016; Aaltonen et al. 2021a). Work prior to the pandemic has found poorer continuity of care for individuals with dementia to be associated with increased rates of emergency department visits, hospitalization and healthcare spending (Amjad et al. 2016). Further, individuals with dementia with greater continuity of care with their physician are less likely to be in acute or urgent care longer than medically necessary (i.e., lower rates of delayed discharge) (Aaltonen et al. 2021b). This was challenged in the first wave of the pandemic when health information on mitigation (e.g., KN95 masks, vaccines) was uncertain and older adults with dementia were particularly vulnerable. Providing resources to assist caregivers in evaluating the importance of medical appointments—determining which ones are critical to keep and where it is feasible to delay or transition to telemedicine—can help maintain continuity of care while mitigating risk. Advanced care plans could also reduce stress for caregivers when urgent needs arise to ensure the individual with dementia's care preferences are attained while abiding by variable health guidelines.

Further, incorporating caregivers into healthcare visits where possible helps to ensure continuity of care. Exceptions for visitation restrictions for care partners of individuals with dementia need to be made, particularly at moderate to later stages of dementia. Caregivers are needed to provide accurate patient history, help to interpret cues from individuals with dementia that a caregiver may best recognize (e.g., agitation, pain) (Kennedy et al. 2020), and help prevent disorientation that may be intensified by the use of personal protective equipment like masks (Gil and Arroyo-Anlló 2021). Help from care partners attending to individuals with dementia in hospital settings can also serve to reduce time limitations of overburdened staff in a chaotic pandemic environment. In hospital settings, care partners may also learn critical bedside information from nursing staff and practice skills for at-home care that may help reduce future readmissions to healthcare settings.

Our study pointed to many benefits of virtual offerings for caregivers, from physically distanced telemedicine appointments to specialized support groups to cultural activities that could be enjoyed from home. However, there were upfront costs in acquiring digital literacy to engage in these activities and concerns around whether the physician was getting the full picture of the individual with dementia's capacity and health. Clear guidance and tech support for older patients would help care dyads reap the benefits of virtual offerings. Further, as care dyads have become more familiar with virtual technologies, more resources offered through these mediums may have ongoing benefit for this population. We also consistently heard from caregivers how useful meal and grocery service delivery applications were. Free or discounted access to such services for older adults and their caregivers could encourage use and support of physical distancing of a vulnerable population.

Overall, in the chaotic early months of the pandemic, individuals with dementia and their caregivers faced increased risk for COVID-19 complications while also dealing with stressors around navigating a pandemic context alongside cognitive, functional, and behavioral challenges associated with a dementia diagnosis. Enhanced virtual training and supports, exceptions where possible for caregivers to attend medical appointments, and guidance around healthcare decision-making (e.g., appointments to prioritize, when a hospital or emergency department visit is medically necessary) would promote health and reduce risk for the care dyad.

Finally, as demonstrated, caregiver management approaches and readiness were associated with healthcare utilization and difficulties in accessing services during the pandemic. These important constructs are understudied in the caregiving literature, yet demonstrate critical associations with outcomes for the care dyad. In particular, interventions that enhance caregiver readiness (e.g., promoting understanding of dementia and its symptoms, building adaptability and comfort in trying new care management approaches) may enhance a necessary skill set for caregivers to confidently manage at-home care and avoid unnecessary urgent healthcare needs in a pandemic environment. This would thereby help to mitigate pandemic exposure and risk, while also improving overall well-being of the care dyad.

Acknowledgements Dr. Leggett is also supported by the Michigan Alzheimer's Disease Research Center (NIAP30AG072931). I would also like to acknowledge the University of Michigan undergraduate research assistants: Elaina Baker, Hannah Lee, Sarah Wallace, Anna Webber, and Tyson Fang— for their support of the qualitative analysis.

References

Aaltonen M, El Adam S, Martin-Matthews A, Sakamoto M, Strumpf E, McGrail K (2021) Dementia and poor continuity of primary care delay hospital discharge in older adults: a population-based study from 2001 to 2016 (in English). J Am Med Dir Assoc 22(7):1484-1492.e3. https://doi.org/10.1016/j.jamda.2020.11.030

Aaltonen M, El Adam S, Martin-Matthews A, Sakamoto M, Strumpf E, McGrail K (2021) Dementia and poor continuity of primary care delay hospital discharge in older adults: a population-based study from 2001 to 2016. J Am Med Directors Assoc 22(7):1484–1492.e3. https://doi.org/10.1016/j.jamda.2020.11.030

Amjad H, Carmichael D, Austin AM, Chang C-H, Bynum JPW (2016) Continuity of care and health care utilization in older adults with dementia in fee-for-service medicare. JAMA Intern Med 176(9):1371–1378. https://doi.org/10.1001/jamainternmed.2016.3553

Ankuda CK, Maust DT, Kabeto MU, McCammon RJ, Langa KM, Levine DA (2017) Association between spousal caregiver well-being and care recipient healthcare expenditures. J Am Geriatrics Soc 65(10):2220–2226. https://doi.org/10.1111/jgs.15039

Alzheimer's Association (2020) Alzheimer's disease facts and figures. Alzheimer's Dementia 16(3):391+

Bakker C, et al (2013) Predictors of the time to institutionalization in young-versus late-onset dementia: results from the needs in young onset dementia (NeedYD) study. J Am Med Directors Assoc 14(4):248–253. https://doi.org/10.1016/j.jamda.2012.09.011

Bronskill SE et al (2022) Trends in health service use for Canadian adults with dementia and parkinson disease during the first wave of the COVID-19 pandemic. JAMA Health Forum 3(1):e214599–e214599. https://doi.org/10.1001/jamahealthforum.2021.4599

Bynum JP, Rabins PV, Weller W, Niefeld M, Anderson GF, Wu AW (2004) The relationship between a dementia diagnosis, chronic illness, medicare expenditures, and hospital use (in English). J Am Geriatr Soc 52(2):187–194. https://doi.org/10.1111/j.1532-5415.2004.52054.x

Le Couteur DG, Anderson RM, Newman AB (2020) COVID-19 through the lens of gerontology. J Gerontol: Ser A. https://doi.org/10.1093/gerona/glaa077

Davis LL, Chestnutt D, Molloy M, Deshefy-Longhi T, Shim B, Gilliss CL, Adapters, strugglers, and case managers: a typology of spouse caregivers.

Qual Health Res 24(11):1492–1500. https://doi.org/10.1177/1049732314548879

de Vugt ME et al (2004) Do caregiver management strategies influence patient behaviour in dementia? Int J Geriatr Psychiatry 19(1):85–92. https://doi.org/10.1002/gps.1044

Elbaz S et al (2021) A systematic review of telemedicine for older adults with dementia during COVID-19: an alternative to in-person health services? (in English). Front Neurol Syst Rev 12. https://doi.org/10.3389/fneur.2021.761965

Feng Z, Coots LA, Kaganova Y, Wiener JM (2014) Hospital and ED use among medicare beneficiaries with dementia varies by setting and proximity to death. Health Affairs 33(4):689–690. https://doi.org/10.1377/hlthaff.2013.1179

Giannouchos TV, Biskupiak J, Moss MJ, Brixner D, Andreyeva E, Ukert B (2021) Trends in outpatient emergency department visits during the COVID-19 pandemic at a large, urban, academic hospital system. Am J Emergency Med 40:20–26. https://doi.org/10.1016/j.ajem.2020.12.009

Gil R, Arroyo-Anlló EM (2021) Alzheimer's disease and face masks in times of COVID-19. J Alzheimers Dis 79:9–14. https://doi.org/10.3233/JAD-201233

Gitlin LN, Rose K (2016a) Impact of caregiver readiness on outcomes of a nonpharmacological intervention to address behavioral symptoms in persons with dementia. Int J Geriatric Psychiatry

Gitlin LN, Rose K (2016b) Impact of caregiver readiness on outcomes of a nonpharmacological intervention to address behavioral symptoms in persons with dementia. Int J Geriatric Psychiatry 31(9):1056–1063. https://doi.org/10.1002/gps.4422

Gitlin LN, Rose K (2014) Factors associated with caregiver readiness to use nonpharmacologic strategies to manage dementia-related behavioral symptoms. Int J Geriatr Psychiatry 29(1):93–102. https://doi.org/10.1002/gps.3979

Harb AA, Chen R, Chase HS, Natarajan K, Noble JM (2021) Clinical features and outcomes of patients with dementia compared to an aging cohort hospitalized during the initial New York City COVID-19 wave. J Alzheimers Dis 81:679–690. https://doi.org/10.3233/JAD-210050

Hinrichsen GA, Niederehe G (1994) Dementia management strategies and adjustment of family members of older patients. Gerontologist 34(1):95–102. https://doi.org/10.1093/geront/34.1.95

Hong S-I, Luo N, Yap P (2013) Maximizing a nurturing care style for persons with dementia: a person-centered analysis. Am J Geriatric Psychiatry 21(10):987–998. https://doi.org/10.1016/j.jagp.2013.01.034

Jones A et al (2021) Impact of the COVID-19 pandemic on home care services among community-dwelling adults with dementia. J Am Med Dir Assoc 22(11):2258-2262.e1. https://doi.org/10.1016/j.jamda.2021.08.031

Kennedy M, et al (2020) The agitated older adult in the emergency department: a narrative review of common causes and management strategies. J Am College Emergency Phys Open 1(5):812–823. https://doi.org/10.1002/emp2.12110

Leggett A (2021) Amanda Leggett: caregiving on the frontline: caring for an individual living with dementia during the COVID-19 pandemic. Columbia Academic Commons, Academic Unit: Northeast Big Data Innovation Hub, Series: COVID-19 Research Lightning Talks. https://doi.org/10.7916/qnk9-fb87

Leggett A, Koo HJ, Park B, Choi H (2022) The changing tides of caregiving during the COVID-19 pandemic: how decreasing and increasing care provision relates to caregiver well-being. J Gerontol: Ser B 77(Supplement_1):S86-S97. https://doi.org/10.1093/geronb/gbac002

Leggett AN, et al (2021) Care challenges due to COVID-19 and mental health among caregivers of U.S. adults with a chronic or disabling condition (in English). Innov Aging 5(3):igab031. https://doi.org/10.1093/geroni/igab031

Liang J, Aranda MP (2023) The use of telehealth among people living with dementia-caregiver dyads during the COVID-19 pandemic: scoping review (in English). J Med Internet Res 25:e45045. https://doi.org/10.2196/45045

Lim J, Griva K, Goh J, Chionh HL, Yap P (2011) Coping strategies influence caregiver outcomes among Asian family caregivers of persons with dementia in Singapore. Alzheimer Dis Assoc Disord 25(1):34–41. https://doi.org/10.1097/WAD.0b013e3181ec18ae

Lo AX et al (2022) COVID-19 hospital and emergency department visitor policies in the United States: impact on persons with cognitive or physical impairment or receiving end-of-life care (in English). J Am Coll Emerg Phys Open 3(1):e12622. https://doi.org/10.1002/emp2.12622

Maust DT, Kales HC, McCammon RJ, Blow FC, Leggett A, Langa KM (2017) Distress associated with dementia-related psychosis and agitation in relation to healthcare utilization and costs (in English). Am J Geriatric Psychiatry: off J Am Assoc Geriatric Psychiatry 25(10):1074–1082. https://doi.org/10.1016/j.jagp.2017.02.025

McClendon MJ, Smyth KA (2013) Quality of informal care for persons with dementia: dimensions and correlates. Aging Mental Health 17(8):1003–1015. https://doi.org/10.1080/13607863.2013.805400

Phelan EA, Borson S, Grothaus L, Balch S, Larson EB (2012) Association of incident dementia with hospitalizations (in English). JAMA 307(2):165–172. https://doi.org/10.1001/jama.2011.1964

Reisberg B (1984) Functional assessment staging (FAST) (ed)

Rimmer A (2020) Covid-19: GPs can stop health checks for over 75s and routine medicine reviews. BMJ 368:m1157. https://doi.org/10.1136/bmj.m1157

Rose KC, Gitlin LN (2017) Background characteristics and treatment-related factors associated with treatment success or failure in a non-pharmacological intervention for dementia caregivers. Int Psychogeriatr

29(6):1005–1014. https://doi.org/10.1017/S1041610217000205
Smith GE, Kokmen E, O'Brien PC (2000) Risk factors for nursing home placement in a population-based dementia cohort (in English). J Am Geriatr Soc 48(5):519–525. https://doi.org/10.1111/j.1532-5415.2000.tb04998.x
Smith GE, O'Brien PC, Ivnik RJ, Kokmen E, Tangalos EG (2001) Prospective analysis of risk factors for nursing home placement of dementia patients (in English). Neurology 57(8):1467–1473. https://doi.org/10.1212/wnl.57.8.1467
Stern Y, et al (1997) Predicting time to nursing home care and death in individuals with Alzheimer disease (in English) JAMA 277(10):806–812. Available https://www.ncbi.nlm.nih.gov/pubmed/9052710.
Tannous J et al (2022) COVID-19 associated disruptions in routine health care of people with mild cognitive impairment or dementia (in English). Alzheimers Dement (Amst) 14(1):e12323. https://doi.org/10.1002/dad2.12323
Truskinovsky Y, Wiemers EE (2023) Paid care among older adults with long-term care needs declined in the first year of COVID-19 while families stepped in. Health Affairs Scholar 1(4):qxad040. https://doi.org/10.1093/haschl/qxad040
Vislapuu M, Angeles RC, Berge LI, Kjerstad E, Gedde MH, Husebo BS (2021) The consequences of COVID-19 lockdown for formal and informal resource utilization among home-dwelling people with dementia: results from the prospective PAN.DEM study. BMC Health Services Res 21(1):1003. https://doi.org/10.1186/s12913-021-07041-8
Wang Q, Davis PB, Gurney ME, Xu R (2021) COVID-19 and dementia: analyses of risk, disparity, and outcomes from electronic health records in the US. Alzheimers Dement 17(8):1297–1306. https://doi.org/10.1002/alz.12296
Watkins DC (2017) Rapid and rigorous qualitative data analysis: the "RADaR" technique for applied research. Int J Qual Methods 16(1):1609406917712131. https://doi.org/10.1177/1609406917712131
Wong BP, Kwok TC, Chui KC, Cheng TS, Ho FK, Woo J (2021) The impact of dementia daycare service cessation due to COVID-19 pandemic (in English). Int J Geriatr Psychiatry, 37(1). https://doi.org/10.1002/gps.5621
Yaffe K et al (2002) Patient and caregiver characteristics and nursing home placement in patients with dementia. JAMA: J Am Med Assoc 287(16):2090–2097. https://doi.org/10.1001/jama.287.16.2090

Amanda N. Leggett is an Assistant Professor in the Institute of Gerontology and Department of Psychology at Wayne State University and an adjunct Assistant Professor in the Department of Psychiatry at the University of Michigan. She received her MS and Ph.D. in Human Development and Family Studies from the Pennsylvania State University and is a Fellow of the Gerontological Society of America. Her research program focuses on the dementia caregiver stress process and mental health in late life. She recently completed both a K01 Career Development Award from the National Institute on Aging and a related grant from the Alzheimer's Association to develop a taxonomy of dementia care management styles and aims to determine how classifying one's caregiving style might be used to target interventions and optimize care. She also has R01 funding from the National Institute on Aging to examine family and friend caregiving networks more broadly and how they impact health and well-being outcomes for persons living with dementia. She is an Associate Editor of *Dementia and Caregiving* for the *Journal Innovation in Aging* and is a ten-year Hospice volunteer.

Laura N. Gitlin Ph.D., FGSA, FAAN, an applied research sociologist and intervention scientist, is distinguished professor, and Dean Emerita of the College of Nursing and Health Professions, Drexel University. She is also the inaugural executive director of its AgeWell Collaboratory which oversees Drexel's Age Friendly University international designation, and partners with over 80 community-based organizations serving racially, ethnically, and culturally diverse communities involving older adults and families. Dr. Gitlin's research focuses on developing, evaluating, implementing and disseminating novel home and community-based interventions to improve quality of life of older adults and/or family members. Her interventions apply a social ecological, health equity and implementation lens to develop and evaluate person-centered and person-directed approaches to support individuals/families. Her interventions are used worldwide. She also is the Chief Scientific Officer of Plans4Care, Inc., a new company she co-founded in 2022 to develop digital solutions to support family caregivers and train healthcare providers in evidence-based programs. She is the co-author of many scientific papers and co-authored books, one of which is forthcoming from Academic Press: *Handbook of Intervention Science: Designing and Implementing for Public Health Impact.*

Cathleen M. Connell is a Professor in the Department of Health Behavior and Health Equity at the University of Michigan School of Public Health. She received her MS and Ph.D. in Human Development and Family Studies from the Pennsylvania State University and completed a postdoctoral fellowship in Cardiovascular Risk Reduction at Washington University in St. Louis. Her research program focuses on dementia caregiving, chronic illness self-management, and pets as a source of support for older adults. She completed therapy dog training and participates in community-based programs that offer pet support.

Natasha Nemmers Ph.D., is a Postdoctoral Research Fellow in the Institute of Gerontology at Wayne State University. She completed her MS and Ph.D. in Gerontology within the Department of Human Development of Family Studies at Iowa State University in 2023. Her research broadly focuses on supporting aging-in-place for older adults with limited social connections. Currently, Dr. Nemmers is investigating living alone as a social determinant of health, and evaluating family and professional care networks among individuals with and without dementia. She also has funding from the Michigan Alzheimer's Disease Research Center to conduct a dyadic mixed-methods study, focusing on the care needs, preferences, and support of persons living alone with cognitive impairment and their care partners.

Wenhua Lai is a Postdoctoral Research Fellow at the Institute of Gerontology at Wayne State University. She earned her Ph.D. in Sociology from Michigan State University and her MA from National Taiwan University. Dr. Lai's research advances understanding of racial and ethnic disparities in cognitive health and biological aging, as well as differences in caregiving networks among minoritized aging populations. Her current work investigates the biopsychosocial pathways linking chronic stress exposure and psychosocial resilience to racial and ethnic disparities in biological aging, with the goal of uncovering mechanisms that drive cognitive health inequities.

Helen C. Kales is Chair of Psychiatry and Behavioral Sciences at University of California, Davis. Prior to this, she was Professor of Psychiatry at the University of Michigan and a Research Investigator in the Center for Clinical Management Research and the Geriatric Research Education and Clinical Center in the Veterans Affairs (VA) Ann Arbor Health System. She is the founding Director of the University of Michigan (UM) Program for Positive Aging (PPA), established in 2009, which moved to the University of California, Davis in 2019. As a fellowship-trained, board-certified geriatric psychiatrist, her research program is directly informed by her clinical work and experiences with patients, families, providers, and systems to diminish the barriers to effective and high-quality care for older patients with mental health issues and/or dementia and their caregivers. Dr. Kales is recognized as a national and international expert in outcomes related to later-life depression, the risks of use of antipsychotic and other psychotropic medications in older adults, and in researching ways to improve dementia care. In recognition of her work in the field of dementia care, she was named to the standing Lancet Global Commission on Dementia Care in 2015. Dr. Kales has had continuous federal funding since 2004, serving as principal investigator on 13 federally-funded grants. In her work with junior investigators, she has served as a mentor to four faculty with National Institute on Aging (NIA) career development funding.

Compliance, Trust, and the Control of COVID-19 in China from 2020 to 2022

11

Katherine A. Mason and Yifeng Troy Cai

Abstract

In China, mass compliance with some of the world's most restrictive infection control measures achieved unparalleled results in the control of COVID-19 during the first two-and-a-half years of the pandemic. This was achieved with the so-called zero-COVID policy, which ended in December 2022. The level of control over the spread of COVID-19 that China was initially able to achieve relied in large part on the mass compliance of its enormous population. In this chapter, we draw upon ethnographic data gathered over a three-year time period in Shanghai, China to suggest some reasons for the ready compliance with zero-COVID that was exhibited by so many Chinese citizens for so long. Rejecting simplistic explanations that attribute popular support for government policies entirely to authoritarian control or collectivist instincts, we show how a combination of self-interest, nationalistic pride, and trust in the central government's competence drove popular compliance with China's policies. When these factors were no longer in place, resistance to government policies grew, culminating in the abandonment of zero-COVID. Our findings offer lessons for future epidemic control within and beyond the Chinese context.

Research funded by

NSF Award #2032407—RAPID (Rapid Response Research Award): Understanding the Process of Social Change through the Transitional Period of the COVID-19 Pandemic.

COVID Information Commons Video Presentation

The COVID Information Commons (CIC) Research Lightning Talk video presentation and individual multilingual transcripts for this project can be found in the Columbia Academic Commons (AC) (Mason and Cai 2022).

K. A. Mason (✉) · Y. T. Cai
Brown University, Providence, RI, USA
e-mail: katemason@brown.edu

Y. T. Cai
e-mail: yifeng_cai@alumni.brown.edu

Keywords

COVID-19 · China · Compliance · Zero-COVID · Trust · Nationalism

11.1 Introduction

In China, mass compliance with some of the world's most restrictive infection control measures achieved unparalleled results in the control of COVID-19 during the first two-and-a-half years of the pandemic (Cai and Mason 2022, 2023; Wang et al. 2020). In January 2020, following an initial outbreak of COVID-19 in the city of Wuhan, the central Chinese

F. D. Hudson (ed.), *The COVID Information Commons*, Sustainable Development Goals Series,
https://doi.org/10.1007/978-3-032-06125-6_11

government instituted a nationwide lockdown to curb the spread of the new virus. The country was almost entirely sealed off for over two months. Public health workers went door to door, taking residents' temperatures and moving those with a fever to government-run quarantine facilities. Guards were stationed at the gates of apartment complexes, keeping outsiders out and residents (Xu et al. 2020). By March 2020, the domestic outbreak had been brought under control.

Following these initial efforts, freedom of movement in China was mostly restored, but close tracking of citizens continued, as did repeated lockdowns of entire city blocks or even entire cities in response to case counts as small as 1. Close contacts were sent to compulsory government quarantine centers for at least 14 days. Everyone in China was assigned a color-coded health pass that tracked their movements using their cell phones.

These measures were impressively successful for a long period of time. Despite accounting for nearly 100% of cases during the first month of the virus' spread, through mid-2022 the People's Republic of China had cumulatively recorded well under 1% of both global cases and global deaths from COVID-19 (Hopkins 2020). Underreporting alone cannot explain these apparent successes (Cai and Mason 2022; Pan et al. 2020; Tian 2021). Barring a major cover-up that somehow rendered a massive disaster invisible both to the general public and to the international scientific community, it appears that China's efforts to control COVID-19 met with considerable success—at least until December 2022 (Zheng et al. 2023). In December 2022, the government abruptly ended its "zero-COVID" policy, resulting in a surge of cases and up to 1.8 million excess deaths nationwide over several months, according to multiple studies (Du et al. 2023; Glanz et al. 2023; Xiao et al. 2023), although the Chinese government reported a total number of just 120,000 deaths as of October 2023 (World Health Organization 2023; Mason and Xie 2024).

The level of control over the spread of COVID-19 that China was able to achieve relied in large part on the mass compliance of its enormous population. In this chapter, we draw upon ethnographic data gathered over a three-year time period in Shanghai, China to offer some explanations for *why* so many Chinese readily complied with zero-COVID policies for so long. Rejecting simplistic explanations that attribute popular support for government policies entirely to authoritarian control or collectivist instincts, we show how a combination of self-interest, nationalistic pride, and trust in the central government's competence drove popular compliance with China's policies. When these factors were no longer in place, resistance to government policies grew, resulting in the abandonment of zero-COVID. Our findings offer lessons for future epidemic control within and beyond the Chinese context.

11.2 Goals and Methods

11.2.1 Goals and Approach

As anthropologists with deep ties to China and a long history of studying epidemic responses there (Mason 2010, 2012, 2016a, b), we set out early in spring 2020 to understand how the COVID-19 pandemic was shaping urban Chinese society in real time. Cai, then a Ph.D. student studying under the mentorship of Mason, was living in Shanghai, China, conducting dissertation research on an unrelated topic, when COVID-19 struck (Cai 2021). Locked down in his home during the first phase of the pandemic in China in early 2020, Cai saw immediately the potential for an ethnographer to closely document the unfolding crisis in real time. He approached Mason about applying for a National Science Foundation (NSF) Rapid Response Research (RAPID) grant to support an ethnographic project based in Shanghai, in collaboration with Mason, that would examine how a persistent state of uncertainty created by pandemic conditions was shaping local daily lives in Shanghai and citizens' plans for the future. After being awarded a RAPID in April 2020, data collection began immediately, with Cai collecting ethnographic data on site in Shanghai, in close contact and conversation with Mason, who was located in the United States (U.S.). Following the first six months of data collection, the authors identified compliance as a central theme and pursued this theme in a more targeted fashion in later stages of data collection and in our analysis.

11.2.2 Study Population

We utilized purposive sampling and a modified cohort design. We selected an initial sample of Shanghai residents ($n = 38$) that was heterogeneous across age, occupation, and what we call "qualitative socio-economic status" (QSES). In April 2022, we added an additional seven participants—three Shanghai residents and four residents of other cities—to gain additional perspectives on the theme of compliance (total $n = 45$). Social media and observational data complemented our interview data. Primary data collection took place June 2020–August 2021, with follow-up interviews conducted April–May 2022 and November–December 2022. The study was approved by Brown University's Institutional Review Board. We created the QSES construct to capture interlocutors' SES as reflected in their subjective narratives (Rubin et al. 2014). During interviews, we collected information on interlocutors' education, housing, employment, income, and household registration (*hukou*). Interviewees were grouped into low, medium, or high QSES based on their own subjective assessments and Cai's familiarity with local social hierarchies, acquired over several years of fieldwork experience in Shanghai.

11.2.3 Setting

Data collection took place primarily in Shanghai, which has a diverse population with residents from nearly every region in China. Follow-up interviews were also conducted with four non-Shanghai residents living in Shenzhen (2), Beijing (1), and Shenyang (1), all of which are also large, migrant-heavy cities that experienced multiple lockdowns. Data were also collected online through social media channels. Our findings primarily reflect the perspectives of Shanghai residents and other urban dwellers and should not be taken to be representative of the full range of experiences in China.

11.2.4 Sampling and Recruitment

The authors began by reaching out to existing contacts in Shanghai to recruit a small purposive seed sample, the members of which then referred additional participants. Efforts were made to balance the sample across age, occupation, and QSES. Thirty-eight Shanghai residents (27 men and 11 women) aged 18–61, who had lived in Shanghai for at least one year, were recruited to participate in 90-min semi-structured interviews (Bernard 2011) conducted between June and August 2020. For a second round of interviews in January–February and July–August 2021, we selected a sub-sample ($n = 10$) of these thirty-eight who were particularly reflective about their experiences. An additional seven participants (five women, two men) were then recruited for a third round of interviews conducted April–May 2022 (third-round $n = 10$, including three participants from the original 38). More women than men were recruited for third-round interviews to help address a sex ratio imbalance in our original sampling. While the sex imbalance remains a limitation of this study, results were consistent across both male and female participants. Finally, a fourth round of interviews was conducted with the 10 third-round participants in November–December 2022, to document responses to the collapse of public support for zero-COVID and the end of the policy in December 2022.

11.2.5 Interviews

Cai conducted all interviews in Mandarin Chinese. Twenty-eight first-round interviews were conducted in a café; ten were conducted by phone. Second-round, third-round, and fourth-round interviews were conducted by phone. Interviews were recorded. We solicited information on a variety of topics related to knowledge consumption, behaviors, and attitudes. Oral consent was obtained at the beginning of each interview. Interviewees received 150 RMB (~23 USD) in compensation. Cai asked consistent questions in each interview as initial probes and then encouraged interviewees to freely discuss COVID-19-related matters. Probes included: What do you think about the government's control measures? How does China's response compare with other countries? Where do you get your information about COVID-19?

11.2.6 Participant Observation

Cai conducted participant observation by traveling to different parts of Shanghai daily during the primary data collection period (June 2020–August 2021). Areas visited covered all major urban areas of Shanghai. Because Shanghai resumed public activities in March 2020, Cai was able to observe venues including gyms, malls, and parks. He wrote field notes about control measures and observed behaviors over the course of the study.

11.2.7 Social Media

Social media was a major source of information about COVID-19 and a place where many people recorded their reactions to the pandemic. Adopting the "walk-through method" (Light et al. 2018), we drew upon the same social media platforms that our interlocutors identified as important and used these platforms the ways they did, to identify and collect content (including articles/posts/images/comments) related to COVID-19. Social media data were collected daily from January through March 2020 on WeChat, the most popular social media platform in China, and periodically thereafter whenever major events (e.g., new outbreaks) occurred. We analyzed posts' structure, genre, and format, to study publicly expressed reactions to control measures. Only publicly available content was collected. Over 50 online articles and 100 + screenshots of social media posts were collected and contributed to our overall analysis. We triangulated social media data with observations and interviews to develop a detailed picture of urban Chinese residents' COVID-19 experiences. Social media collection was ongoing through December 2022.

Particularly during lockdowns, social media platforms were our interlocutors' main sources of information about COVID-19, and these platforms significantly shaped people's perceptions of the pandemic and behavioral responses (Lathifah et al. 2021). It is important to note that Chinese social media is not simply a channel for state-controlled propaganda. When collecting social media data, we observed creative ways in which people used pictures, euphemisms, or Virtual Private Networks (VPNs) to counteract censorship efforts, and they frequently posted opinions that contradicted official statements. Still, our interlocutors knew that information they accessed through social media was heavily mediated by both political and market forces.

11.2.8 Analysis

Interview recordings were transcribed verbatim. Material from the primary data collection period (June 2020–August 2021), including interview transcripts, field notes, images, and online articles, was imported into NVivo (v.13) for coding. Cai generated a codebook, which Mason revised. A research assistant (RA) then conducted a first round of coding. Relevant responses from all interviews were extracted and reviewed to identify broad themes. A second round of coding was conducted by the RA to identify subthemes under the broad theme of "compliance". Responses were sorted by participant to examine chronological changes and consistencies over time. Data from interviews, observations, and social media were triangulated to produce as nuanced a representation as possible of common ideas and behaviors. Additional rounds of interview and social media materials were coded as they came in.

11.3 Findings

11.3.1 Self-Protection and Fear of Death

The primary reason for most of our interlocutors' compliance with strict control measures was fear of COVID-19 infection and subsequent death if they failed to comply. That fear created an emotional link between compliance and self-interest that lasted long after the initial lockdowns ended and was continually reinforced by gruesome imagery emerging from countries where control measures were looser.

Trapped indoors in the early days of the pandemic, hundreds of millions of people had nothing to do but browse social media and watch news. Messages of fear and danger were relayed early and often, and sensational images and videos out of Wuhan's overwhelmed hospitals went viral. The impression for our interviewees in these early months—an impression borne out by high fatality rates in Wuhan and later in Italy, Ecuador, and the U.S.—was that COVID equaled death.

In response to Cai's question about why he eagerly complied, one interviewee explained: *"Fear. ... we don't want our family or ourselves to contract it. This is just normal. ...*

You pay for your own choice. We all do, eventually. You can choose not to wear a mask. But it will eventually lead to a large-scale outbreak ... Can you ensure that you won't be infected, that your most personal interests won't be impaired? Your purpose of taking other people's interests into consideration is to benefit yourself in the end. Eventually this is about taking care of ourselves. It's not about grand ideological beliefs." For this participant, complying with large-scale control measures was a matter of self-protection. One had to care about protecting others in order to protect oneself and one's family. Failing to comply would result in personal harm. This participant rejected the idea that there may be a tradeoff between public health goals and individual interests.

Even those who experienced substantive harms from COVID-19 control measures made similar assertions. A barber whose shop closed for four months early in the outbreak could not afford his rent and had to move in with a friend. Even after the barbershop resumed operations, business was only half what it was before the lockdown. Still, this participant expressed strong support for China's strict measures. The lack of deaths in China compared with other countries provided all the evidence he needed: *"I think [the measures] are necessary, so we can guarantee everyone's safety, our national safety. So we can effectively control the pandemic. ... That's why we proactively comply. ...If you lose your life, what's the point of freedom?"*.

11.3.2 Conscious Indifference to Transparency

Chinese citizens were not naïve about the nature of the data they were provided. Far from being "brainwashed" (Wu 2017)—as is often portrayed in Western media (Ge 2012; Thokmay 2019)—our interlocutors displayed what we came to think of as "conscious indifference to transparency." They were keenly aware that at least some of the information about COVID reported in the Chinese media were likely poor approximations of the "truth". Yet this was not of concern to them. As long as society was stable and they were able to stay healthy and safe, the accuracy of statistical data was beside the point.

As one participant explained in an interview:

> Participant: It's not about credibility. I trust [the leaders] whether they are credible or not.
>
> Cai: You don't care much whether the information is true?
>
> Participant: No, I don't. Listen, does it matter if we care about truth? We are laobaixing [ordinary people], it's not our job to care. This is vague and lofty. What's important is that we are well-fed ... For laobaixing, anything else is empty talk.

Another participant explained this attitude in this interchange with Cai:

> Participant: Are 100,000 new cases different from 50,000? Does it really matter to you? Either way, you'll have to comply with preventive measures to protect yourself. As for how accurate the statistics are, I don't think that's important. It shouldn't influence your behavior.

Cai: Why isn't it important that statistics are 100% accurate?

Participant: Because … they're never 100% accurate. Some people might get sick and recover quickly and were never counted. [Whether] 50,000, 50,100, or 51,000, it doesn't impact your individual interests at all … Sometimes the state reports a smaller number not because they are trying to hide anything, but because some of the cases weren't confirmed yet. Plus, sometimes you just must learn to interpret state-reported data. Data sharing is a craft. … Take the Wuhan situation early on … True, the situation might be awful. But what do you expect the government to say? To admit that, "yes, our hospital system collapsed and ran out of places to hospitalize patients?" For what? To produce more panic? We both know the state should not and could not have said anything like that. At a time when we already face disaster, would you really choose to publicize negative news?

For this participant, there were three reasons full transparency should not be a goal. First, whether the data were entirely accurate did not change the overall situation, in his view. So long as people understood the situation was serious, the exact number of cases should not impact anyone's actions. Second, this participant pointed to the impracticality of achieving total accuracy in data reporting during a rapidly evolving outbreak. Finally, he argued that officials were wise to withhold certain information, because revealing the full scale of devastation would cause panic. Those we interviewed did not trust fellow citizens to remain calm in the face of bad news any more than government officials did. But they *did* trust their government to ensure stability. To protect this stability—and thus, ultimately, themselves—government officials and citizens alike agreed that it was better for ordinary citizens to *not* have full information.

Another interviewee similarly told us:

I always believe in the government because they are the authority. The authority has social responsibility for the news and data it releases and puts a lot of thought behind it. People on social media are free to say whatever they want because they don't have to be responsible for society. As for whether it's 100% accurate and whatnot, you can make your own judgment. But the most important things are the policies and directions.

These comments highlight another reason our interviewees found total transparency to be unnecessary and undesirable: just because data were "true" did not mean they should be shared. For this participant, full transparency about what happened in Wuhan was irresponsible and potentially harmful (Mason 2020). Government-generated data might not technically be true, but our interlocutors suggested they were the product of thoughtful decisions made in the interest of responsibly promoting a healthy and stable environment. Information posted by the public on social media, on the other hand, might be neither "true" *nor* responsible.

Such comments mirrored language frequently used in government statements claiming that national development strategies were beyond ordinary people's ability to understand (Mason 2016a). The underlying message was clear: trust us, we have your best interests at heart. For our interlocutors, this message was not empty propaganda. During the first two years of the pandemic, it was supported by considerable evidence: society remained stable and personal safety was maintained. With this lived experience, our interlocutors overwhelmingly expressed confidence that the state did have their best interests at heart.

11.3.3 Nationalistic Pride

Even as they cited self-interest as the primary motivating factor for their actions, interviewees also ascribed their support for their government to a sense of responsibility and collective consciousness rooted in Chinese history and culture (Mason 2016b). Nationalistic rhetoric common on social media similarly reflected the entanglement of these two dynamics.

The slogan "don't cause more trouble to our nation" (*bugei guojia tianluan*) is a case in point. It emerged on social media in January 2020 in response to those few who publicly pushed back against government policies, and quickly made its way into the repertoire of everyday expressions. In the first half of 2020, social media was filled with videos of noncompliant individuals being reprimanded by police officers, security guards, and neighborhood committee members for not wearing masks or for refusing to stay indoors. In the comments section of videos, people left comments such as "Punish them hard" or "Arrest them and lock them up." Among our interviewees there was little pushback against this stigmatizing approach.

Even in the face of prolonged quarantine and uneven distribution of food and other supplies, impatience with the noncompliant persisted among many of our interviewees. In an interview in April 2022, when a strict two-month quarantine was in place in Shanghai and popular frustrations with the zero-COVID policy were boiling over, one interviewee told Cai in a follow-up interview:

> It's always these lice causing trouble. When the government sends them to isolation centers, they scream like a pig being slaughtered and complain on social media. If the government doesn't send them to quarantine, they are whining again, saying that the government leaves them to fend for themselves. These lice are selfish.[1]

Prior to the summer and fall of 2022, the most serious period of popular discontent to emerge in response to COVID-19 occurred during the first months of the pandemic, following the death of Li Wenliang, a Wuhan physician who blew the whistle about COVID's initial appearance in December 2019. Li was detained by authorities and made to deny his claims. Anger erupted on social media as citizens demanded to know what was happening in Wuhan and why Li's claims had not been taken seriously. The unrest culminated in Li's death from COVID on February 7, 2020. Li's mask-wearing image became the symbol of calls for justice amidst perceived government cover-up and inaction.

The crisis of trust in government that Li's death unleashed did not last, however. By the time COVID-19 began spreading widely in the rest of the world, anger had dissipated and was replaced with nationalistic pride. Indeed, according to surveys on trust in government conducted in March 2020 in over 170 countries, China's government ranked among the "most trusted" by its citizens (Fetzer et al. 2020; Rieger and Wang 2022). According to the 2021 Democracy Perception Index Report (DeVeaux and Dölitzsch 2020), China similarly ranked second on citizens' trust in governmental responses to COVID (U.S. ranked forty-second) (Rieger and Wang 2022).

Analysis of social media materials suggests that a multi-pronged deflection of blame made this trust possible. First, blame for COVID's emergence was redirected from the central government to local officials. This is an established tactic of the central Chinese government, which often seeks to distinguish a caring, trustworthy central government from incompetent, corrupt local governments (Su et al. 2016; Chen 2017). The central government declared Li a "national martyr" (XinhuaNews 2020) and punished local officials who mistreated him (Beijing News 2020). These moves—which came alongside the rapid containment of COVID-19 inside China—were enough to tamp down most criticism, at least in urban areas.

Punitive measures against local officials continued throughout the pandemic. Every time a regional outbreak took place, or a local response was bungled, local officials were removed from their positions to set examples for others and reassure the public that action was being taken. These measures redirected discontent onto local officials. While local responses to COVID-19 were critiqued, those critiques rarely extended to the national response.

By spring 2020, public sentiment also coalesced around the idea that COVID-19 originated outside of China. Social media rumors swirled with claims that SARS-CoV 2 was found in Spanish sewage systems in mid-2019 (Chavarria-Miró et al. 2020). Additional rumors suggested that the U.S. Army base Fort Detrick may have been responsible for releasing the virus (Zhang and Chen 2021), and that U.S. military personnel brought the virus with them when participating in the 2019 Military World Games

[1] Lice (虱子, shi zi) is a common derogatory term in Chinese that refers to people who rely on others to support them without producing any social benefit in return.

in Wuhan (Chinese Communist Youth League 2021). By June 2020, most of our interviewees in China felt certain that COVID-19 originated outside of China.

Finally, media reports about COVID-19 in other countries instilled an image of Western governments as incompetent and irresponsible (CCTV News 2022). Supporting the Chinese government was equated with protecting the Chinese nation from the dangers created by untrustworthy Western governments, which allowed a deadly threat to spread (Mason 2016b).

11.4 Summary of Themes

While different people's precise reasons for complying with the zero-COVID policy differed, almost all our interviewees agreed that tangible results were what mattered. Trust and pride in the nation were based on a "job well-done." Pushing back against suggestions that he had been forced to comply, one interviewee told us, "*Foreign media always assume that what we ordinary people say are taught by the government. But this is simply false. We ordinary people have deep lived experiences of the government's actions.*"

This is not to say that government power is not important. The contemporary Chinese state continues to hold immense power, and most Chinese still usually comply with state requirements even when they don't want to Zuo (2022). Fear of public shaming and subsequent consequences also plays an important role in shaping public opinion and actions (Zhao and Burgess 2021).

In arguing that our interlocutors are making thoughtful decisions about whether to willingly support state control measures, we thus do not suggest that their choices are limitless or that they are making purely rational decisions unmediated by outside influences. Like citizens of any country, Chinese are heavily influenced by the information they access, and are constrained by the laws of the land where they live and the consequences for violating them. Still, they make reasoned decisions based on the information they have. When that information changes, their perspectives change too.

11.4.1 The End of Zero-COVID

Through mid-2022, our findings suggested that willing compliance among urban Chinese was associated with calculated decisions to make sacrifices that they were confident would pay off in the form of long-term personal and national health and stability. These were priorities that they trusted were shared by a state capable of achieving its goals. In the latter part of 2022, however, our interlocutors came to feel that this trust had been betrayed. Not only did the state no longer seem to have their—or their country's—best interests at heart, but it no longer was clear that the state was even capable of protecting those interests. This doubt became most prevalent after Omicron became the dominant SARS-CoV-2 variant in China and the country's stringent lockdown measures failed to keep the increasingly infectious virus at bay. One of our interlocutors commented during the Shanghai lockdown in April 2022:

> Many people have been locked in their home for over a month. But there have still been over twenty thousand new infections every day... It is time the state should critically reflect if lockdown was an appropriate response to Omicron at all (Cai and Mason 2023).

When people no longer believed that their national government was competently, or with their best interests in mind, protecting them or their country from harm, the willing compliance that had characterized China's COVID-19 response since early 2020 began to crumble (Yeung et al. 2022). Their doubt in the government's ability to deliver what it had promised produced dynamic discussions and online debates about power, authority, class, and the legitimacy of the state, resulting eventually in highly unusual protests in major cities, including Shanghai, in late 2022 (Che and Chien 2022).

If the state was no longer capable of protecting either their own or their country's best interests, then what did that mean for the last three years of people's lives? A dismal collective sentiment emerged among some of our interlocutors and on social media: after three years of sacrificing personal freedoms, financial gains, and social connectivity to win the "war on COVID," the war seemed to be lost. Or worse, it might never have been winnable in the first place.

A long-time interlocutor expressed deep disappointment in the state's failures in a follow-up interview:

> At first, I supported it because we didn't know much about the virus. I thought that because we were so strict, our life could get back to normal faster than foreign countries. But now, we know that the virus isn't as deadly as we thought, and everyone [overseas] lives a normal life without strict control measures. But we are still in lockdown every once in a while. ... I feel that our efforts and sacrifices were in vain.

11.5 Conclusions and Recommendations

Successful containment of COVID-19 depended in large part on Chinese citizens' willingness to actively support government actions. In the first two years of the pandemic, our interlocutors overwhelmingly believed that complying with state directives would benefit themselves, their families, and their country—and so they readily complied even with the most stringent COVID-19 control restrictions.

While China's authoritarian government played a role in achieving this result, it did not—at least in the first two years of the pandemic—do so primarily through coercion. Instead, it controlled the narrative about COVID 19 and supported that narrative with effective actions. That it was able to do this largely without resistance for close to three years speaks to many Chinese citizens' trust in, and satisfaction with, the central government's ability to protect them and their families. The deaths of millions of people around the world served as a powerful warning of what could happen to individuals if collective control of COVID-19 was not achieved.

Our study provides insights into the relationship between compliance and data. Despite the unreliable nature of data provided to the public, most urban Chinese willingly complied with the state's strict control measures. They did this because they trusted that the state was acting based on the knowledge it had, and that it was using this knowledge to protect them. They did not need to know everything the government knew to trust in the path set for them. Key to this trust was the observation that the control measures being implemented were compatible with their own individual interests. When this compatibility changed, attitudes changed too. While fear of punishment may have been able to force China's population into compliance for a time, in the long term an unwilling populace is much harder to control than a willing one. The Chinese government knew this, and it ended zero-COVID when both the SARS-CoV-2 virus, and public opposition to the policy, became impossible to contain.

Our findings have policy implications that go beyond China. While we certainly do not suggest that governments release inaccurate data or implement the exceptionally draconian control measures that China used, we do suggest there are useful lessons to be learned from the Chinese experience. While authoritarian power and the threat of punishment no doubt played a role in China's successes, this dynamic may not be necessary when other components are in place. Few of our interlocutors mentioned fear of the authorities as a major motivating factor. Instead, they complied because the evidence before them suggested it was in their interest to do so, and because—for a time at least—they had something currently lacking in the U.S. context: trust in the competence and good intentions of their national government.

We thus offer three recommendations based on China's experience with COVID-19 that U.S. public health officials might consider implementing to encourage compliance with disease containment efforts during future pandemic events: (1) provide clear, simple, easy to follow directions from sources the public trusts; (2) rally nationalistic pride as part of a cohesive, nationwide pandemic response; and (3) tie public health benefits closely to the personal interests of individuals. The Chinese government successfully rallied public support through these three methods—none of which require authoritarian rule to accomplish.

While trust in government and its institutions is lacking in many spheres in the U.S. today, our findings suggest that if such trust is restored, major public health benefits may follow. On the other hand, as China's experiences in late 2022 show, without public trust, even the strictest control measures will eventually fail. Restoring trust in institutions thus should be the cornerstone of any future pandemic response.

Acknowledgements Portions of the text previously appeared in two published articles by the authors: Cai, Yifeng Troy and Katherine A. Mason. 2022 Why They Willingly Complied: Ordinary People, the Big Environment, and the Control of COVID-19 in China. *Social Science and Medicine* 309: 115239, published by Elsevier; and Cai, Yifeng Troy and Katherine A. Mason. 2023 "Our Sacrifices Were in Vain": Zero-COVID and the Betrayal of Trust. *HAU: Journal of Ethnographic Theory* 13(2): 308–314 (Cai and Mason 2023), published by University of Chicago Press. The original published text has been edited for this publication.

In addition to the NSF Award #203407 that supported this work, Mason acknowledges the support of the Population Studies and Training Center, supported by National Institutes of Health (NIH) Center grant P2C HD041020. The authors also would like to thank Claire Lin, Karina Bao, and Kexin Zhao, who contributed to this work with transcription, coding, and organizational support.

References

Bernard HR (2011) Research methods in anthropology: Qualitative and quantitative approaches. Rowman & Littlefield, New York

Cai YT (2021) Male sex work in China: Digital technology and its emerging role. In: The Routledge handbook of male sex work, culture, and society. Routledge

Cai YT, Mason KA (2023) "Our sacrifices were in vain": zero-COVID and the Betrayal of Trust. HAU: J Ethnogr Theory 13:308–314

Cai YT, Mason KA (2022) Why they willingly complied: ordinary people, the big environment, and the control of COVID-19 in China. Soc Sci Med 309:115239. https://doi.org/10.1016/j.socscimed.2022.115239

Chavarria-Miró G, Anfruns-Estrada E, Guix S, et al (2020) Sentinel surveillance of SARS-CoV-2 in wastewater anticipates the occurrence of COVID-19 cases. medRxiv 2020.06.13.20129627. https://doi.org/10.1101/2020.06.13.20129627

Che C, Chien AC (2022) Memes, puns and blank sheets of paper: China's creative acts of protest. The New York Times. https://www.nytimes.com/2022/11/28/world/asia/china-protests-blank-sheets.html

Chen D (2017) "Supervision by public opinion" or by government officials? Media criticism and central-local government relations in China. Modern China 43:620–645

Chinese Communist Youth League (2021) US journalist says US athletes at Wuhan Military Games may be COVID-19 "patient zero", more and more clues are pointing to the US. In: Baidu News. https://baijiahao.baidu.com/s?id=1707571587138941012&wfr=spider&for=p. Accessed 21 Nov 2021

DeVeaux F, Dölitzsch C (2020) Democracy perception index. https://allianceofdemocracies.org/democracy-perception-index. Accessed 4 Nov 2022

Du Z, Wang Y, Bai Y, et al (2023) Estimate of COVID-19 deaths, China, December 2022–February 2023. Emerging Infectious Diseases J 29(10) (CDC). https://doi.org/10.3201/eid2910.230585

Fetzer T, Witte M, Hensel L, et al (2020) Measuring worldwide Covid-19 attitudes and beliefs. Warwick

Ge Q (2012) China's brainwashed youth. Foreign Policy

Glanz J, Hvistendahl M, Chang A (2023) How deadly was China's Covid wave? The New York Times. https://www.nytimes.com/interactive/2023/02/15/world/asia/china-covid-death-estimates.html

Johns Hopkins (2020) Johns Hopkins Coronavirus Resource Center. https://coronavirus.jhu.edu. Accessed 4 Nov 2022

Lathifah A, Tiani R, Mazida F (2021) Public perception of information about Covid 19 on social media Instagram: an netnography study of visual information. In: E3S web of conferences, vol 317, p 5015. https://doi.org/10.1051/e3sconf/202131705015

Light B, Burgess J, Duguay S (2018) The walkthrough method: an approach to the study of apps. New Media Soc 20:881–900

Mason KA (2010) Becoming modern after SARS. Battling the H1N1 pandemic and the politics of backwardness in China's Pearl River Delta. BEHEMOTH-J Civil 3:8–35

Mason KA (2012) Mobile migrants, mobile germs: migration, contagion, and boundary-building in Shenzhen, China after SARS. Med Anthropol 31:113–131

Mason KA (2016a) The correct secret: discretion and hypertransparency in Chinese biosecurity. Focaal 2016:45–58

Mason KA (2016b) Infectious change: reinventing Chinese public health after an epidemic. Stanford University Press, Stanford, CA

Mason KA (2020) Divergent trust and dissonant truths in public health science. Can science and technology save China. Cornell University Press, Ithaca, NY, pp 95–114

Mason KA, Xie J (2024) Seesaw precarity: journaling anxious hope on a Chinese University campus during covid-19. Cult Med Psychiatry 48:66–90. https://doi.org/10.1007/s11013-024-09846-8

Mason K, Cai Y (2022) Kate Mason and Yifeng Cai: understanding the process of social change through the transitional period of the COVID-19 pandemic. Columbia Academic Commons, Academic Unit: Northeast Big Data Innovation Hub, Series: COVID-19 Research Lightning Talks. https://doi.org/10.7916/9pk7-0y43

CCTV News (2022) 西方抗疫“躺”出四大灾难， 带给中国四点启示 [The West’s anti-epidemic efforts brought four major disasters, and four lessons to China]. https://news.cctv.com/2022/04/30/ARTIuVhq9lHJPxyua4M0hurj220430.shtml. Accessed 19 Aug 2024

Pan A, Liu L, Wang C et al (2020) Association of public health interventions with the epidemiology of the COVID-19 Outbreak in Wuhan, China. JAMA 323:1915–1923. https://doi.org/10.1001/jama.2020.6130

Rieger MO, Wang M (2022) Trust in government actions during the COVID-19 crisis. Soc Indic Res 159:967–989. https://doi.org/10.1007/s11205-021-02772-x

Rubin M, Denson N, Kilpatrick S et al (2014) “I Am Working-Class”: subjective self-definition as a missing measure of social class and socioeconomic status in higher education research. Educ Res 43:196–200. https://doi.org/10.3102/0013189X14528373

Su Z, Ye Y, He J, Huang W (2016) Constructed hierarchical government trust in China: formation mechanism and political effects. Pac Aff 89:771–794

Thokmay K (2019) The brainwashed: a generation of Chinese millennials. Asia Times

Tian W (2021) How China managed the COVID-19 pandemic. Asian Economic Papers 20:75–101. https://doi.org/10.1162/asep_a_00800

Wang L, Yan B, Boasson V (2020) A national fight against COVID-19: lessons and experiences from China. Aust N Z J Public Health 44:502–507. https://doi.org/10.1111/1753-6405.13042

World Health Organization (2023) WHO COVID-19 Dashboard. https://covid19.who.int. Accessed 13 Oct 2023

Wu AX (2017) Brainwashing paranoia and lay media theories in China: the phenomenological dimension of media use (and the self) in digital environments. Media Cult Soc. https://doi.org/10.1177/0163443717734409

(2020) 武汉警方对李文亮医生被训诫一案相关责任人作出处理 [Wuhan Police punished personnel related to Dr. Li Wenliang Case]. In: Beijing News. https://www.bjnews.com.cn/detail/158462617015389.html. Accessed 24 Apr 2022

Xiao H, Wang Z, Liu F, Unger JM (2023) Excess all-cause mortality in China after ending the zero COVID policy. JAMA Netw Open 6:e2330877. https://doi.org/10.1001/jamanetworkopen.2023.30877

XinhuaNews (2020) 14 Hubei pandemic frontliners recognized as Martyrs. In: Xinhua News. http://www.xinhuanet.com/english/2020-04/02/c_138941398.htm

Xu T-L, Ao M-Y, Zhou X et al (2020) China’s practice to prevent and control COVID-19 in the context of large population movement. Infect Dis Poverty 9:115. https://doi.org/10.1186/s40249-020-00716-0

Yeung J, Gan N, Xiong Y, Beijing Bureau (2022) Xi’an lockdown brings heartbreak and dysfunction as political pressure to contain outbreak grows. https://www.cnn.com/2022/01/06/china/xian-lockdown-zero-covid-intl-hnk/index.html. Accessed 9 Jul 2024

Zhang Y, Chen W (2021) There are many doubts about the US military base and this university. If we want to trace the source of the new coronavirus, we must investigate them. http://k.sina.com.cn/article_6145283913_16e49974902001gs4d.html. Accessed 19 Aug 2024

Zhao I, Burgess A (2021) China brings back public shaming, parading alleged people-smugglers through the streets for breaking COVID-19 rules. ABC News. https://www.abc.net.au/news/2021-12-30/china-public-shaming-covid-rule-breakers-people-smugglers/100731450

Zheng L, Liu S, Lu F (2023) Impact of National Omicron Outbreak at the end of 2022 on the future outlook of COVID-19 in China. Emerging Microbes Infections 12:2191738. https://doi.org/10.1080/22221751.2023.2191738

Zuo Y (2022) Shanghai police: violating pandemic regulations will be punished. In: Xinmin News. https://wap.xinmin.cn/content/32140581.html. Accessed 24 Apr 2022

Katherine A. Mason is an Associate Professor of Anthropology at Brown University. She is a medical anthropologist whose research addresses issues in population health, bioethics, mental health, and global health in China and the U.S. Her first book, *Infectious Change: Reinventing Chinese Public Health after an Epidemic,* was published by Stanford University Press in 2016. Her work has appeared in *American Ethnologist, American Anthropologist, Medical Anthropology Quarterly, Social Science and Medicine,* and *The Journal of Infectious Diseases,* among other venues. Mason is co-founder of the Pandemic Journaling Project (PJP), an online platform that collects and records longitudinal journals of ordinary people’s COVID-19 experiences, and she co-leads two extramurally funded research studies that build on the PJP platform. She is also currently working on a book on perinatal mood and anxiety disorders. Mason received her Ph.D. in Social Anthropology from Harvard University in 2011 and also holds a Bachelor’s degree in Biology from Yale University.

Yifeng Troy Cai earned his Ph.D. in Anthropology from Brown University, where he also received his Master of Public Health degree. His major research interests include digital anthropology, medical anthropology, queer theory, and urban studies. His dissertation, "Taming Desires: Precarious Intimacy and Queer Urban Sociality in Digital China," is based on three years of in-person and virtual fieldwork in China. It focuses on urban gay men's experiences of intimacy, pleasure, and competition in an increasingly digitized China. Prior to Brown, he received training from the University of Pennsylvania, University of Chicago, and Renmin University of China. His research has been supported by the National Science Foundation (NSF), the Wenner-Gren Foundation, the American Council of Learned Societies, and the Luce Foundation, among others. His work has appeared in *Social Science and Medicine*, *Medical Anthropology Quarterly*, and *HAU: Journal of Ethnographic Theory*, among others. Cai currently lives in Shanghai, China where he works as an independent scholar and runs a college consulting firm.

Part IV
COVID and Human Behavior

12 Computational Cognitive Theory of Attitudes, Social Information, and Behavior During Pandemics

Peter Pirolli, Archna Bhatia, Kathleen M. Carley, Bonnie J. Dorr, Christian Lebiere, Brodie Mather, Konstantinos Mitsopoulos, Don Morrison, Mark Orr, Tomek Strzalkowski, and Choh Man Teng

Abstract

Human culture, social interactions, the information ecology, and psychology are crucial elements of the behavioral environment shaping the emergence, transmission, response, and mitigation of infectious diseases. In this chapter, we describe computational theories, models, and methods developed to understand and predict the dynamics of human behavior and the surrounding information environment in the United States of America (U.S.A.) during the COVID-19 pandemic. This builds upon and combines research on computational cognitive models, natural language processing, network science, social media, and mass

Research funded by

NSF Award #2033390—RAPID (Rapid Response Research Award): Improving Computational Epidemiology with Higher Fidelity Models of Human Behavior;
NSF Award #2200112—Predictive Intelligence for Pandemic Prevention (PIPP) Phase I: Computational Theory of the Co-evolution of Pandemics, (Mis)information, and Human Mindsets and Behavior;

The Office of the Director of National Intelligence (ODNI), Intelligence Advanced Research Projects Activity (IARPA), via 2020–20092500001.

COVID Information Commons Video Presentations

The COVID Information Commons (CIC) Research Lightning Talk video presentations and individual multilingual transcripts for this project can be found in the Columbia Academic Commons (AC) (Pirolli 2021, 2023).

P. Pirolli (✉) · A. Bhatia · B. Mather · K. Mitsopoulos · M. Orr · C. M. Teng
Florida Institute of Human and Machine Cognition, Pensacola, FL, USA
e-mail: ppirolli@ihmc.org

A. Bhatia
e-mail: abhatia@ihmc.org

B. Mather
e-mail: bmather@ihmc.org

K. Mitsopoulos
e-mail: kmitsopoulos@ihmc.org

M. Orr
e-mail: morr@ihmc.org

C. M. Teng
e-mail: cmteng@ihmc.org

K. M. Carley · C. Lebiere · D. Morrison
Carnegie Mellon University, Pittsburgh, PA, USA
e-mail: kathleen.carley@cs.cmu.edu

C. Lebiere
e-mail: cl@cmu.edu

D. Morrison
e-mail: dfm2@cmu.edu

B. J. Dorr
University of Florida, Gainesville, FL, USA
e-mail: bonniejdorr@ufl.edu

T. Strzalkowski
Rensselaer Polytechnic University, Troy, NY, USA
e-mail: tomek@rpi.edu

F. D. Hudson (ed.), *The COVID Information Commons*, Sustainable Development Goals Series,
https://doi.org/10.1007/978-3-032-06125-6_12

media analysis. One aim is to make predictions about transmission-mitigating behavior, such as mask wearing and social distancing, in relation to regional population variations in demographics, psychographics, politics, weather, social media, news, and perceptions of ongoing pandemic transmission rates. We discuss the specific data pipelines employed, the models, and empirical evaluations. Insights concerning observed behavioral and psychological phenomena are discussed as well as the opportunities for developing population-scale psychological models and information-environment simulators for impactful events.

Keywords

ACT-R · COVID-19 · Cognition · Non-pharmaceutical interventions

12.1 Introduction

Throughout history, people have modulated their behavior to mitigate pandemics (Christakis 2020; Jones et al. 2021; Leech et al. 2022). The responses of public officials to the COVID-19 pandemic were guided, in part, by epidemiological models that have an abundance of very large uncertainties and large regional heterogeneity (Flaxman et al. 2020; Redlin 2022) around the human response and the resulting epidemiological effects of non-pharmaceutical interventions (NPIs) and vaccine guidance (Aviral et al. 2022; Cramer et al. 2022; Ferguson et al. 2020; Ioannidis et al. 2022; Yang et al. 2023). These uncertainties and imprecision have been attributed, in part, to a failure to address the crucial role of social and psychological-behavioral mechanisms (Spiliopoulos 2022; Vardavas et al. 2021). The National Academies of Sciences, Engineering, and Medicine (Brossard et al. 2020) emphasize the importance of psychological science to the mitigation of the spread of COVID-19. Human behavior plays a crucial role in mediating viral transmission (West et al. 2020).

An empirical illustration of the interplay of human behavior and viral transmission (Weitz et al. 2020; Ianni and Rossi 2021; Pirolli et al. 2021) is in Fig. 12.1, which presents the relationship of the effective reproduction number, R_t on day t and R_{t+7} at a seven-day lag for the first three waves of COVID-19 (pre-vaccination rollout; 4/24/2020–3/31/2021). The effective reproduction number is the average number of people infected by an infected case. Many things might control the value of R_t at any given time, including the proportion of non-susceptible hosts in the population or mitigating behaviors such as mask wearing or social distancing (to the extent they are effective). The signature temporal phenomenon observed in virtually all regions is the damped oscillation pattern of the effective reproductive number, R_t. At the beginning of a pandemic, there is a rapid increase in R_t, followed by an asymmetric decline, followed by oscillations around $R_t = 1$. This oscillation is like that produced by a Proportional-Integral-Derivative control system, often used in engineering solutions for industrial control systems. However, in the case of pandemics, it is mainly natural biology, psychology, and behavior that produces interventions (e.g., mask wearing) in proportional response to the state of the system (e.g., R_t), although there may be lags between changes in the system and awareness of them, between the awareness of the system state and the response, and between the response and effecting a change. We argue that signature phenomena in Fig. 12.1 emerge, in large part, from the complexities and adaptability of human social-psychological mechanisms, driven by past knowledge and experience, and by perception of, and cognition about, information from the physical and digital environments. This chapter provides a summary of our early explorations of computational theory to explain and model the social-psychological mechanisms that produce decisions and behaviors that mitigate pandemics.

In this chapter, we describe our research on computational models of human behavior during the COVID-19 pandemic. Our focus (Pirolli et al. 2021, 2024, 2020, 2022; Mitsopoulos et al. 2023; Orr et al. 2023) is on modeling beliefs, attitudes, intentions, and behavior that are

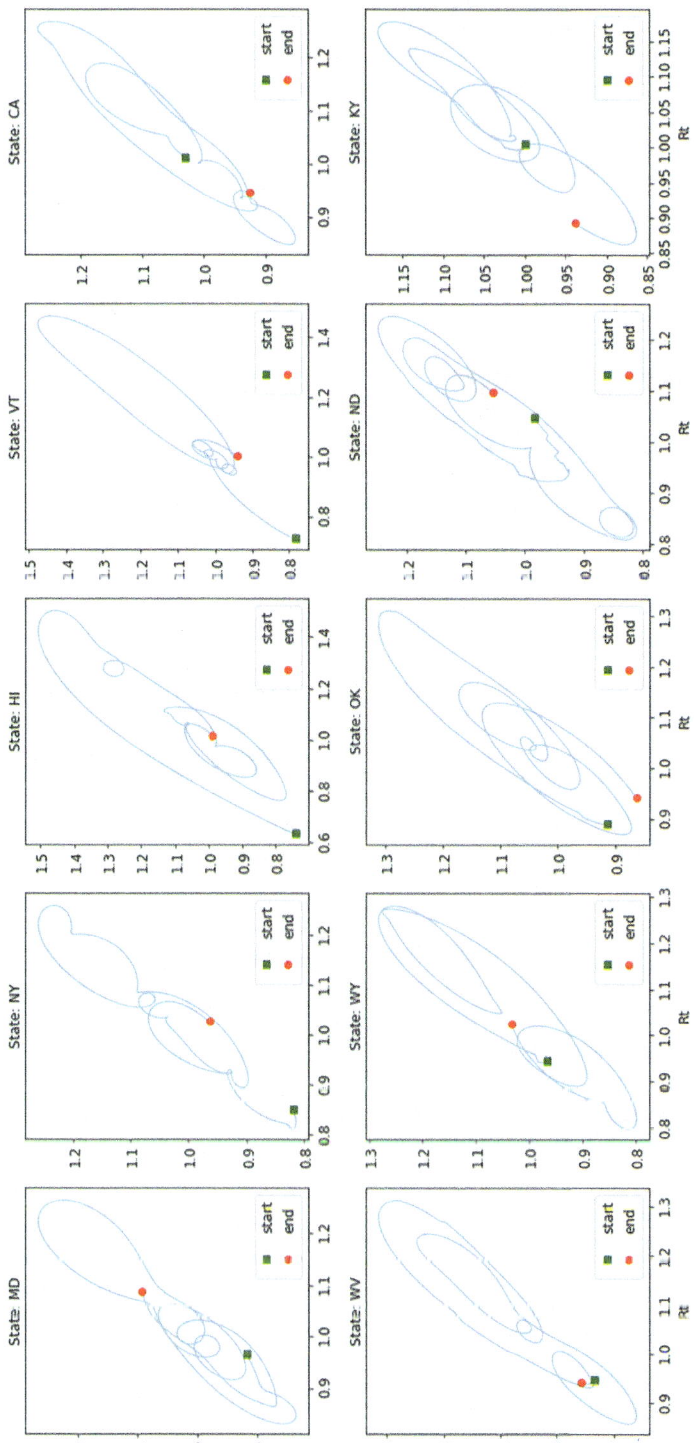

Fig. 12.1 Dampened oscillation of R_t. The values of R_t are plotted against R_{t+7}, where t is in days

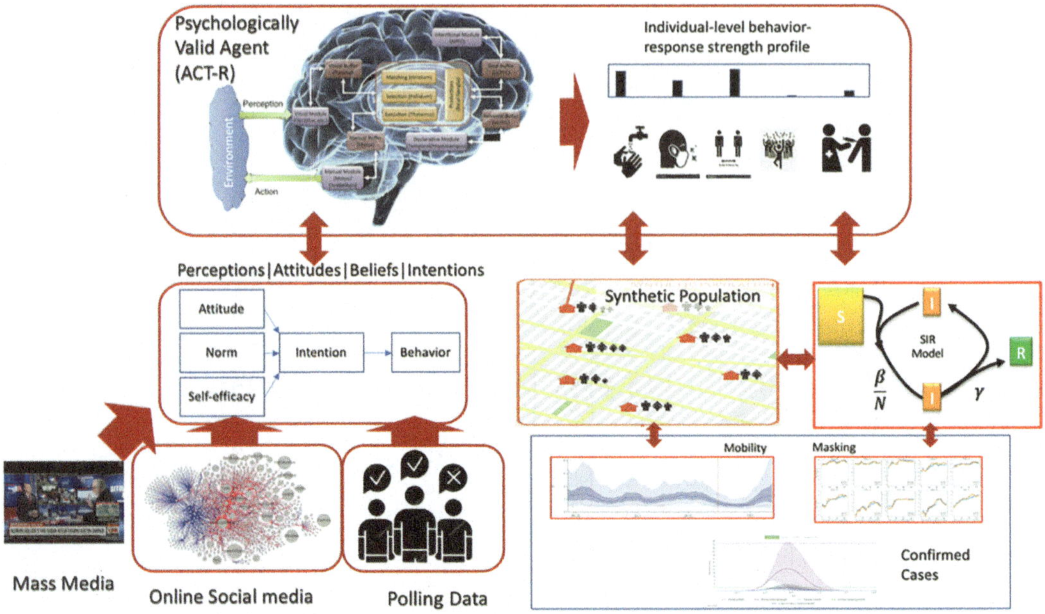

Fig. 12.2 Overview of the psychological valid agents agent approach

assumed to influence the transmission of infectious disease (e.g., COVID-19). Our approach (Fig. 12.2) uses Psychologically Valid Agents (PVAs) implemented in the Adaptive Control of Thought-Rational (ACT-R) cognitive architecture (Anderson et al. 2004), with input drivers induced from heterogeneous sources including online media such as Twitter/X and mass media that provide indicators of pandemic awareness, beliefs, and attitudes (Pirolli et al. 2020). PVAs are used to model behavior change, such as social distancing and mask wearing in response to dynamics in the physical and information environments and are designed to represent individuals or homogeneous segments of a population of interest. Datasets are used to extract cognitive content (e.g., perceptions, attitudes, beliefs, and intentions) and initialize PVAs implemented in the ACT-R cognitive architecture. Given decision contexts, the PVAs simulate the choice of behaviors, which predict behavior-response profiles (e.g., for hand washing, social distancing, vaccination). The PVAs can be run in agent-based models or abstracted into SIR models (Susceptible (S), Infected (I), and Recovered (R)) to simulate effects on disease progression.

12.2 Computational Cognitive Models for Psychologically Valid Agents (PVAs)

Our PVA simulations are based on a variety of data sources including case rate, death rate, survey data, regional personality data, as well as analyses derived from cable news, and social media. Online data sources were used to induce norms, attitudes, and intentions that represent people in different regions over time. These PVAs were grounded in the mechanisms and mathematics of the ACT-R cognitive architecture described below, an architecture for which the theoretical mechanisms have been validated in various human experimental contexts. Notably, aside from seeding the PVAs with data available at the beginning of the pandemic, our simulations were run with zero or only a few adjusted cognitive architecture parameters. In other words, the models should be highly generalizable because of the constraints of the cognitive architecture.

PVAs can address systems of behavior from the level of individuals up through population-scale agent-based models (ABMs). Our PVA simulations are tested in multiple ways. For instance, we can use the PVAs to generate and test predictions about differences in the dynamics of behavior across regional populations arising from the natural experiments across US regions that differ in public health interventions, policies and the public's mindsets. Validation involves simulating dynamic behavioral responses to the infectious agent dynamics—specifically mask wearing and social distancing. We have developed a robust pipeline for modeling *Regional* PVAs (Pirolli et al. 2024), have merged PVAs, agent-based models, and Large Language Models to create Generative PVAs that can simulate richer human behavior in counterfactual simulations (Mitsopoulos et al. 2023).

The abundance of regionally organized (e.g., state, county) demographic, psychographic, epidemiological, behavioral, and information environment data makes it feasible to validate such models. These data are used to initialize agents and provide time-series inputs representing the pandemic context (e.g., local transmission rates). The PVAs iteratively assess the current context and make decisions (e.g., wear a mask or not) over discrete time steps (e.g., every day). They are thus capable of predicting various regional time-series data, e.g., the United States (U.S.) county- or state-level daily mobility patterns or daily mask wearing.

12.2.1 ACT-R Core Mechanisms

Our approach builds on decades of work on a computational cognitive architecture, ACT-R. ACT-R (Anderson 2007) has evolved since the 1990s to address a wide variety of experimental results on problem solving, decision-making, memory, learning, cognitive skill acquisition, perception, and attention, as well as the fine-grained time course of neural processes. The theory has had a close relationship to advances in neuroscience and artificial intelligence (AI), and consequently can be related in many ways to general psychology and behavior, the brain, neural processing, or AI approaches. The theory has been successfully applied to a variety of domains including computer tutoring systems (Anderson et al. 1990; Corbett et al. 1995; Koedinger et al. 1997), human–computer interaction (Pirolli and Durso 1999; Pirolli 2007), language learning (Pavlik and Anderson 2008, 2005), and health behavior change (Pirolli 2016a).

The ACT-R theory specifies modules that process different kinds of cognitive content, coordinated through a centralized *procedural module* controlled by procedures called *productions*. Each module has been associated with a specific brain region. Declarative knowledge and experience in long-term *declarative memory* are represented formally as sets of attribute-value pairs called *chunks* (Simon 1974; Miller 1956) with *activation* levels governing retrieval probability. Subsymbolic mechanisms in ACT-R govern retrieval and processing of chunks and productions, reflecting neural-like mechanisms defined by a set of dynamic equations that capture processes such as learning, forgetting, and blending of knowledge and experiences. There is a continuous processing cycle of perceiving the world, setting goals, retrieving memories, moving information across modules, making decisions, and executing actions. New memories and habits are continuously learned and their subsymbolic parameters updated.

Table 12.1 presents a subset of the core ACT-R subsymbolic mechanisms. The first six rows of the table capture the dynamics of declarative memory: how the level of activation of chunks in memory relates to the probability of their retrieval at any given time, how activation levels are increased by repeated experiences (including messaging, communications, observations, etc.), or decay with time (forgetting), and how blended retrievals of past experiences produce aggregate values over similar experiences. Blending over similar past experiences, and partial matching to information about the current context, provide a basis for mechanistically understanding how past behavior will generalize to novel situations.

Table 12.1 Key subsymbolic mechanisms in the core ACT-R architecture

Mechanism	Equation	Description
Activation	$A_i = B_i + S_i + P_i + \varepsilon_i$	B_i: Base-level activation reflects the recency and frequency of use of chunk i S_i: Spreading activation reflects the effect that context has on the retrieval process P_i: Partial matching reflects the degree to which the chunk matches the cueing context ε_i: Noise generated from quasi-normal distribution with parameter value s
Base-level	$B_i = ln\left(\sum_{j=1}^{n} t_j^{-d}\right) + \beta_i$	n: The number of presentations for chunk i t_j: The time since the *jth* presentation d: A decay rate, $0 < d < 1$ β_i: A constant offset
Spreading activation	$S_i = \sum_k \sum_j W_{kj} S_{ji}$	**k:** *index of buffers used to represent context* **j:** index of chunks which are in the slots of the chunk in buffer k W_{kj}: Amount of activation from sources j in buffer k S_{ji}: Strength of association from sources j to chunk i
Partial matching	$P_i = \sum_k PM_{ki}$	**P:** Match scale parameter which reflects the weight given to the similarity M_{ki}: Similarity between the value k in the retrieval specification and the value in the corresponding slot of chunk i
Declarative retrievals	$p(i) = \frac{e^{\frac{A_i}{s}}}{\sum_j e^{\frac{A_j}{s}}}$	$p(i)$: The probability that chunk i will be recalled A_i: Activation strength of chunk i $\sum A_j$: Activation strength of all eligible chunks j s: Chunk activation noise parameter
Blended retrievals	$V = \text{argmin}_j \sum_i p(i)\left(1 - \text{Sim}_{ij}\right)^2$	$p(i)$: Probability of retrieval of chunk i Sim_{ij}: Similarity between compromise value j and actual value i
Utility learning	$U_i(n) = U_i(n-1)$ $+\alpha[R_i(n) - U_i(n-1)]$	$U_i(n-1)$: Utility of production i after its n-1st application $R_i(n)$: Reward production receives for its nth application $U_i(n)$: Utility of production i after its nth application α: temporal discount
	$s(i) = \frac{e^{U_i/\tau}}{\sum_j e^{U_j/\tau}}$	$s(i)$: Probability that production i will be selected U_i: Expected utility of the production determined by the utility equation U_j: Expected utility of the competing productions j

12.2.2 Instance-Based Learning

Our PVAs often rely on a decision-making modeling approach called *Instance-Based Learning* (IBL) (Gonzalez et al. 2003; Lebiere et al. 2016). IBL employs ACT-R to model human decisions that are based on experiences stored in declarative memory. Declarative memory stores multi-attribute instances of decision contexts, actions, outcomes, and rewards. In a current decision context, these experiences are retrieved and blended to adapt and make decisions according to the dynamics in Table 12.1. Such models are effective in variable and uncertain domains. The mathematical foundations of IBL lie in Statistical Learning Theory (Vapnik 1998). Specifically, IBL is a linear smoother which is a non-parametric instance-based learning function approximator. Therefore, IBL can implement various types of learning algorithms. These include Supervised Learning, with applications in regression and classification, and Reinforcement Learning, which facilitates utility-based learning for habitual behavior and with additional modules (e.g., goal or imaginal buffers) it can support goal-driven behavior. As a nonparametric instance-based learning model, IBL does not require a typical training phase like parametric models do. Instead, it keeps the "training data" within its memory repository, allowing it to adapt dynamically to new situations. To accommodate large datasets, IBL computations can be vectorized and parallelized, enhanced by techniques such as approximate k-nearest neighbors for efficient scalability. This capacity for handling large-scale data, dynamic adaptation to new situations without training, combined with a cognitively plausible approach to decision-making through approximate expectations, makes IBL particularly suitable for modeling evolving human behavior in large-scale multi-agent systems.

The PVA modeling approach integrates three levels of modeling within the IBL framework. Memory chunks encode features of the context, action, and outcome (including possibly reward) for a behavior. Behaviors such as mask wearing or social distancing are driven by prevalence of social norms in a context-free process. Attitude-driven reactions are context-sensitive to perceptions of the pandemic environment (especially whether the situation is getting better or worse) and driven by affective reactions such as "hope" and "fear." Decision-making under risk and uncertainty is driven by a combination of risk messaging and direct experience, and reflects sampling dynamics, e.g., risk aversion and confirmation bias. The three levels of reactive, affective, and reflective decision-making are combined in a single IBL model.

12.2.3 Health Behavior Change

The core ACT-R mechanisms provide a basis for explaining behavior change (Pirolli et al. 2022; Pirolli 2016a, b, 2018; Orr et al. 2024a, b). ACT-R is a dual-system theory (Kahneman 2011): First, a deliberative or controlled goal-striving process motivates and guides seminal attempts at behavior in the relevant contexts. Second, habit learning and strengthening processes form new habits through repeated practice, and habitual behaviors are executed without effortful, controlled, goal-striving.

Figures 12.3, 12.4, and 12.5 are simulations of synthetic ACT-R agents formalizing key social-psychological processes that are implemented in PVAs (Pirolli et al. 2022). Attitudes are assumed (Ajzen 1991) to be an *expectancy-value* assessment, such that an attitude toward a behavior is proportional to the strength of beliefs about outcomes and their values (utilities). Expectancy-value formulations are pervasive in the social and behavioral sciences, and psychological process theories of attitude change and behavior change. Attitudes can be perturbed or modified socially via experiences, messaging, exposure to social norms, and other social contexts. Figure 12.3 illustrates synthetic PVA *expectancy-value* learning and perturbation by messaging (Pirolli et al. 2022).

The concept of *self-efficacy* (Bandura 1998, 1977) is pervasive in the theoretical and applied literature on health behavior change (Michie et al. 2013; Samdal et al. 2017). Self-efficacy is the individual's belief in their capacity to execute behaviors necessary to produce specific performance attainments. As self-efficacy increases (e.g., through experience and mastery of behavior) then people are more likely to engage in the behavior when intended. Related to self-efficacy are psychological processes defined in Motivational Intensity Theory (Silvestrini et al. 2022). When self-efficacy is low relative to a difficult task, then it is likely to be judged as being more effortful, so high motivated effort can be compensatory—up to a threshold (Vancouver 2008; Vancouver et al. 2008). The ACT-R theory (Pirolli et al. 2022; Pirolli 2016a) assumes that self-efficacy and intentional effort are the result of memory processes. IBL mechanisms blend past experiences similar to the goal to judge self-efficacy and judge the level of effort required to achieve a behavior, and the individual will put in the effort if it is less than a threshold. If the behavior is performed, then a new instance memory is learned, and new successful experiences will tend to improve self-efficacy with repeated experience. Figure 12.4 depicts another synthetic simulation illustrating the dynamics of self-efficacy, intentional effort, and probability of performing the behavior.

Figure 12.5 shows how polarization evolves from small initial differences and subsequent reaction to messaging. Behavior-change decisions involve judgments about safe and risky options. As has been shown empirically, the probability of engaging with an NPI behavior can differ initially between groups having different initial beliefs and those differences can be amplified over time (Gollwitzer et al. 2020). Figure 12.5 illustrates this amplification process. The chunks in this PVA model reflect

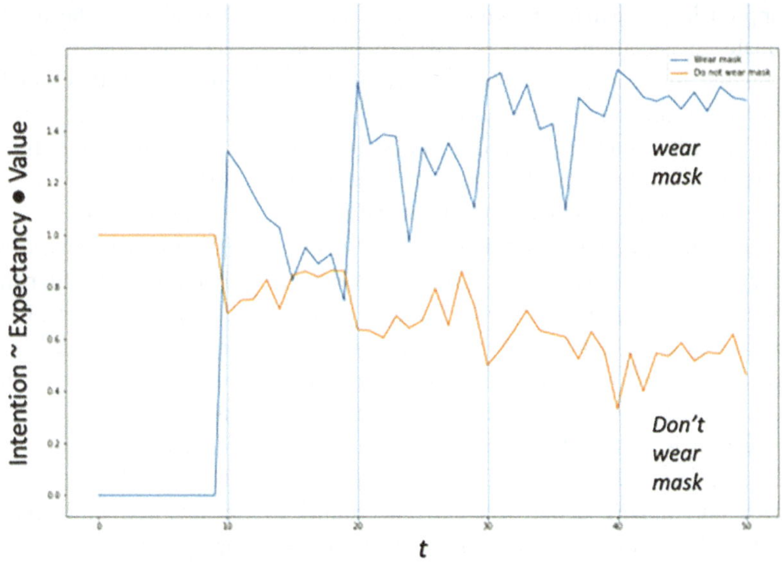

Fig. 12.3 Dynamics of subjective intentions for mask wearing or not wearing masks for a synthetic PVA with messaging promoting masks at time $t = 10, 20, 30, 40$ (purely synthetic simulation runs)

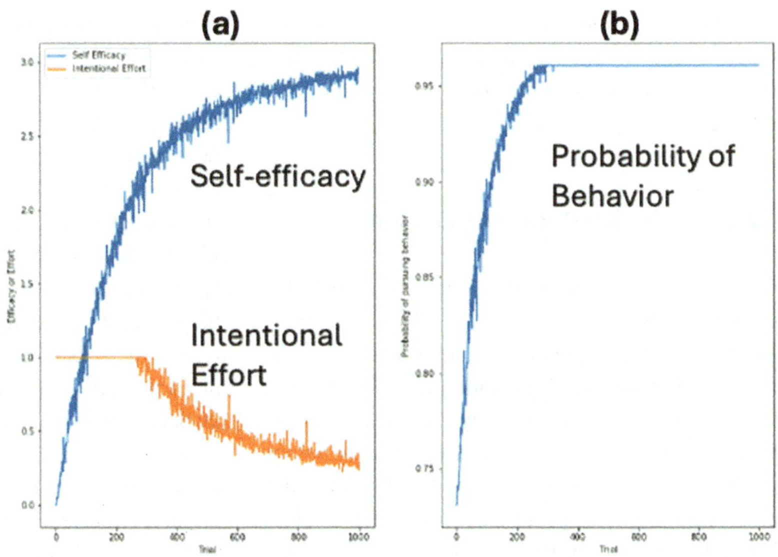

Fig. 12.4 **a** The dynamics of self-efficacy and intentional effort over time, **b** The impact on the probability of pursuing a behavior (over the same period for the same PVA)

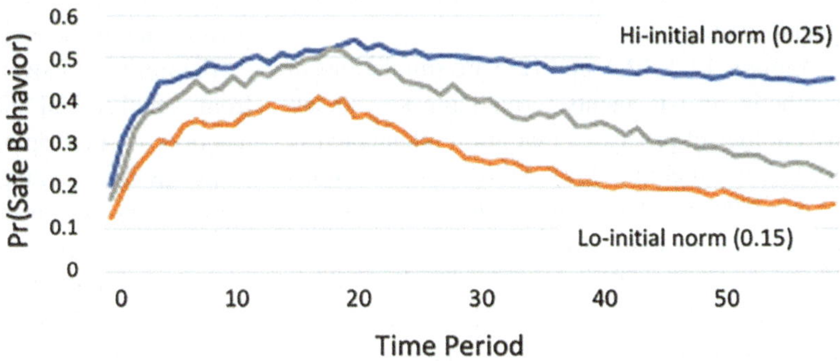

Fig. 12.5 Synthetic simulation of the evolving polarization of behavior adoption in response to messaging (Pirolli et al. 2022)

the relative payoff values of the two options of wearing or not wearing a mask, including both outcomes of the riskier option (basically usual, unencumbered life vs. illness, and potentially death) as well as the single outcome of the safe option (the annoyance of wearing a mask). The relative memory activations of those options reflect the initial messaging propagated in social and mass media, gradually supplemented by personal experience. A key characteristic of this process of decision-making under risk is that later decisions are influenced by the sampling from memory of options that occurred early in the process, itself driven by the initial presentation of the options (Lebiere et al. 2007; Erev et al. 2010). An important consequence is to amplify initial differences.

12.2.4 Using Social Media to Infer Stances and Attitudes

In our modeling, we supplement a wide variety of public data (described below) with novel data induced from analysis of Twitter/X. We (Pirolli et al. 2021) have developed two kinds of stance detection techniques: coarse-grained and fine-grained. These can be mapped to attitude and norm representations in PVA models that vary by region and over time. They can also be analyzed independently (Pirolli et al. 2021; Uyheng et al. 2023; Mather et al. 2021) as responses to unfolding events (local or national news; mandates) and as predictors of real-world behavior (i.e., mask wearing).

12.2.4.1 Coarse-Grained Stance Detection from Twitter/X

For coarse-grained stances (Pirolli et al. 2021; Uyheng et al. 2023), tweets were selected based on COVID-19-related keywords in multiple languages for specified periods using the Twitter/X API. Machine learning algorithms were then used to label these tweets by location country, and then again within country then by state. Machine learning algorithms were then used to add additional labels classifying the source as a bot, a news agency, or a government actor. Additional classifiers can be used, when feasible, to identify city location.

Coarse-grained stance analysis is partly based on identifying a set of pro-facemask and anti-facemask hashtags. Then using a label propagation algorithm (Kumar and Carley 2019) available in ORA-PRO (Carley 2017) each tweet is labeled, and each tweeter is labeled being either Contra − 1 (anti-facemask) or Pro + 1 (pro-facemask), along with the confidence in this label. Tweets not containing hashtags are neutral. Using a threshold, stance x confidence values for tweets are classified as Pro- or Con-mask. For each individual, stance is determined as the average of the stance x confidence across the individual's tweets. The time series of regional weighted volumes and proportions of these pro- and con-mask stance data can be used as variables input to the Regional PVA models.

Overall, there are more Pro-facemask tweets and individuals than there are Con-facemask tweets and individuals in the data. Detailed analysis of dynamics of expressed Pro and Con stances, and the relations between online stances and observed behavior (mask wearing) are explored in detail elsewhere (Uyheng et al. 2023). For the purposes of PVA modeling, coarse-grained stances provide empirical indicators—a kind of instantaneous informal attitude survey—about the population expressing themselves on Twitter. The daily volumes of Pro and Con tweets appear to show location-specific spikes that are often correlated with a local event. For instance, the early spike in the Sacramento data (Fig. 12.6) aligns with the two-year anniversary of the police shooting of Stephon Clark in Sacramento, the resulting marches, and tweets about "no masks" at the marches. The tweet volumes also show spikes that are correlated with national and state events. Examples of Pro and Con tweet dynamics are illustrated in Fig. 12.6 for four major cities in California for 2020.

12.2.4.2 Fine-Grained Stances

We also apply a Natural Language Processing (NLP) pipeline for detecting and analyzing fine-grained stances (Mather et al. 2021) for rapid discovery of contrasting views in conversations.

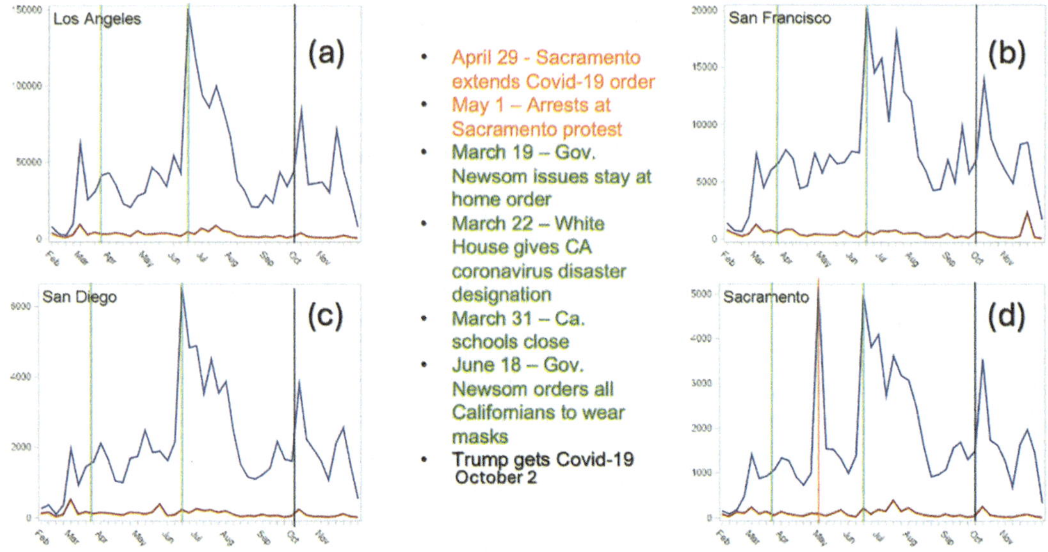

Fig. 12.6 California metro Pro/Anti mask wearing Tweet volumes (blue, top curve = Pro; red, lower curve = Anti), by week in 2020, and national, state, local events, in **a** Los Angeles, **b** San Francisco, **c** San Diego, and **d** Sacramento

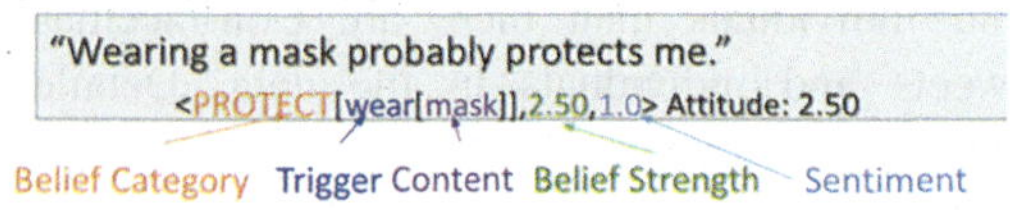

Fig. 12.7 Fine-grained stance analysis of a tweet

In online settings, different groups champion their own beliefs while attacking others who hold opposing beliefs, causing extreme polarization. For example, messages composed as a direct attack against anti-mask groups convey beliefs (and associated belief strengths) that masks do provide protection, that social distancing does not restrict freedom, and that it's important to show solidarity. A fine-grained stance (Fig. 12.7) is defined as a proposition, such as wear (a mask), coupled with a task-specific belief category such as PROTECT (to capture a belief such as "wearing masks provides protection") or RESTRICT (to capture a belief such as "mask mandates limit freedom."). These are taken together with a sentiment toward this proposition, ultimately yielding an overall attitude toward the topic at hand (e.g., "mask wearing"). This pipeline can be applied to the same geo-located Twitter data described above. Figure 12.8 illustrates a mapping of fine-grained stance representations of attitudes and beliefs induced from Twitter (onto chunk representations used in the ACT-R cognitive model (Pirolli et al. 2021)).

12.3 Data

PVA Data Lake Data from a variety of sources have been collected and merged for development of PVAs (Pirolli et al. 2024). The merged data contain predictor variables for *static* regional variables and *time-series* regional variables (U.S. State and County) capturing day-by-day COVID-19 statistics and behavior rates (mask wearing, social distancing, mobility, and vaccination). Some of the variables are subjected to dimensionality reduction (e.g., weather data) to induce a smaller set of latent features, and features are downselected based on standard regression techniques (Fig. 12.9). The sources included data aggregated by the Johns Hopkins Coronavirus Resource Center (n.d.) (Killeen et al. 2020), regional Big 5 personality statistics (Ebert et al. 2022), and a fused data set used in a study of predictors of social distancing behavior (Gollwitzer et al. 2020). For each county and/or state, we combined survey data from the COVIDCast (Reinhart et al. 2021) and the COVID States

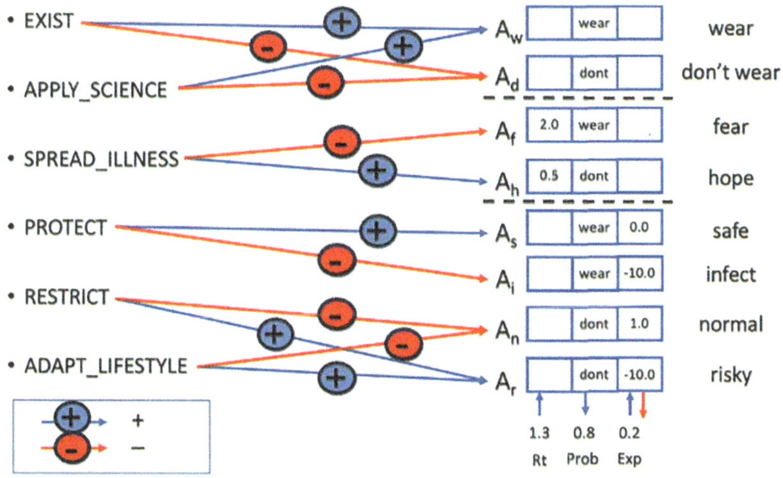

Fig. 12.8 Mapping of belief category attitudes from Stance Detection (left) to chunk activations and relative chunk frequencies in ACT-R (right) (Pirolli et al. 2021)

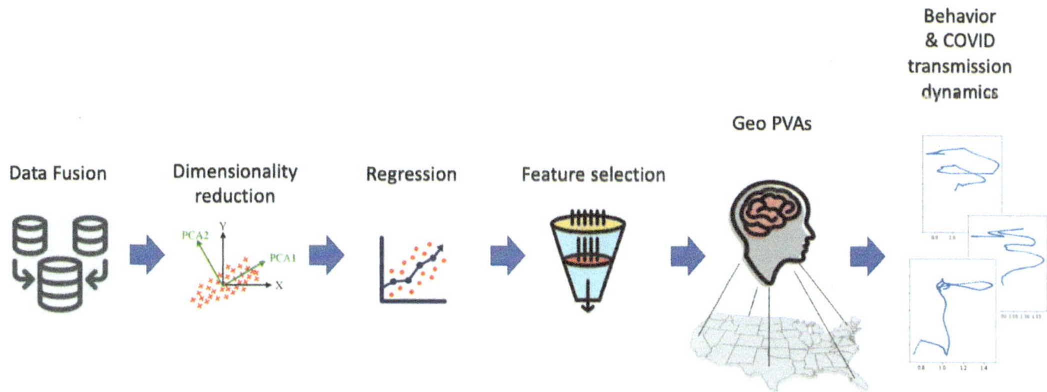

Fig. 12.9 Overview of the data analysis and modeling pipeline

(Lazer et al. 2021) projects. R_t data at the state and county levels were downloaded from covidestim.org, a project of the Yale School of Public Health. Mask wearing data at the state level was downloaded from the COVID States website. The time-series data is for every day for the pre-vaccination rollout period of 4/24/2020 to 3/31/2021. The mask wearing data compiled by the COVID States project was captured in survey waves deployed every three weeks. We have also added daily statistics of COVID-19 news topics from the Stanford Cable TV News Analyzer, and analyses of Twitter discussions of mask wearing (Uyheng et al. 2023).

12.4 Regional PVAs for Mask Wearing in 50 States Over Three Waves of COVID-19

One innovative aspect of the PVA approach to modeling population-scale COVID-19 behavior (e.g., mask wearing, social distancing) is the fusion of cognitive architecture mechanisms with cognitive content learned from big data and large-scale natural language processing (Pirolli et al. 2021). This provides a way of generating decision-making and behaviors in PVAs that vary due to differences in cognitive traits and cognitive states. These Regional PVAs offer a unique opportunity

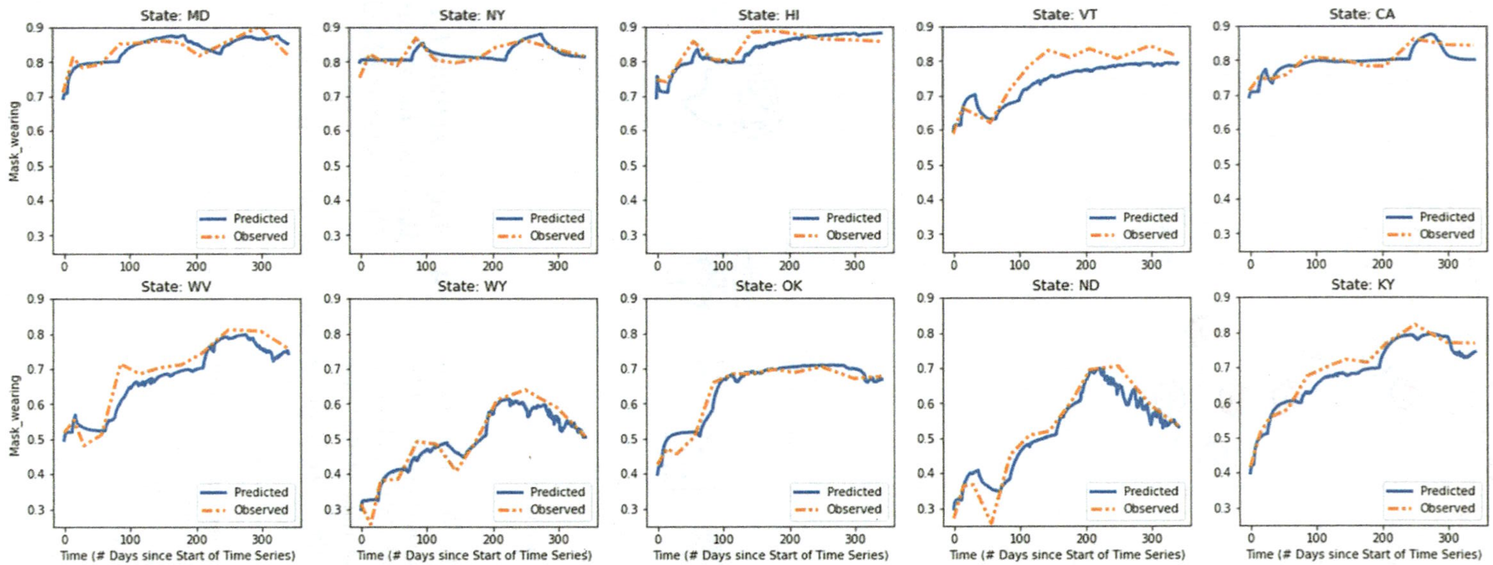

Fig. 12.10 Observed proportion of mask wearing and proportion predicted by R-PVA for 10 U.S. states over the first three waves of COVID-19. The top row shows the five states with the lowest proportion voting for Trump in 2016 and the bottom row shows the five states with the highest proportion voting for Republican Presidential Candidate Trump in 2016, which is the dominant predictor of state-wise differences

to explore demographic or psychographic factors related to behavior, using a variety of methods.

Leveraging the COVID-19 data repositories described above, we have created Regional PVAs (R-PVAs) in an ABM framework to simulate the behavior of regional U.S. populations during the pre-vaccination phase of COVID-19. Figure 12.10 (Pirolli et al. 2024) is from R-PVAs simulations for behaviors in the 50 states of the U.S. across the first three waves of COVID-19. Most of the details of this model are presented elsewhere (Pirolli et al. 2024), but the results presented here are enhanced by inclusion of inputs from mass media analysis and the social media coarse-grained and fine-grained stances. The R-PVAs predict mask wearing based on selected predictor variables and the R_t time lag suggested by the Granger causality analysis. The data were aggregated at the state level and the time series covered the pre-vaccine rollout period of three waves of COVID-19 over 4/24/2020 to 3/31/2021. For the full 50-state, three-wave, dataset we achieve very good model fits [$R^2 = 0.93$]. Moreover, we capture signature phenomena that appear in the oscillating dynamics of mask wearing and the effective reproduction number (e.g., Fig. 12.11). Figure 12.11 is similar to the phase space diagrams of Fig. 12.1, but includes a notable "learning effect" in that mask wearing shows increased response levels with each new wave, which is the result of self-efficacy mechanisms in our models.

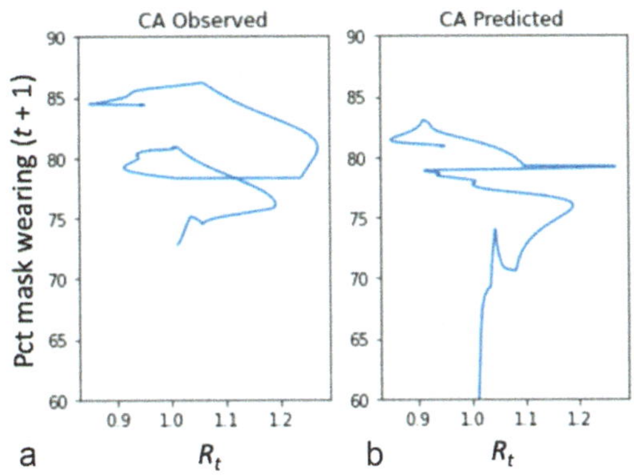

Fig. 12.11 **a** Observed and **b** PVA-predicted relationship between effective transmission number (R_t) at time t days and percent mask wearing at time $t + 7$ days over the first three waves of COVID-19. The general phenomena is an oscillation around $R_t = 1$ combined with a learning effect toward higher masking over three waves of COVID-19

12.5 Conclusions and Insights

In the infectious disease policy space, the use of mechanistic or mathematical epidemiological models is supported by technological and scientific advances in simulation, mathematics, and statistics, as well as integration with synthetic or dynamic data structures (social networks, administrative data, and industry data) (Mooij et al. 2023; Jalali et al. 2021). The latest addition to this arsenal is artificial intelligence (Arik et al. 2021; Brownstein et al. 2023). The results are sophisticated tools that provide current state estimates and future-looking scenario estimates related to a host of policy questions along standard epidemiological outcomes (e.g., case counts, hospitalizations, case-fatalities, etc.). This kind of policy modeling sometimes attempts to represent features of stylized human or economic behavior into mechanistic epidemiological models (Weitz et al. 2020; Manfredi and d'Onofrio 2013; Fenichel et al. 2011; Espinoza et al. 2021; Verelst et al. 2016), but such attempts are not derived from realistic, validated models of human behavior or measures of human beliefs, attitudes, intentions, and behaviors, thus narrowing their application potential. This lack of empirical grounding and theoretical constraints is why, when policy makers are confronted with the mitigation of human behavior directly as an outcome (e.g., when concerned with population-level behavior change such as overcoming vaccine hesitancy), policy decisions rely on survey instruments for surveillance and past/current state assessment of beliefs, attitudes, intentions, and behavior. The limitation is that scenario modeling and forecasts are notably absent and rightly so as the technological and scientific development has yet to happen to afford such capability.

Advances in computational cognitive modeling can serve as a foundation for predictive and causal-explanatory models of human behavior in response to pandemics. Such models are needed for improved infectious disease control. They can help shape the course of human behavior response in more precise and less burdensome ways and aid in the development of technologies for scenario modeling of behavioral response to mitigation efforts. Our approach of developing models based on a unified theory of human cognition promises generalizability and deeper understanding of at-scale interactions between human behavior and pathogens and we recommend the approach of modeling the intricacies of the human mind as a foundation for future success in mitigating pandemics.

Acknowledgements The writing of this chapter has been supported by the U.S. National Science Foundation under Grant #2200112. Portions of the research reported herein were supported by National Science Foundation Grant #2033390 and the Office of the Director of National Intelligence (ODNI), Intelligence Advanced Research Projects Activity (IARPA), via 2020-20092500001.

References

Ajzen I (1991) The theory of planned behavior. Organ Behav Hum Decis Process 50:179–211. https://doi.org/10.1016/0749-5978(91)90020-T

Anderson JR et al (1990) Cognitive modelling and intelligent tutoring. Artif Intell 42:7–49. https://doi.org/10.1016/0004-3702(90)90093-F

Anderson JR et al (2004) An integrated theory of mind. Psychol Rev 11(4):1036–1060. https://doi.org/10.1037/0033-295X.111.4.1036

Anderson JR (2007) How can the human mind occur in the physical universe? Oxford University Press, Oxford, UK. https://doi.org/10.1093/acprof:oso/9780195324259.001.0001

Arik SO, et al (2021) A prospective evaluation of AI-augmented epidemiology to forecast COVID-19 in the U.S.A. and Japan. NPJ Digit Med **4**(1):146. https://doi.org/10.1038/s41746-021-00511-7

Aviral C et al (2022) Accuracy of US CDC COVID-19 forecasting models, p 22274097. https://doi.org/10.3389/fpubh.2024.1359368

Bandura A (1977) Self-efficacy: toward a unifying theory of behavioral change. Psychol Rev 82(2):191–215. https://doi.org/10.1037/0033-295X.84.2.191

Bandura A (1998) Self-efficacy: the exercise of control. W.H. Freeman, New York

Brossard D, et al (2020) National academies of sciences, engineering, and medicine: encouraging adoption of protective behaviors to mitigate the spread of COVID-19: strategies for behavior change. In: E. National Academies of Sciences, and Medicine 2020 (ed). The National Academies Press, Washington, DC. https://doi.org/10.17226/25881

Brownstein JS et al (2023) Advances in artificial intelligence for infectious-disease surveillance. N Engl J Med 388(17):1597–1607. https://doi.org/10.1056/NEJMra2119215

Carley KM (2017) ORA: a toolkit for dynamic network analysis and visualization. In: Alhajj R, Rokne J (eds) Encyclopedia of social network analysis and mining. Springer. https://doi.org/10.1007/978-1-4939-7131-2_309

Christakis NA (2020) Apollo's arrow. Little, Brown Spark, New York, NY

Corbett AT, Anderson JR, O'Brien AT (1995) Student modelling in the ACT Programming Tutor. In: Nichols PD, Chipman SF, Brennan RL (eds) Cognitively diagnostic assessment. Lawrence Erlbaum Associates, Hillsdale, NJ, pp 19–41

Cramer EY et al (2022) Evaluation of individual and ensemble probabilistic forecasts of COVID-19 mortality in the United States. Proc Natl Acad Sci U S A 119(15):e2113561119. https://doi.org/10.1073/pnas.2113561119

de Mooij J et al (2023) A framework for modeling human behavior in large-scale agent-based epidemic simulations. SIMULATION 99(12):1183–1211. https://doi.org/10.1177/00375497231118489

Ebert T, Gebauer JE, Brenner T, Bleidorn W, Gosling SD, Potter J, Rentfrow PJ (2022) Are regional differences in psychological characteristics and their correlates robust? Applying spatial-analysis techniques to examine regional variation in personality. Perspect Psychol Sci 17(2):407–441. https://doi.org/10.1177/1745691621998326. Epub 2021 Oct 26 PMID: 34699736

Erev I et al (2010) A choice prediction competition: Choices from experience and from description. J Behav Decis Mak 23:15–47. https://doi.org/10.1002/bdm.683

Espinoza B et al (2021) Asymptomatic individuals can increase the final epidemic size under adaptive human behavior. Sci Rep 11(1):19744. https://doi.org/10.1038/s41598-021-98999-2

Fenichel EP et al (2011) Adaptive human behavior in epidemiological models. Proc Natl Acad Sci 108(15):6306–6311. https://doi.org/10.1073/pnas.1011250108

Ferguson NM, et al (2020) Impact of non-pharmaceutical interventions (NPIs) to reduce COVID- 19 mortality and healthcare demand. https://doi.org/10.25561/77482

Flaxman S, et al (2020) Imperial College Report 13: estimating the number of infections and the impact of non-pharmaceutical interventions on COVID-19 in 11 European countries. Imperial College, London. https://doi.org/10.25561/77731

Gollwitzer A et al (2020) Partisan differences in physical distancing are linked to health outcomes during the COVID-19 pandemic. Nat Hum Behav 4(11):1186–1197. https://doi.org/10.1038/s41562-020-00977-7

Gonzalez C, Lerch JF, Lebiere C (2003) Instance-based learning in dynamic decision making. Cogn Sci 27:591–635. https://doi.org/10.1207/s15516709cog2704_2

Ianni A, Rossi N (2021) SIR-PID: a proportional–integral–derivative controller for COVID-19 outbreak containment. Physics 3(3):459–472. https://doi.org/10.3390/physics3030031

Ioannidis JPA, Cripps S, Tanner MA (2022) Forecasting for COVID-19 has failed. Int J Forecast 38(2):423–438. https://doi.org/10.1016/j.ijforecast.2020.08.004

Jalali MS et al (2021) Evolution and reproducibility of simulation modeling in epidemiology and health policy over half a century. Epidemiol Rev 43(1):166–175. https://doi.org/10.1093/epirev/mxab006

Jones JH, et al (2021) Summary report: NSF predictive intelligence for pandemic prevention (PIPP) "Human" workshop. Human PIPP

Kahneman D (2011) Thinking, fast and slow. Farar, Straus and Giroux, New York

Killeen BD, et al (2020) A county-level dataset for informing the united states' response to COVID-19. arXiv:2004.00756. https://doi.org/10.48550/arXiv.2004.00756

Koedinger KR et al (1997) Intelligent tutoring goes to the big city. Int J Intell Educ 8:30–43

Kumar S, Carley KM (2019) Tree LSTMs with convolution units to predict stance and rumor veracity in social media conversations. In: Proceedings of the 57th conference of the association for computational linguistics, pp 5047–5058. https://doi.org/10.18653/v1/P19-1498

Lazer D, et al (2021) The COVID states Project #26: trajectory of COVID-19-related behaviors. https://doi.org/10.31219/osf.io/37jbe

Lebiere C, Gonzalez C, Martin M (2007) Instance-based decision making model of repeated binary choice. In: 8th International conference on cognitive modeling. Ann Arbor, Michigan, USA. Ann Arbor, MI

Lebiere C, et al (2016) Cognitive models of prediction as decision aids. In: Reitter D, Ritter FE (eds) Proceedings of 2016 international conference on cognitive modeling (ICCM-2016). Penn State, University Park, PA

Leech G et al (2022) Mask wearing in community settings reduces SARS-CoV-2 transmission. Proc Natl Acad Sci U S A 119(23):e2119266119. https://doi.org/10.1073/pnas.2119266119

Manfredi P, d'Onofrio A (2013) Modeling the interplay between human behavior and the spread of infectious disease. Springer, New York. https://doi.org/10.1007/978-1-4614-5474-8

Mather B et al (2021) A general framework for domain-specialization of stance detection. In: The international FLAIRS conference proceedings. https://doi.org/10.32473/flairs.v34i1.128457

Michie S et al (2013) The behavior change technique taxonomy (v1) of 93 hierarchically clustered techniques: building an international consensus for the reporting of behavior change interventions. Ann Behav Med 46(1):81–95. https://doi.org/10.1007/s12160-013-9486-6

Miller GA (1956) The magical number seven plus or minus two: some limits on our capacity for processing information. Psychol Rev 63:81–97

Mitsopoulos K, et al (2023) Psychologically-valid generative agents: A novel approach to agent-based modeling in the social sciences. In: 2023 AAAI fall symposium. AAAI Press, Arlington, VA. https://doi.org/10.1609/aaaiss.v2i1.27698

Orr M, et al (2023) A 10-year prospectus for mathematical epidemiology. Front Psychol 14. https://doi.org/10.3389/fpsyg.2023.986289

Orr M, et al (2024a) A comparison of two memory models of attitude retrieval. In: Proceedings of the annual meeting of the cognitive science society, vol 46, Retrieved from https://escholarship.org/uc/item/78k2q3pz

Orr M, et al (2024b) A comparison of frequency effects in two attitude retrieval models. In: Proceedings of the international conference on cognitive modeling proceedings

Pavlik PI, Anderson JR (2005) Practice and forgetting effects on vocabulary memory: An activation-based model of the spacing effect. Cogn Sci 29:559–586. https://doi.org/10.1207/s15516709cog0000_14

Pavlik PI, Anderson JR (2008) Using a model to compute the optimal schedule of practice. J Exp Psychol Appl 14(2):101–117. https://doi.org/10.1037/1076-898X.14.2.101

Pirolli P (2016a) A computational cognitive model of self-efficacy and daily adherence in mHealth. Transl Behav Med 6(4):1–13. https://doi.org/10.1007/s13142-016-0391-y

Pirolli P (2016b) From good intentions to healthy habits: towards integrated computational models of goal striving and habit formation. Conf Proc IEEE Eng Med Biol Soc 2016:181–185. https://doi.org/10.1109/EMBC.2016.7590670

Pirolli P (1999) Cognitive engineering models and cognitive architectures in human-computer interaction. In: Durso FT et al (eds) Handbook of applied cognition. Wiley, West Sussex, England, pp 441–477

Pirolli P, Lebiere C, Orr M (2022) A computational cognitive model of behaviors and decisions that modulate pandemic transmission: expectancy-value, attitudes, self-efficacy, and motivational intensity. Front Psychol 13:981983. https://doi.org/10.3389/fpsyg.2022.981983

Pirolli P (2007) Information foraging: a theory of adaptive interaction with information. Oxford University Press, New York. https://doi.org/10.1093/acprof:oso/9780195173321.001.0001

Pirolli P, et al (2018) Scaffolding the mastery of healthy behaviors with fittle+ systems: evidence-based interventions and theory. Human–computer interaction,

pp 1–34. https://doi.org/10.1080/07370024.2018.1512414
Pirolli P (2021) Peter Pirolli: improving computational epidemiology with higher fidelity models of human behavior. Columbia Academic Commons, Academic Unit: Northeast Big Data Innovation Hub, Series: COVID-19 Research Lightning Talks. https://doi.org/10.7916/5cj7-vz72
Pirolli P (2023) Peter Pirolli: computational theory of the co-evolution of pandemics, (mis)information, and human mindsets and behavior. Columbia Academic Commons, Academic Unit: Northeast Big Data Innovation Hub, Series: COVID-19 Research Lightning Talks. https://doi.org/10.7916/kqt2-xn68
Pirolli P, et al (2020) Cognitive modeling for computational epidemiology. In: Thomson R, et al (eds). 2020 International conference on social computing, behavioral-cultural modeling & prediction and behavior representation in modeling and simulation (SPB-BRIMS 2020). Springer, Washington, DC
Pirolli P, et al (2021) Mining online social media to drive psychologically valid agent models of regional covid-19 mask wearing, in social, cultural, and behavioral modeling. Thomson R et al (eds). Springer, Cham, pp 46–56. https://doi.org/10.1007/978-3-030-80387-2_5
Pirolli P, et al (2024) Computational modeling of regional dynamics of pandemic behavior using psychologically valid agents PREPRINT. R. Square (ed). https://doi.org/10.21203/rs.3.rs-4189570/v1
Redlin M (2022) Differences in NPI strategies against COVID-19. J Regul Econ 62:1–23. https://doi.org/10.1007/s11149-022-09452-9
Reinhart A et al (2021) An open repository of real-time COVID-19 indicators. Proc Natl Acad Sci 118(51):e2111452118. https://doi.org/10.1073/pnas.2111452118
Johns Hopkins Coronavirus Resource Center (JHUa) (n.d). Medicine
Samdal GB et al (2017) Effective behaviour change techniques for physical activity and healthy eating in overweight and obese adults; systematic review and meta-regression analyses. Int J Behav Nutr Phys Act 14(1):42. https://doi.org/10.1186/s12966-017-0494-y
Silvestrini N et al (2022) An integrative effort: Bridging motivational intensity theory and recent neurocomputational and neuronal models of effort and control allocation. Psychol Rev. https://doi.org/10.1037/rev0000372
Simon HA (1974) How big is a chunk? Science 183:482–488. https://doi.org/10.1126/science.183.4124.482
Spiliopoulos L (2022) On the effectiveness of COVID-19 restrictions and lockdowns: Pan metron ariston. BMC Public Health 22(1):1842. https://doi.org/10.1186/s12889-022-14177-7
Uyheng J, Robertson DC, Carley KM (2023) Bridging online and offline dynamics of the face mask infodemic. BMC Digital Health 1(1):27. https://doi.org/10.1186/s44247-023-00026-z
Vancouver JB (2008) Integrating self-regulation theories of work motivation into a dynamic process theory. Hum Resour Manag Rev 18(1):1–18. https://doi.org/10.1016/j.hrmr.2008.02.001
Vancouver JB, More KM, Yoder RJ (2008) Self-efficacy and resource allocation: support for a nonmonotonic, discontinuous, model. J Appl Psychol 93(1):35–47. https://doi.org/10.1037/0021-9010.93.1.35
Vapnik VN (1998) Statistical learning theory. Wiley, Hoboken, NJ
Vardavas R et al (2021) Modeling infectious behaviors: the need to account for behavioral adaptation in COVID-19 models. Policy Complex Sys 7(1):21–32
Verelst F, Willem L, Beutels P (2016) Behavioural change models for infectious disease transmission: a systematic review (2010–2015). J R Soc Interface. 13(125). https://doi.org/10.1098/rsif.2016.0820
Weitz JS et al (2020) Awareness-driven behavior changes can shift the shape of epidemics away from peaks and toward plateaus, shoulders, and oscillations. Proc Natl Acad Sci 117(51):32764. https://doi.org/10.1073/pnas.2009911117
West R et al (2020) Applying principles of behaviour change to reduce SARS-CoV-2 transmission. Nat Hum Behav 4(5):451–459. https://doi.org/10.1038/s41562-020-0887-9
Yang C, Lapp L, Tebbutt SJ (2023) Overcoming COVID-19 vaccine hesitancy hurdles. Lancet 402(10408):1129–1130. https://doi.org/10.1016/S0140-6736(23)01425-3

Peter Pirolli is a Senior Research Scientist at the Institute for Human and Machine Cognition (IHMC). His research combines cognitive science, artificial intelligence, and human–computer interaction, with applications in human-AI teaming, mobile health, sensemaking, information foraging, and measuring and enhancing cognitive capabilities. His recent work is aimed at computational theories and models of information flows, human behavior, and the transmission and evolution of pandemic pathogens. He has a doctorate in cognitive psychology from Carnegie Mellon University in 1985, and a BSc in psychology and anthropology from Trent University. He has been elected as a Fellow of the National Academy of Inventors, the American Association for the Advancement of Science, the American Psychological Association (Div 3 and Div 21), the Association for Psychological Science, the National Academy of Education, and the ACM Computer–Human Interaction Academy.

Archna Bhatia is a Research Scientist at the Institute for Human and Machine Cognition (IHMC), where her work integrates artificial intelligence with speech and natural language processing to address real-world challenges in health and other domains. Her research includes developing systems, such as mobile applications for word retrieval; technologies for detecting and monitoring stress and its impact on individuals' performance; technologies for tracking Amyotrophic Lateral Sclerosis (ALS) progression, and methods for predicting human behaviors like physical exercise through speech and language analysis and information extraction. She contributes to cognitive modeling and cybersecurity by leveraging natural language processing to generate outputs that inform these areas. Her work also aims to advance semantic parsing and natural language understanding. Dr. Bhatia earned her PhD in Linguistics at the University of Illinois at Urbana-Champaign, followed by postdoctoral fellowships at the University of Colorado at Boulder and the Language Technologies Institute at Carnegie Mellon University. She contributed significantly to developing the foundation of the ask detection system at IHMC collaborating with Dr. Bonnie Dorr and Brodie Mather.

Kathleen M. Carley is a Professor in the College of Computer Science at Carnegie Mellon University (CMU), and the Director of the Center for Informed Democracy and Social CyberSecurity (IDeaS). She received her PhD in Sociology from Harvard University, SB degrees from Massachusetts Institute of Technology, and an HD from the University of Zurich. She received the USGIF (United States Geospatial Intelligence Foundation) Academic Award at GEOINT 2018 for her work on geo-spatially enabled dynamic-network analytics, the Lifetime Achievement Award from the Sociology and Computers Section of the American Sociological Association (ASA) in 2001, and the Simmel Award for advances in social networks from the International Network for Social Network Analysis (INSNA) in 2011. Her research combines cognitive science, sociology, and computer science to address complex social issues. Her pioneering research led to the areas of computational social science, dynamic-network analysis, and social cybersecurity, and to new theories and methods related to disinformation, hate speech, extremism, counter-terrorism, insider threats, health analytics, social media and elections, and COVID-19. She has developed technologies used worldwide, including a computational linguistic tool for sentiment, social network, semantic networks, and cues extraction (NetMapper), and an agent-based dynamic-network model of information diffusion (Construct). Her Organization Risk Analyzer (ORA) system supports geo-spatial, high-dimensional and dynamic-network analytics and visualization.

Bonnie J. Dorr is a Professor in the Department of Computer and Information Science and Engineering at the University of Florida, where she directs the Natural Language Processing & Culture Laboratory (NLP&C). She is also an affiliate of the Institute for Human and Machine Cognition (IHMC) and the Florida Institute for National Security. She is former program manager of the Defense Advanced Research Projects Agency (DARPA) Human Language Technology programs and Professor Emerita at University of Maryland. Dorr is a recognized leader in artificial intelligence and natural language, specializing in machine translation and cyber-aware language processing. Her research explores neural-symbolic approaches for accuracy, robustness, and explainable outputs. Applications include cyber-event extraction for detecting and mitigating attacks, detecting influence campaigns, and building interpretable models. She is a NSF Presidential Early Career Awards for Scientists and Engineers (PECASE) Fellow, a Sloan Fellow, and a Fellow of the Association for the Advancement of Artificial Intelligence (AAAI), Association for Computational Linguistics (ACL), and Association for Computing Machinery (ACM).

Christian Lebiere is a Research Professor in the Psychology Department at Carnegie Mellon University (CMU). He received his PhD from the CMU School of Computer Science where he studied connectionist models and was the co-developer of the Cascade-Correlation neural network deep learning algorithm. Since 1991, he has worked on the development of the ACT-R cognitive architecture and was co-author of the 1998 book *The Atomic Components of Thought*. Most recently he has been involved with the specification of the Common Model of Cognition, a community-wide effort to consolidate and formalize the scientific progress resulting from the 50-year research program in cognitive architectures. He is a founding member of the Biologically Inspired Cognitive Architectures Society, of the International Conference on Cognitive Modeling, and of the *Journal of Artificial General Intelligence*. His main research interests are cognitive architectures and their applications to psychology, artificial intelligence, human–computer interaction, decision-making, network science, cognitive robotics and human–machine teaming.

Brodie Mather is a Research Scientist at the Institute for Human and Machine Cognition (IHMC) doing interdisciplinary research spanning areas such as artificial intelligence, cybersecurity, cognitive architectures, social psychology, natural language processing, and more. Brodie's research also explores inherently interpretable AI which provides explainable outputs for better understanding and trust. Brodie earned his PhD in Intelligent Systems at the University of West Florida in September 2023. As co-inventor of ask and stance detection systems at IHMC with Dr. Bonnie Dorr, he has designed tools that aid in individual understanding through the processing of textual messages. The ask detection system identifies "asks" that a social engineer makes of a potential victim in order to actively defend against attacks. Stance detection allows for the automatic extraction of beliefs that individuals hold and their attitudes toward those beliefs. His work in these areas has also produced tools for understanding influence campaign indicators. Brodie continues to pursue similar research as well as the intersection of these areas.

Konstantinos Mitsopoulos is a Research Scientist at the Institute for Human and Machine Cognition (IHMC) where his work integrates cognitive science, psychology, and machine learning to understand human decision-making, in order to identify patterns that can improve artificial intelligence systems. He earned a BSc in Physics from the University of Athens, an MSc in Machine Learning from University College London, and a PhD in Computational Cognitive Science from Birkbeck, University of London. Prior to his current role, he was a Project Scientist at Carnegie Mellon University's Robotics Institute and a postdoctoral researcher in Carnegie Mellon's Psychology Department, where he worked in developing autonomous agents for effective human–machine collaboration and cognitive architectures for explaining deep reinforcement learning models and computational epidemiology.

Don Morrison is a Senior Research Programmer both in Carnegie Mellon University's (CMU) Psychology Department and in its Social and Decision Sciences Department, focusing on work supporting research involving cognitive models and related experimentation. He has connected models written using the ACT-R cognitive architecture to simulations in a variety of domains, including anesthesia, robotics, cybersecurity, intelligence analysis, search and rescue, and network science activities. He is the author both of PyACTUp and of PyIBL, ACT-R inspired frameworks used for a variety of modeling experiments both at CMU and other institutions. He previously worked in CMU's Human–Computer Interaction Institute developing CogTool, a predictive user interface prototyping tool built around an ACT-R cognitive model of the user. Prior to joining CMU he worked on and managed a variety of commercial software and hardware projects, including Lisp Language implementation, computer algebra, database tools, video editing, workflow automation, and financial simulations. He earned his Bachelor of Science (SB) in Physics from the Massachusetts Institute of Technology and his MS in Computer Science from the University of Utah.

Mark Orr joined the Institute for Human and Machine Cognition (IHMC) as a Senior Research Scientist in July 2024. He is a contributing faculty member to the joint University of West Florida - IHMC Intelligent Systems and Robotics PhD program. He previously was a professor at the Biocomplexity Institute at the University of Virginia. Orr was originally trained as a cognitive psychologist at the University of Illinois at Chicago. Orr received augmentation to this training with postdoctoral fellowships in computational modeling (Carnegie Mellon University), neuroscience (Albert Einstein College of Medicine), and epidemiology/complex systems (Columbia University). Over the past decade, he has become heavily involved in understanding dynamic processes and drivers of risky behavior and decision-making, primarily in a public health context, at the scale of the individual and populations for various sponsors, including the Defense Advanced Research Projects Agency (DARPA), Intelligence Advanced Research Projects Activity (IARPA), National Science Foundation (NSF), National Institutes of Health (NIH) and Army Research Institute (ARI).

Tomek Strzalkowski is the Tetherless World Constellation Chair, Professor of Cognitive Science and Computer Science (http://tw.rpi.edu/) at Rensselaer Polytechnic Institute (RPI), where he directs the Language and Cognition Artificial Intelligence Lab (lacailab.cogsci.rpi.edu). He is also a visiting scientist at the Institute for Human and Machine Cognition (IHMC). Prior to joining RPI in 2019, he was Director of the Institute for Informatics, Logics, and Security Studies (ILS) and Professor of Computer Science at SUNY Albany. Between 1994 and 2000 he was a Principal Scientist at GE R&D (General Electric Research & Development) in Niskayuna, New York, heading the Natural Language Group. Before coming to GE, he was research faculty at the Courant Institute of New York University. Professor Strzalkowski has done research in computational linguistics and sociolinguistics, information retrieval, question-answering, human–computer dialog, serious games, social media analytics, formal semantics, and reversible grammars. He serves on the Editorial Board of the *Journal of Natural Language Engineering* and is a fellow of the Association for Computational Linguistics.

Choh Man Teng is a Research Scientist at the Institute for Human and Machine Cognition (IHMC). She received a Ph.D. in computer science from the University of Rochester. Her research interests include natural language understanding, uncertain reasoning, machine learning and knowledge representation and reasoning. Some of her recent work is focused on computational cognitive modeling of human behavior and information dissemination.

13 RAPID: Media Exposure, Objective Knowledge, Risk Perceptions, and Risk Management Preferences of Americans Regarding the Novel Coronavirus Outbreak

Branden B. Johnson

Abstract

A six-wave longitudinal panel survey of a United States (U.S.) online opportunity sample panel occurred between February 28, 2020, with confirmed U.S. SARS CoV-2 viral infections < 50, and April 13, 2021, a few months after general public vaccinations began. Each survey wave included the same baseline questions—e.g., COVID-19 attention, media use, risk perceptions, psychological distance, COVID-19 knowledge, behavioral intentions, policy support, trust in decision-makers—to control for variation within and across waves, with other variables in specific waves—e.g., demographics, conspiracist thinking, anti-Chinese prejudice, science attitudes. Among other findings, risk perceptions, trust, behavior, and policy support varied over time. Results supported behavioral motivation (higher judged risk now means greater protective action later) and accuracy (lower reported protection now means higher risk perception now). However, results strongly refuted risk reappraisal (greater protective action now meant higher, not lower, risk perceptions later), perhaps reflecting continuing uncertainties about COVID-19 dynamics and protectiveness of policies and behaviors. Risk perceptions for the U.S. and the world, and affective risk perceptions (e.g., concern, dread) were more strongly associated with behavior and policy support for mask-wearing, avoiding public gatherings, and vaccination than was personal risk perception. These results have implications for future pandemic risk management and communication practice.

Research funded by

NSF Award #2022216—RAPID (Rapid Response Research Award): Media Exposure, Objective Knowledge, Risk Perceptions, and Risk Management Preferences of Americans Regarding the Novel Coronavirus Outbreak;
NSF Award #2411612—RAPID (Rapid Response Research Award): Media Exposure, Objective Knowledge, Risk Perceptions, and Risk Management Preferences of Americans Regarding the Novel Coronavirus Outbreak.

COVID Information Commons Video Presentations

The COVID Information Commons (CIC) Research Lightning Talk video presentations and individual multilingual transcripts for this project can be found in the Columbia Academic Commons (AC) (Johnson 2020, 2022a).

B. B. Johnson (✉)
Decision Research Center, Oregon Research Institute, (formerly at Decision Science Research Institute), Springfield, OR, USA
e-mail: bjohnson@ori.org

F. D. Hudson (ed.), *The COVID Information Commons*, Sustainable Development Goals Series,
https://doi.org/10.1007/978-3-032-06125-6_13

Keywords

Risk perceptions · Behavioral intentions · Policy support · Media effects · Cultural biases · COVID-19

13.1 Goal of the Research

The Protective Action Decision Model (PADM) (Lindell and Perry 2004, 2012; Lindell 2018) is at the core of our analyses, which focus particularly on its predictors of behavioral intentions. These predictors include measures such as "threat perception" of SARS-CoV-2, and "protective action perception" of the potential effectiveness of, and ability to enact, these behaviors, concepts which overlap with those in other behavioral models, as well as "stakeholder perception" measures such as trust that other models lack. These models are often applied to hazards for which agency is deemed mostly restricted to individuals (e.g., excessive sun tanning), or to hazards where the responsibility for protection may be more widely shared, but a given study may focus on individuals' propensity to protect themselves. As impersonal organizations and institutions can play a major role in management of infectious disease epidemics (e.g., by imposing quarantines or providing vaccination opportunities), as for many other hazards (e.g., nuclear power accidents), incorporating stakeholder perception into one's measures can greatly enhance the model's empirical value.

13.2 Results

For simplicity, we first summarize the overall methods, and then report results by publications, as each tackles separate aspects of project aims. Details of methods are available in the publications specified and listed in the references.

13.2.1 Overall Methods

A six-wave longitudinal panel survey recruited an opportunity sample of Americans from the Prolific online panel (http://www.prolific.com/) over 14 months:

- $N_1 = 2004$, February 28–29, 2020
- $N_2 = 1613$, April 27–May 6, 2020
- $N_3 = 1184$, August 5–13, 2020
- $N_4 = 1026$, October 12–21, 2020
- $N_5 = 866$, January 22–February 11, 2021
- $N_6 = 1019$, March 25–April 13, 2021.

The first wave was launched on the same day that the first community transmission case of COVID-19 in the United States (U.S.)—that is, a case not travel-associated—was officially announced by the Centers for Disease Control and Prevention (CDC). At the time, there were fewer than 50 confirmed cases in the nation. The last wave in this study occurred when COVID-19 vaccines had been available to the general public, starting with those over 65, for close to three months. At that time, 26% of Americans had completed their first vaccination. This study was labeled exempt by the Decision Research Institutional Review Board (IRB), meaning it posed no more than minimal risk to participants. Participant consent was obtained with the Prolific panel's privacy and participation agreement.

Everyone answering a given wave was recruited for the next wave, except for the final wave to which all Wave 1 respondents were invited. To control for attention, all multiple-wave analyses omitted people claiming vaccination in Waves 1–4, when vaccines were unavailable excluding clinical trial participants (which included none of our Wave 4 respondents).

Compared to 2020 U.S. Census estimates for U.S. adults 18+, respondents to Wave 1 were slightly less female, perhaps older (Census median age estimates include children, while this was an adult-only sample, so the medians are difficult to compare), more non-Hispanic White, far better educated, and somewhat lower in income. From a political perspective, 50%

of respondents reported being Democrats, 15% Republicans, and 35% independent or undeclared political partisanship, while 61% reported holding liberal political ideology, and 20% conservative ideology.

Attrition is inevitable in a longitudinal panel study, but the critical issue is whether it changes findings. We compared demographics and risk perceptions of those who completed all six waves (38%) to those who dropped out after Wave 1, including 271 who returned in Wave 6. Gender and political stance did not differ, but the college-educated and non-Hispanic whites were more likely to persist through the survey period, and younger people were less likely. Only two of seven risk perception types showed significant differences between enduring and dropout groups (see Sect. 13.2.4 "Classifying risk perception measures" below). We believe there is no substantive attrition effect for our findings.

On measures used, readers are referred to the published papers, but we explain some briefly in the following sections for clarity. We refer to behavioral "intentions" below for simplicity, but these measures also included actual self-reported behavior. Participants reported whether "My household" never considered taking this action, is considering it, decided against taking this action, decided to take this action, has taken this action, or has taken this action and will continue to take this action as needed. For most behavioral analyses, we omitted the against-action respondents to provide a strictly ordinal measure.

13.2.2 Early Behavioral Responses to COVID-19

Partial models[1] of behavioral intentions were tested (Johnson and Mayorga 2021) using Wave 1 data only from the February 28–29, 2020 time period, thus assessing how well the models worked very early in what had not yet been recognized by most Americans as the emergent pandemic. Behavioral intentions were presumed to be predicted by action perceptions, stakeholder perceptions, and threat perceptions. Action perceptions and stakeholder perceptions were based on factual knowledge of transmission pathways as well as subjective knowledge. Threat perceptions were influenced by knowledge indicators, following news about COVID-19 in the U.S. and China, and perceived psychological distance. The behaviors were hand-washing, mask-wearing, avoiding travel to infected areas in the U.S. or elsewhere, avoiding large public gatherings (formal or informal), and avoiding Asians (an ineffective preventive behavior that some people were reported by mass media as doing in the U.S.).

Structural equation models had only moderate fit by comparative fit index (CFI = 0.91–0.94) and Tucker-Lewis index (TLI = 0.89–0.93) statistics, if good fit by root mean square error of approximation (RMSEA) statistics (RMSEA 90% confidence intervals = 0.06–0.07), with variance explained ranging from 12% (hand-washing) to 28% (mask-wearing). Action and threat perceptions increased intentions, while trust in stakeholders reduced them; psychological distance and factual knowledge reduced threat perceptions, while subjective knowledge increased threat perceptions; factual knowledge increased trust, and subjective knowledge decreased trust; and, news-following increased both factual and subjective knowledge, but particularly U.S. news-following increased subjective knowledge. Variations in the strength of these relationships across behaviors, and the moderate power of the models, may reflect the outlier status of Wave 1—well before official emergency or pandemic declarations—observed in later analyses of data from all six waves, summarized below.

13.2.3 Comparing Culture Effects in China Versus U.S.

Inspired by this study, data were collected in China (n = 646) May 21–June 10, 2020, and compared to data from the second wave of the U.S. survey conducted April 27–May 6, 2020

[1] "Partial" versus our proposed model, as some of its content was not measured in Wave 1.

(Yuan et al. 2024). The model proposed that a 12-item measure of cultural biases (Jenkins-Smith and Smith 1994) which had been advanced in grid/group cultural theory (Thompson et al. 1990)—comprising hierarchist (e.g., valuing order), individualist (liberty), egalitarian (equality), and fatalist (good luck) biases—would affect behavioral intentions via the three PADM mediators (threat, action, and stakeholder perceptions).

Across the two nations, cultural effects were mainly indirect, and in both countries, greater hierarchical bias increased intentions by mediation through action perceptions of response efficacy and self-efficacy. The mediation by action perceptions varied across nations for other cultural biases and by the behavior involved. Trust in government was a significant mediator only in China, while threat perception was a significant mediator only in the U.S. The cultural bias scales, which, like most such scales, were developed with North American samples, were usually more reliable measures in the U.S. than in the Chinese sample. However, fatalism was more reliable in the Chinese sample, further underlining the challenges and necessity of replicating studies in multiple societies, rather than assuming that findings from one place can be generalized elsewhere.

13.2.4 Classifying Risk Perception Measures

It does not take great imagination to assume that different risk perception items would factor separately. But that premise is belied by several considerations. First, it is rare for empirical studies to include diverse measures of risk perception. They may distinguish judged susceptibility from judged severity in health behavior models, or might distinguish cognitive from affective measures in risk analysis studies, but that is limited diversification. Thus, if one does not use multiple measures in one's empirical work, one cannot test that premise. Second, the distinction is often presumed, so that no factoring of the responses is conducted to confirm that these conceptual differences appear empirically, or that all items that ostensibly reflect a given latent concept in fact factor together. Third, recent efforts to formally categorize different risk perception measures do not agree with each other on either the overall conceptual structure of such measures or which items belong in which category of risk perception [e.g., compare Ferrer et al. 2016; Walpole and Wilson 2021].

Tests were conducted of how diverse measures of COVID-19 risk perception would factor across the six waves of the U.S. survey, and the relative association between these classes of risk perception measures on the one hand, and the risk responses of protective behavioral intentions and policy support on the other hand (Johnson and Kim 2023a). The risk perception items included two items on personal perceived risk from the coronavirus: one item was conditional on non-action, protective behavior not already taken or expected ("if you or your family doesn't do anything new to protect yourself"), while the other was conditional on action ("do anything new"). It is typical in the scientific literature when asking about personal risk to use non-conditional language, without mentioning behavior. But non-conditional risk perception measures will be misleading in testing the relationship between risk perception and self-protective behavior, because the two measures overlap (Brewer et al. 2004). For example, imagine a non-conditional question about personal risk related to COVID-19, with one respondent expecting that they will wear a mask more often when away from home, and another person expecting to spend more time unmasked in large public gatherings. Not only is behavior included in both the "perception" and the behavioral responses, leading to overestimates of the correlation of the two measures, but these respondents are also answering different questions, further confounding good science. Besides these two personal risk situations, non-conditional collective risk perceptions of the coronavirus were elicited for the risk to the U.S. and the world. Affective risk perceptions were elicited for concern, dread (defined for respondents as meaning "in terror of, or fear intensely"), and affect (good versus bad feelings) about the coronavirus. The perceived likelihood of a large outbreak of coronavirus infections in the U.S. in the next five years was elicited during Wave 1 only, because at later waves there were COVID-19 cases throughout the U.S. The final risk perception items were expected U.S. infections and deaths, and expected duration of the outbreak, elicited during Waves 2–6.

Most of these risk perception items factored as would have been expected from their content: four factors reliable across waves comprised 1) the two personal risk perception items, 2) the two collective risk items (U.S., global), 3) concern and dread, two of the three affective items, and 4) expected infections and death, two of the three severity items. However, the affect item did not load with the other affective items as expected, the duration item did not load with the other severity items as expected, and the likelihood item did not load with any other item in Wave 1. Thus, face validity is not a guarantee of associations between risk perception items.

Behavioral intentions and policy support for mask-wearing, avoidance of large public gatherings, and vaccination were regressed against these five (personal, affective, affect, collective, likelihood—Wave 1) or six (personal, affective, affect, collective, severity, duration—Waves 2–6) risk perception indicators. Collective risk perceptions most strongly predicted both intentions and support, followed by affective/affect risk perceptions.

These results indicate that researchers cannot rely on the face content of risk perception items as an indicator of their similarity. Furthermore, the usual assumption that self-protective behavior must be associated only, or at least primarily, with personal risk perception cannot be taken for granted. Because multiple regression analysis includes common variance among predictors, it cannot be ruled out that some associations of behavioral intentions and policy support with personal risk perception were partly reflected in other predictors' effects, particularly through the affective and affect measures. But this caveat does not offset the fact that, despite common variance, associations of behavior and policy support with collective risk perceptions were far stronger than for any other risk perception. Perhaps these findings are unique to a hazard like COVID-19, given that risks for self and for others are mutually reinforcing. But they also underline the importance of multi-dimensionality (i.e., not just asking about personal risk perceptions) in designing risk perception research, even when trying to explain personal protective actions.

13.2.5 Examining Predictors of Risk Perception Measures

The temporal dynamics of risk perceptions were examined over the 14-month period of the project, as well as their associations with diverse potential predictors (Johnson et al. 2023). The focus was only on predictors elicited in every wave, thus excluding predictors such as political ideology that were measured once or occasionally over the six waves. In general, risk perceptions increased from February 2020 to April 2021, but with varying trends across personal, collective, affective, affect, severity, and duration measures. Factors in baseline risk perceptions in Wave 1 were examined, plus how different respondents' trends in risk perception over time differed. Baseline ratings were associated with how immediate the threat seems (temporal distance), how likely the threat would affect people like oneself (social distance), and with following United States news about the pandemic. Inter-individual trend differences were shaped most by temporal distance, whether local coronavirus infections were accelerating their upward trend from official data, and subjective knowledge about viral transmission. The gap in risk perception magnitude between people high and low on the predictor consistently widened or narrowed over these six waves. While the narrowing might be explained as reversion to the mean, the widening is harder to explain. Associations of subjective knowledge and risk trend with risk perceptions could change signs, e.g., from positive to negative, over time.

13.2.6 Testing Hypotheses About Risk Perception-Protective Behavior Associations

Authors of a 2004 paper (Brewer et al. 2004) clarified a then-emerging literature on health behavior about such associations. They argued that cross-sectional studies can only test an "accuracy" hypothesis: e.g., people with higher risk perceptions at a specific time (e.g., Time 1) will exhibit low protective behavior and/or high risky behavior at that same time. Yet cross-sectional data are often wrongly interpreted as able to test two other hypotheses, which are only testable longitudinally. One is the "behavioral motivation" hypothesis, that high risk perception at Time 1 increases protective behavior at some later Time 2, while the "risk reappraisal" hypothesis is that protective behavior at Time 1 reduces risk perception at Time 2.

These hypotheses were tested for six behaviors (Johnson and Kim 2023b): hand-washing, mask-wearing, avoiding travel to infected areas, avoiding large public gatherings, vaccination, and (for waves 2–6) social isolation at home. Accuracy and behavioral motivation hypotheses were supported for both behaviors and intentions, excluding a few behaviors and waves (particularly in February–April 2020, when the pandemic was new in the U.S.). The risk reappraisal hypothesis was contradicted—protective behavior at one wave increased risk perception later—perhaps reflecting continuing uncertainty about efficacy of COVID-19 protective behaviors and/or that dynamic infectious diseases may yield different patterns than the chronic diseases that have dominated the limited empirical testing of these hypotheses.

13.2.7 Affect as a Mediator of Culture's Effects on Risk Responses

Understanding risk perceptions, behavioral intentions, and policy support has included attention to such predictors as affect (how people feel about the threat) and culture (e.g., the ways of life and associated values that people hold dear). Two recent models have suggested that affect mediates the culture-risk response relationship, but differ on whether the mediator is affect toward the hazard, such as negative feelings about COVID-19, or affect toward a proposed action (e.g., negative feelings about a specific COVID-19 policy, such as mandatory masking or vaccination). Prior research using data on Americans' responses to the Zika virus had concluded that both models fit with the data, but "solution aversion" or negative affect toward a proposed action was a stronger influence on risk perceptions or policy support (Johnson 2022b). Solution aversion, the latter mediator, accounted for more associations of culture with coronavirus risk perceptions (personal risk, collective risk, risk severity) than did dread, the indicator of hazard affect used here (Johnson and Kay 2025). The same pattern occurred for behavioral intentions (mask-wearing, avoiding large public gatherings, vaccination) and policy support (i.e., mask mandates, public gathering bans, vaccination mandates). Statistically significant direct effects were rare, mainly for people holding an egalitarian cultural bias (which values equality). High-egalitarian respondents had higher COVID-19 responses, in that they perceived more risk, intended more protective behavior, and supported protective policies more. Indirect effects occurred for egalitarians, political conservatives, and individualists who value freedom, with the latter two groups downplaying COVID-19 risks and voicing lower protective intentions and policy support. Hierarchists who value order and fatalists who value good luck exhibited no significant direct or indirect effects.

13.2.8 Media Content Effects

COVID-19 coverage of health-focused protective behaviors in the *New York Times*, *Wall Street Journal*, and *USA Today*, and in Twitter posts, was analyzed for the February 2020-April 2021 survey period, including 10,528 news articles and 209,261 tweets (Shao et al. 2025). There was a two-way interactive agenda flow between legacy media and social media platforms, particularly in controversial topics like vaccination. Exposure to media coverage was associated with survey respondents seeing more benefits from certain behaviors, but these benefit perceptions did not always translate into behavioral adoption. For example, while one's exposure to these specific media outlets increased perceived benefits of vaccination by reducing health risks to both one's household and to vulnerable people in one's community, it was not consistently linked across waves to higher intentions to get vaccinated.

13.2.9 Other Studies

The work cited above has been reported in peer-reviewed publications. Other manuscripts are under review or in development; we briefly describe some of them below:

Media Exposure and Psychological Distance Cross-sectional U.S. data were used in an earlier study to conclude that conservative media use increased psychological distance from COVID-19, which then decreased susceptibility and severity perceptions. Distance, low perceived susceptibility, and low perceived severity then reduced intentions to be vaccinated or get a booster shot (Hmielowski et al. 2023). This paper uses the current longitudinal data to better test the causal implications of this model, including non-vaccination as well as vaccination intentions.

Factors in Intentions to Avoid Asians Such avoidance is not effective at preventing coronavirus infection, but anecdotal reports and survey responses suggested that some Americans thought it was a means of protection. Wave 1 data show that conservative political ideology increased anti-Asian prejudice, conspiracist thinking, and perceived household and community efficacy of avoiding Asians. Yet with Asian-avoidance intentions as the outcome, there were no longer direct effects of political ideology or community efficacy.

Stability of Cultural Biases There is a long history of both qualitative and quantitative study of cultural biases' role in human attitudes and behavior. However, there are many unresolved theoretical and empirical questions in this area, including whether people should or do have fixed cultural biases—e.g., they support one bias, such as individualism, all of their lives—or they have varying biases, across time, between domains of life (e.g., in the home versus work), or have mixed views (e.g., they might support both hierarchism and fatalism simultaneously). The COVID-19 survey collected cultural bias responses in Waves 2, 5, and 6, with intervals of about two to 11 months. At the individual level, classifying someone as supporting a given bias occurred only when the person agreed with each one of the three items that comprised each cultural bias scale. Results differed somewhat by waves compared, and whether the focus was only on people who supported only one bias, or supported more than one bias; for individual classification, the latter group was classified by the bias they supported the most. Overall, 85% - 95% of respondents in a comparison of two waves at a time were classified the same way, with this proportion not varying consistently with the length of the gap in time between waves. When the focus was on the 650 people who answered these questions at all three waves, a third of those who supported only a single cultural bias were consistently classified across all three waves, while half of those who supported one bias more than any other were consistently classified. Many of those who were not classified as "consistent" were not necessarily inconsistent, but rather their equal support for two or more cultural biases, or other reasons, prevented them from being classified as having a single bias. Thus, stability of cultural biases, as measured by these scales, seems more common than not.

Temporal Variation in Prediction of Behavioral and Policy Stances Section 13.2.2 reported on early behavioral responses of Americans (i.e., February 2020) using the limited number of variables available in that first survey (Johnson and Mayorga 2021). We are now analyzing data from all six waves to apply the central predictors of the PADM (Lindell 2018)—risk perceptions, behavior/policy perceptions (e.g., efficacy at reducing risk; value threat), and stakeholder perceptions (e.g., trust in institutions; attitudes toward science)–as predictors of behavior/behavioral intentions and of policy support, to see whether there is temporal variation in patterns of association across 14 months in which COVID-19 incidence changed, compliance with protective recommendations changed, vaccines became available to the general public, and federal administrations changed.

13.3 Recommendations Regarding Mitigation, Response, and Recovery from Future Pandemics

Practical inferences from a single study with an opportunity sample (if a diverse one) from a single nation should be cautious. The following suggestions are offered with those caveats in mind. Some general remarks are offered, followed by a few more specific recommendations.

First, both researchers and practitioners need to be far more aware of and responsive to the temporal dynamics of pandemics. This is exhibited most explicitly in our longitudinal analyses, in which Wave 1 responses in late February 2020, when there were few confirmed cases of SARS-CoV-2 infection in the U.S., and no emergency or pandemic had yet been declared, did not track with those in waves two months or more later. This may explain why the variance explained in Wave 1 behavioral intentions was relatively low, and only modest fit of the model occurred, though the model itself could be wrong or incomplete (Johnson and Mayorga 2021). But we observed this temporal variation as well in changing associations of risk perceptions with predictors asked in every wave (Johnson et al. 2023), and in the failure of our sample to adapt to the pandemic in the sense of reappraising the risk downward after having adopted protective behaviors, as instead their risk perceptions increased (Johnson and Kim 2023b). Furthermore, we observed in unpublished data that we were unable to establish a coherent cross-temporal measure of our sample's factual knowledge about COVID-19 even when we excluded Wave 1. Clearly, the facts changed in both scientific understanding and public health policies, including the spring 2020 communication debacle about whether mask-wearing protected the wearers or others, due to official concerns about mask supplies. But it seems citizens' grasp of what is known also was variable, as reflected in spring 2020 open-ended responses to questions about mask-wearing during Wave 1 of this study. Practitioners need to be prepared in their operational planning and baseline messaging plans, even before a pandemic occurs, to expect that scientific knowledge will change, public policies, actions, and messages will change, and citizen "factual knowledge" will change. Efforts that do not account for that multi-dimensional change well in advance are less likely to succeed.

Second, practitioners should not retain assumptions that personal risk perceptions are the only, or even the most important, factors in public compliance with public health advice. This is pointed up by two findings here. One study concerned the stronger effect of collective and affective risk perceptions than personal risk perceptions on both protective behavioral intentions and policy support (Johnson and Kim 2023a), which converges with some efforts by public health officials to invoke collective risk as a reason to take protective action in the U.S. The other concerns the associated roles of action perceptions and stakeholder perceptions on behavioral intentions (Johnson and Mayorga 2021). For example, beliefs about COVID-19 vaccine safety and efficacy, and trust in the U.S. President, were among critical factors in intention and actual action to get vaccinated (Johnson 2023). While models of health behavior have long emphasized both perceived risk and perceived self-efficacy and response efficacy of protective actions, public health communications must emphasize a wider range of influential factors *and* get more granular about the details of these issues. For example, the well-intentioned early message to avoid hoarding the most protective masks until front-line health personnel have enough was not clear enough about officials' concern that otherwise everyone's health will be threatened. In another example, it is important for practitioners to determine which specific affective reactions are driving public action or inaction, because not all negative emotions about an infectious agent are necessarily identical or follow similar temporal pathways (Johnson and Kim 2023a).

Third, this research found that one long-held "behavioral motivation" premise—that higher personal risk perceptions at one time are associated with more self-reported protective behavior later—held true for COVID-19 over six survey waves and 14 months (Johnson and Kim 2023b). Although adding experimental evidence would further justify drawing causal inferences, this finding bolsters the argument for risk communications that bolster risk perceptions, with the caveats that collective and affective risk perceptions are even stronger influences on behavioral intentions than personal risk perceptions (Johnson and Kim 2023a), and that higher risk perceptions not associated with high self-efficacy can undermine action. However, the same data revealed that the reverse "risk reappraisal" hypothesis—that taking protective action now reduces later risk perceptions—was not supported, with perceived risk continuing to rise over time. It is unclear whether COVID-19 presented a unique case in this respect, or this unexpected result is typical of pandemics but not other health issues, or it signals that this specific hypothesis needs to be assessed anew for each health problem. There were certainly unusual aspects to the COVID-19 event that might have altered expected risk reappraisal dynamics: e.g., a novel coronavirus, ever-changing policies, and messages by a public health establishment often at odds with each other or with political leaders, and a public that could never be sure how protective its actions or those of its leaders might be. We simply argue that the risk reappraisal hypothesis should not be treated as gospel by public health officials regarding future pandemics, nor should the behavioral motivation hypothesis, despite its supporting evidence here.

Acknowledgements I appreciate the collaborations on this project with Marcus Mayorga, Ph.D., Decision Science Research Institute, who was co-PI on the original National Science Foundation (NSF) proposal but left the project after data collection; Byungdoo Kim, Ph.D., Norwegian Institute of Technology; Cameron Kay, Ph.D., Union College; Meng Yuan, Ph.D., Chongqing University; and Brendon Swedlow, Ph.D., JD, Northern Illinois University. I also appreciate ongoing collaborations with currently anonymous colleagues on yet-to-be-published papers, as well as the willingness of our respondents to answer multiple surveys over 14 months during a daunting pandemic.

References

Brewer NT, Weinstein ND, Cuite CL, Herrington J (2004) Risk perceptions and their relation to risk behavior. Ann Behav Med 27(2):125–130. https://doi.org/10.1207/s15324796abm2702_7

Ferrer A, Klein WM, Persoskie A, Avishai-Yitshak A, Sheeran P (2016) The tripartite model of risk perception (TRIRISK): Distinguishing deliberative, affective, and experiential components of perceived risk. Ann Behav Med 50(5):653–663. https://doi.org/10.1007/s12160-016-9790-z

Hmielowski JD, Matthews A, Chu H (2023) Going the distance for COVID-19: relationships among news use, psychological distance, risk perceptions, and behavioral intentions. Sci Commun 45(4):460–484. https://doi.org/10.1177/10755470231188694

Jenkins-Smith HC, Smith WK (1994) Ideology, culture, and risk perception. In: Coyle DJ, Ellis RJ (eds) Politics, policy and culture. Westview, Boulder, CO, pp 17–32. https://doi.org/10.4324/9780429302503

Johnson BB (2020) Branden Johnson: media exposure, objective knowledge, risk perceptions, and risk management preferences of americans regarding the novel coronavirus outbreak. Columbia Academic Commons, Academic Unit: Northeast Big Data Innovation Hub, Series: COVID-19 Research Lightning Talks. https://doi.org/10.7916/kffs-ty33

Johnson BB (2022) Branden Johnson: media exposure, objective knowledge, risk perceptions, and risk management preferences of americans regarding the novel coronavirus outbreak. Columbia Academic Commons, Academic Unit: Northeast Big Data Innovation Hub, Series: COVID-19 Research Lightning Talks. https://doi.org/10.7916/fcse-sf04

Johnson BB (2022b) Affect toward the policy option versus the hazard differentially mediates cultural effects on Americans' Zika risk perceptions and policy support: comparing the solution aversion-based model and the affect heuristic-cultural cognition theory model. Hum Ecol Risk Assess 28(3–4):281–315. https://doi.org/10.1080/10807039.2022.2048354

Johnson BB, Kay CS (2025) Affect mediates culture's effects on COVID-19 risk perceptions, behavioral intentions, and policy support among Americans. Health Risk Soc 27(1–2):1–26

Johnson BB, Kim B (2023a) COVID-19 risk perception measures: factoring and prediction of behavioral intentions and policy support. J Risk Res 26(11):1191–1212. https://doi.org/10.1080/13669877.2023.2264301

Johnson BB, Kim B (2023b) Cross-temporal relations of conditional risk perception measures with protective actions against COVID-19. Soc Sci Med 324:115867. https://doi.org/10.1016/j.socscimed.2023.115867

Johnson BB, Mayorga M (2021) Americans' early behavioral responses to COVID-19. Hum Ecol Risk Assess 27(7):1733–1746. https://doi.org/10.1080/10807039.2021.1884842

Johnson BB, Mayorga M, Kim B (2023) Americans' COVID-19 risk perceptions and risk perception predictors changed over time. J Risk Res 26(7):815–835. https://doi.org/10.1080/13669877.2023.2208149

Johnson BB (2023) Factors in intention to get the COVID-19 vaccine change over time: Evidence from a two-wave U.S. study. Health Risk Soc 25(3–4):151–179. https://doi.org/10.1080/13698575.2023.2173727.

Lindell MK (2018) Communicating imminent risk. In: Rodriguez H, Trainor J, Donner W (eds) Handbook of disaster research. Springer, New York, pp 449–477. https://doi.org/10.1007/978-3-319-63254-4_22

Lindell MK, Perry RW (2004) Communicating environmental risk in multiethnic communities. Sage, Thousand Oaks, CA. https://doi.org/10.4135/9781452229188

Lindell MK, Perry RW (2012) The protective action decision model: theoretical modifications and additional evidence. Risk Anal 32(4):616–632. https://doi.org/10.1111/j.1539-6924.2011.01647.x

Shao A, Chen K, Johnson BB, Miranda S, Xing Q (2025) Ubiquitous news coverage and its varied effects in communicating protective behaviors to American adults in infectious disease outbreaks: time-series and longitudinal panel study. J Med Internet Res 27:e64307. https://www.jmir.org/2025/1/e64307. https://doi.org/10.2196/64307

Thompson M, Ellis R, Wildavsky A (1990) Cultural theory. Westview, Boulder, CO. https://doi.org/10.4324/9780429501180

Walpole HD, Wilson RS (2021) A yardstick for danger: developing a flexible and sensitive measure of risk perception. Risk Anal 41(11):2031–2045. https://doi.org/10.1111/risa.13704

www.prolific.com

Yuan M, Johnson BB, Mayorga M, Swedlow B (2024) Explaining compliance with COVID-19 regulation in China and the United States: cultural biases, political trust, and perceptions of risk and protective actions. J Publ Policy 44(2):284–326. https://doi.org/10.1017/S0143814X23000429

Branden B. Johnson conducts risk perception and risk communication research for environmental, technological, natural, and social hazards as a Senior Scientist, Decision Research Center, Oregon Research Institute, a nonprofit research firm in Springfield, Oregon. Besides his research on COVID-19 responses, in the last ten years he has studied (among other topics) U.S. public responses to Ebola and Zika, improving cost–benefit analysis estimates of the social benefits of prolonging one life, assessing recovery times for lifeline infrastructures after the Cascadia subduction zone earthquake, improving risk perception measures, and how citizens respond to disputes among scientists. He earned a BA in Environmental Values and Behavior from University of Hawai'i-Manoa, and a MA in Environmental Affairs and Ph.D. in Geography from Clark University. He was an assistant professor of geography (University of Pittsburgh) and assistant and associate (tenured) professor of science, technology, and society (Michigan Technological University), plus adjunct professor of human ecology, of geography, and of public health at Rutgers University. Dr. Johnson served in the New Jersey Department of Environmental Protection's Division of Science, Research and Technology for 20 years, researching topics from black bears to outdoor air pollution to wetlands, and in its regulatory Bureau of Safe Drinking Water for five years.

Communication of COVID-19 Information by Decision Makers in the U.S. Southeast

14

Kathryn Corvey, Janna R. Willoughby, Ryan D. Williamson, Mina Becker, and Kelly Dunning

Abstract

Decision makers play a crucial role in communicating information to the public to garner collective action in response to crises like the COVID-19 pandemic, but face a critical knowledge gap on effective strategies for communicating scientific recommendations within vulnerable communities. This chapter discusses how decision makers communicated COVID-19 information in the three lowest-income states in the United States (U.S.) Southeast: Alabama, Mississippi, and Louisiana. This study examined what COVID-19 information was communicated to the public, how this information was communicated, and the successes and challenges experienced by decision makers. The study used semi-structured interviews of decision makers and thematic analyses. Data show that decision makers played diverse roles in disseminating information. Communication was tailored to local populations, and trusted local figures were used to enhance effectiveness. Institutional response varied, with coordination levels affecting communication success. Misinformation and public distrust presented significant challenges. Decision makers and the public experienced burnout due to the constant flow of information and evolving guidance. Our study suggests that, to improve future crisis communication, pre-crisis preparation is essential. This includes identifying and training communicators, clarifying communication channels, and targeting vulnerable populations. These steps will ensure effective communication of scientific information, promoting collective action and better public health outcomes.

Research funded by

NSF Award #2032133—RAPID (Rapid Response Research Award): A participatory study of how policy-makers and marginalized communities in the American South consume and act on scientific information in the context of COVID-19.

COVID Information Commons Video Presentation

The COVID Information Commons (CIC) Research Lightning Talk video presentation and individual multilingual transcripts for this project can be found in the Columbia Academic Commons (AC) (Corvey 2021).

K. Corvey (✉) · M. Becker
University of Alabama at Birmingham, Birmingham, AL, USA
e-mail: kcorvey@uab.edu

M. Becker
e-mail: mbecker1@uab.edu

J. R. Willoughby
Auburn University, Auburn, AL, USA
e-mail: jwilloughb@auburn.edu

R. D. Williamson · K. Dunning
University of Wyoming, Laramie, WY, USA
e-mail: ryan.williamson@uwyo.edu

K. Dunning
e-mail: kelly.dunning@uwyo.edu

F. D. Hudson (ed.), *The COVID Information Commons*, Sustainable Development Goals Series,
https://doi.org/10.1007/978-3-032-06125-6_14

Keywords

Coronavirus disease 2019 · COVID-19 · Public policy · Communications · Qualitative

14.1 Goal of the Research

COVID-19 presents unprecedented health policy challenges, as this disease was a community-wide outbreak that required non-pharmaceutical interventions. Even before the COVID-19 crisis, the Centers for Disease Control and Prevention (CDC) recommended community-based interventions, such as social distancing, to "buy time," decrease the peak number of cases, and reduce mortality by lowering the total number of cases. Social distancing guidance suggests decision makers leverage *collective action*, or the interactive engagement of stakeholders and the public, to mitigate pandemics. Collective action requires individuals to adopt behaviors that benefit society—in this case, reducing transmission of SARS-CoV-2, the causative agent of COVID-19—but are personally costly (e.g., staying home over periods of time and social distancing) (Centers for Disease Control and Prevention 2007).

Collective action is especially important—and especially difficult—in vulnerable communities. The Southeastern United States (U.S.) contains many such communities. Alabama, Mississippi, and Louisiana have among the nation's highest prevalence of adult obesity. They also have the highest age-adjusted death rates for chronic respiratory conditions and diabetes (Centers for Disease Control and Prevention 2020a, b, c, d). These conditions are related to higher risk of serious complications from COVID-19 (Centers for Disease Control and Prevention 2020d). These three states are also among the poorest in the U.S. (U.S. Department of Commerce 1993), with approximately two million residents living in poverty at the height of the pandemic (U.S. Census 2022). This amplifies vulnerability to the pandemic (Centers for Disease Control MMWR 2020). Extensive public health literature links poverty and negative community health outcomes (Brennan-Olson et al. 2015; Erhunmwunsee et al. 2012; Li et al. 2015; Sandel et al. 2016; Vuuren et al. 2014).

Decision makers play a crucial role in interpreting and communicating information to the public to signal (or compel) collective action in a crisis, but they face a critical knowledge gap on the most effective strategies for communicating scientific recommendations within our most vulnerable communities. We aim to fill this gap by examining how decision makers responded to COVID-19 in the three lowest-income states in the U.S. Southeast: Alabama, Mississippi, and Louisiana. This study will benefit society by helping decision makers improve and hone messaging to compel more people to act faster and more meaningfully when collective action is required to lessen the impacts of crises.

State and local decision makers from the states of Alabama, Mississippi, and Louisiana participated in this study. One-on-one semi-structured interviews were conducted remotely, using Zoom video conferencing technology, between December 2020 and June 2021. Thematic analysis of the qualitative data revealed eight major themes that represent the considerations for effective communications made by the study decision makers. Themes include which COVID-19 information to communicate, institutional response, targeting messaging, trust in science, trust in government, misinformation, effectiveness of messaging, and messaging burnout.

14.2 Results

14.2.1 Characteristics of Decision Makers

Seventeen decision makers were interviewed, including local government employees and elected officials, state and regional public health officials, an employee of a non-governmental organization involved in community health efforts, university officials, local and regional business association employees, and employees of various media outlets. Respondents were contacted through a snowball process, where respondents were asked for contact details of other respondents. Respondents were interviewed with their response data coded iteratively using a first cycle of an in vivo grounded theory approach followed by a second cycle in which these codes were distilled into themes (Saldana 2021). A grounded theory approach works for in vivo coding and works well for this case, especially for examining stakeholder approaches to communication during a pandemic.

Decision makers filled various roles in responding to the COVID-19 pandemic. A primary role of decision makers was gathering, adapting, and communicating COVID-19 information and representing the interests of their respective constituents during decision-making around communication efforts. Decision makers noted that their primary roles revolved around gathering information about COVID-19 from national and state-level sources, assessing the credibility and appropriateness of that information, and deciding upon communication tactics to disseminate that information to their constituents. To disseminate information, most decision makers reported using traditional media outlets, such as newspapers, television, and radio, as well as social media outlets. Several decision makers were involved in sharing information via websites. These websites were often existing community or organizational websites, though some decision makers reported creating new websites, or online "resource centers" to serve their constituents.

Many decision makers were also responsible for communicating and coordinating with other agencies. Several described the difficulty in navigating existing channels of communication whose weaknesses were exposed during the pandemic response. For example, regional public health officials reported having some degree of autonomy (from their respective state health departments) in communicating with the public. These decision makers described some regional offices being more comfortable with exercising this autonomy than others, which caused some miscommunications to occur. In general, this was described in the context of institutional leadership issues, in which clear guidance about what types of information should be communicated and when was sometimes lacking.

Most decision makers were also part of task forces dedicated to COVID-19 response. Several of them—primarily local government employees and elected officials, as well as state and regional public health officials—described how they had been involved in Unified Command structures activated after states of emergency were declared. Separate task forces were also established in many cities, consisting of the mayor, city council, and local law enforcement, fire and rescue, hospitals, non-profits, and faith-based organizations. As the pandemic progressed, decision makers representing local and regional business associations began serving on task forces focused on economic recovery.

Nearly all decision makers reported that their pre-pandemic roles and responsibilities had changed during the first year of the COVID-19 response. For some, their pre-pandemic job roles remained unchanged and COVID-related responsibilities were simply added to them. Many discussed expanded working hours and work intensity. Others were required to transition to a more communications-based role as they became more important sources of information for their constituents than they had been previously.

14.2.2 Information Communicated

A primary consideration of decision makers was what types of COVID-19 information should be communicated to constituents. Early in the pandemic, decision makers decided to focus on scientific information, including virus traits, daily case numbers, risk factors, and behaviors (such as hand washing, social distancing, etc.) that could be used to mitigate the spread of the virus. As vaccines became available, the focus of communications shifted to emphasize where and how to access the vaccine.

Most decision makers communicated data from their state health department or the CDC. In general, decision makers felt that state-level data messaging was more unified than federal messaging. However, some decision makers reported dissatisfaction with state health departments for various reasons. These decision makers were more likely to communicate data they had received from local physicians, hospitals, or academics, which they felt was timelier and more accurate. They also felt that this information had already been somewhat tailored to their local constituents and was therefore easier to pass along.

Decision makers discussed whether, why, and how they altered information from state health departments or the CDC prior to communicating the information to their specific constituents. Many decision makers passed on information without alteration. Those who did alter information noted that, while they did not change the actual information, they did alter the presentation of the information. This included making the language more accessible and easily digestible for the average constituent. Some felt that the bureaucratic language used by governmental agencies was a barrier to communication in their communities. Decision makers also reported tailoring the presentation of information to appeal to their local populations, especially those deemed high risk for infection and high severity of illness.

Information was communicated via numerous and varied mediums, including print and online news, e-newsletter, direct mail, text messaging, radio, podcast, television, live stream, e-mail, organizational websites, social media, posters, flyers, and t-shirts. Many decision makers utilized their main organizational website as a hub for more detailed information while using other mediums to communicate abbreviated information. Radio was mentioned as a particularly useful medium, as it did not rely on complex technology or graphic design and was more widely accessible among rural populations. Social media was viewed as a rapid and convenient medium for disseminating information, though many decision makers noted that both correct and incorrect information spreads quickly via social media platforms, an observation that is supported by other research demonstrating that falsehood spreads farther and faster than truth on social media platforms (Vosoughi et al. 2018).

14.2.3 Institutional Response

Decision makers faced high levels of uncertainty, not only due to the evolving nature of the pandemic, but also because of the changing nature of their roles and the roles of their organizations. The evolving situation demanded timely and actionable messaging, requiring many decision makers to rely on new communication technologies, presenting some with a steep learning curve. To complicate matters, decision makers struggled with uncertainty over how long the pandemic would last. Several described organizational turmoil as timeline estimates changed from several days to weeks, and then to months. As understanding of the virus evolved, guidance for preventing the spread of the virus evolved as well. Decision makers were cognizant that the public may interpret changing guidance as ineptitude on the part of those communicating guidance rather than a normal process of adapting public administration based on new information.

The decision of when and how to communicate information to the public was difficult for many decision makers. In the early stages of the pandemic, some reported waiting for the federal government to initiate communication of information to the public. As frustration grew with the perceived lack of leadership from the federal government, many decision makers decided to rely instead on their state governments and public health departments for leadership, a relationship that would continue throughout the pandemic. Some local organizations decided to be proactive in communicating guidance, driven by frustration with governmental agencies. Others decided that it was less important to be the first to convey a message, and instead waited until they could communicate more complete and clear information. These decisions reflected the tension between the desire to provide timely information and providing the most accurate information.

The evolving nature of the pandemic highlighted both the importance of and limitations of existing institutional coordination. Decision makers representing agencies and organizations that had high preexisting levels of coordination reported greater ease and confidence in their communication roles. For example, decision makers at various governmental levels in Louisiana reported fairly high levels of confidence in their communication decisions, as high degrees of coordination were necessary to effectively respond to the natural disasters that regularly affected the state. For these decision makers, channels of communication had already been established and trust had been built between responding agencies and the public. Rapid access to high-quality data was also facilitated through these existing relationships.

Other decision makers reported high levels of discord among organizations and agencies within their state, leading to disjointed and sometimes conflicting communications. Some felt that COVID messaging was influenced by factors other than science, namely financial interests or fear of liability. The perceived top-down administration of state departments of health was also cited by decision makers in other organizations as hampering effective communication and vaccine rollout, once vaccines became available. Some decision makers at the local level felt that state health departments were out of touch with reality "on the ground" and the hierarchy did not allow for flexibility in response decisions. Furthermore, some local decision makers described feeling that state health department employees seemed to make decisions that supported their continued employment and institutional structure over the well-being of the public.

Institutional coordination issues were exacerbated by a general lack of resources within all three states, including limited workforce capacity. In state agencies, employees were asked to take on COVID-related responsibilities in addition to their regular work duties. As agencies were already under-resourced prior to the pandemic, these responsibilities were extraordinarily burdensome to employees. As all three states were also medically underserved prior to the pandemic (Health Resources Services Administration 2020), states struggled to recruit employees with the needed skill sets and levels of expertise to effectively respond to the pandemic.

14.2.4 Targeting Messaging

Most decision makers reported targeting their messaging in some way, usually based on various demographics of their respective populations. Many discussed the importance of targeting vulnerable populations, or those who were predicted to be most adversely affected by COVID-19. These included individuals residing in rural areas, those with limited access to high-speed internet, recent immigrants, those for whom English was not their primary language, the elderly, those with chronic health conditions, low-income individuals, and those with low educational attainment. While some decision makers reported proactive outreach to vulnerable populations, others noted that vulnerable populations were only identified and targeted after disparities in COVID-19 outcomes began to arise. Targeting was accomplished both through tailoring the method of communication, such as allowing for conference calls that were easier for the elderly to navigate compared to Zoom meetings, and adapting messages themselves to specific audiences, such as providing tailored messages to populations about their increased risk of contracting COVID-19.

Many decision makers also discussed the importance of tailoring the messenger, believing that communications would likely be more effective if the messenger relaying information to certain subpopulations was someone who was perceived to be trustworthy among that group. For example, several decision makers reported partnering with African American ministers and barbershop employees to communicate with African American populations rather than relying on state health department employees. In Louisiana, the National Guard was utilized for communications as well as for logistical coordination with the public during mass vaccination events. Decision makers in Louisiana reasoned that residents of the state were accustomed to the presence of the National Guard during natural disaster response situations, and that the National Guard may therefore be the most effective means by which to communicate information and coordinate vaccination sites. In general, decision makers felt that the more "local" the messenger was to the target population, the more receptive those populations would be to the message.

14.2.5 Trust in Science

Decision makers believed that distrust among the public was created as guidance evolved over the course of the pandemic. Changing guidance, though necessary as understanding of the COVID-19 virus expanded, created distrust in both "government science" (institutions such as the CDC and National Institutes of Health (NIH), as well as government-funded university research) and local messengers. Decision makers also felt that an anti-science sentiment was fostered by President Donald Trump (Webb and Kurtz 2022) and other leaders who publicly countered guidance from scientists and medical doctors and promoted scientifically untested prevention and treatment methods. This issue, they felt, was compounded by relatively low educational attainment among the populations they served. Decision makers who had not previously been responsible for conveying scientific information reported great discomfort with this type of communication, particularly epidemiological and statistical information. They admitted to several miscommunications due to this discomfort and felt that this led to credibility issues with the public. Conversely, several decision makers did have a medical or scientific background and believed that this background led to greater trust in their messaging, as they had become known in their communities for this knowledge and the public felt that their guidance regarding COVID-19 could be trusted. To combat growing distrust, many decision makers attempted to distill guidance about prevention and treatment into recommendations that would not necessarily change as new knowledge emerged about the virus.

14.2.6 Trust in Government

In addition to distrust in science, decision makers felt that the public had significant levels of distrust in government more generally, an observation supported by national Gallup polls finding low levels of trust in government (Gallup 2023). This distrust was most often associated with politically conservative beliefs, an observation supported by subsequent research demonstrating differences in trust by political affiliation (Justwan and Williamson 2022), as well as rurality and low levels of education. Several decision makers working for local and state governments or health departments reported significant pushback from self-identified conservative individuals who expressed distrust in what they perceived to be liberal government leaders and agencies. Some expressed frustration with local leaders who seemed to take advantage of these political sentiments by further politicizing the response to the pandemic. To appeal to conservative constituents, some decision makers tailored messaging to focus on the preservation of individual freedom and benefits to the individual as opposed to benefits that would accrue to the larger community from various behavioral interventions. For example, messages highlighted the decrease in individual risk from mask wearing, rather than the benefits to the community from slowing the spread of the virus.

Trust in government may have also been complicated by the historical experiences between the government and certain racial and ethnic groups. Several decision makers described the negative impact of historical events, including the United States Public Health Service Untreated Syphilis Study at Tuskegee (Centers for Disease Control and Prevention 2023), on African Americans' trust in government. Many decision makers, including state health department employees, cited this historical distrust as an explanation for poorer COVID-19-related health outcomes among African American communities. However, African American decision makers in our study felt that this belief among state health department employees was used as an excuse for not providing equitable resources to African American communities during the pandemic, leading to poorer health outcomes. In their view, African Americans had long since moved on from the mistrust that arose following events such as the Tuskegee study.

14.2.7 Misinformation

Nearly all decision makers felt that misinformation was the primary barrier to public acceptance of behavioral changes necessary to slow the spread of COVID-19, including mask wearing, social distancing, and vaccination. Decisions had to be made whether to try to correct misinformation or not. Some tried to combat misinformation by pushing accurate information about COVID-19 to the public rapidly, before misinformation could take root. To complement efforts to target populations through tailoring of messengers, some decision makers focused on getting accurate information to these messengers as quickly as possible so that this information could be shared prior to misinformation circulating on the internet. Eventually, many decision makers decided that misinformation was so pervasive that it could not be effectively countered, so chose to focus on disseminating accurate information without addressing misinformation.

Many decision makers felt that the spread of misinformation was amplified through social media. Those who worked in more traditional media or press roles explained that their typical response to the spread of misinformation during political speeches or other means, which was to publicly call out the inaccurate information, was not effective for misinformation that was spread more rapidly through social media. The sheer number of social media platforms also proved overwhelming to decision makers, as it became increasingly difficult to combat misinformation across multiple platforms.

14.2.8 Effectiveness of Messaging

The effectiveness of messaging was a primary concern of all decision makers. They expressed concern about whether messaging would lead to behavior changes or adherence to guidance. Several discussed factors that they felt improved the effectiveness of their messaging. Many mentioned the importance of charisma. Charismatic local leaders, celebrities, university football coaches, and other sports figures were credited with more effective messaging. Decision makers also emphasized the importance of "we're all in this together" messages and communication methods that highlighted this message, including virtual town halls and other group-based communications. Honesty and transparency, about both successes and failures, was also highly valued.

Despite holding strong beliefs about what comprised effective messaging, many decision makers noted the difficulty of measuring effectiveness. The number of social media posts and shares, website downloads, number of persons signing up for text messages, and other similar metrics were used by most communities to gauge effectiveness. These points of engagement were compared to the rate of disease spread, number of COVID-19 tests completed, and number of vaccinations, to determine which messaging was most effective. Despite these efforts, decision makers admitted that the effectiveness of their messaging was likely impacted by the public's ability to respond, as supply chain, transportation, financial, and other issues hampered the ability of some populations to make behavioral modifications or follow testing and vaccination guidance.

14.2.9 Messaging Burnout

Decision makers experienced burnout during the pandemic, often related to the perception that, despite caring deeply about providing high-quality actionable information to their constituents, these constituents rapidly lost interest or chose not to follow guidance being provided to them. Several spoke of the frustration caused by ever-increasing morbidity and mortality statistics, despite their best efforts to communicate prevention information to the public. Decision fatigue also contributed to messenger burnout. Decision makers described the difficulty of making numerous high-stakes decisions daily, for months. Many worked within organizations that were not designed to support this rapid decision making.

Like the burnout experienced by decision makers, burnout among the public was sensed as well. Decision makers noted that the public rapidly began to express burnout over the constant stream of COVID-19 information being directed at them. As the pandemic wore on, the public seemed less and less responsive to COVID messaging. To overcome this burnout, decision makers attempted to convey information in creative ways, reemphasizing necessary precautions to slow the spread of the virus. Some acknowledged information overload as a related issue and made efforts to streamline and simplify their communications.

14.3 Recommendations

Themes identified through this study inform recommendations for decision makers that will improve communication of scientific information to spur collective action during future crises. Many of these recommendations are actions that should be implemented prior to future crises, when urgency is at a low ebb. The recommendations are focused on improving communication systems, as well as the capacity of the individuals within those systems.

One recommendation is to identify those persons who will be responsible for communications. Critical response time was lost during the early stages of the pandemic as the communication roles of decision makers were clarified. Those who will be responsible for communicating scientific information during future crises should be identified prior to these crises occurring. Roles and responsibilities should be clarified such that response times can be shortened. This may occur alongside more routine emergency response planning.

Once those who will be responsible for communications have been identified, those individuals should receive training in the communication of scientific information. Many of the decision makers in this study described difficulty in communicating scientific information, including medical guidance and statistical information, particularly in this context since stay-at-home orders, based on the best available science at the time, had a large impact on people's lives. Those responsible for communication should also receive training in how to address and counter misinformation, if appropriate. Training should be based on the best available evidence on effective scientific messaging.

Similar to issues caused by decision makers' uncertainty over their individual roles, delays and miscommunication also occurred due to confusion regarding intraorganizational and inter-organizational channels of communication. These channels should be evaluated and clarified prior to future crises, such that roles and responsibilities are understood at each level of the response system. Emphasis should be placed on the respective roles of local and state health departments, and how communications flow through these organizations to other agencies and individuals with a role in communication. Thought should also be given to nontraditional sources of scientific information, including chambers of commerce, university systems, and others who serve specific subsections of the public.

In addition to clarifying intra- and inter-organizational channels of communications, various appropriate platforms for communication should be identified. Most decision makers described the necessity of utilizing varied platforms to reach targeted subpopulations. Familiarity with newer forms of technology, including social media platforms, should be fostered. However, a need to re-familiarize with and/or maintain use of older technologies was also noted. For example, several decision makers described the necessity of using older conference call technology in addition to web-based conferencing technology such as Zoom so that those who were unfamiliar with newer technologies were not neglected.

Target populations should also be identified and studied prior to future crises, as certain platforms may be more effective for different subpopulations. Many decision makers described targeting known vulnerable populations. However, several previously unknown or misunderstood vulnerable populations were uncovered through the pandemic. Agencies and organizations should reconsider which subpopulations among their constituents should be specifically targeted should a similar crisis arise in the future. The unique needs of these subpopulations should be considered and efforts should be made to increase the cultural competency of decision makers to improve the effectiveness of communications.

Finally, in preparation for future crises, effort should be made to further understand the role of trust in spurring collective action. Known mistrust should be addressed. Evaluations of mistrust that arose during the COVID-19 pandemic should also be performed. Our findings suggest that deep mistrust in science and in government could hamper collective action and the effective response to future crises.

While the above recommendations focus on efforts that should occur prior to future crises, recommendations did emerge for scientific communication during crises as well. Firstly, communication expectations should be managed when responding to crises. Many decision makers felt that, in hindsight, they would have had fewer miscommunications and misunderstandings had they framed the pandemic response as an evolving situation from the start. They felt this may have supported the perception that changing guidance was due to an evolving understanding of COVID-19 rather than contradictions of previous guidance. Relatedly, several decision makers noted their efforts to provide guidance that would be likely to remain stable over time when possible. In the context of COVID-19, behaviors such as hand washing, mask wearing, and social distancing would fall into this category.

14.4 Conclusion

In conclusion, this research highlights the significant role played by decision makers in Alabama, Mississippi, and Louisiana in combating the COVID-19 pandemic through effective communication strategies. By conducting in-depth interviews, the study identified key themes and best practices for disseminating scientific information to diverse and vulnerable populations. Decision makers had to navigate unique challenges such as misinformation, fluctuating public trust, and institutional coordination issues while adapting their messaging to resonate with specific community needs. The reliance on both traditional and new communication platforms underscored the importance of versatile and culturally competent strategies in promoting public health measures.

The results reveal that clear identification of communication roles, pre-crisis training in scientific communication, and improved inter-organizational coordination are critical steps towards enhanced pandemic response efforts. Furthermore, targeting messaging to vulnerable populations and repairing historical mistrust are vital for achieving effective collective action. Understanding that guidance must evolve with emerging scientific knowledge, decision makers are advised to manage communication expectations and prioritize stable, consistent public health messages. This study's insights offer valuable recommendations for improving communication systems and fostering public trust, ultimately aiming to mitigate the impacts of future health crises.

References

Brennan-Olson S, Williams LJ, Holloway KL et al (2015) Small area-level socioeconomic status and all-cause mortality within 10 years in a population-based cohort of women: data from the Geelong Osteoporosis Study. Prev Med Rep 2:505–511. https://doi.org/10.1016/j.pmedr.2015.05.011

Centers for Disease Control and Prevention (2007) Interim pre-pandemic planning guidance: community strategy for pandemic Influenza mitigation in the United States- early, targeted, layered use of nonpharmaceutical interventions. https://stacks.cdc.gov/view/cdc/11425. Accessed 31 July 2024
Centers for Disease Control and Prevention (2020a) New adult obesity maps. https://web.archive.org/web/20240503025458/, https://www.cdc.gov/obesity/data/prevalence-maps.html
Centers for Disease Control and Prevention (2020b) Stats of the states: chronic lower respiratory disease mortality. https://www.cdc.gov/nchs/pressroom/sosmap/lung_disease_mortality/lung_disease.htm
Centers for Disease Control and Prevention (2020c) National diabetes statistics report. https://web.archive.org/web/20240224010233/, https://www.cdc.gov/diabetes/data/statistics-report/index.html?CDC_AA_refVal=, https://www.cdc.gov/diabetes/data/statistics/statistics-report.html
Centers for Disease Control and Prevention (2020d) Coronavirus disease 2019 (COVID-19). https://web.archive.org/web/20240221014229/https://www.cdc.gov/coronavirus/2019-ncov/your-health/understanding-risk.html?CDC_AA_refVal=https%3A%2F%2Fwww.cdc.gov%2Fcoronavirus%2F2019-ncov%2Fneed-extra-precautions%2Findex.html
Centers for Disease Control and Prevention (2023) The U.S. Public Health Service Untreated Syphilis Study at Tuskegee. https://www.cdc.gov/tuskegee/index.html Accessed 31 July 2024
Centers for Disease Control MMWR (2020) Preliminary estimates of the prevalence of selected underlying health conditions among patients with coronavirus disease 2019-United States, February 12-March 28, 2020. MMWR Morbidity & Mortality Weekly. Rep. 69. https://doi.org/10.15585/mmwr.mm6913e2
Corvey K (2021) Katie Corvey: a participatory study of how policy-makers and marginalized communities in the American South consume and act on scientific information in the context of Covid-19. Columbia Academic Commons, Academic Unit: Northeast Big Data Innovation Hub, Series: COVID-19 Research Lightning Talks. https://doi.org/10.7916/3vd4-7y86
Erhunmwunsee L, Joshi MBM, Conlon DH, Harpole DH (2012) Neighborhood-level socioeconomic determinants impact outcomes in nonsmall cell lung cancer patients in the Southeastern United States. Cancer 118:5117–5123. https://doi.org/10.1002/cncr.26185
Gallup (2023) Trust in Government. https://news.gallup.com/poll/5392/trust-government.aspx Accessed 31 July 2024
Health Resources & Services Administration (2020) Workforce shortage areas. https://bhw.hrsa.gov/workforce-shortage-areas Accessed 31 July 2024
Justwan F, Williamson RD (2022) Trump and trust: examining the relationship between claims of fraud and citizen attitudes. PS-Polit Sci Polit 55(3):462–469. https://doi.org/10.1017/S1049096522000221
Li X, Sundquist J, Zoller B, Sundquist K (2015) Neighborhood deprivation and lung cancer incidence and mortality: a multilevel analysis from Sweden. J Thorac Oncol Off Publ Int Assoc Study Lung Canver 10:256–263. https://www.jto.org/article/S1556-0864(15)32327-3/fulltext
Saldana J (2021) The coding manual for qualitative researchers. SAGE Publications Ltd.
Sandel M, Faugno E, Mingo A et al (2016) Neighborhood-level interventions to improve childhood opportunity and lift children out of poverty. Acad Pediatr 16:S128-135. https://doi.org/10.1016/j.acap.2016.01.013
U.S. Census (2022) Poverty in the United States: 2021. https://www.census.gov/content/dam/Census/library/publications/2022/demo/p60-277.pdf
U.S. Department of Commerce (1993) Poverty in the United States- changes between the censuses. https://web.archive.org/web/20250308082127/, https://www.census.gov/library/publications/1993/demographics/sb93-15.pdf
Vosoughi S, Roy D, Aral S (2018) The spread of true and false news online. Science 359(6380):1146–1151. https://doi.org/10.1126/science.aap9559
Van Vuuren CL, Reijneveld SA, Van der Wal MF, Verhoeff AP (2014) Neighborhood socioeconomic deprivation characteristics in child (0–18 years) health studies: a review. Health Place 29:34–42. https://www.sciencedirect.com/science/article/abs/pii/S1353829214000793?via%3Dihub
Webb RM, Kurtz L (2022) Politics v. science: How President Trump's war on science impacted public health and environmental regulation. Prog Mol Biol Transl Sci 188(1):65–80. https://doi.org/10.1016/bs.pmbts.2021.11.006

Kathryn Corvey is Director of the BS in Public Health program and Assistant Professor in the Department of Health Policy and Organization at the University of Alabama at Birmingham (UAB) School of Public Health. She received her Bachelor of Arts degree in International Affairs from Kennesaw State University and her MPH and DrPH from UAB. She teaches undergraduate courses in Public Health Practice, Public Health Systems, Profession of Public Health, and Capstone. Many of these courses incorporate service-learning and/or undergraduate research. She has also taught graduate-level courses in Health Policy and Public Health Management and Evaluation. Her primary area of research is public health pedagogy. Other research interests include health insurance coverage and adequacy, health care utilization and costs, children with special health care needs, public health systems and practice, and public health communication.

Janna R. Willoughby is an Assistant Professor at Auburn University, where she leads a dynamic research group focused on evolutionary genetics and population dynamics. Her research investigates the mechanisms of various evolutionary processes, focusing on the evolution of knowledge in human communities and the genetic diversity and disease dynamics of wild animal populations. Through these efforts, her work has important applications for both conservation and human health. Dr. Willoughby's research spans a diverse array of species—including caribou, kangaroo rats, mosquitoes, and songbirds—integrating genomics, ecological modeling, and field-based approaches to address pressing challenges in planetary health. She actively collaborates with government agencies, non-profits, and interdisciplinary teams to develop innovative solutions that bridge ecological and evolutionary perspectives. Dr. Willoughby has authored numerous peer-reviewed publications and regularly contributes to national and international scientific initiatives.

Ryan D. Williamson is an Assistant Professor of political science at the University of Wyoming. He received his doctorate from the University of Georgia and his undergraduate degree from the University of Alabama at Birmingham. Before joining the faculty at University of Wyoming, he served as an American Political Science Association Congressional Fellow in the U.S. Senate Committee on Rules and Administration and as a governance fellow with the R Street Institute. His research focuses on American politics, with a particular emphasis on elections, legislative politics, and institutional development. He has published articles in journals such as the *Journal of Politics*, *Political Research Quarterly*, *State Politics & Policy Quarterly*, and *Electoral Studies*. He has also recently published a book, *Nationalized Politics*, with Oxford University Press.

Mina Becker holds a Master's in Clinical Laboratory Science from the University of Alabama at Birmingham's School of Health Professions and a Bachelor of Science in Biomedical Science from Auburn University. While at Auburn University, she served as a research assistant in the Department of Forestry and Wildlife and a learning assistant in the College of Science and Mathematics. As an undergraduate research assistant, Mina contributed to the National Science Foundation (NSF)-funded RAPID project for grant #2032133 by collecting and summarizing articles for the systematic review of science-based messaging during the COVID-19 pandemic.

Kelly Dunning is the Timberline Professor of Outdoor Recreation at the University of Wyoming Haub School of Environment and Natural Resources. Her lab, the Wildlife & Wilderness Recreation Lab, looks at issues at the heart of public policy, conservation, and human well-being. Her interdisciplinary lab is interested in policy responses to scientific phenomena across a wide range of topics and settings.

Communicating Crises Collaboratively: Employing a Journalist-Scholar Partnership to Implement and Evaluate COVID-19 Communication

15

Natasha Strydhorst, Scott Burg, and Asheley R. Landrum

Abstract

Our COVID-19 research, communication, and evaluation project leveraged an existing National Science Foundation (NSF)-funded project team to research and evaluate public knowledge gaps and public-facing COVID-19 communication strategies as a multidisciplinary group of journalists, scholars, and evaluators. Using the established researcher-practitioner partnership, the project explored how COVID-19 coverage could most effectively engage as well as educate young audiences about viral transmission and its prevention, and which journalistic practices could best be applied to pandemic communication and translated to other media outlets. The collaboration resulted in identification of top COVID-19 misinformation topics and journalistic stories to address them, two research reports for general audiences, and two peer-reviewed articles (with other studies in process), as well as a series of process evaluation reports detailing the collaboration's challenges and opportunities. All of these have been made publicly available at the project website (https://www.kqed.org/crackingthecode) and on https://informalscience.org. Crisis communication—including through future pandemic scenarios—may benefit from developing similar collaborative researcher-practitioner partnerships to

Research Funded by

NSF Award #1811019 DRL*—Collaborative Research: Influencing Millennial Science Engagement;

NSF Award #1810990 DRL*—Collaborative Research: Influencing Millennial Science Engagement;

NSF Award #2028473—RAPID**: Collaborative Research: Influencing Young Adults' Science Engagement and Learning with COVID-19 Media Coverage;

NSF Award #2028469—RAPID**: Collaborative Research: Influencing Young Adults' Science Engagement and Learning with COVID-19 Media Coverage.

*Division of Research on Learning in Formal and Informal Settings.
**Rapid Response Research Award.

COVID Information Commons Video Presentation

The COVID Information Commons (CIC) Research Lightning Talk video presentation and individual multilingual transcripts for this project can be found in the Columbia Academic Commons (AC) (Landrum 2021).

N. Strydhorst (✉)
Morgridge Institute for Research, University of Wisconsin-Madison, Madison, WI, USA
e-mail: nstrydhorst@morgridge.org

S. Burg
Rockman Et Al Cooperative, Inc., San Francisco, CA, USA
e-mail: scott@rockman.com

A. R. Landrum
Arizona State University, Tempe, AZ, USA
e-mail: asheley.landrum@asu.edu

F. D. Hudson (ed.), *The COVID Information Commons*, Sustainable Development Goals Series,
https://doi.org/10.1007/978-3-032-06125-6_15

leverage the expertise of journalists and researchers to produce, evaluate, and refine public-facing crisis coverage in the teeth of substantial, evolving, and immediate threats. This chapter details our collaborative model, summarizes the journalistic, academic, and evaluative deliverables it produced, and outlines findings of proposed practices and partnership dynamics for future crisis communication scenarios.

Keywords

Crisis communication · COVID-19 pandemic · Infodemic · Research-practice partnership · Interdisciplinary task force

15.1 Introduction

The global COVID-19 pandemic, far beyond merely forcing a pivot in our group's research and practice, upended commerce, education, industry, and healthcare around the world. It was the first global pandemic in a century. It was also arguably the first accompanied by a polarized, polluted, social-media-saturated, (mis)information environment that was quickly dubbed an "infodemic" (Cinelli et al. 2020)—a setting utterly alien to the media backdrop of the last global pandemic, in 1918.

This maelstrom inevitably posed challenges to researchers and journalists alike, as well as the publics which they both aimed to understand and inform. Speed and volume presented particular challenges: the pace and quantity with which (mis)information arose and piled up necessitated rapid and efficient communication to audiences. Meanwhile, the novelty of the virus and pandemic presented challenges and opportunities for researchers investigating the infodemic milieu and the effects of efforts by media organizations in the midst of the crisis.

As the pandemic and infodemic presented these interlocking challenges and opportunities to journalists and communication scholars, our group leveraged an existing journalist/communication researcher partnership and a framework initially developed and refined as part of our initial National Science Foundation (NSF) grant (see below), to apply in the COVID-19 era. This framework enabled us to develop a novel public communication strategy and deployment plan, a process for evaluating its impact and effectiveness—as well as the partnership's—and a tentative roadmap for applying such collaborations to future societal challenges.

The large-scale collaboration featured in this chapter commenced in 2018 with a \$3 million NSF grant to develop a research-practice partnership consisting of KQED, a San Francisco-based National Public Radio (NPR) and Public Broadcasting System (PBS) member station; the Science Communication and Cognition Research Group, then at Texas Tech University; the evaluation, research, and consulting firm Rockman et al. Cooperative; and—in the early stages—Yale University. The project, "Cracking the Code: Influencing Millennial Science Engagement" (NSF awards #1810990 and #1811019) was developed to investigate means of drawing more diverse audiences into science media, and ended up spanning over three years. Research on science curiosity indicated swathes of "missing audiences"—members of the public who have high science curiosity, yet engage less frequently than expected with science media (Kahan et al. 2017). Research in the early phases of the grant found science-curious women were a potential missing audience for KQED's online Deep Look videos, and were in the midst of exploring awe-inspiring narratives' ability to spur engagement with science news coverage (e.g., Landrum et al. 2022) when the pandemic struck. KQED's science reporting abruptly shifted to account for the torrent of pandemic information—and counter the accompanying torrent of misinformation—while pandemic restrictions thwarted planned studies in the psychophysiology lab.

Needing to be agile, the team leveraged its existing collaboration acumen to shift the research and practice foci onto the unfolding crisis, applying for and obtaining a 2020 NSF Rapid Response Research collaborative grant for Advancing Informal Science, Technology, Engineering and Mathematics (STEM) Learning

(RAPID/AISL) (NSF awards #2028473 and #2028469) to develop and evaluate COVID-19 communication, along with the merits and challenges of the collaboration itself (NSF 2021). Both the original NSF grant in 2018 (NSF award #1810990) and the additional RAPID/AISL grant included funding to retain an evaluator from Rockman et al. Cooperative to methodically evaluate the collaborative model's strengths and weaknesses as a research-practice partnership (RPP).

15.2 Goals

In theory, science communication scholars are invested in uncovering findings with broad applicability to real-world science communication. In practice, there often yawns a gulf between this theory-grounded work and the audiences—science journalists and other communicators of science—intended to benefit (Gerber et al. 2020). Much academic research is ensconced behind paywalls that confine its access to fellow academics. There is an interest barrier as well: many practitioners do not see academic studies as relevant or interesting to their on-the-ground work. Even with increasing volumes of open-access publications, science journalists are rarely privy to academic findings about their field, while academics may have little communication with practitioners outside of intentional, broadly disseminated—though still rare (Radcliffe and Lewis 2021)—models such as The Conversation.[1] One overarching project goal, therefore, was to establish and evaluate a collaborative model for conducting and applying science communication research to audience-focused practice.

In a nutshell, the overarching project goal—even before the emergence of COVID-19—was to "develop and advance best practices for engaging and informing diverse millennial audiences on science content and information, through a combination of rigorous research methodology and professional science media knowledge and experience" (Burg 2022a, par. 8). When the pandemic emerged, forcing a research and practice pivot, the RAPID/AISL grant project introduced a narrower goal: producing and researching effective strategies of informing, engaging, and educating young audiences about viral transmission and prevention. This involved producing enormous amounts of science news coverage—necessitated by the speed and volume of the pandemic and accompanying infodemic—and studying the effectiveness of various communication strategies, as well as COVID-19 knowledge and knowledge gaps among audiences. Since public audiences were a key focus for the researchers and journalists alike, a particularly strong objective of the evaluation side was to "inform project participants, stakeholders and the general public on progress and research findings through regular ongoing reports" (Burg 2022b, par. 4).

In addition to the overarching research and practice goals, the project aimed to evaluate the interdisciplinary researcher/journalist partnership model itself. An independent evaluator was therefore retained throughout to observe, evaluate, and disseminate findings on the strengths and weaknesses, opportunities, and challenges presented by the model. This aspect provided a launching point to hypotheses about the RPP's translatability to other contexts (e.g., different news agencies and research groups, non-academic environments, non-pandemic vs. pandemic collaborations).

15.3 Results

As a three pronged effort, this collaboration: (i) produced public-facing science communication about the pandemic; (ii) investigated the effectiveness of that communication among its intended audience; and (iii) evaluated the promises and pitfalls of this RPP model, with implications for future academia/journalism bridging endeavors.

[1] *The Conversation* is an online publication that aims to spotlight the work of experts among public audiences. It does this by approaching (or receiving pitches from) academic researchers, pairing them with journalists, and collaboratively producing public-facing stories of the scientific research to be made freely available to general publics (The Conversation 2023).

15.3.1 Broad Findings

The studies this collaboration embarked upon began with research questions posed by researchers and practitioners about how to engage missing audiences—that is, how to reach audiences who were curious about science but not substantively engaging with science content. With the advent of the pandemic, new questions, methods, and studies emerged to address the infodemic crisis accompanying the viral one. The collaboration began investigating content produced by KQED's science news team and, more broadly, misinformation topics and consequent knowledge gaps among audiences. The social media platform X (then Twitter) conversations from March to October 2020 teased out circulating misinformation about COVID-19. This included suggestions that the pandemic was a hoax, COVID-19 was comparable to influenza, wearing face masks was hazardous and could cause CO_2 poisoning, and that certain foods (such as garlic) could combat or kill the virus.

The team conducted studies on masking messages to combat the misinformation that COVID-19 was no more hazardous than the flu and that wearing masks posed health risks. One study (Landrum and Huxster 2021) presented participants with an infographic of how droplets travel from a sneeze or cough with and without masks. KQED subsequently incorporated the infographic into social media materials, investigating whether the graphic as still image, Graphics Interchange Format (GIF), or video encouraged more engagement from audiences. It was determined that videos were the most enticing. Additional studies investigated (mis)perceptions among publics via survey (Opat and Landrum 2021) and in-depth interviews (Strydhorst and Landrum 2022) documenting misperceptions. All the research reports produced were disseminated to public audiences via a blog post written by the researchers and practitioners to highlight the findings in public-friendly language—these posts can be found at the project website kqed.org/crackingthecode (KQED 2019). Additionally, evaluation reports were made publicly available via Medium blog posts, as well as KQED's website and social media platforms.

Identifying circulating misperceptions presented communication targets for KQED. Further, the rapid-fire pace necessitated by the pandemic context meant journalists had to adjust their reporting to produce content that was: "More thoughtful, relevant, and actionable for communities under duress…. This required a more acute understanding of their audience, a place where science communication research could be most helpful. Insights gained through… research studies conducted during the pandemic enabled the science news staff to think differently about their work and relationships with diverse audiences" (Burg 2022c, par. 14).

Evaluation of the collaborative model turned up other valuable insights as well. On the researchers' part, entering the project steeped in published literature, it was valuable to encounter the media landscape from the perspective of those working directly in it. What media professionals are most keen to uncover about their audiences, and why, are salient targets for science communication research. The collaborators learned from and about each other and their respective vocational worlds, in addition to producing content to distribute throughout those worlds. In this, both groups had to learn something relatively novel to each of them: how to write effectively for social media contexts.

There were challenges as well. While the team shared broad research and practice goals, the expectations of research's applicability and timeliness to influence the day-to-day reporting of science in practice tended to vary between researchers and journalists. The reporting styles understandably varied, to best serve the different audiences each was accustomed to addressing, as did the timeframes within which those addresses were expected. Researchers and journalists are also driven by different types of professional and organizational incentives. Academic rewards are predicated on the number of research-based publications, a time-consuming, carefully vetted activity, which enhances opportunities for promotion and tenure. Journalists work in a totally different, more bottom-line-oriented environment, where funding to support work comes not from applying for grants, but from advertisers or the private sector, and where value is judged in terms of volume and brevity, and meeting tight deadlines. The research results were neither as conclusive nor as timely as practitioners had hoped, though still relevant to ongoing practice. From the academic perspective, the project was never designed to mirror marketing research; the goal was to observe trends with applicability for the field as a whole, rather than develop strategies particular to KQED's content and audience.

Project participants suggested that occasional discussions of the research-practice relationship were insufficient: it would have been preferable to develop and maintain an immersive, continual reflective practice to center the theory and application, research and practice, linkages, and implications.

15.4 Deliverables

In addition to the insights and cross-disciplinary experience gained, the collaboration produced materials for general, academic, and mixed audiences throughout the project.

Journalistic Deliverables KQED maintained a live blog to answer audience questions about COVID-19 in real time, an innovation at that time. The format had never been used by KQED before, but was a natural extension of the audience engagement mission, enabling science reporters to communicate directly and instantaneously with audience members in a reciprocal format. They also reached diverse audiences via radio broadcasts, daily news coverage on their website, Spanish translations of disaster news coverage, and social media materials reporting on the pandemic and the science behind viral transmission, mask-wearing, vaccines, social distancing, and other salient topics. This coverage was informed by the science communication scholarship conducted as a part of the collaboration.

The KQED website hosted—and continues to as of this 2024 writing—reports co-written by journalists and academics detailing project findings at a general audience level (KQED 2019). These reports detail the project's scope and findings, offering takeaways from audience studies ("Deep Look"), COVID-19 research, and the national surveys conducted, collaborative best practices, and the project evaluation.

Academic Deliverables The studies conducted through the project produced deliverables for general audiences and domain researchers, and practitioners. Two research reports, which are hosted on KQED's project website, were written for general audiences to inform the science curious and the general public (KQED 2019). Two peer-reviewed articles with deeper research context and results were published with open access to be accessible to academics, journalists, and the general public alike. Further papers are in progress.

The first research report (Landrum and Huxster 2021) details a United States (U.S.) nationally representative survey that enrolled 2871 participants and investigated the effects of presenting a consensus message about masks' utility and an infographic illustrating the message. The study found that, while political party affiliation was the greatest predictor of participants' policy preferences and COVID-19 attitudes, viewing the infographic was associated with greater participant agreement that masks were efficacious and that there was a scientific consensus to that effect (regardless of whether participants also viewed the consensus message text). The second national survey (Opat and Landrum 2021) enrolled 3006 individuals and investigated misperceptions about germs and vaccines, generally and the COVID-19 virus and vaccines, specifically. Particularly prevalent misperceptions captured at that time included the beliefs that antibiotics and hydroxychloroquine were effective preventatives and/or treatments for COVID-19, that COVID-19 was caused by a bacterial infection, that vaccines cause autism, and that viruses are able to replicate without a live host.

An in-depth interview study (Strydhorst and Landrum 2022), enrolling 27 participants from Texas, New Jersey, and California, investigated which (mis)perceptions were top-of-mind for individuals at that point in the pandemic: the summer of 2021. While most participants were found to hold at least an elementary understanding of COVID-19, many relayed that members of their networks clung to persistent misperceptions—and some participants held unconfirmed and false pandemic beliefs. Among these were perceptions that COVID-19 was developed as a bioweapon and that the speed with which vaccines became available suggested that their safety was suspect. A survey study probed 3276 participants' media consumption habits and beliefs in false COVID-19 and vaccination claims (Strydhorst et al. 2023). Using Ad Fontes' media bias and reliability ratings (Ad Fontes Media 2023) for participants' news sources, the study found that a media diet's average bias and reliability was a significant predictor of participants' false beliefs about both COVID-19 and vaccination.

Evaluative Deliverables The evaluation component, produced by including a professional evaluator in the project, yielded insights into both the challenges and opportunities presented by a collaborative researcher-practitioner partnership.

Findings indicated that some of the most substantial challenges were posed by differing reporting timescales and visions of findings' applicability (narrowly targeted to KQED's audience and conclusively determined vs. broadly applicable to U.S. audiences and tentatively arrived at). The strengths of the collaboration included providing a tangible and productive link between researchers and practitioners, crucially connecting academics to the audiences their research questions concerned and journalists to the research that could inform their work to reach those audiences. Both researchers and practitioners involved in the project developed greater understanding of and appreciation for the other's work styles, vernacular, professional incentives (e.g., tenure versus job promotion), audiences, and methods for developing consensus. Creative tensions led to collaborative problem solving.

The project was also successful in producing content for general audiences and academics alike, and bridged the frequent vernacular gap by producing reports collaboratively, making them linguistically and literally accessible. The findings, in addition to informing the project work, were made publicly available via an open-access 14-part Medium blog series (Burg 2022d) and a three-part best practices report series published on the project website (KQED 2019). The academic deliverables were additionally published with open access to preserve their availability to all audiences.

15.5 Recommendations

Despite the evident applicability of researcher-practitioner partnerships in the realm of science communication, these formal collaborations between science communication scholars and practicing journalists remain uncommon, but we are hopeful that the positive results of our work will inspire future multidisciplinary collaborations to benefit the public good. There are, of course, challenges to implementing such RPPs: work styles, vernaculars, and timescales may be frequent friction points, along with non-overlapping bureaucratic sources of red tape (such as Institutional Review Boards (IRBs) at academic institutions and legal departments at news outlets). Even so, this collaboration was a fruitful one, producing science content for general audiences, as well as publications in the academic sphere and a roadmap of considerations for future RPP endeavors.

To make best use of researchers' and journalists' expertise in a collaboration of this nature, our findings recommend the following model (Fig. 15.1) as a launching point for future RPPs—particularly in a pandemic context, for which time is of the essence.

In this model, journalists identify issues in practice and formulate questions to address them, at which point researchers and journalists collaboratively discuss the questions and relevant literature. Researchers then take the discussion points and existing literature to develop hypotheses and design a study to investigate them. Another round of collaborative design work follows with researchers and journalists,

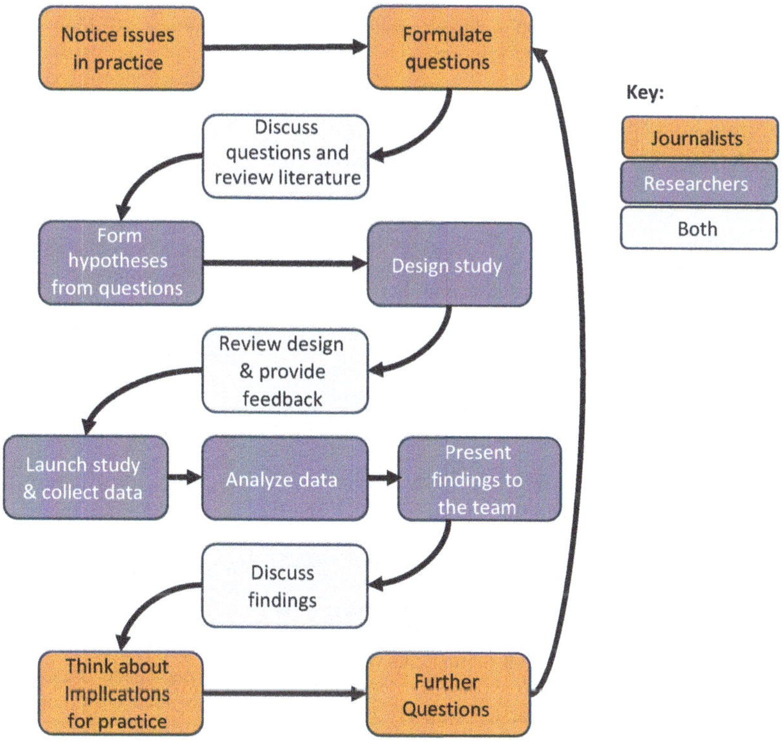

Fig. 15.1 The stepwise process of enacting a researcher-journalist research-practice partnership, identifying the roles of journalists (in orange) and researchers (in purple)

after which researchers launch the study, collect and analyze data, and present the findings to the team. At this stage, journalists and researchers discuss the findings, then journalists consider the implications for their practice and develop further questions, at which point the process begins again.

This project identified both friction points and opportunities from which future RPPs can take note. A collaboration that begins with the understanding that workflows, timelines, vocabularies, and expectations are likely to be divergent between the parties rather than discovering these frictions along the way—is positioned to adjust expectations and adapt to one another's systems.

This project highlights the value research-practice partnerships bring to the pandemic communication context. The nimbleness of journalism and research methodology of communication science presents opportunities to rapidly inform public audiences, assess communications, and adapt strategies to expand and improve upon public engagement with research in a crisis context. It is our hope that this collaborative model can inform future endeavors when next public health crisis strikes.

References

Ad Fontes Media (2023) Methodology. https://adfontesmedia.com/methodology/

Burg S (2022a) Cracking the code: a science media—research collaboration: a three year case study. https://medium.com/cracking-the-code-a-science-media-research/cracking-the-code-a-science-media-research-collaboration-df77742d5aa9. Accessed 25 July 2024

Burg S (2022b) Cracking the code: a science media—research collaboration: reporting and dissemination: reaching audiences. https://medium.com/cracking-the-code-a-science-media-research/cracking-the-code-a-science-media-research-collaboration-d96eeb13de5a. Accessed 25 July 2024

Burg S (2022c) Cracking the code: a science media—research collaboration: collaborating during a pandemic. https://medium.com/cracking-the-code-a-science-media-research/cracking-the-code-a-science-media-research-collaboration-bd00571166. Accessed 25 July 2024

Burg S (2022d) Cracking the code: a science media—research partnership for improving the quality of

science communication. https://medium.com/cracking-the-code-a-science-media-research. Accessed 25 July 2024
Cinelli M, Quattrociocchi W, Galeazzi A, Valensise CM, Brugnoli E, Schmidt AL, Zola P, Zollo F (2020) Scala A (2020) The COVID-19 social media infodemic. Sci Rep 10:16598
Gerber A, Gabriel M, Lorenz L, Lorke J, Merten W, Metcalfe J, Müller B, Warthun N (2020) Science communication research: an empirical field analysis. E. Innovare. https://sciencecomm.science/app/uploads/2020/05/Research_Field_Analysis__Science_Communication__2020__public.pdf
Kahan DM, Landrum AR, Carpenter K, Helft L, Jamieson KH (2017) Science curiosity and political information processing. Pol Psy 38(S1):179–199. https://doi.org/10.1111/pops.12396
KQED (2019) Cracking the code. https://www.kqed.org/crackingthecode
Landrum A (2021) Asheley Landrum: collaborative research: influencing young adults' science engagement and learning with COVID-19 media coverage. Columbia Academic Commons, Academic Unit: Northeast Big Data Innovation Hub, Series: COVID-19 Research Lightning Talks. https://doi.org/10.7916/gcd8-ff33
Landrum AR, Janét K, Opat K, Akin H (2022) When science journalism is awesome: measuring audiences' experiences of awe from reading science stories. J Pract 15:1–17. https://doi.org/10.1080/17512786.2022.2116724
Landrum AR, Huxster JK (2021) Mask messaging for COVID-19: examining the effectiveness of a scientific consensus message versus an explanatory infographic. RAPID preliminary report 2. KQED.org. https://www.kqed.org/about/16011/mask-messaging-for-covid19. Accessed 20 July 2024
NSF (2021) Award abstract # 2028473 RAPID: collaborative research: Influencing young adults' science engagement and learning with COVID-19 media coverage. U.S. National Science Foundation. https://www.nsf.gov/awardsearch/showAward?AWD_ID=2028473&HistoricalAwards=false. Accessed 20 July 2024
Opat K, Landrum AR (2021) Germ knowledge: reporting during an infodemic: RAPID preliminary report 1. KQED.org. https://www.kqed.org/about/16059/germ-knowledge-reporting-during-infodemic. Accessed 20 July 2024
Radcliffe D, Lewis SC (2021) The datafication of journalism: strategies for data-driven storytelling and industry-academia collaboration. In: Bounegru L, Gray J (eds) The data journalism handbook: towards a critical data practice. Amsterdam University Press, Amsterdam, pp 314–330. https://doi.org/10.2139/ssrn.3907053
Strydhorst N, Landrum AR (2022) Charting cognition: mapping public understanding of COVID-19. PUS 31:5. https://doi.org/10.1177/09636625221078462
Strydhorst N, Morales-Riech J, Landrum AR (2023) Exploring partisans' biased and unreliable media consumption and their misinformed health-related beliefs. Misinfo Rev 4:5
The Conversation (2023) About US the conversation. https://theconversation.com/us/who-we-are

Natasha Strydhorst Ph.D. is an affiliate researcher in the University of Wisconsin-Madison Department of Life Sciences Communication and a Rita Allen Foundation Civic Science Fellow at the Morgridge Institute for Research. Her research interests include uncertainty communication, (dis)trust in experts, and narrative science communication. Natasha's research investigates the nexus of science communication practice and scholarship, seeking ways to shore up the links between scientific research and general audiences. Her background is in scholastic science communication, and she has written science-centered stories for Yale School of Medicine, Yale Center for Clinical Investigation, and Hope College. Natasha's research work has been published in *Science Communication, Public Understanding of Science*, and the Harvard Kennedy School *Misinformation Review*. She earned her Ph.D. in Science Communication from Texas Tech University, her M.S. in Science Journalism from Boston University, and her B.A. in Environmental Studies and Writing from Calvin University.

Scott Burg is a Senior Research Principal at Rockman et al. Cooperative (www.rockman.com). He has a formal background in adult education, instructional systems design, health services, and qualitative research. Scott has served as external evaluator on a wide range of projects funded by agencies such as the National Science Foundation (NSF), National Institutes of Health (NIH), National Oceanic and Atmospheric Administration (NOAA), and the National Academies, focusing on formal and informal education in the sciences and health services. Projects include studies on science communication methods for public audiences, research-practice partnerships, interdisciplinary STEM programs in post-secondary settings, environmental research in indigenous communities, and climate change education and resiliency programs. Scott has also conducted process evaluations exploring partnerships and capacity building in non-profit organizations, and studies assessing diverse eye care delivery models to rural and urban populations in such countries as Guatemala, India, and Peru. Scott has a B.A. in Literature/Creative Writing from the University of Illinois and a Master's in Library Science (MLS) from the University of Maryland (https://www.linkedin.com/in/scott-burg-1a77365/).

Asheley R. Landrum Ph.D. is an associate professor at the Walter Cronkite School of Journalism and Mass Communication and a Senior Global Futures Scientist at Arizona State University. Her research explores how values and worldviews shape information selection and processing, particularly focusing on how these dynamics develop from childhood to adulthood. Asheley is a fellow of the American Association for the Advancement of Science, and she served on the National Academies of Science, Engineering, and Medicine's consensus panel on "Understanding and Addressing Misinformation about Science." Her work has received funding from the National Science Foundation (NSF) and Templeton Religion Trust. Asheley is dedicated to bridging research and practice. She has collaborated with journalistic organizations such as KQED public media, PBS Digital Studios, and CNN's Anderson Cooper 360. These partnerships ensure that her research informs scholarly discourse and real-world communication strategies. Asheley earned her Ph.D. in Psychological Sciences from the University of Texas at Dallas, specializing in children's trust in others as sources of information (part of their social-cognitive development). She then completed postdoctoral fellowships at the University of Louisville and the Annenberg Public Policy Center at the University of Pennsylvania, solidifying her expertise in the intersection between social-cognitive psychology and public communication.

Concurrent Hazards and Resilience: COVID-19 in Hurricane-Prone Areas

16

Alka Sapat, Diana Mitsova, Ann-Margaret Esnard, and Karen D. Sweeting

Abstract

This chapter focuses on how concurrent hazards, specifically COVID-19 and hurricanes, impact individual and household coping and adaptive capacities. The National Science Foundation (NSF) funded the underlying research. As the pandemic spread and ebbed and waned through 2020 and the first half of 2021, individuals and households faced multiple challenges with health risks, precarious housing conditions, economic hardship, and multiple concurrent hurricanes. In our research, we employed the theoretical lens of subjective resilience and adaptive and coping capacity to understand individual cognition of risks and perceptions of recovery from the pandemic, particularly for those living in hurricane-prone areas facing the potential of multiple hazards. We collected individual-level data via surveys and other secondary data to explore the factors affecting these subjective perceptions of resilience and recovery, such as capacities to adapt and cope, risk perceptions, communication sources, and levels of physical and social vulnerabilities. The results of our research help fill gaps in knowledge that can improve contingency planning and policymaking to deal with multiple hazards and concurrent crises.

Keywords

COVID-19 · Concurrent hazards · Resilience · Hurricanes · Recovery · Policy

Research Funded by

NSF Award #2028968—RAPID*: Health, Housing, and Hazards: COVID-19, Subjective Resilience, Vulnerabilities, and Policy Evolution in Hurricane Prone Counties.
*Rapid Response Research Award.

COVID Information Commons Video Presentation

The COVID Information Commons (CIC) Research Lightning Talk video presentation and individual multilingual transcripts for this project can be found in the Columbia Academic Commons (AC) (Sapat 2021).

A. Sapat (✉) · D. Mitsova
Florida Atlantic University, Boca Raton, FL, USA
e-mail: asapat@fau.edu

D. Mitsova
e-mail: dmitsova@fau.edu

A.-M. Esnard
Georgia State University, Atlanta, GA, USA
e-mail: aesnard@gsu.edu

K. D. Sweeting
University of Rhode Island, South Kingstown, RI, USA
e-mail: ksweeting@uri.edu

16.1 Introduction

This chapter focuses on how concurrent hazards, specifically COVID-19 and hurricanes, impact individual and household coping and adaptive

F. D. Hudson (ed.), *The COVID Information Commons*, Sustainable Development Goals Series,
https://doi.org/10.1007/978-3-032-06125-6_16

capacities. The underlying research was funded by the National Science Foundation (NSF). The global COVID-19 pandemic led to over 1,205,773 deaths in the United States (U.S.) and over 84,540 deaths in the state of Florida for the period January 2020—September 2024 (Covid data tracker 2024). In the first year, by August 2021, there were ~600,000 deaths in the U.S. and over 37,000 deaths in the state of Florida. As the pandemic spread and ebbed and waned through 2020 and the first half of 2021, individuals and households in Florida faced multiple challenges that included health risks, precarious housing conditions, and economic hardship, along with multiple concurrent hurricanes, evident in the 2024 hurricane season and the devastating impact of these hurricanes. In the U.S., these risks were further compounded by government missteps at the beginning of the pandemic and throughout 2020, which led to the abdication of federal responsibility to the state level, leading to a fragmented policy response and considerable uncertainty. In this NSF-funded research project, we employed the theoretical lens of subjective resilience and adaptive and coping capacity to understand individual cognition of risks and perceptions of recovery from the pandemic, particularly for those living in hurricane-prone areas facing the potential of multiple hazards. We collected individual-level data via surveys and other secondary data to explore the factors affecting these subjective perceptions of resilience and recovery, such as capacities to adapt and cope, risk perceptions, communication sources, and levels of physical and social vulnerabilities.

16.2 Goals

The major goals of this COVID-related project were to capture ephemeral and time-bound data on the subjective perceptions of resilience of individuals and households, including their coping and adaptive capacities. At the height of the COVID-19 pandemic, individuals and households faced multiple challenges, including health risks, precarious housing conditions, and exposure to weather and climate hazards. This occurred within the context of rapidly evolving policy mandates and short-term measures (such as moratoriums on housing evictions) that were being implemented to address the uncertainties stemming from the growing and increasingly widespread pandemic.

16.3 Research Questions

The main research questions were: (1) How did the COVID-19 pandemic impact the subjective perceptions of the resilience of individuals and households living in areas still recovering from the 2016–2018 hurricane seasons?; and (2) To what extent did rapidly evolving and often fragmented and ambiguous federal, state, and local policy responses to the COVID-19 pandemic affect the coping and adaptive capacities of individuals and households, particularly those with intersectional vulnerabilities (such as being a renter, recovering from past hurricanes, being elderly, and having minority status)?

16.3.1 Hypothesis

Our main hypothesis was that the subjective resilience of individuals and households (i.e., their coping and adaptive capacities) would be affected by the following factors: (i) Individual and household cumulative and intersectional vulnerabilities (including health, housing, and economic needs, and past and potential hazard exposure); (ii) the immediate impact of temporary policies designed to reduce those public health, housing, and other vulnerabilities; and, (iii) spatial/regional vulnerabilities, access to infrastructure and services, and social capital.

Our study areas included counties in south and central Florida and the Panhandle, which were still recovering from Hurricanes Michael (2018) and Irma (2017), and which saw an influx of displaced persons from Puerto Rico after Hurricane Maria (2017). We subsequently included five counties in Texas after the impacts of the ice storm in February 2021 (during the project period) to better understand the impacts of dual and compound hazards and levels of household resilience, coping, and adaptive capacities.

16.4 Data Collection

To address our research questions and test our hypotheses, we undertook several research activities during the one-year project period from May 2020 to April 2021, which included primary data collection from three repeated cross-sectional household surveys in Florida and one survey in Texas to assess the impact of the pandemic on subjective perceptions of resilience, recovery, and decision-making.

The first survey was sent out in July 2020 to align with the hurricane season, and when policies, such as lockdowns and mask mandates, were already in effect and being implemented to address the fallout of the pandemic. The second survey was sent out in January 2021, after hurricane season and when vaccines had started to become available. The third survey was sent out in April 2021, a year after policy responses to the pandemic began. The first and second surveys were conducted via the Internet, cell phones, and landlines (the latter mode was used to help us reach older populations), and the third survey was a fully online survey. In addition to the surveys sent out in Florida, a separate survey was sent to five counties in Texas that experienced an ice storm in February 2021 during the pandemic. The surveys were administered through a company called Dynata, which used a variety of digital sources to recruit participants. The surveys were administered in both English and Spanish. The data was collected by Florida Atlantic University's Business and Economics Polling Initiative (BEPI).

16.5 Main Findings

Overall, the findings from our research show that the initially slow and then fractured U.S. response was accompanied by exponential surges in cases and deaths across states and counties. The politicization of pandemic mitigation measures impeded actions to slow the spread of COVID-19. Differences also existed across Florida counties and across different populations in their responses to the pandemic. *The main characteristics of the survey respondents and some key insights are given below.*

16.5.1 Florida Household Survey 1

Survey 1 included thirty-one questions focused on five key topics: (i) risk perception, adaptive and coping capacity; (ii) COVID information sources and policies; (iii) access and mobility; (iv) exposure to past hurricanes; and (v) economic hardship and social capital. A total of 1392 people over the age of 18 responded to the survey conducted in July 2020. Figure 16.1 below provides the self-reported demographics of respondents.

Some of the key findings of this survey were that the most trusted sources of information about the COVID-19 pandemic were: news and media organizations (34%); official government agencies (31%); family doctors (20%); social media (9.3%); and community (6.3%). In terms of the level of preparedness for a potential hurricane during the pandemic, the results indicated that most respondents were very well prepared (28.2%) and mostly prepared (36.1%). Some indicated that they were only somewhat prepared (26.9%) or not prepared at all (6.6%). The survey responses suggest that the majority of respondents learned to adjust to disruptions created by the pandemic—a great deal (36.5%), a lot (28.1%), a little (21.8%), and not at all (8.1%). At that point in time, respondents had varying perceptions of the timeline for returning to normal pre-COVID life. Responses were as follows: 3–6 months (17.3%); 6 months–1 year (30.7%); 1–2 years (26.6%); more than 2 years (8.9%); never (6.3%).

16.5.2 Florida Household Survey 2

Survey 2 sought to understand the extent to which households learned to adapt to the disruptions of

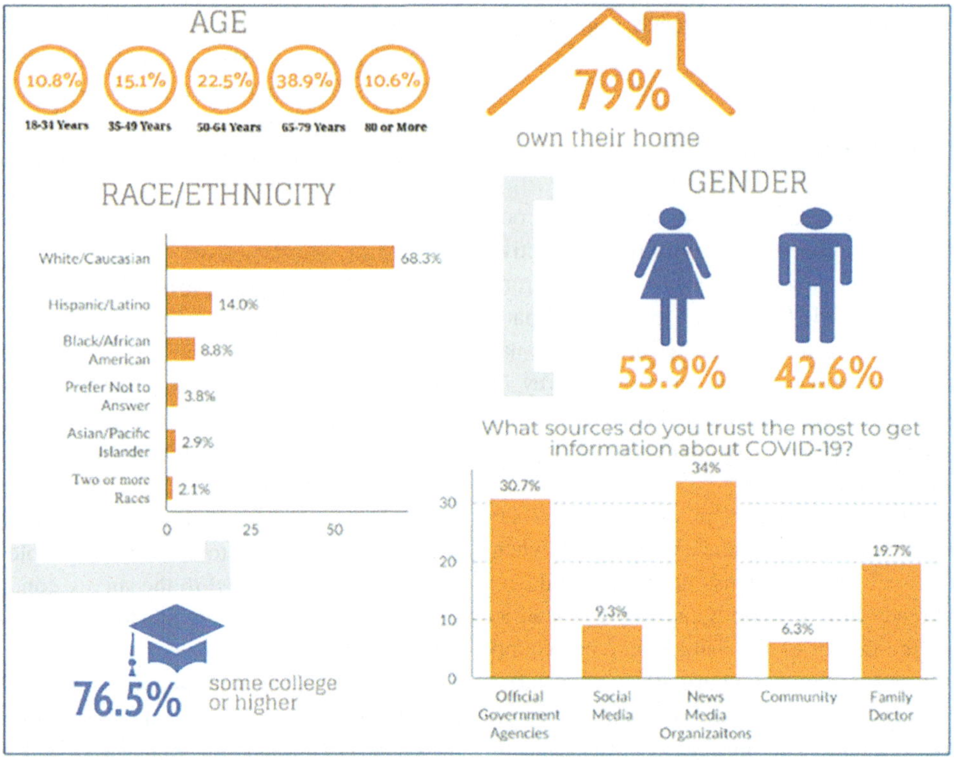

Fig. 16.1 Demographic data (Florida household survey 1)

everyday life caused by the pandemic. A total of 1320 respondents participated in the survey. The demographic profile of the survey respondents (Fig. 16.2) was similar to that in the first survey, allowing for some comparisons. The findings from Survey 2 indicated that 83.8% owned their own home, and around 80% of the respondents had a college education, which is similar to the findings from Survey 1. However, sources of information that Survey 2 respondents relied on were mostly scientists, such as Anthony Fauci (39%) and other scientists (22.8%).

By January 2021, the vaccine was starting to become available, and a majority of respondents (54.4%) indicated they would be willing to take a Food and Drug Administration (FDA) approved vaccine (Fig. 16.3). To manage the spread of COVID-19, respondents also indicated strong support for stricter COVID-19 measures (59.8%).

Figure 16.4 shows a comparative summary of the responses to the adaptive capacity (adjustment) question between Survey 1 and Survey 2. Only 8% of the respondents in both surveys indicated that they had not learned how to adjust their daily activities to the new reality imposed by the COVID-19 restrictions. Another 22% (Survey 1) to 25% (Survey 2) reported that they had learned how to adjust a little. Over 50% of the respondents in both surveys said that they had adjusted their daily activities to COVID-19 either a lot or a great deal.

16.5.3 Florida Household Survey 3

With insight from 1402 survey respondents, the third survey focused on assessing the impact of the COVID-19 pandemic on Florida households. The survey aimed to gather data to help us understand

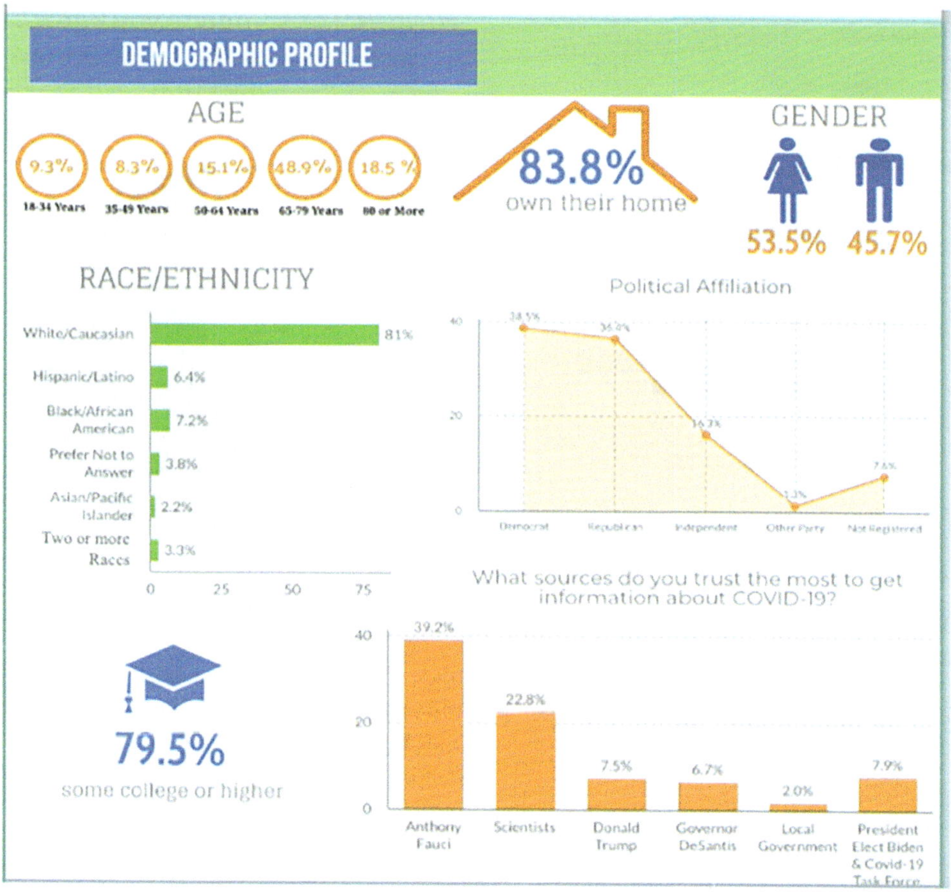

Fig. 16.2 Demographic data (Florida household survey 2)

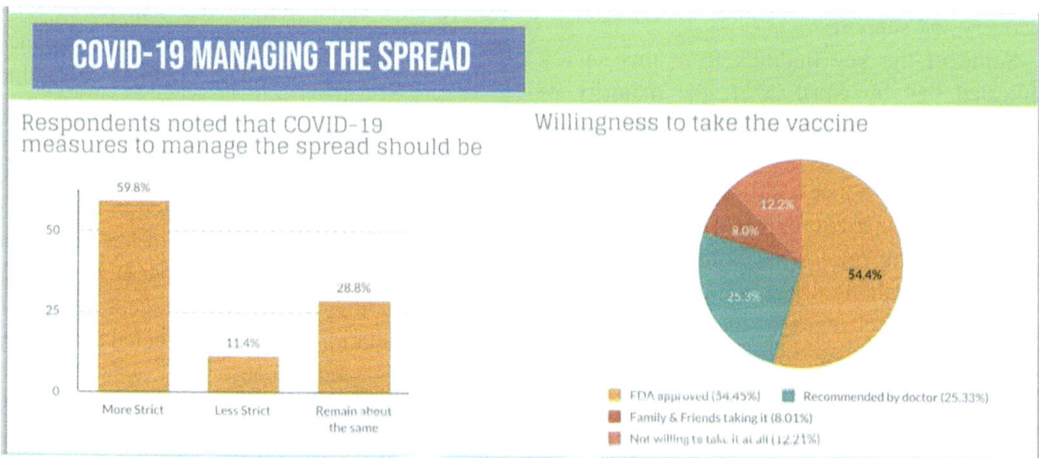

Fig. 16.3 Managing the COVID-19 spread (survey 2)

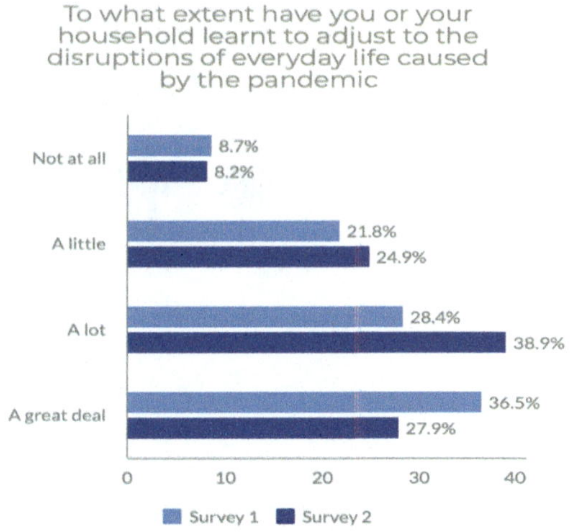

Fig. 16.4 Comparison of adaptive capacity between the respondents to survey 1 and survey 2

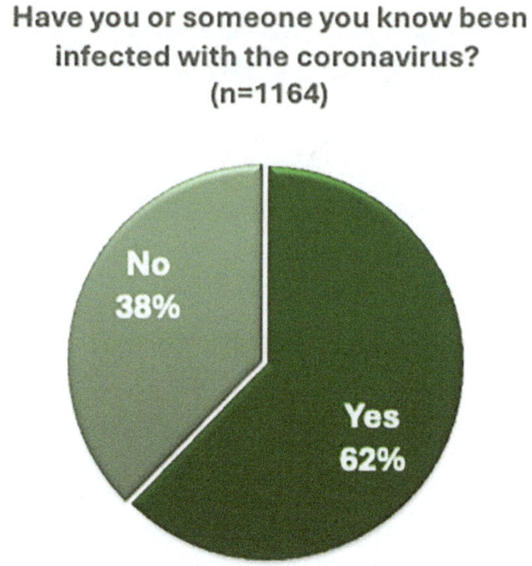

Fig. 16.5 Personal impact of COVID (survey 3)

the social and economic effects of the pandemic, including changes in financial situations, housing stability, and employment status. We focused on questions that garnered insights into adjustments to daily life, remote work, and challenges faced in working from home. The survey asked about engagement in preventive activities, vaccine uptake, hesitancy, and perceived societal benefits of vaccination. Demographic data allowed for the analysis across different age groups, income levels, and racial/ethnic backgrounds, revealing disparities and guiding targeted interventions for recovery and support.

Some of the key findings from this survey indicated that by April 2021, the majority of respondents had learned to adjust to the disruptions caused by the COVID-19 pandemic either a great deal (35%) or a lot (36%) (Fig. 16.5). This question captured the spectrum of resilience and coping strategies households employed in response to challenges like remote work, social distancing, and changes in daily routines. The responses highlight the extent to which people were able to navigate the disruptions and maintain some semblance of normalcy amid the ongoing crisis of the pandemic.

To gather further insight into the personal and familial impact of COVID-19, we asked respondents if they or someone they knew had contracted the virus (Fig. 16.5). A large majority of the respondents (62%) indicated that either they or someone they knew had contracted COVID-19, and Fig. 16.6 shows that by April 2021, the majority of respondents were vaccinated (66%) or planned to get vaccinated (12%).

16.5.4 Texas Household Survey

The focus of the Texas household survey was to understand households' experience with the process of applying for disaster assistance following a winter storm in February 2021 that impacted five Texas counties (Brazoria, Dallas, Galveston, Harris, and Jefferson) during the time of the COVID-19 pandemic. Insights were gained from primary survey data collected in May 2021 from 196 respondents, 18 years of age and older [for additional insights, see (Sapat et al. 2023)]. The five counties were chosen because they had diverse populations, had been affected previously by disasters, and because they were affected by both Winter Storm Uri and the COVID-19 pandemic.

With respect to the pandemic, by the end of February 2021, Harris County accounted for 17.43% of cases and 12.17% of deaths, and Dallas County accounted for 10.09% of cases

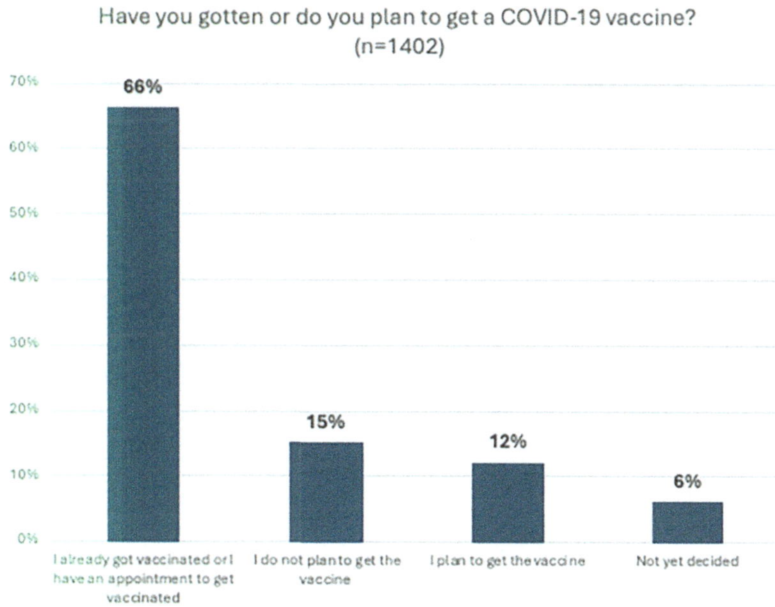

Fig. 16.6 Public sentiment regarding COVID-19 vaccine (survey 3)

and 8.17% of deaths in Texas. Jefferson, Galveston, and Brazoria counties had 1–1.5% cases and deaths (0.88–1.25%) (Texas COVID-19 data 2022). Winter Storm Uri severely compounded problems for Texas residents. It was the deadliest winter storm since 1993 (City of Austin 2021), killing 246 people and leaving close to 4.5 million homes and businesses without power at its peak (Texas Health and Human Services News Updates 2021; Busby et al. 2021). There were approximately $90 billion (Sullivan 2021) to $295 billion in estimated economic losses (Hobby School of Public Affairs at the University of Houston 2021), and only a small proportion of those losses ($20 billion of the estimated $90 billion) were covered by insurance (Sullivan 2021). The storm had other cascading impacts on those at medical risk, as well as resulting in infrastructure damage to water and transportation systems (Doss-Gollin et al. 2021; Lee et al. 2022).

Figure 16.7 displays the descriptive statistics of the respondents. A majority of the respondents owned their own home (65.3%), had some college or higher (62.3%), 51% identified as White/Caucasian, 27% as Black/African American, 15.3% as Hispanic or Latino, and 4.6% as Asian/Pacific Islander. Respondents were evenly split on gender (49.5% each male and female), and a majority were between 18 and 49 years old (82.2%), while 17.8% were 50 years and older.

Most respondents applied for individual assistance from the Federal Emergency Management Agency (FEMA) (46.9%), business assistance (20.9%), and disaster unemployment assistance (26%). Individual assistance applications were likely the highest, as ~68% of respondents experienced damage to their residence. In terms of facing compound disasters, we found that over 61% found it difficult to cope with the storm in the middle of a pandemic, and over 69% found the process of applying for government assistance somewhat difficult to extremely difficult. Several respondents found the process time-consuming (35.7%) and too complex (34.2%), while some reported that they did not receive adequate help to navigate the process (20%).

Figure 16.8 provides information on the sources of information for disaster relief and assistance that respondents found were most helpful. The responses indicate that respondents found information from the state government most helpful (38.8%), followed by information provided by local governments (33.7%) and FEMA (26.5%). Non-governmental sources of information included the internet/social

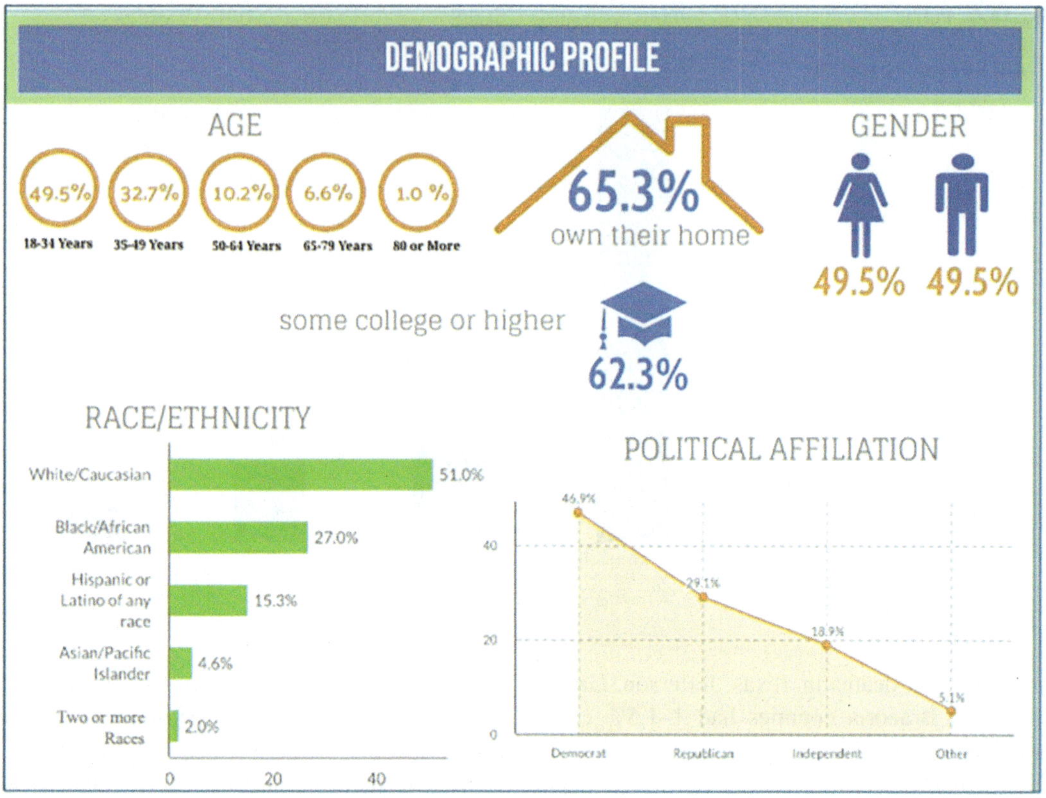

Fig. 16.7 Demographic profile of respondents (Texas survey)

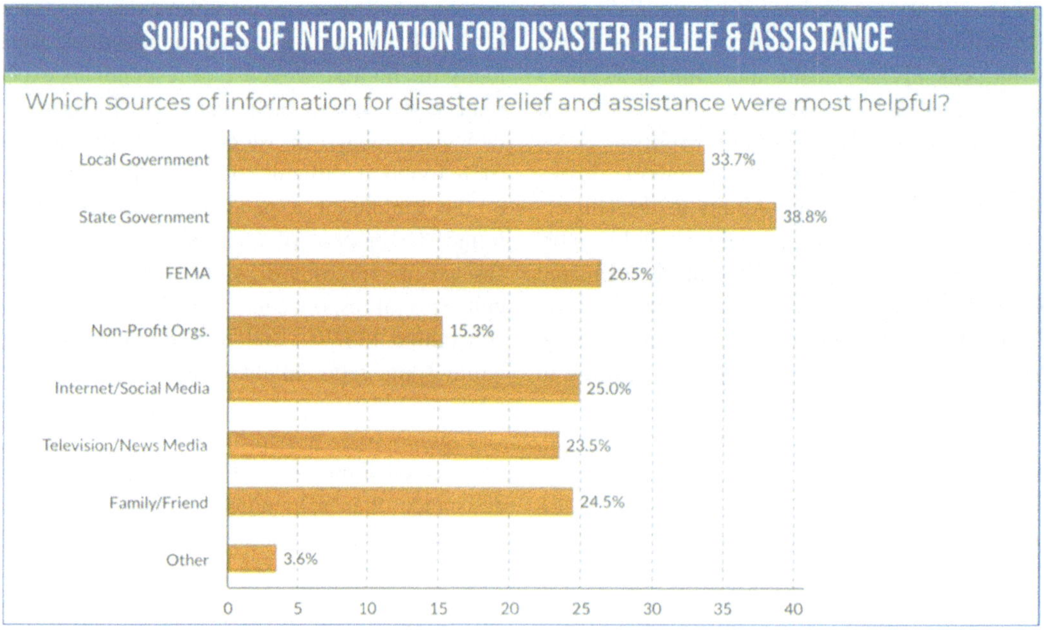

Fig. 16.8 Sources of information for disaster relief and assistance (Texas survey)

media, TV/news media, family and friends, and non-profit organizations.

16.6 Policy Recommendations

The research findings from these surveys offer practical applications for pandemic preparedness and disaster management beyond our study areas, specifically for socially vulnerable populations with respect to housing, sheltering, and evacuation in hazard-prone areas. By focusing on the impacts of the pandemic on socially vulnerable populations, this study highlights the importance of focusing policy efforts and planning practices on some less well-understood aspects of resilience.

Our findings highlight specific policy responses that policymakers could adopt to help individuals and households recover. For instance, our results show that receiving assistance from a food bank and getting a temporary job were the most important financial assistance measures that helped respondents have a more positive perception of recovery and helped them cope. While short-term measures such as stimulus checks can be useful, governments need to enable long-term income-generating measures. Enabling and funding civil society organizations to distribute food and other assistance during a pandemic can help in providing assistance more quickly to affected individuals. Effective leadership is also key to ensuring better compliance and improving recovery outcomes for individuals and households. Given the heavy health and economic toll of the pandemic, effective leadership that enables people to envision and sense a return to normalcy can lead to faster recovery from the pandemic.

Overall, the results of our research help fill gaps in knowledge that can improve contingency planning and policy-making to deal with multiple hazards and concurrent crises and contribute to theory-building on several topics, including: (i) factors influencing adaptive and coping capacities; and (ii) the impacts of policy fragmentation and ambiguity in addressing intersectional and cumulative vulnerabilities related to housing, health, and hazards.

References

Busby JW et al (2021) Cascading risks: understanding the 2021 winter blackout in Texas. Energy Res Soc Sci 77:102106. https://doi.org/10.1016/j.erss.2021.102106

City of Austin (2021) City of Austin and Travis county winter storm uri after-action report and improvement plan technical report. https://www.austintexas.gov/sites/default/files/files/HSEM/Winter-Storm-Uri-AAR-and-Improvement-Plan-Technical-Report.pdf

Center for Disease Control and Prevention [CDC] (2024) Covid data tracker. https://covid.cdc.gov/covid-data-tracker/#maps_deaths-total

Doss-Gollin J et al (2021) How unprecedented was the February 2021 Texas cold snap? Environ Res Lett 16(6):064056. https://doi.org/10.1088/1748-9326/ac0278

Hobby School of Public Affairs at the University of Houston (2021) The winter storm of 2021. https://uh.edu/hobby/winter2021/storm.pdf

Lee CC, Maron M, Mostafavi A (2022) Community-scale big data reveals disparate impacts of the Texas winter storm of 2021 and its managed power outage. Hum Soc Sci Commun 9(1):1–2. https://doi.org/10.1057/s41599-022-01353-8

Sapat A (2021) Health, housing, and hazards: COVID-19, subjective resilience, vulnerabilities, and policy evolution in hurricane prone counties. In: Columbia academic commons, academic unit: Northeast big data innovation hub, series: COVID-19 research lightning talks. https://doi.org/10.7916/w67b-e514

Sapat A et al (2023) Concurrent disasters: perceived administrative burdens and household coping capacities. Public Admin Rev 83(5):1202–1220. https://doi.org/10.1111/puar.13637

Sullivan BK (2021) Most of Texas deep freeze $90 billion in losses avoidable. https://www.insurancejournal.com/news/southcentral/2021/02/25/602807.htm

Texas Department of State Health Services (2022) Texas COVID-19 data. https://www.dshs.texas.gov/coronavirus/additionaldata/

Texas Health and Human Services News Updates (2021) Winter storm-related deaths. https://www.dshs.texas.gov/sites/default/files/news/updates/SMOC_FebWinterStorm_MortalitySurvReport_12-30-21.pdf. Accessed 23 Nov 2021

Alka Sapat is a Professor and Director at the School of Public Administration at Florida Atlantic University. Her expertise encompasses disaster management, public policy processes, resilience assessment, and methodology. She has been involved in numerous initiatives, including National Science Foundation (NSF) and Department of Homeland Security (DHS)—funded projects on building code regulations, disaster-induced population displacement, post-disaster housing, the role of the Haitian diaspora and Non-Governmental Organizations (NGOs) in disaster recoveryport recovery, and critical infrastructure resilience. Her work has been published in the *Journal of Public Administration Research and Theory*, *Natural Hazards Review*, *Public Administration Review*, the *American Review of Public Administration*, *Risk Analysis*, *Natural Hazards*, the *International Journal of Disaster Risk Reduction*, and other scholarly venues. She has served on the Florida State Disaster Housing Task Force, the Governor's Hurricane Conference committee, and local committees on post-disaster housing initiatives. She is the coauthor of the 2014 book *Displaced by Disasters: Recovery and Resilience in a Globalizing World*, and co-editor of the 2017 book *Coming Home after Disaster: Multiple Dimensions of Housing Recovery*.

Diana Mitsova is a Professor and Chair of the Department of Urban and Regional Planning at Florida Atlantic University and John DeGrove Eminent Scholar in Growth Management and Development. Her research has been funded by the National Science Foundation (NSF), the National Institutes of Health (NIH), the Florida Sea Grant, and the Florida Department of Emergency Management. Her research focuses on topics related to infrastructure disruptions and community resilience, climate adaptation, and the health effects of disasters and climate change. She is a co-author of *Geospatial Applications for Climate Adaptation Planning* (2019, Routledge), which examines the fundamentals of using geospatial technologies in climate vulnerability assessments and mitigation and adaptation strategies.

Ann-Margaret Esnard is a Distinguished University Professor of Public Management and Policy at Georgia State University. She has been involved in several NSF-funded interdisciplinary research initiatives, including long-term recovery, population displacement from catastrophic disasters, school recovery after disasters, and community resilience. Her research is dedicated to advancing our understanding of disparate rates and patterns of post-disaster recovery, the complexities of community and household recovery in hurricane-prone communities, household perceptions of recovery, and drivers and implications of population displacement. In addition to numerous academic journal publications and book chapters, she is the co-author of the 2014 book *Displaced by Disasters: Recovery and Resilience in a Globalizing World*, co-editor of the 2017 book *Coming Home after Disaster: Multiple Dimensions of Housing Recovery*, and co-author of the 2019 book *Geospatial Applications for Climate Adaptation Planning*.

Karen D. Sweeting is an Associate Professor in the Department of Political Science at the University of Rhode Island. Her research focuses on enhancing public service delivery for vulnerable, minoritized, and marginalized communities, with a strong emphasis on fostering a humanistic and justice-oriented approach to public service. With over two decades of professional experience in public service prior to transitioning to academia, Dr. Sweeting brings a deep, practice-informed perspective to her scholarship, focusing on issues of equity, inclusion, identity, and ethics.

Part V
COVID and Education

17 Evaluating the Costs of School and Daycare Closure on Family Well-Being and Child Development During the COVID-19 Pandemic

Joshua K. Hartshorne and Mengguo Jing

Abstract

When the COVID-19 pandemic struck, our research group set out to understand the effect of isolation on child development with the help of National Science Foundation (NSF) RAPID award #2030106. Our major results were unexpected. We found plenty of negative effects of pandemic-era policy on families, but they were driven primarily by loss of childcare support, rather than isolation or COVID itself. These results were repeated over and over across many large, national datasets, including both surveys and passive data. These results are particularly striking in the context of a pandemic where children were at low risk of illness and had minimal impact on disease transmission. That is, the closure of schools and daycares came at enormous societal cost and with little benefit. In the future, policymakers should consider this cost-benefit ratio more carefully. In the event the cost-ratio benefit still motivates closure of schools and daycares, proactive measures should be taken to address the known harms.

Keywords

COVID-19 · Screen time · Mental health · Child development · Families

Research Funded by

NSF Award #2030106—RAPID*: Collaborative Research: A "Citizen Science" approach to COVID-19 social distancing effects on children's language development.
*Rapid Response Research Award.

COVID Information Commons Video Presentation

The COVID Information Commons (CIC) Research Lightning Talk video presentation and individual multilingual transcripts for this project can be found in the Columbia Academic Commons (AC) (Hartshorne 2022).

J. K. Hartshorne (✉)
Communication Sciences and Disorders, MGH Institute of Health Professions, Boston, MA, USA
e-mail: joshua.hartshorne@hey.com

M. Jing
Crane Center for Early Childhood Research and Policy, The Ohio State University, Columbus, OH, USA

17.1 Introduction

The COVID-19 pandemic was famously difficult for parents and children. It forced schools across the world to shut down in March 2020, disrupting the educations of roughly 55.1 million students in 124,000 schools across the United States (U.S.) alone (Week 2020). The impacts permeated almost every aspect of family life, with broad effects on both parents and children (Racine et al. 2022; Trott et al. 2022; Whaley and Pfefferbaum 2023). According to a national U.S. survey, just months after the onset of the COVID-19 pandemic in March 2020, 27% of parents reported worsening mental health for themselves

F. D. Hudson (ed.), *The COVID Information Commons*, Sustainable Development Goals Series,
https://doi.org/10.1007/978-3-032-06125-6_17

and 14% reported increased behavioral problems in their children (Patrick et al. 2020). Children's screen time increased by 52% from before the COVID-19 pandemic (Madigan et al. 2022), and students' test scores in fall 2021 were significantly lower than pre-pandemic levels, especially for students from historically disadvantaged backgrounds (Goldhaber et al. 2022).

While it is unreasonable to expect pandemics to be cost-free for families or anyone else, the obvious question is whether we could have avoided some of these harms to families and mitigated the others (Whaley and Pfefferbaum 2023; Muldrew et al. 2022; Viner et al. 2021; Esposito and Principi 2020; Roos et al. 2021; Moss et al. 2023, 2022). Answering this question requires a better understanding of what caused those harms (Whaley and Pfefferbaum 2023; Muldrew et al. 2022). Were they the outcome of policy decisions such as school closures, or were they consequences of people being sick and scared? That is, perhaps the direct impacts of the pandemic on health and stress were sufficiently strong that while school closures and social distancing did nothing to promote family flourishing, they did not make it much worse.

While there has been a mountain of research on these questions, most of it is correlational (e.g., Aiano et al. 2021; Buja et al. 2021; Stage et al. 2021; Walsh et al. 2021). In particular, the effect of school closures is subject to confounding and collinearity from other concurrent interventions making it difficult to isolate the effects of school closures alone. Whether the strictness of social distancing policies is correlated with various metrics of child development is informative but not determinative. It is famously difficult to disentangle correlations between complex social phenomena. For instance, the fact that middle children are more likely to go to jail has proved to be less a result of middle child criminality per se but because children in large families are more likely to go to jail for a variety of reasons, including the fact that large families tend to have low socioeconomic status (Hartshorne et al. 2009; Rahav 1980; Cundiff 2013; for a similar example, see D'Onofrio et al. 2010). By analogy, pandemic-era public policies were not distributed randomly but rather are themselves the result of cultural differences across polities—cultural differences that themselves are known to affect child development and family flourishing.

In our studies reviewed below, we used natural experiments to tease out the causal relationship between pandemic-era social distancing policies and familial flourishing. Lockdown and unplanned homeschooling are believed to have been two major stressors among the many pandemic-related challenges faced by families. While much of the work to date has focused on the direct effect of school closures on student learning (e.g., Goldhaber et al. 2022; Contini et al. 2021; Blainey and Hannay 2021; Engzell et al. 2021), our focus was on how the resulting increase in childcare burden on parents affected family flourishing. Critically, we sought to disentangle the effects of school shutdowns from other losses of support and other COVID-related stressors (for a broad discussion of such stressors, see Whaley and Pfefferbaum 2023).

To measure familial flourishing, we focused on parent mental health and child recreational screen time. There were three primary reasons. First, as a practical matter, we were able to obtain reasonably comprehensive data about both that allowed us to address a number of confounds.

Second, both factors are of significant independent interest. Poor parent mental health is obviously a problem both for parents and for children and is a known risk factor for any number of poor outcomes (Smith 2004; Mensah and Kiernan 2010). More broadly, mental health in the general population has been steadily declining for decades, putting both society and individuals in a less-than-ideal position to weather additional shocks. Children's recreational screen time has been a topic of persistent concern in child development and care, since it is linked with risks across a range of developmental and health outcomes, such as attention deficits, social and behavioral concerns, cognitive difficulties, language delays, sleeping disorders, and obesity (Calvert and Wilson 2011). Pandemic-era increases in children's recreational screen time were well-documented and the topic of extensive public and scientific discussion and concern (Bergmann et al. 2022; Hedderson et al. 2023; Eales et al. 2021; Richtel 2021; McClain 2022). Note also that poor parent mental health is itself known to be associated with increased child recreational screen time (Seguin et al. 2021; Stienwandt et al. 2022).

Third and finally, understanding the etiology of children's recreational screen time during the pandemic informs a long-running debate. While there is no doubt that recreational screen time is *associated* with poor developmental outcomes, causality is debated. Public policy has been predicated on the assumption that screen time displaces more enriching but less attractive activities (Roberts et al. 1993; Lizandra et al. 2019; Chen et al. 2020) and thus has focused on training parents to control children's access to screens (Hill et al. 2016; Organization et al. 2019). However, there is some reason to believe that that screen time is more of a default backup activity when more enriching activities are not available (Oh and Vukina 2022). In that case, policy experts may be doing no more than haranguing parents about something outside their control. Indeed, given growing evidence that screen time is actually beneficial in some circumstances (e.g., Jing et al. 2023; Kirkorian et al. 2016; Zack and Barr 2016; Egami et al. 2024), they may be doing more harm than good.

17.2 Overview of the Studies

As reviewed above, the fact that child recreational screen time had increased dramatically and parental mental health declined precipitously is well established and was clear even early in the pandemic. The question is whether this (a) could have been avoided, or (b) mitigated.

The answer depends on causation. If, for instance, parents were deliberately using recreational screen time to address children's anxiety, a potential strategy would be to provide parents and children with more effective and healthful soothing techniques, like mindfulness meditation. If, instead, recreational screen time increased because many schools provided students with laptops which were then repurposed for recreational uses, we might want to consider more effective device controls.

In the research reviewed in this chapter, we tested the hypothesis that increases in child recreational screen time and declines in parental mental health were both direct consequences of the closure of schools and daycares, mediated by parenting burden: the more pressure on parents' time, the more children utilized recreational screen time and the worse the parents felt. While there is some evidence that unmet childcare needs are associated with degraded parenting (e.g., Roos et al. 2021; Baldus et al. 2022), evidence to date is primarily correlational, again raising questions about causation.

We investigated the causal relationship between childcare burden and familial flourishing in two related natural experiment studies we conducted. The first study considered increases in recreational screen time and decreases in parent mental health early in the pandemic. We find that both changes were tightly time-locked to the closure of schools and daycares. The second study considered what happened when schools reopened, finding that this reversed the negative effects of the pandemic. In both studies, the causal mechanism appears to be time pressure: the more parents were childcare-burdened, the worse the problems. We also were able to rule out a number of potential confounds.

We describe the two studies in the next two sections. Both studies focused exclusively on the United States. We conclude with lessons and strategies for future pandemics.

17.3 Study 1: Lessons from Closures

Our first study (Hartshorne et al. 2021) focused on the early days of the pandemic in 2020. Our approach was to get temporally fine-grained measures of our outcome variables (parent mental health and child recreational screen time) and see how closely time locked they were to temporally fine-grained measures of potential predictors (COVID rates, mobility data, school lockdown policies, and parent childcare burden). If, for instance, changes in our outcome measures tracked changes in COVID rates rather than mobility data, that would suggest parent mental health and child recreational screen time were driven directly by disease: perhaps sick people (or their friends and relatives) were stressed out and watching TV.

At the time we conducted this study, the reported studies were not sufficiently fine-grained to answer this question. Some initial data suggested that viewership of child-related TV channels increased in the immediate wake of the pandemic (O'Reilly 2020). In a mid-April survey of members of a parent advocacy organization, 49% of parents ($N > 3000$) reported their children spending more than 6 h per day online, with 26% reporting more than 8 h, compared to an estimated 8 and 4%, respectively, before the pandemic (ParentsTogether 2020). Small studies in Canada, China, and South Korea showed similar patterns (Moore et al. 2020; Xiang et al. 2020; Guan et al. 2020). The most fine-grained data came from SuperAwesome, a kid-safe advertising service, which reported kids' weekday internet traffic in the US and the UK approximately doubled in mid-March—roughly coinciding with school closures—though the organization noted that their data did not clearly distinguish recreational screentime from remote schooling (SuperAwesome 2020).

While the above results were suggestive of a sharp change somewhere in March or April 2020, many things were happening in those months, so more systematic analysis was required. In Study 1, we and our colleagues set out to address this gap.

17.3.1 Method

We were able to obtain eight large data sets that provided good geographic and temporal resolution. Below, we briefly describe the samples we analyzed:

- Kaiser Family Foundation Health Tracking Poll: Four nationally representative surveys (March 11–15, March 23–30, May 13–18, July 14–19, 2020) including a total of 1310 parents and 3626 non-parents (Kirzinger et al. 2020; Hamel et al. 2020a, 2020b).
- AP-NORC: Two nationally representative surveys (March 26–29, July 16–20, 2020) including a total of 587 parents (Associated Press-NORC Center for Public Affairs Research 2020a, b).
- NORC COVID-19 Household Impact Survey: Three national surveys with regional oversamples (April 20–26, May 4–10, May 30–June 8, 2020) including a total of 5996 parents and 19,273 non-parents (Wozniak et al. 2020).
- Understanding America: A tracking poll that re-surveyed the same 2331 parents every two weeks from March 10 through July 20, 2020 (Understanding America Study 2020; Alattar et al. 2018).
- KidTalk Survey: Survey data collected through our own KidTalk project, with responses from 1120 U.S.A.-based parents who responded between April 16 and September 11, 2020, with 94% of responses coming by June 15 (Hartshorne et al. 2021).
- ReelGood: Weekly state-by-state data on average number of initiations of child-related streaming content, provided by ReelGood, a streaming search app, from December 30, 2019 to July 19, 2020 (Hartshorne et al. 2021).

In addition, we obtained fine-grained school closure dates from MCH Strategic Data (http://www.mchdata.com/); daily state-level COVID rates from the New York Times COVID tracker (New York Times 2020); daily, geographically localized estimates of time spent at home from the Google Community Mobility Reports (Aktay et al. 2020); and state-by-state population estimates from the 2019 U.S. Census report (U.S. Census Bureau 2019).

17.3.2 Results and Discussion

For full statistical results, see the original paper on "Screen time as an index of family distress" (Hartshorne et al. 2021). Key findings are depicted in two interactive figures, available online (jkhartshorne.shinyapps.io/Spaghetti and jkhartshorne.shinyapps.io/Heatmap). To summarize, the many datasets show a consistent pattern: a sharp increase in child recreational screen time and decline in parent mental health in late March 2020. This was essentially unrelated to actual COVID case rates, either in aggregate or on a state-by-state level.

However, the spikes in recreational screen time and parents reporting poor mental health both coincided with lockdown measures: school closures, increases in time spent at home, and decreases in time spent at work. Notably, while school closures and changes in mobility were initially coupled, schools remained closed even though mobility numbers started recovering almost immediately, even in states that officially maintained social distancing policies. Critically, our outcome measures remained roughly stable after the initial spike—tracking school closures and not mobility measures.

Tellingly, not only did parents report a sharp uptick in time spent on childcare, there was a dose-dependent relationship with our outcome measures: the more extra time a particular parent reported spending on childcare, the worse their mental health and the more the children engaged in recreational screen time. Strikingly, parents who reported having no change in childcare (presumably because they were already caring for their children exclusively at home) also appeared to be largely unaffected with respect to our outcome measures.

Thus, it seems likely that the problem was not school and daycare closures per se but the strain those closures put on parents. Families fared better or worse depending on how great the resulting strain was. In contrast, the disease itself had no measurable impact, nor did non-school-related social distancing.

17.4 Study 2: Lessons from Reopenings

While Study 1 tied increased child recreational screen time and decreased parent mental health specifically to school and daycare closures (rather than work-at-home policies, unemployment, fluctuating COVID rates, etc.), losses are different from gains, and just because losing childcare hurts families does not mean increasing childcare will help them. In the second study, we thus focused on the effects of school reopenings during the COVID-19 pandemic on child recreational screen time and parental mental health. This study capitalized on the fact that schools and daycares reopened at different times across different regions, providing a natural experiment for examining how returning to in-person schooling affected these family outcomes.

17.4.1 Method

We made use of three pre-existing datasets as well as two waves of our own KidTalk survey:

- Adolescent Brain and Cognitive Development (ABCD) study: Survey data of 5953 adolescents and their parents, collected as part of a large longitudinal study (Jernigan et al. 2020).
- Understanding America COVID Panel: A tracking poll that re-surveyed the same 2694 adults (Understanding America Study 2020; Alattar et al. 2018). We used weeks 11 through 28.
- Census Household Pulse Survey: Conducted every two weeks on a national scale. We included data from September 2, 2020 to May 24, 2021, for a total of 270,265 respondents (Fields et al. 2020).
- KidTalk: A survey conducted through our KidTalk app, with 162 respondents in Wave 2 and 159 in Wave 3 (Hartshorne et al. 2021).

These data captured family experiences as schooling formats varied, allowing for a cross-sectional comparison of families whose children were attending in-person schooling, virtual schooling, or a hybrid of both. The key outcome measures were changes in child recreational screen time and parental mental health, assessed through surveys administered to parents across different schooling conditions. The richness of the data also allowed us to control for a number of potential confounding factors, such as local COVID-19 prevalence, loss of employment, and whether parents were working from home.

17.4.2 Results and Discussion

For full statistical results, see the original paper (Hartshorne 2020). The findings were consistent across multiple datasets, showing that parents whose children returned to in-person schooling reported significantly better mental health, and significantly lower childcare burden, compared to those whose children remained in virtual or hybrid schooling. The results suggested that the reduction in childcare burden played a key role in this improvement.

Similarly, children who returned to in-person schooling exhibited significantly lower levels of recreational screen time compared to those in virtual or hybrid models. The reduction in screen time was most pronounced in families where children were fully back to in-person schooling.

Interestingly, the study found that hybrid schooling did not significantly alleviate parental stress or reduce screen time compared to fully virtual schooling. This suggests that hybrid models, which often require complex scheduling and additional supervision, may not provide enough relief to parents compared to fully in-person schooling.

The study controlled for several potential confounds, such as employment status and work-from-home policies, and found that the positive effects of school reopenings were robust even when these factors were taken into account. Other factors that could not explain the results were COVID-19 prevalence, employment status, and parental remote work. This strengthens the conclusion that schooling format, and the related childcare burden, was a primary driver of the observed outcomes.

Thus, the driving factor behind the improvements in both parental mental health and child screen time was the alleviation of parental childcare burden. When children were home for virtual learning, parents had to take on more childcare responsibilities, leading to increased stress and reliance on screen time as a way to occupy children. In contrast, the return to in-person schooling provided parents with much-needed relief, reducing the time pressure they faced and improving their overall mental health.

This study provides strong evidence that reopening schools had a direct, positive impact on both parent mental health and child screen time, primarily by reducing the childcare burden that parents faced during school closures. The reduction in childcare demands allowed parents to better manage their own mental health, and children benefited from reduced screen time as they returned to structured learning environments. The findings suggest that policymakers should prioritize keeping schools and daycares open during future crises, as the availability of childcare is critical to maintaining family well-being. Moreover, the limited impact of hybrid schooling raises important questions about the effectiveness of such models in reducing parental stress.

In summary, the second study demonstrates that the negative effects of school closures were largely reversible once schools reopened, reinforcing the importance of childcare in supporting family flourishing during times of crisis.

17.5 Lessons and Suggestions for the Future

School and daycare closures came at considerable cost even without taking into account children's lost learning and parents' lost professional advancement. That said, the costs of public policy have to be measured against the expected gains. With respect to COVID-19, with its low incidence in young children and low rates of transmission by children—and given that private schools in the U.S. and many school systems abroad were able to offer in-person instruction with low case rates—it is increasingly clear that school and daycare closures did little or nothing to lessen spread of the disease (Walsh et al. 2021; Donohue and Miller 2020; Wu et al. 2022; Chen et al. 2022; Neil-Sztramko et al. 2024). Moreover, the fact that public schools remained closed while many private schools successfully opened appears to have driven a significant exodus of families from public education, exacerbating inequalities and heightening demographic pressures on public school systems (Lieberman 2023; Dee 2023; Dee et al. 2023; Azevedo et al. 2022).

However, this will not be true of all pandemics. Some diseases are more severe in children or transmit more readily in children than adults (Goldstein et al. 2018; Zheng et al. 2022; Fang et al. 2014). Indeed, even more recent variants of COVID-19 are more readily transmitted by children than were earlier variants (Zhu et al. 2023). In some cases, it may nonetheless be possible to keep schools and daycares open by using strategies that became familiar during the pandemic: cohorting students, proactive testing, quarantine protocols, using outdoor learning, or improving hygiene and ventilation procedures (e.g., McGee et al. 2021; Colosi et al. 2022; Burke et al. 2021).

In cases where daycare and school closures are nonetheless judged to be worth the cost, it is still possible to spread that cost more equitably. For instance, the increased childcare burden might be somewhat ameliorated by subsidizing (socially distanced) services, such as grocery and meal delivery. While less is currently known about the learning efficacy and disease-prevention success of home learning pods, it would be worth exploring in advance whether wide-spread, official support for such programs might be helpful. Providing affected parents with targeted professional development opportunities post-pandemic could help address long-term costs borne by parents. Announcing those opportunities in advance might help alleviate stress. Our point here is not to make specific policy suggestions but rather to suggest that there should be a policy.

That said, there is one policy that we would caution against: attempting to train parents to control children's screen time. The studies described above are yet more evidence that screen time is a symptom, not a disease. Moreover, it is a very difficult symptom to intervene on directly. If screens are the babysitter of last resort, by definition there is no additional resort to turn to. In short, diatribes against screens may, at best, stress out parents (Wolfers et al. 2024) and, at worst, take away a valuable coping device. Indeed, one large correlational study showed positive effects of video-gaming on mental health during the pandemic (Egami et al. 2024).

The silver lining of the COVID-19 pandemic is that we have learned a great deal. Now we need to put that knowledge into practice.

Acknowledgements We also thank collaborators on Study 1, including Yi Ting Huang, Pablo Martín Lucio Paredes, Kathleen Oppenheimer, Parker T. Robbins, and María Daniela Velasco.

References

Aiano F, Mensah AA, McOwat K, Obi C, Vusirikala A, Powell AA, Flood J, Bosowski J, Letley L, Jones S et al (2021) COVID-19 outbreaks following full reopening of primary and secondary schools in England: cross-sectional national surveillance, November 2020. Lancet Reg Health Europe 6:120. https://doi.org/10.1016/j.lanepe.2021.100120

Aktay A, Bavadekar S, Cossoul G, Davis J, Desfontaines D, Fabrikant A, Gabrilovich E, Gadepalli K, Gipson B, Guevara M et al (2020) Google COVID-19 community mobility reports: anonymization process description (version 1.0). arXiv preprint arXiv:2004.04145

Alattar L, Messel M, Rogofsky D (2018) An introduction to the understanding America study internet panel. Soc Sec Bull 78:13

Associated Press-NORC Center for Public Affairs Research (2020a) The July 2020 AP-NORC center poll. https://apnorc.org/wp-content/uploads/2020/07/topline_release2.pdf

Associated Press-NORC Center for Public Affairs Research (2020b) The March 2020 AP-NORC center poll. https://apnorc.org/wp-content/uploads/2020/04/march2020_topline.pdf

Azevedo JP, Gutierrez M, Hoyos R, Saavedra J (2022) The unequal impacts of COVID-19 on student learning. In: Primary and secondary education during COVID-19: disruptions to educational opportunity during a pandemic, pp 421–459. https://doi.org/10.1007/978-3-030-81500-4_16

Baldus C, Franz S, Thomasius R (2022) Help needs among parents and families in times of the COVID-19 pandemic lockdown in germany. Int J Environ Res Public Health 19(21):14159. https://doi.org/10.3390/ijerph192114159

Bergmann C, Dimitrova N, Alaslani K, Almohammadi A, Alroqi H, Aussems S, Barokova M, Davies C, Gonzalez-Gomez N, Gibson SP et al (2022) Young children's screen time during the first COVID-19 lockdown in 12 countries. Sci Rep 12(1):2015. https://doi.org/10.1038/s41598-022-05840-5

Blainey K, Hannay T (2021) The impact of school closures on autumn 2020 attainment. RS Assess Hodder Educ 503:48

Buja A, Paganini M, Cristofori V, Baldovin T, Fusinato R, Boccuzzo G, Cocchio S, Coretti S, Rebba V, Parpinel M (2021) Opening schools and trends in SARS-CoV-2 transmission in European countries. Int J Public Health 66:1604076. https://doi.org/10.3389/ijph.2021.1604076

Burke A, Moore S, Molyneux L, Lawlor A, Kottwitz T, Yurich G, Sanson R, Andersen O, Card B et al (2021) Children's wellness: outdoor learning during COVID-19 in Canada. Educ North 4:28

Calvert SL, Wilson BJ (eds) (2011) The handbook of children, media and development. Wiley-Blackwell, West Sussex

Chen B, Bernard JY, Padmapriya N, Ning Y, Cai S, Lança C, Tan KH, Yap F, Chong Y-S, Shek L et al (2020) Associations between early-life screen viewing and 24 hour movement behaviours: findings from a longitudinal birth cohort study. Lancet Child Adolescent Health 4(3):201–209. https://doi.org/10.1016/s2352-4642(19)30424-9

Chen F, Tian Y, Zhang L, Shi Y (2022) The role of children in household transmission of COVID-19: a systematic review and meta-analysis. Int J Infect Dis 122:266–275

Colosi E, Bassignana G, Contreras DA, Poirier C, Boëlle P-Y, Cauchemez S, Yazdanpanah Y, Lina B, Fontanet A, Barrat A et al (2022) Screening and vaccination against COVID-19 to minimise school closure: a modelling study. Lancet Infect Dis 22(7):977–989. https://doi.org/10.1016/s1473-3099(22)00138-4

Contini D, Di Tommaso ML, Muratori C, Piazzalunga D, Schiavon L (2021) The COVID-19 pandemic and school closure: learning loss in mathematics in primary education. In: Technical report, IZA discussion papers. https://docs.iza.org/dp14785.pdf

Cundiff PR (2013) Ordered delinquency: the "effects" of birth order on delinquency. Person Soc Psychol Bull 39(8):1017–1029. https://doi.org/10.1177/0146167213488215

D'Onofrio BM, Singh AL, Iliadou A, Lambe M, Hultman CM, Grann M, Neiderhiser JM, Lichtenstein P (2010) Familial confounding of the association between maternal smoking during pregnancy and offspring criminality: a population-based study in Sweden. Arch Gen Psych 67(5):529–538. https://doi.org/10.1001/archgenpsychiatry.2010.33

Dee TS (2023) Where the kids went: nonpublic schooling and demographic change during the pandemic exodus from public schools. Teach College Rec 125(6):119–129

Dee TS, Huffaker E, Phillips C, Sagara E (2023) The revealed preferences for school reopening: Evidence from public-school disenrollment. Am Educ Res J 60(5):916–940

Donohue JM, Miller E (2020) COVID-19 and school closures. JAMA 324(9):845–847

Eales L, Gillespie S, Alstat RA, Ferguson GM, Carlson SM (2021) Children's screen and problematic media use in the United States before and during the COVID-19 pandemic. Child Develop 92(5):866–882. https://doi.org/10.1111/cdev.13652

Egami H, Rahman MS, Yamamoto T, Egami C, Wakabayashi T (2024) Causal effect of video gaming on mental well-being in japan 2020–2022. Nat Hum Behav 12:1–14. https://doi.org/10.1038/s41562-024-01948-y

Engzell P, Frey A, Verhagen MD (2021) Learning loss due to school closures during the COVID-19 pandemic. Proceed Natl Acad Sci 118(17):2022376118. https://doi.org/10.1073/pnas.2022376118

Esposito S, Principi N (2020) School closure during the coronavirus disease 2019 (COVID-19) pandemic: an effective intervention at the global level? JAMA Pediat 174(10):921–922. https://doi.org/10.1001/jamapediatrics.2020.1892

Fang Y, Wang S, Zhang L, Guo Z, Huang Z, Tu C, Zhu BP (2014) Risk factors of severe hand, foot and mouth disease: a meta-analysis. Scand J Infect Dis 46(7):515–522

Fields J, Hunter-Childs J, Tersine A, Sisson J, Parker E, Velkoff V, Logan C, Shin H (2020) Design and operation of the 2020 household pulse survey. Accessed 6 Jan 2021

Goldhaber D, Kane TJ, McEachin A, Morton E (2022) A comprehensive picture of achievement across the COVID-19 pandemic years: examining variation in test levels and growth across districts, schools, grades, and students. In: Working paper no. 266–0522. National center for analysis of longitudinal data in education research (CALDER). https://eric.ed.gov/?id=ED620384

Goldstein E, Nguyen HH, Liu P, Viboud C, Steiner CA, Worby CJ, Lipsitch M (2018) On the relative role of different age groups during epidemics associated with respiratory syncytial virus. J Infect Dis 217(2):238–244. https://doi.org/10.1093/infdis/jix575

Guan H, Okely AD, Aguilar-Farias N, Pozo-Cruz B, Draper CE, El-Hamdouchi A, Florindo AA, Jáuregui A, Katzmarzyk PT, Kontsevaya A, Löf M, Park W, Reilly JJ, Sharma D, Tremblay MS, Veldman SLC (2020) Promoting healthy movement behaviours among children during the COVID-19 pandemic. Lancet Child Adolesc Health 4(6):416–418. https://doi.org/10.1016/S2352-4642(20)30131-0

Hamel L, Kearney A, Kirzinger A, Lopes L, Muñana C, Brodie M (2020a) KFF health tracking poll—May 2020. https://www.kff.org/report-section/kff-health-tracking-poll-may-2020-methodology/

Hamel L, Kearney A, Kirzinger A, Lopes L, Muñana C, Brodie M (2020b) KFF health tracking poll—July 2020. https://www.kff.org/report-section/kff-health-tracking-poll-july-2020-methodology/

Hartshorne JK, Salem-Hartshorne N, Hartshorne TS (2009) Birth order effects in the formation of long-term relationships. J Individ Psychol 65(2):156–176

Hartshorne JK, Huang YT, Paredes PML, Oppenheimer K, Robbins PT, Velasco MD (2021) Screen time as an index of family distress. Curr Res Behav Sci 2:100023. https://doi.org/10.1016/j.crbeha.2021.100023

Hartshorne JK (2020) Just give them childcare: the COVID-19 pandemic as a natural experiment in parenting practices (submitted)

Hartshorne J (2022) A 'citizen science' approach to COVID-19 social distancing effects on children's language development. In: Columbia academic commons, academic unit: northeast big data innovation hub, series: COVID-19 research lightning talks. https://doi.org/10.7916/b6sq-d991

Hedderson MM, Bekelman TA, Li M, Knapp EA, Palmore M, Dong Y, Elliott AJ, Friedman C, Galarce M, Gilbert-Diamond D et al (2023) Trends in screen time use among children during the COVID-19 pandemic, July 2019 through August 2021. JAMA Netw Open 6(2):2256157–2256157. https://doi.org/10.1001/jamanetworkopen.2022.56157

Hill D, Ameenuddin N, Reid Chassiakos YL, Cross C, Hutchinson J, Levine A, Boyd R, Mendelson R, Moreno M, Swanson WS et al (2016) Media and young minds. Pediatrics 138(5):2591. https://doi.org/10.1542/peds.2016-2591

http://www.mchdata.com/

Jernigan TL, Brown SA, Dale AM, Tapert SF, Sowell ER, Herting M, Laird A, Gonzalez R, Squeglia L, Gray K et al (2020). Study 1041: adolescent brain cognitive development study® (ABCD) data release: COVID rapid response research (RRR) survey first data release (surveys #1, 2, and 3). Accessed 23 Jan 2021

Jing M, Ye T, Kirkorian HL, Mares M-L (2023) Screen media exposure and young children's vocabulary learning and development: a meta-analysis. Child Develop 94(5):1398–1418. https://doi.org/10.1111/cdev.13927

jkhartshorne.shinyapps.io/Spaghetti and jkhartshorne.shinyapps.io/Heatmap

Kirkorian HL, Choi K, Pempek TA (2016) Toddlers' word learning from contingent and noncontingent video on touch screens. Child Develop 87(2):405–413. https://doi.org/10.1111/cdev.12508

Kirzinger A, Kearney A, Hamel L, Brodie M (2020) KFF health tracking poll—early April 2020: the impact of coronavirus on life in America. https://www.kff.org/coronavirus-COVID-19/poll-finding/kff-coronavirus-poll-march-2020/

Lieberman M (2023) More young kids opted for private school after COVID hit. Educ Week. https://www.edweek.org/policy-politics/more-young-kids-opted-for-private-school-after-covid-hit/2023/12

Lizandra J, Devís-Devís J, Valencia-Peris A, Tomás JM, Peiró-Velert C (2019) Screen time and moderate-to-vigorous physical activity changes and displacement in adolescence: a prospective cohort study. Eur J Sport Sci 19(5):686–695. https://doi.org/10.1080/17461391.2018.1548649

Madigan S, Eirich R, Pador P, McArthur BA, Neville RD (2022) Assessment of changes in child and adolescent screen time during the COVID-19 pandemic: a systematic review and meta-analysis. JAMA Pediat 176(12):1188–1198. https://doi.org/10.1001/jamapediatrics.2022.4116

McClain C (2022) How parents' views of their kids' screen time, social media use changed during COVID-19. Pew Research Center. https://www.pewresearch.org/short-reads/2022/04/28/how-parents-views-of-their-kids-screen-time-social-media-use-changed-during-covid-19/

McGee RS, Homburger JR, Williams HE, Bergstrom CT, Zhou AY (2021) Model-driven mitigation measures for reopening schools during the COVID-19 pandemic. Proceed Natl Acad Sci 118(39):2108909118. https://doi.org/10.1073/pnas.2108909118

Mensah FK, Kiernan KE (2010) Parents' mental health and children's cognitive and social development: families in England in the millennium cohort study. Soc Psych Psych Epidemiol 45:1023–1035. https://doi.org/10.1007/s00127-009-0137-y

Moore SA, Faulkner G, Rhodes RE, Brussoni M, Chulak-Bozzer T, Ferguson LJ, Mitra R, O'Reilly N, Spence JC, Vanderloo LM, Tremblay MS (2020) Impact of the COVID-19 virus outbreak on movement and play behaviours of Canadian children and youth: a national survey. Int J Behav Nutr Phys Activity 17(1):85. https://doi.org/10.1186/s12966-020-00987-8

Moss SJ, Lorenzetti DL, FitzGerald EA, Smith S, Harley M, Tutelman PR, Birnie K, Mizen SJ, Anglin MC, Stelfox HT et al (2022) Strategies to mitigate the impact of the COVID-19 pandemic on child and youth well-being: a scoping review protocol. BMJ Open 12(7):062413. https://doi.org/10.1136/bmjopen-2022-062413

Moss SJ, Mizen SJ, Stelfox M, Mather RB, FitzGerald EA, Tutelman P, Racine N, Birnie KA, Fiest KM, Stelfox HT et al (2023) Interventions to improve well-being among children and youth aged 6–17 years during the COVID-19 pandemic: a systematic review. BMC Med 21(1):131. https://doi.org/10.1186/s12916-023-02828-4

Muldrew DH, Fee A, Coates V (2022) Impact of the COVID-19 pandemic on family carers in the community: a scoping review Health Soc Care Commun 30(4):1275–1285. https://doi.org/10.1111/hsc.13677

Neil-Sztramko SE, Belita E, Traynor RL, Hagerman L, Akaraci S, Burnett P, Kostopoulos A, Dobbins M (2024) What is the specific role of schools and daycares in COVID-19 transmission? A final report from a living rapid review. Lancet Child Adoles Health 2:7. https://doi.org/10.1016/s2352-4642(23)00312-7

New York Times (2020) Coronavirus (COVID-19) data in the United States. https://github.com/nytimes/covid-19-data

O'Reilly L (2020) With schools shut, kids linear TV ratings are on the up. Digiday. https://digiday.com/media/schools-shut-kids-linear-tv-ratings/

Oh SE, Vukina T (2022) Estimating parental demand for children's screen time in a model of family labor supply. Int Adv Econ Res 28(3):69–89. https://doi.org/10.1007/s11294-022-09854-7

Organization WH et al (2019) Guidelines on physical activity, sedentary behaviour and sleep for children under 5 years of age. World Health Organization. https://iris.who.int/bitstream/handle/10665/325147/WHO-NMH-PND-2019.4-eng.pdf

ParentsTogether (2020) Survey shows parents alarmed as kids' screen time skyrockets during COVID-19 crisis. https://parents-together.org/survey-shows-parents-alarmed-as-kids-screen-time-skyrockets-during-covid-19-crisis/

Patrick SW, Henkhaus LE, Zickafoose JS, Lovell K, Halvorson A, Loch S, Letterie M, Davis MM (2020) Well-being of parents and children during the COVID-19 pandemic: a national survey. Pediatrics 146(4):824. https://doi.org/10.1542/peds.2020-016824

Racine N, Eirich R, Cooke J, Zhu J, Pador P, Dunnewold N, Madigan S (2022) When the bough breaks: a systematic review and meta-analysis of mental health symptoms in mothers of young children during the COVID-19 pandemic. Infant Mental Health J 43(1):36–54. https://doi.org/10.1002/imhj.21959

Rahav G (1980) Birth order and delinquency. Br J Criminol 20(4):385–395

Richtel M (2021) Children's screen time has soared in the pandemic, alarming parents and researchers. The New York Times, New York. https://www.nytimes.com/2021/01/16/health/covid-kids-tech-use.html

Roberts DF, Henriksen L, Voelker DH, Van Vuuren D (1993) Television and schooling: displacement and distraction hypotheses. Austr J Educ 37(2):198–211

Roos LE, Salisbury M, Penner-Goeke L, Cameron EE, Protudjer JL, Giuliano R, Afifi TO, Reynolds K (2021) Supporting families to protect child health: Parenting quality and household needs during the COVID-19 pandemic. PLoS ONE 16(5):0251720. https://doi.org/10.1371/journal.pone.0251720

Seguin D, Kuenzel E, Morton JB, Duerden EG (2021) School's out: parenting stress and screen time use in school-age children during the COVID-19 pandemic. J Affect Disord Rep 6:100217. https://doi.org/10.1016/j.jadr.2021.100217

Smith M (2004) Parental mental health: disruptions to parenting and outcomes for children. Child Family Soc Work 9(1):3–11. https://doi.org/10.1111/j.1365-2206.2004.00312.x

Stage HB, Shingleton J, Ghosh S, Scarabel F, Pellis L, Finnie T (2021) Shut and re-open: the role of schools in the spread of COVID-19 in Europe. Philos Trans Royal Soc B 376(1829):20200277. https://doi.org/10.1098/rstb.2020.0277

Stienwandt S, Cameron EE, Soderstrom M, Casar M, Le C, Roos LE (2022) Family factors associated with hands-on play and screen time during the COVID-19 pandemic. In: Child and youth care forum. Springer, New York, pp 1–25. https://doi.org/10.1007/s10566-021-09668-4

SuperAwesome (2020) Everyone is a kids and family brand now: data, observations and recommendations for companies interacting with kids and families during lockdown. In: Technical report. https://www.superawesome.com/everyone-is-a-kids-and-family-brand-now/

Trott M, Driscoll R, Iraldo E, Pardhan S (2022) Changes and correlates of screen time in adults and children during the COVID-19 pandemic: a systematic review and meta-analysis. EClinicalMedicine 48:1452. https://doi.org/10.1016/j.eclinm.2022.101452

U.S. Census Bureau (2019) State population totals and components of change: 2010–2019. https://www.census.gov/data/tables/time-series/demo/popest/2010s-state-total.html#par_textimage_1574439295,%202019%20estimates

Understanding America Study (2020) Understanding America study methodology. Brookings Institution

Viner RM, Bonell C, Drake L, Jourdan D, Davies N, Baltag V, Jerrim J, Proimos J, Darzi A (2021) Reopening schools during the COVID-19 pandemic: governments must balance the uncertainty and risks of reopening schools against the clear harms associated with prolonged closure. Arch Dis Childhood 106(2):111–113. https://doi.org/10.1136/archdischild-2020-319963

Walsh S, Chowdhury A, Braithwaite V, Russell S, Birch JM, Ward JL, Waddington C, Brayne C, Bonell C, Viner RM et al (2021) Do school closures and school reopenings affect community transmission of COVID-19? A systematic review of observational studies. BMJ Open 11(8):053371. https://doi.org/10.1136/bmjopen-2021-053371

Week E (2020) Map: coronavirus and school closures. https://www.edweek.org/leadership/map-coronavirus-and-school-closures-in-2019-2020/2020/03

Whaley GL, Pfefferbaum B (2023) Parental challenges during the COVID-19 pandemic: psychological outcomes and risk and protective factors. Curr Psych Rep 25(4):165–174. https://doi.org/10.1007/s11920-023-01412-0

Wolfers LN, Nabi RL, Walter N (2024) Too much screen time or too much guilt? How child screen time and parental screen guilt affect parental stress and relationship satisfaction. Media Psychol 54:1–32. https://doi.org/10.1080/15213269.2024.2310839

Wozniak A, Willey J, Benz J, Hart N (2020) COVID impact survey: version 1.3, 2.2, and 3. National Opinion Research Center. https://www.covid-impact.org/results

Wu JT, Mei S, Luo S, Leung K, Liu D, Lv Q, Liu J, Li Y, Prem K, Jit M et al (2022) A global assessment of the impact of school closure in reducing COVID-19 spread. Philos Trans Royal Soc A 380(2214):20210124

Xiang M, Zhang Z, Kuwahara K (2020) Impact of COVID-19 pandemic on children and adolescents' lifestyle behavior larger than expected. Prog Cardiovasc Dis 10:13. https://doi.org/10.1016/j.pcad.2020.04.013

Zack E, Barr R (2016) The role of interactional quality in learning from touch screens during infancy: context matters. Front Psychol 7:1264. https://doi.org/10.3389/fpsyg.2016.01264

Zheng Z, Warren JL, Shapiro ED, Pitzer VE, Weinberger DM (2022) Estimated incidence of respiratory hospitalizations attributable to RSV infections across age and socioeconomic groups. Pneumonia 14(1):6. https://doi.org/10.1186/s41479-022-00098-x

Zhu Y, Xia Y, Pickering J, Bowen AC, Short KR (2023) The role of children in transmission of SARS-CoV-2 variants of concern within households: an updated systematic review and meta-analysis, as at 30 June 2022. Eurosurveillance 28(18):2200624. https://doi.org/10.2807/1560-7917.es.2023.28.18.2200624

Dr. Joshua K. Hartshorne is an Associate Research Professor in Communication Sciences and Disorders at the Massachusetts General Hospital (MGH) Institute of Health Professions. He earned a BA in Mathematics at Oberlin College and a Ph.D. in Psychology from Harvard University. His work falls at the intersection of psychology, linguistics, and computer science, and focuses on understanding what makes human children such phenomenally efficient learners, particularly when it comes to language. He is best-known for his research into critical periods for learning a second language. He is also known for pioneering massive online experiments and citizen science approaches to understanding human behavior. Dr. Hartshorne's work has been published in leading journals, including *Science*, *Nature Human Behavior*, *Psychological Science*, and *Cognition*. He is a recipient of the National Science Foundation (NSF) Graduate Fellowship, the National Defense Science & Engineering Graduate Fellowship, a National Institutes of Health National Research Service Award (NIH NRSA), and an NSF CAREER Award.

Dr. Mengguo Jing is currently a research senior associate at the Crane Center for Early Childhood Research at The Ohio State University. She earned a Ph.D. in Human Development and Family Studies from the University of Wisconsin-Madison. A developmental and educational psychologist, Dr. Jing studies early childhood development and learning in digital contexts. Her research program aims to leverage digital media and technology in promoting children's well-being from a cognitive development perspective. Her current work focuses on understanding the cognitive processes (i.e., attention, symbolic representation, intuitive reasoning) that underlie children's learning from screen media. In this work, she also designs and tests media features that support early learning in domains such as vocabulary, math and spatial reasoning, and science learning. In another line of inquiry, she studies the role of social media in adolescent development, with a particular emphasis on underrepresented communities. Dr. Jing's work has been published in leading academic journals across psychology, education, and human–computer interaction, including *Child Development*, *Journal of Experimental Child Psychology*, and *Computers in Human Behaviors*.

Impact of COVID-19 on Science Education: Evidence from California

18

Niu Gao, Kathy DiRanna, and Maria Chang Fay

Abstract

The COVID-19 pandemic continues to disrupt public education and affect the academic, social, and emotional well-being of students. The long-term underinvestment in science education, combined with the unprecedented disruptions caused by COVID-19, heightened the challenges facing students and school staff in schools, and are likely to constrain science education for years to come. In this chapter, we summarize the impact of COVID-19 on science education. Science education became a lower priority in the 2020–2021 school year and key efforts to implement new science standards were delayed. Staffing shortages, teacher burnout, lack of dedicated funding for science education, and an emphasis on math and English were key barriers to prioritizing science education. Districts provided limited support for science education during the pandemic, and most did not prioritize science education in their recovery plans. In this chapter, we outline key steps for policymakers to support equitable recovery in science education.

Keywords

COVID-19 · Science education · Next generation science standards · Science literacy

18.1 Introduction

In 2020, 44 states serving 71% of United States (U.S.) students had education standards influenced by the *Framework for K-12 Science Education and/or the Next Generation Science Standards (NGSS)* (National Science Teaching Association 2021). If implemented effectively, NGSS have the potential of catalyzing improvements in science education, increasing scientific literacy, and strengthening the global competitiveness of the American labor force (National

Research Funded by

NSF Award #2128789—RAPID*: A Researcher-Practitioner-Partnership to Assess the Impact of COVID-19 Recession on NGSS Implementation. *Rapid Response Research Award.

COVID Information Commons Video Presentation

The COVID Information Commons (CIC) Research Lightning Talk video presentation and individual multilingual transcripts for this project can be found in the Columbia Academic Commons (AC) (Gao 2023).

N. Gao (✉)
American Institutes for Research, Sacramento, CA, USA
e-mail: ngao@air.org

K. DiRanna
WestEd, San Francisco, CA, USA
e-mail: Kdirann@wested.org

M. C. Fay
San Francisco Exploratorium, San Francisco, CA, USA

F. D. Hudson (ed.), *The COVID Information Commons*, Sustainable Development Goals Series,
https://doi.org/10.1007/978-3-032-06125-6_18

Research Council 2014). Early studies raised challenges in science course sequencing, the quality of NGSS-aligned instructional materials, teacher preparedness, assessment alignment, local capacity building, and instructional shifts in classrooms (Banilower et al. 2013; Penuel et al. 2015; Reiser 2013; Trygstad et al. 2013; Krajcik et al. 2014; Tyler et al. 2018; Gao et al. 2018; Haag and Megowan 2015). California was one of the early implementation states, and in 2019, the new NGSS-aligned assessment, the California Science Test, became operational.

The COVID-19 pandemic changed the landscape abruptly. Our review of a random sample of 280 California public school districts in April 2020 showed that only 45% of districts provided home learning resources for science. Only 56% of districts aligned their science instruction with NGSS during distance learning in fall 2020 (Gao et al. 2020). Macias et al. (2022) surveyed 324 K-8 science teachers from 25 states and identified additional challenges. The pandemic and extended school closures had a disproportionate impact on students of color, English learners, and students with disabilities (Hamilton et al. 2020; Garet et al. 2020; Gao et al. 2021). Because many states were at the early implementation stage prior to spring 2020, concerns grew that the COVID-19 pandemic may stall or even reverse some of the early progress schools had made, by shifting resources further away from science education and constraining districts' decisions for several years. In 2017–2018, 60% of California districts reported that science is a priority in their districts, but the number decreased to 52% during 2020–2021 (Gao et al. 2018, 2020). In addition, only 17% of districts considered science education as a "high priority" in fall 2020; 10% reduced the number of science courses, and 64% believed the financial constraints would have a negative impact on science education (Gao et al. 2018).

In this chapter, we summarize the impact of COVID-19 on science education drawing on survey data from California school districts and science teachers. We describe district plans to support science education, from purchasing new instructional materials to providing professional learning. We report on the results of in-depth interviews conducted with school districts, science education organizations, and statewide policymakers to identify policy levers that may support an equitable science education post-pandemic and beyond.

18.2 Data and Method

We drew on three primary data sources:

Statewide Survey of School Districts: We launched a statewide survey of California public school districts in fall 2021. A total of 213 (of 1034) districts responded to the survey, and these districts serve 50% of the state's K-12 population. There was no significant difference in respondent vs. nonrespondent districts' shares of high-need or low-income students. Most of the differences were driven by district size and locale: large districts and urban districts were more likely to respond to our survey. In addition to the district survey, we also designed a (smaller) set of questions that were specific to science teachers. The California Science Educator Association (CASE) shared the teacher survey with its members and 124 science teachers completed the questions. More than 90% of the teachers were middle school or high school science teachers. The average teaching experience was 16 years, and 74% of the teachers had a master's degree or higher. Because of the small sample, we caution that the findings are not generalizable to all science teachers across California.

Local Control Accountability Plans (LCAPs): All districts are required to prepare an LCAP, which is a 3-year plan that describes their goals, actions, and expenditures to support positive student outcomes. Districts were required to develop their LCAPs for the 2021–2024 school years, and 858 (of 1034) posted them online. We downloaded and analyzed all publicly available LCAPs ($N = 858$) to examine districts' plans to support science education.

Interviews with Key Educational Partners: We conducted semi-structured interviews with all county regional leads ($N = 9$), high-need districts with varying size, region, and locale codes ($N = 10$), NGSS early implementation districts ($N = 9$), statewide science organizations ($N = 7$), and statewide policymakers ($N = 8$).

18.3 Impact of COVID-19 on Science Education

18.3.1 COVID-19 Derailed Science Education in Most Districts

Science became a lower priority in 62% of the districts during the 2020–2021 school year when most districts were in distance learning (Fig. 18.1). There was no significant difference by district's share of low-income or racially minoritized students. However, the impact was less severe in rural communities: nearly half of rural districts said science had the same priority, and just 40% said science became a lower priority vs. the average of 62% across all districts. Interviews with science educators in two large rural counties suggested that lower population density, remote location, and ample outdoor space made it easier to continue science education during the pandemic.

Before the pandemic, California districts were making progress in NGSS implementation, but the progress stalled during the pandemic. The state defines three phases of NGSS implementation: (1) the awareness phase represents an introduction to the California NGSS, the initial planning of systems implementation, and establishment of collaborations; (2) the transition phase is the concentration on building foundational resources, implementing needs assessments, establishing new professional learning opportunities, and expanding collaborations between all stakeholders; and (3) the implementation phase expands new professional learning support, fully aligns curriculum, instruction, and assessments, and effectively integrates these elements across the field (California Department of Education 2014). During the 2020–2021 school year, 94% of districts were implementing NGSS, representing a significant increase from 2018 where 78% of districts were in the implementation stage (Gao et al. 2018). Implementation was uneven across grades, with more progress made in K-8 grades than in high schools.

To implement NGSS effectively, schools need to align local curriculum (such as adopting standards-aligned instructional materials, and developing/aligning new science course sequences for middle and high schools), provide high quality training to teachers and administrators, support teachers to align classroom instruction, and develop/align local science assessments (National Research Council 2014). Implementation across those key stages of implementation

Fig. 18.1 Science became a lower priority in most districts during the 2020–2021 school year (*Source* Authors' calculations) *Note* Sample includes 213 respondent districts. High-poverty districts: at least 75% of students are eligible for free/reduced price lunch. High-minority districts: at least 75% of students are Black or Latino

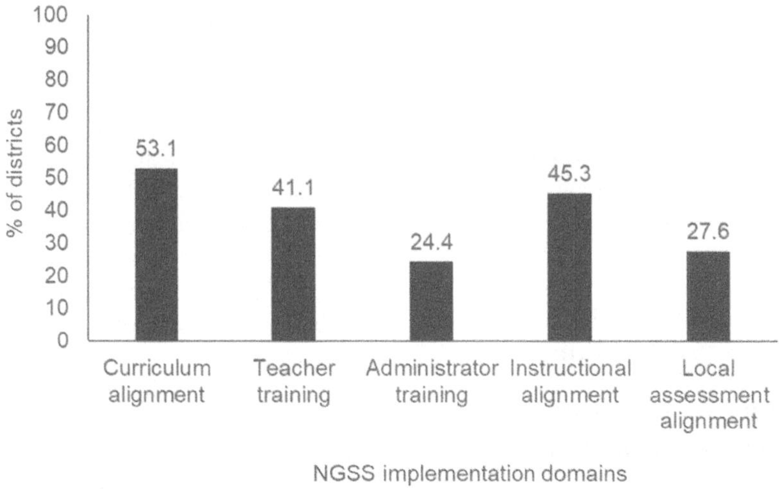

Fig. 18.2 NGSS progress was greatest in aligning local curriculum before spring 2020 (*Source* Authors' calculations). *Note* Sample includes 213 respondent districts. High-need districts: at least 55% of students are high-need (i.e., low-income, English learners or foster youth). High-poverty districts: at least 75% of students are eligible for free/reduced price lunch. High-minority districts: at least 75% of students are Black or Latino (*Source* Authors' calculations)

was uneven, with more districts in the process of aligning curriculum. Most districts had not yet started the process to provide teacher training, shift classroom instruction, align local assessments or train administrators (Fig. 18.2). Among districts that had not adopted new standards-aligned textbook or course models, most said the pandemic delayed their process.

The pandemic also affected teachers' instructional practices. Nearly 70% of teachers reported using more free online lessons or resources during COVID-19, and 40% used more open education resources, paid online resources, and online units that students study at their own pace. However, most of these resources were not fully aligned with NGSS, raising concerns about the quality of these instructional resources.

At the onset of the pandemic, nearly half of teachers said their instruction involved primarily reviewing content rather than teaching new content (Hamilton et al. 2020; Besecker et al. 2020). In contrast to spring, in fall 2020 most districts were focusing their remote content on new material (59%), while most of the remainder were teaching new material and reviewing past content (Gao et al. 2021). This represents a marked improvement from spring 2020, when nearly half of teachers focused mostly or exclusively on reviewing content (Hamilton et al. 2020). During the 2020–2021 school year, 90% of science teachers taught new content; however, most of them covered <50% of the intended curriculum, especially in high schools. Close to a quarter of teachers said 75–100% of students completed their assignments, while 51% of teachers said over 50% of students in their class completed the assignments. In elementary grades, however, nearly a quarter of teachers said less than 25% of students completed the assignment. When asked about the factors contributing to the noncompletion, most teachers said student engagement was the biggest factor (56%). Other important factors included parental involvement (46%), home environment (45%), social-emotional support (27%), academic support (22%), and no internet/device (11%).

Professional development for teachers remains highly inadequate, as only 43% of teachers received adequate training for transition to distance learning. Less than 20% of teachers received adequate training to support English learners and 25% received adequate training

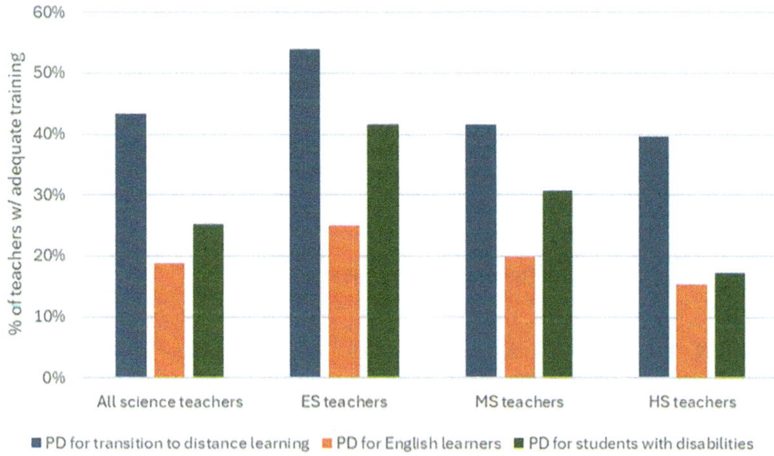

Fig. 18.3 Most teachers did not receive adequate professional development (PD) during COVID-19 (*Source* Authors' calculations) *Note* Sample includes 124 science teachers

Table 18.1 Most teachers did not always align their instructions with NGSS during COVID-19

	All science teachers (%)	ES teachers (%)	MS teachers (%)	HS teachers (%)
Engage in phenomenon or problem solving	18	50	29	8
Use 3 dimensions to explain phenomena	19	30	37	8
Use EP&C to explain phenomena	10	20	18	4
Meaningful learning sequence	38	50	58	24
Negotiate current understanding	20	20	34	12
Equitable science using universal design for learning (UDL) and social-emotional learning (SEL)	19	20	24	16

Source Authors' calculations. *Note* Sample includes 124 science teachers

for students with disabilities (Fig. 18.3). Inadequate professional development affects more high school (HS) teachers than elementary (ES) and middle school (MS) science teachers.

In our survey, we asked teachers how often they engaged in NGSS aligned instruction and the results are summarized in Table 18.1. Less than 20% of teachers always engaged in relevant phenomena or used three dimensions to explain phenomena or solve problems, 38% engaged students in a meaningful, logical learning sequence, and 20% provided student-centered instruction in which students review prior knowledge to negotiate current understanding. Very few teachers used environmental principles and concepts (EP&C) to explain phenomena. The alignment is higher in K-8 grades (elementary school and middle school).

18.3.2 Districts Provided Limited Programs to Support Science Learning During COVID-19

Supplemental instructional materials to support science learning were provided in 60% of the districts during distance learning, and services and programs to support social-emotional learning were provided in 40% (Fig. 18.4). Compared to math/ELA (English Language Arts), the range of programs the school districts provided for science learning was more limited. For example, 89% of districts assessed student learning in math/ELA, while only 28% did so for science. Nearly two-thirds (65%) of districts provided small group instruction for math/ELA, compared to 25% in science. Close to half

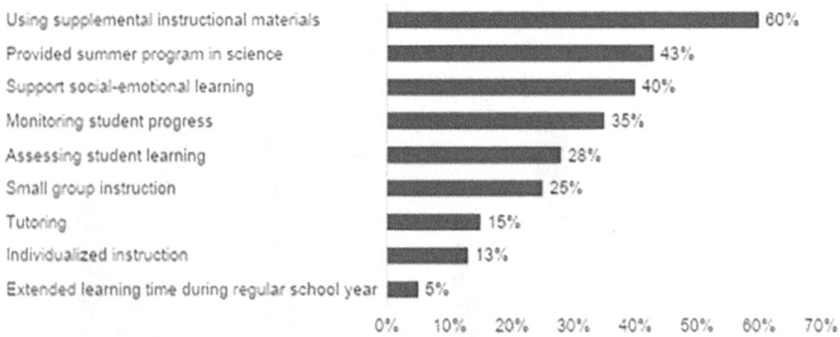

Fig. 18.4 Percent (%) of districts providing science learning support during the 2020–2021 school year (*Source* Authors' calculations) *Note* Sample includes 213 respondent districts. High-need districts: at least 55% of students are high-need (i.e., low-income, English learners or foster youth). High-poverty districts: at least 75% of students are eligible for free/reduced price lunch. High-minority districts: at least 75% of students are Black or Latino

of districts (48%) provided extended learning opportunities or individualized instruction in math/ELA while very few districts did so for science (Gao et al. 2021). Summer programs in science were provided in 43% of districts, and most of these programs prioritized students who were academically behind.

18.3.3 Challenges in High-Need Districts

Our survey data do not show much significant or meaningful difference by districts' share of high-need students. One possible explanation is that most districts did not prioritize science education during COVID-19, so the resources dedicated to science education are limited across all districts. On the other hand, however, emerging evidence shows that the pandemic has left historically under-served students, including Black, Latino, and low-income students, and English learners, further behind. We interviewed ten high-need districts, where at least 55% of students are high-need (meaning low-income, English learners, or foster youth), in spring 2022 to understand how they handled science education during COVID-19 and the challenges facing those districts. Those districts are diverse in terms of enrollment size, locale, and region.

In contrast to the survey results, nine of the ten high-need districts which participated in the interviews reported adoption of NGSS-aligned curriculum prior to the outbreak of COVID-19. This difference can likely be attributed to sample bias due to nonresponse. Although most districts have adopted NGSS-aligned curriculum, they were in different stages of implementation. Participants at two of the districts, one of which was part of the Early Implementation Initiative, reported having started the process 6 years ago in 2016. One district was in its third year of implementation. The rest were in the early stages of adoption, one is still in the piloting phase while the remaining five districts are in the first or second year of implementation. One participant reported their district as just beginning the adoption process this current school year in 2021. The districts that were in the earlier stages of implementation reported a disruption of that process due to COVID-19. Similar to the survey results, there was a higher proportion of adoption in grades K-8 than in high schools. Of the five districts with high schools, two participants reported their districts as just starting the process for high school NGSS adoption.

Participants identified staff shortages, funding, and the emphasis of ELA and math over science as barriers to prioritizing science learning. Staff shortages presented two separate issues: (1) the inability to find substitute teachers to allow teachers to have release time for professional training and (2) a lack of single-subject credentialed science teachers to teach middle and high school science. Funding was less of an issue for most districts; however, many indicated the need for designated funding for teacher training, science materials (initial purchase and replacement), and funding for science teachers on special assignment (TOSA) positions or science teacher leaders. During remote learning and the return to in-person schooling, there was a focus on ELA and math. This led to fewer instructional minutes for science, particularly in the elementary grades. This was less true for the two districts who used an integrated model for science instruction, however. In addition to these three barriers, the level of teacher burnout prevented many districts from pushing any new initiatives that prioritized science.

There isn't a clear picture of the amount of learning loss that occurred during remote instruction. Only one of the participants reported their district administered the California Science Test (CAST) during the 2020–2021 school year. The middle and high schools had assessments embedded in their curriculum but there wasn't a district-wide reporting of performance. All participants reported that there was learning loss due to the reduction in instructional minutes and the transition to remote learning, particularly at the elementary level; however, there was no actionable data specific to science. When asked if the districts had plans to address unfinished learning in science, only three participants reported their districts as having anything in place. The additional supports included before- and after-school tutoring, additional support in the classroom (second teacher or tutor hired from an outside party), and summer Science, Technology, Engineering and Mathematics (STEM)-based learning opportunities.

The one aspect that stands out is the way districts were using science as a lever for English language development. Participants from three of the five districts with a large English Learner (EL) population reported using science content as a means to teach English comprehension. This integrated model has ensured that EL students are receiving ELA instruction without sacrificing their access to science content.

18.4 District Recovery Plans

Our analysis of district responses during the 2020–2021 school year has shown the profound impact of the pandemic on science education, from teacher instruction to curriculum adoption. This raises the urgency of prioritizing science education as schools recover from the pandemic. However, few districts did so. Only 27% of districts said science is a high priority in their district's recovery plans, whereas more than 80% said math/ELA is a high priority (Fig. 18.5). The lack of prioritization means that fewer resources will be devoted to support science teaching and learning, raising concerns about whether students can fully recover from the pandemic learning disruptions.

To explore districts' plans to support science education in the following years, we downloaded and analyzed the local control accountability plan (LCAP) for all districts. We used natural language processing algorithms to analyze the LCAPs (e.g., text classification), and the results are summarized in Fig. 18.6. Close to half of districts plan to adopt, develop or purchase standards-aligned instructional materials; 38% plan to provide teacher training; 32% set student performance goals in the California Science Test (CAST); 25% plan to provide extended learning opportunities such as after-school and enrichment programs. Other

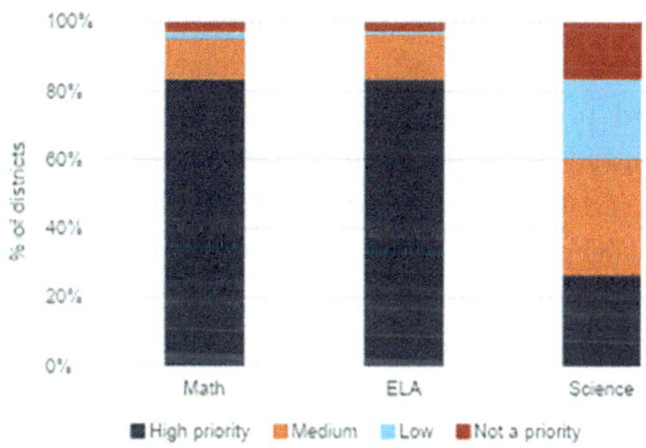

Fig. 18.5 Less than 30% of districts said science is a high priority in their recovery plans. (*Source* Authors' calculations) *Note* Sample includes 213 respondent districts. High-need districts: at least 55% of students are high need (i.e., low-income, English learners, or foster youth). High-poverty districts: at least 75% of students are eligible for free/reduced price lunch. High-minority districts: at least 75% of students are Black or Latino

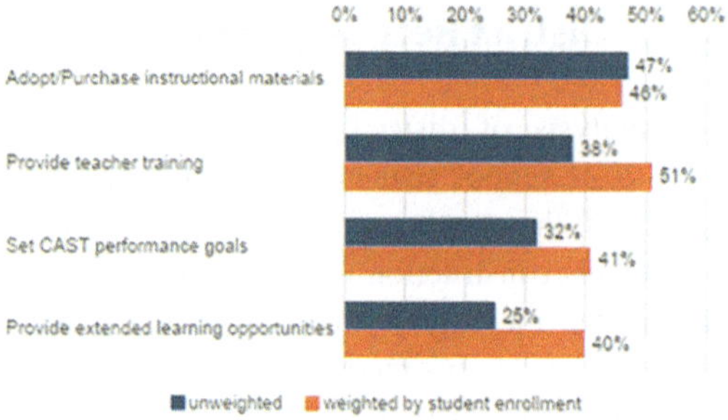

Fig. 18.6 Districts' plans to support science education (*Source* Authors' calculations). *Note* Sample includes the universe of 2021–2024 LCAPs ($N = 858$). 14% of LCAPs did not mention science education or NGSS

uncommon programs include allocating dedicated staff (such as TOSA or coordinator) to support science teachers (14%), purchasing science lab equipment and materials (14%), and using teacher surveys to measure and monitor local implementation (13%).

18.5 Conclusion

The Next Generation Science Standards (NGSS) describe a new vision for science teaching and learning that, if implemented well, could catalyze improvements in science literacy and increase the global competitiveness of the U.S. labor force (National Research Council 2014). Before the pandemic, most California districts were in the early stages of implementing the new science standards, and some districts were headed toward full implementation. However, COVID-19 caused major disruptions to California NGSS implementation in most districts. Science became a lower priority in most districts, and major NGSS implementation activities were delayed. Districts provided limited support for science education, and only 27% made science education a high priority in their recovery plans. Taken together, these findings suggest that the pandemic disruptions may affect science learning for years to come. To help students recover, we recommend that the state include science in its district accountability requirements to encourage districts to prioritize science education and dedicate resources to support California NGSS implementation. We also recommend that the state provide more evidence-based strategies to support science learning recovery. As schools recover from the COVID-19 pandemic, continued commitment from state, district, and school leaders, and teachers will be required to support an effective and equitable recovery in science too.

References

Banilower ER, Smith PS, Weiss IR, Malzahn KA, Campbell KM, Weis AM (2013) Report of the 2012 national survey of science and mathematics education. Horizon Research, Inc.

Besecker M, Thomas A, Daley G (2020) Student engagement online during school facilities closures: an analysis of L.A. Unified secondary students' Schoology activity from March 16 to May 22, 2020. Independent Analysis Unit, Los Angeles Unified School District

California Department of Education (2014) Next generation science standards systems implementation plan for California. https://www.cde.ca.gov/ci/pl/ngssintrod.asp

Gao N (2023) Impact of COVID-19 on science education: early evidence from California. In: Columbia academic commons, academic unit: northeast big data innovation hub, series: COVID-19 research lightning talks. https://doi.org/10.7916/z22v-0c20

Gao N, Adan S, Lopes L, Lee G (2018) Implementing the next generation science standards: early evidence from California. Public Policy Institute of California

Gao N, Lafortune J, Hill L (2020) Who is losing ground with distance learning? Public Policy Institute of California

Gao N, Hill L, Lafortune J (2021) Distance learning strategies in California schools. Public Policy Institute of California

Garet M, Rickles J, Bowdon J, Heppen J (2020) National survey on public education's coronavirus pandemic response. American Institutes for Research

Haag S, Megowan C (2015) Next generation science standards: a national mixed-methods study on teacher readiness. School Sci Math 115(8):416–426. https://doi.org/10.1111/ssm.12145

Hamilton L, Kaufman J, Diliberti M (2020) Teaching and leading through a pandemic: key findings from the American educator panels spring 2020 COVID-19 surveys. RAND Corporation

Krajcik J, Codere S, Dahsah C, Renee B, Mun K (2014) Planning instruction to meet the intent of the next generation science standards. J Sci Teach Educ 25(2):157–175. https://doi.org/10.1007/s10972-014-9383-2

Macias M, Iveland A, Tyler B, White MS (2022) Teaching K-8 science through distance learning: specific challenges and successes during the COVID-19 pandemic. WestEd

National Research Council (2014) Guides to implementing the next generation science standards. National Academies Press. https://doi.org/10.17226/18802

National Science Teaching Association. 2021. K-12 Science Standards Adoption.

Penuel WR, Harris CJ, DeBarger AH (2015) Implementing the next generation science standards. Phi Delta Kappan 96(6):45–49

Reiser BJ (2013) What professional development strategies are needed for successful implementation of the next generation science standards? Center for K-12 assessment and performance management, educational testing service

Trygstad PJ, Smith PS, Banilower ER, Nelson MM (2013) The status of elementary science education: are we ready for the next generation science standards? Horizon Research

Tyler B, Britton T, Iveland A, Nguyen K, Hipps J (2018) Engaged and learning science. How students benefit from next generation science standards teaching? WestEd. https://www.wested.org/resource/engaged-and-learning-science/

Niu Gao is a principal researcher at the American Institutes for Research (AIR), where she specializes in K-12 educational policy. Her research concerns the evaluation of policies, programs, and practices to improve the educational opportunities for students. Her recent work focused on understanding the far-reaching impact of the COVID-19 pandemic on students and families, and she is leading a research-practice-partnership with the California Department of Education to improve pandemic recovery efforts. Her work has been supported by the U.S. Department of Education, the National Science Foundation (NSF), the Administration for Children and Families, among others. She received the New Scholars Award from the Association for Education Finance and Policy and the research-practice-partnership award from the California Education Research Association. Before joining AIR, she was a senior fellow at the Public Policy Institute of California. She received her PhD in educational policy and master's in economics, both from Florida State University.

Kathy DiRanna is the Statewide Director Emeritus of WestEd's K-12 Alliance, a professional learning organization focused on improving science education for K-12 through content, pedagogy, and leadership. She served as a Principal Investigator/Program Director for several NSF-funded projects, and helped shape California's science reform efforts for the past 33 years including facilitating the California's Science Expert Panel for the state's adoption of Next Generation Science Standards (NGSS), serving on the state's STEM Task Force, the State NGSS Implementation Plan and the California Science Framework. Ms. DiRanna directed the California NGSS K-8 Early Implementation Initiative (2014–2020) to assist eight districts and two charters in implementing NGSS and serving as lighthouse Local Education Agencies (LEAs). She was a major contributor to the state's NGSS Roll Outs and Science Community of Practice. Ms. DiRanna received WestEd's Paul Hood Individual and Team Award for making significant contributions to the field; Learning Forward's Susan Loucks Horsley Award for building learning communities; California Science Teachers' Margaret Nicolson Award for distinguished service to science education; and the National Science Education Leadership Association's Outstanding Leadership in Science Education Award for making significant contributions to the field. Upon retirement, WestEd awarded Ms. DiRanna with a Lifetime Achievement award for her quality work in science education.

Maria Chang Fay studiess the incorporation of antiracist pedagogy into secondary science classrooms. She is a native Californian and received her bachelor degrees from the University of California, Berkeley, and her master's degree from the University of California, Davis. Dr. Fay received her doctorate in Science and Agricultural Education from the University of California, Davis' School of Education. She currently works with the San Francisco Exploratorium and University of San Francisco to support science teacher coaches and mentors. She worked as a coach for novice secondary science teachers for 7 years through the San Francisco Exploratorium's Leadership Program. Prior to that, she worked as a math specialist and resource specialist assistant at an elementary school in the Oakland Unified School District. She also taught middle school science for 8 years in the East Bay. Dr. Fay earned her Multiple Subject Teaching Credential from Holy Names University and completed her coursework for her Single Subject Credential in foundational science at the University of California, Los Angeles.

COVID's Impact on STEM Learning in High School: Lessons Learned

19

Jennifer Hamilton and Debbie Kim

Abstract

As a result of the COVID-19 pandemic, schools across the United States (U.S.) closed their doors in spring 2020, disrupting the learning environments of a generation of students. Student experiences during this turbulent time varied greatly, with disparities in access to effective instruction largely dependent on socioeconomic status. We report on findings from a nationally representative youth survey conducted in spring 2020. The survey asked students about how COVID-19 impacted their access to Science, Technology, Engineering and Math (STEM) content, their learning, and goals for the future. Survey results illustrate ongoing themes in education—traditionally marginalized students had far less productive remote learning experiences than their higher-resourced peers. In other words, the pandemic exacerbated existing inequalities in student access to effective high school instruction and, by extension, access to the STEM pipeline. We turn to social capital theory and Bourdieu's concept of hysteresis to consider the contextual factors that shaped students' STEM-related experiences during COVID. We assert that social capital played a central role in whether and how access to learning tools were taken up and encourage educators to consider the full range of factors (e.g., students' sense of self efficacy, access to adults who facilitate online learning) that shape students' progress along their STEM trajectories.

Keywords

COVID-19 · Education · STEM trajectory · Remote learning · Access

This study is supported by the National Science Foundation under Grant No. 2030436. The study is entitled "Pandemic Learning Loss in U.S. High Schools: A National Examination of Student Experiences." Any opinions, findings, and conclusions or recommendations expressed in this material are those of the author(s) and do not necessarily reflect the views of the National Science Foundation.

Research Funded by
NSF Award #2030436—RAPID*: Pandemic Learning Loss in U.S. High Schools: A National Examination of Student Experiences.

*Rapid Response Research Award.

COVID Information Commons Video Presentation

The COVID Information Commons (CIC) Research Lightning Talk video presentation and individual multilingual transcripts for this project can be found in the Columbia Academic Commons (AC) (Hamilton and Kim 2022).

J. Hamilton (✉) · D. Kim
NORC, University of Chicago, Chicago, IL, USA
e-mail: Hamilton-Jennifer@norc.org

D. Kim
e-mail: debbie.h.kim@gmail.com

F. D. Hudson (ed.), *The COVID Information Commons*, Sustainable Development Goals Series,
https://doi.org/10.1007/978-3-032-06125-6_19

19.1 Introduction

As a result of the COVID-19 pandemic, schools across the United States (U.S.) closed their doors in spring 2020, disrupting the learning environments of a generation of students. Student experiences during this turbulent time varied greatly, with disparities in access to effective instruction largely dependent on socioeconomic status. While America has long been struggling with diversifying and strengthening the Science, Technology, Engineering and Mathematics (STEM) pipeline, the pandemic had the opposite effect—by exacerbating the existing inequalities in student access to effective high school instruction and, by extension, access to the STEM pipeline.

To more fully understand the impact of the pandemic on high school students, and longitudinally, the country's future STEM workforce, the National Science Foundation (NSF) funded NORC at the University of Chicago (NORC) to conduct a nationally representative survey of American youth. The goal was to understand how COVID-19 impacted their access to STEM content, their learning, and goals for the future. We asked which students had the opportunity to participate in distance learning during spring 2020, as well as which students did not have this opportunity, and why. The study also asked high schoolers about their plans after high school, and the extent to which those plans had changed due to COVID.

In this chapter, we present findings from this survey. What we learned supports what other studies have uncovered—access to broadband was uneven, uptake was thin, and many students returned to in-person schooling underprepared. And, most importantly, low-income students had far less productive remote learning experiences than their high-income peers. In other words, COVID-19 exacerbated existing inequalities in education and pushed some ugly truths into the spotlight.

We turn to social capital theory and Bourdieu's concept of hysteresis (Strand and Lizardo 2016) to consider the contextual factors that shaped students' STEM-related experiences during COVID. While the era of fully remote learning may be behind us, we can still extract important lessons from that time.

19.2 Literature Review

It is estimated that COVID-19-related school closures affected 55.1 million students across 124,000 public and private schools in the United States in the spring of 2020 (Map: Coronavirus and School Closures 2020). In most cases, traditional classroom-based instruction was replaced with a mix of live and self-paced learning delivered through a variety of online platforms such as Blackboard, Google Classroom, Schoolify, and Canvas (Gross and Opalka 2020; Hamilton et al. 2020; Harris et al. 2020). These closures persisted into the following academic year, with 17 of the 20 largest school districts in the United States starting fall 2020 with remote-only instruction (School Districts' Reopening Plans 2020).

The majority of schools provided at least some remote instruction during the last months of the 2019–2020 school year (Lake and Dusseault 2020). However, teachers generally had little previous experience with online instruction, and many families' lack of appropriate devices/Wi-Fi meant that students' access to quality instruction was highly variable. In fact, the Los Angeles Unified School District in California reported that nearly 15% of high school students never logged on to its online learning system in the spring of 2020, and an additional 25% logged on only infrequently (Burke 2020). Nationwide, just 9% of teachers reported that their students regularly completed their online learning activities (Hamilton et al. 2020). Parents during this time were understandably worried. Approximately one-quarter of parents felt their child was underprepared for school in the coming fall. These worries were more prevalent for non-white and low-income households (Saavedra et al. 2020).

These worries turned out to be well-founded. Research indicates that school closures led to substantial learning loss in math and science among high school students. For instance, students experienced significant setbacks in mathematics during school closures, with disadvantaged students disproportionately affected (Kuhfeld et al. 2020). It is not unreasonable to conclude that these inequities will not only have severe, long-term repercussions for this cohort of students as individuals, but will also have major consequences for the future of the STEM workforce and the economic success of the Nation.

19.2.1 The STEM Pipeline to College and Career

The STEM pipeline and the factors that support a student along the pipeline are complicated. Students must achieve specific milestones as they progress through the STEM pipeline, with failure at any point effectively ending that trajectory. Meeting these milestones include hitting academic markers (e.g., graduating high school, success in advanced course work) as well as the presence of internal (e.g., a sense of self-efficacy in STEM subject areas) and external supports (e.g., access to high-quality instruction).

For instance, research has demonstrated that high-quality instruction, characterized by engaging and effective teaching methods and a rigorous curriculum, increases student achievement in STEM subjects (Tai et al. 2006). Such instruction challenges students and better prepares them for more advanced studies (Sadler and Tai 2007). The provision of high-quality, engaging instruction also boosts students' understanding and appreciation of STEM fields and sense of self-efficacy in STEM, all of which are critical for pursuing STEM careers (Sadler and Tai 2007; Tyson et al. 2007). In short, high-quality instruction is connected with students' long-term perceptions of the value of STEM and their sense of place within the STEM field.

In addition, students' attitudes and performance set the stage for their success along the STEM trajectory. Strong performance in math and science courses as well as taking advanced coursework, such as advanced placement (AP) classes, prepare students for the rigors of STEM degrees and careers (Tyson et al. 2007). Students who plan early and express aspirations for STEM careers are more likely to follow through on these plans. This early commitment to a STEM pathway increases the likelihood of taking the necessary preparatory courses and seeking relevant experiences (Sadler and Tai 2007). For instance, students who plan to attend college are more likely to pursue STEM careers and take the necessary steps in high school to qualify for STEM programs (Maltese and Tai 2011).

19.2.2 Theoretical Framework

In order to study the impact of COVID-19 on high school students' STEM trajectories, we draw on two complementary theoretical frameworks—social capital theory and Bourdieu's notion of hysteresis.

19.2.2.1 Social Capital Theory

The concept of social capital has a rich and intellectually profound history, providing valuable framing and insight into the ways individuals, groups, and their social contexts interact to shape life trajectories over time. At its core, social capital theory states that people's behaviors can be understood in relation to the social context they are behaving within (Bourdieu and Passeron 1990; Lin 2001). Social capital theory conceptualizes people's social worlds as a durable network of three key elements:

1. Relationships, referring to social ties through group membership and associations that can help generate benefits for individuals and social groups
2. Trust, referring to people's faith in others as well as confidence in their political institutions, and
3. Norms, referring to forms of social support and collective efficacy.

This network establishes norms and values of a social class and, by doing so, may provide people within that class power and opportunity over those outside of the network (Bourdieu

1986; Makridis and Wu 2021). Social capital is consciously and unconsciously formed and enacted between and across networks (Dika and Singh 2002) and consists of tangible and intangible resources.

For instance, in the context of the COVID-19 pandemic, we might have seen students' social capital work for them or against them as they strove to adjust to the new way of learning. They may have had access to tangible resources, such as pre-existing broadband internet and laptops at home, and intangible resources, such as the knowledge of and confidence to reach out to online tutors to enhance their online learning experiences. In this example, the students' social capital helped bolster them through the remote learning period, providing them with valuable information and tools to move through the pandemic with more success.

Social capital theory provided us with a theoretical baseline for understanding the status quo, pre-COVID-19, as well as a rich lens for understanding what happened when COVID hit and students pivoted quickly to online learning. Students' contexts (e.g., home, school, peer) shaped their experiences with remote schooling in a broad range of ways. Remote learning likely magnified the importance of pre-existing resources such as adults who could assist or support online learning, a quiet place to work, and broadband and technology access and expertise. In this situation, social capital played an important supportive role. For instance, if a student's family was in communication with other families, they may have exchanged information that could support a student's remote learning needs (e.g., passing on tips for accessing websites, creating "pods" for learning). Indeed, during COVID, student engagement in remote learning was shaped by factors such as pre-existing access to technological resources, their parents' communications with their schools, and their parents' communications with classmates' parents (Makridis and Wu 2021). In other words, resources available within their contexts as well as tacit resources that existed between individuals mattered when COVID hit.

19.2.2.2 Bourdieu's Notion of Hysteresis

Social capital theory helped us understand students' lives pre-COVID-19 and the ways their contexts interacted with their schooling experiences. However, COVID-19 provided a unique case of abrupt change and disruption to the norm. As such, we turn to Bourdieu's notion of hysteresis to help conceptualize the impacts of the COVID pandemic at a system level.

Hysteresis occurs when there is a major, abrupt change—often spurred by a crisis or catastrophe—in which one's current environments are mismatched or disconnected from emerging social conditions (Strand and Lizardo 2016): "It feels like living in a different time, carrying risks and opportunities" (Graham 2020, p. 451). Hysteresis can occur when upwardly mobile students have relied upon schools to equip them with classroom studying skills when their parents could not do so (Perna 2004); this would mean that, during the COVID-19 pandemic and the transition to remote learning, students who succeeded within schools faced challenges for which they might be unequipped at home. Bourdieu even stated about such situations, "The hysteresis effect is proportionately greater for agents who are more remote from the educational system and who are poorly or only vaguely informed about the market in educational qualifications" (Bourdieu 2000, p. 142).

In these times of transition or disruption, individuals and systems do not respond simultaneously or identically, but rather in disparate and different ways. And those responses often magnify and exacerbate existing inequalities, since the highest resourced people are the first to head for new positions (Bourdieu 1986). Indeed, students who are poorest in economic, cultural, and social capital are likely to suffer disproportionately in unprecedented situations, such as COVID and remote learning.

Taken together, social capital theory and the concept of hysteresis provided complementary lenses for understanding how students' existing networks, resources, and trust in their institutions interacted with a sudden change in their schooling environment. We argue that the COVID pandemic brought an abrupt change that exacerbated existing inequalities by making certain forms of social capital matter even more.

19.3 Methodology and Results

To understand the impact of school closures and the abrupt switch to remote instruction on American high school students, NORC developed and fielded a national survey through its long-standing AmeriSpeak Teen Panel. This panel is a nationally representative sample of 2306 students aged 13–17, and closely reflects Census population benchmarks (Fig. 19.1).

a. Race/Ethnicity

Race/Ethnicity	n	%
White	1261	53
Hispanic	331	24
Black	219	13
2+ race	102	5
Asian	109	4
Other	14	1

b. Family Income

Family Income	n	%
<=$34,999	409	29
$35,000-74,999	583	30
$75,000-124,999	594	25
$125,000+	450	16

c. Gender

Gender	n	%
Male	981	51
Female	1055	49

Fig. 19.1 Survey demographic raw frequency counts,**a** race/ethnicity, **b** family income, **c** gender

The survey was administered just before the end of the spring 2020 semester and asked high school students about a range of topics related to their experiences during remote learning. These topics included their schools' responses to COVID, technology access, barriers and facilitators for remote learning, how they spent their time at home, concerns about returning to school, STEM course enrolment and engagement pre-COVID-19 and while in quarantine, perceptions of their STEM learning during quarantine, and any changes in their future plans in the face of the pandemic.

For analysis, NORC calculated panel weights by incorporating the appropriate probability of selection, then incorporating nonresponse and making ratio adjustments to population benchmarks. NORC then ran descriptive analyses with a focus on crosstabs between groups of interest, along with t-tests to determine significance.

19.3.1 Analysis/Discussion/ Evidence

Our survey results clearly demonstrate that COVID-19 impacted the most important milestones shaping high school students' STEM trajectories. Disparities already existed between high- and low-income students and COVID pushed these groups apart even further. COVID also affected students' sense of confidence in their schools and in their own future success in STEM subjects. These changes in their sense of self-efficacy were further compounded by changes in their plans post-high school and in their engagement in important steps along the STEM trajectory (e.g., taking and passing AP exams).

19.3.1.1 Existing Disparities

Prior to COVID-19, disparities already existed between low-income high school students and their high-income peers. According to our survey, 88% of high-income students were getting As and Bs just prior to COVID-19, compared to only 56% of low-income students. Similarly, only 21% of low-income students were enrolled in three

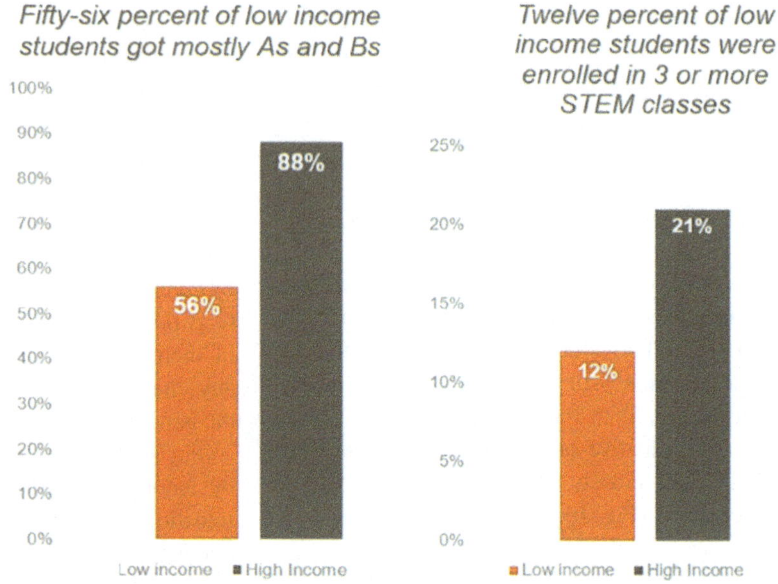

Fig. 19.2 Disparities just prior to COVID-19 disruption

or more STEM courses, compared to 33% of high-income students (both statistically significant differences, see Fig. 19.2).

Moreover, students' responses about their pre-COVID-19 engagement in schooling (e.g., considering dropping out) and in STEM in general (e.g., considering a STEM career) also supported the notion that there were existing disparities in engagement. Pre-COVID-19, 15% of low-income students had considered dropping out of school, while only 5% of their high-income peers did so. Additionally, only 36% of low-income students had considered a STEM career, while 56% of high-income students had done the same (both statistically significant at the $p < 0.05$ level).

Given these disparities going into the COVID disruptions, the additional interruptions during COVID can only have made the pursuit of a STEM career more difficult. Our survey results indicate that COVID-19 affected low-income students much more negatively than their more affluent counterparts, exacerbating the existing disparities.

19.3.1.2 Impact of COVID-19 on the STEM Pipeline

The STEM pipeline has many interconnected factors that influence students' entry into the STEM career pipeline. These include: (1) access to high-quality instruction, (2) enrolment in AP classes/exams, and (3) their post high school plans.

Access to High-Quality Instruction

The sudden pivot to online learning impacted all students. In an open-ended question about obstacles faced during remote learning, students expressed concerns about access to the internet at home or limited access to technological resources. They also noted the challenges that changes in their learning environment brought—their home was too loud or chaotic, specialized instruction was no longer available to them, and online learning was less engaging (Fig. 19.3).

While these challenges were felt across the board, we observed more distinct differences in the ways high- versus low-income students experienced instruction during lockdown. Twenty-two percent of high-income students reported that their instruction was "excellent during lockdown" compared to 13% of students in the low-income category (statistically significant). In terms of time spent engaged in learning, 33% of high-income students reported spending more than four hours per day doing schoolwork, compared to 25% of students in the low-income category.

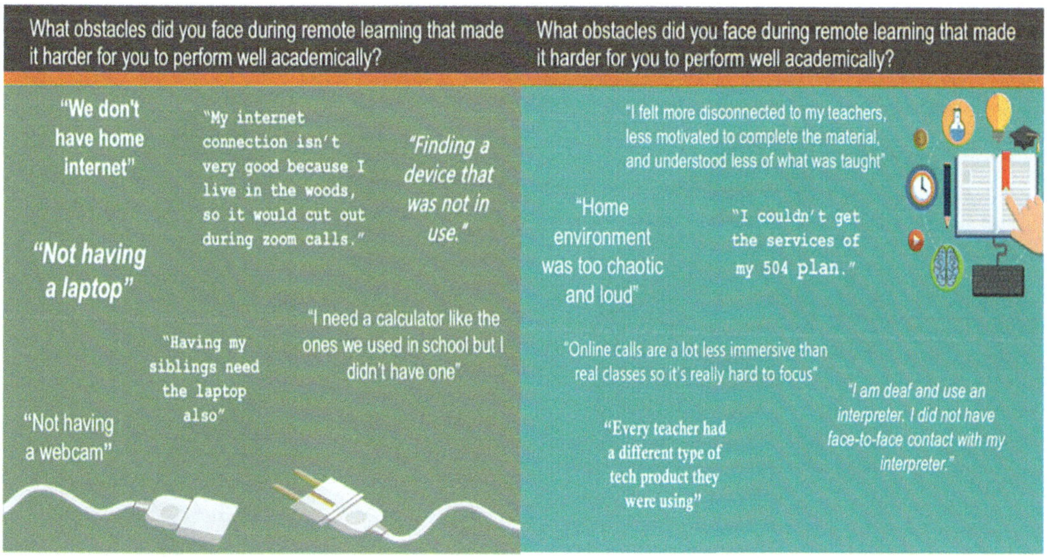

Fig. 19.3 Obstacles during remote learning

We also observed important differences in students' perceptions of their own performance and their sense of confidence in their schooling. Figure 19.4 displays findings from a survey question about whether students were worried about their STEM education. About one-third of students reported being "extremely" or "moderately" worried, with 83% of low-income students being worried and only 67% of high-income students feeling the same way (a statistically significant difference).

These differences in students' sense of confidence in their schooling were also evident in other findings. Sixty-four percent of high-income students felt completely or mostly prepared for the upcoming fall semester, compared to 46% of low-income students. In addition, 10% of low-income students thought their grades in STEM subjects would decline because of the COVID disruption while only 3% of high-income students thought the same.

These findings are particularly important because personal interest and confidence in one's abilities in STEM subjects are shown to be critical predictors of students pursuing STEM careers (Wang and Degol 2013). In addition, a sense of trust in your political institution (in this case, your school) to support you in your ambitions is an important form of social capital and plays a critical role in shaping students' futures.

Enrolment in Advanced Placement (AP) and Exam Taking

Students' engagement and progress in their overall STEM trajectory involves not only a sense of confidence in themselves and their futures, but also taking concrete steps along the way. One important concrete step is enrolling in AP courses, and then taking the exam and receiving a grade of three or higher ("qualified"). Looking at the overall results, we find that, for most STEM subjects (AP Biology, AP Chemistry, and AP Physics) half of all enrolled students did not take the exam.[1] For AP Calculus, the numbers were substantially higher—for Calculus BC, 66% took the exam; and for Calculus AB, 70% took the exam. However, when we look at these data by economic status, we can see that students from less affluent homes took the exam in much lower numbers. For example, for AP Physics there is

[1] *The College Board worked to ensure that every student who wanted to take the test and couldn't, was provided with internet and a laptop (with the help of generous funders). They provided videos on YouTube of expert teachers going through the materials necessary for the test. And they shortened the tests themselves. The College Board also waived the cancellation fees.*

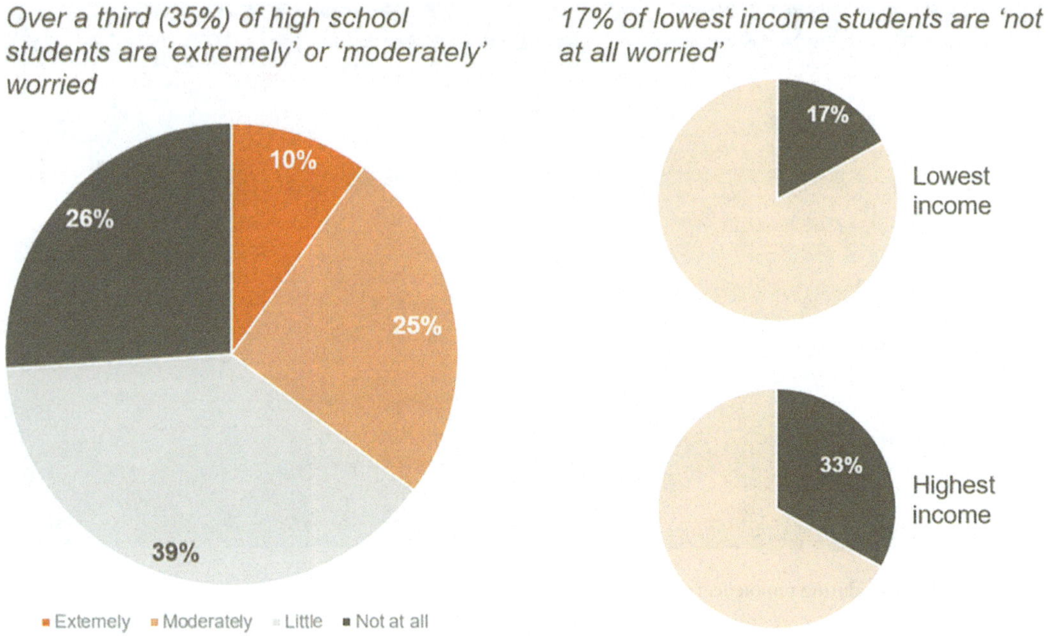

Fig. 19.4 Students "worried" about their STEM education

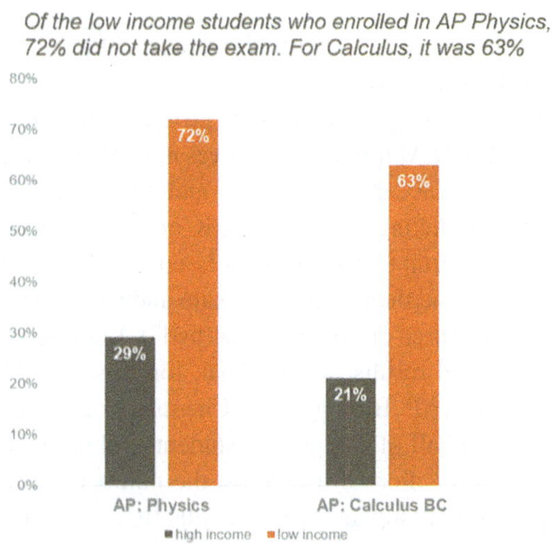

Fig. 19.5 Advanced placement (AP) test taking by income category

a difference of 43% in exam taking between the high- and low- income students, and a 42% difference for AP Calculus BC (both statistically significant) (Fig. 19.5).

The bottom line is that for motivated low-income students who were engaged in higher level STEM coursework, the educational disruptions of 2020 resulted in them not taking the AP exams that would help them get into college and pursue a STEM career.

When asked why they did not take the exam, students' responses were very similar. Some reported that the exams were not offered in their school; while others stated that they were not prepared enough and so were afraid to take them. Still others indicated that, since taking the exams was not required, they decided not to take them. Such steps count for students' progress along the STEM trajectory, with research demonstrating that engaging in the advanced coursework and test taking set students up for success in their STEM careers (Tai et al. 2006).

Post-high School Plans

Another factor that influences a student's STEM trajectory is their post-high school plans. Overall, there was no change in the proportion of teens planning to attend a 4-year college, and more teens were planning to attend community college posthigh school (Fig. 19.6). In addition, there was an increase in the proportion of students planning to take a part-time job after graduation.

	Pre-COVID	Post-COVID
Work Part-Time	30%	22%
Work Full-Time	13%	18%
Community College	20%	25%
4-Year College	50%	50%

Fig. 19.6 Students' postsecondary plans

		Low Income	High Income	Stat Sig	Pre-Post % change Low Income	Pre-Post % change High Income
Work Part Time	Pre	35%	18%	*	-15%	-3%
	Post	20%	15%			
Work Full Time	Pre	13%	10%		+11%	-3%
	Post	24%	7%	*		
Community College	Pre	18%	12%		+6%	-1%
	Post	24%	11%	*		
4-Year College	Pre	36%	69%	*	-2%	+4%
	Post	34%	73%	*		

Fig. 19.7 Students' postsecondary plans by income level

However, when looking at subgroups, the narrative shifts. Plans for the most affluent students did not change too much as a result of COVID-19 (with a 4% increase in high-income students planning to attend a 4-year college), while students in the low-income category who had previously anticipated attending a 4-year college changed their minds (a decrease of 2%). Conversely, we see an 11% increase of low-income students looking for a full-time job out of high school (Fig. 19.7). This increase may be due to financial pressures brought to households through the COVID-19 crisis. (Note: Stat Sig is Statistically Significant).

19.4 Conclusions and Recommendations

Research has long demonstrated that students from higher socioeconomic backgrounds with access to better resources are more likely to plan for and attend college, thereby increasing their chances of entering STEM fields (Perna 2004). Our survey indicates that the educational disruptions from COVID may have made it even more difficult for these students to attend college—via a reduction in access to STEM coursework and instruction, not taking AP exams, and reporting being less likely to attend a 4-year college post-graduation.

This study provides valuable and unprecedented insight on both students' personal experiences and engagement with STEM learning during the pandemic, and the actual change in their STEM learning pre- and post-COVID. Findings indicate what most of the education world already knew—that COVID-19 negatively impacted low-income students more than high-income students. However, it provides additional insight into the intricacies of how and why this differential impact occurred.

These survey findings have important implications as we look to the future. During COVID, districts and schools did everything they could to ensure all students received instruction. In rural areas, teachers drove out to students' homes to drop off hardcopy materials (Kim et al. 2021). Other schools made laptops available to students and tried to provide basic technical support. Despite these efforts, access to instruction was uneven across socioeconomic groups. The variation in access was exacerbated by students' differences in pre-existing resources and ability to leverage resources in productive ways. In other words, the differences in students' social capital were magnified as a state of hysteresis was introduced. During such times, individuals and groups respond differently, working within their existing contexts. During COVID-19, we observed this in differences in the use and adoption of broadband internet at home, and in the ways families were able to support their students in accessing and understanding online instruction.

The era of COVID-19-prompted remote learning may be behind us. However, online and digital learning tools continue to develop and expand, including the fast-moving technology of artificial intelligence (AI). These tools hold enormous potential for learning and, because they can be disseminated relatively easily, also hold potential for providing access to high-quality instruction. However, as we see from this study, while access is an obstacle it is not the only issue. Social capital also plays a central role in whether and how access to learning tools are taken up and used. These elements (e.g., resources to access online learning tools, adults who can help students understand how to use the learning tools, trust in an institution to go to for help or support, peers who also engage in the same tool) create the context where students either work hard to actualize their potential or have that potential easily available to harness.

19.5 Future Research and Final Thoughts

Our survey captured real-time data at a distinct moment in time. Future research could include a follow up with the AmeriSpeak Teen Panel survey sample. We would ask questions about whether or not they went to college, their sense of confidence in the schooling they received post-COVID, whether or not they are majoring in a STEM field, and questions centred around the contextual supports they received upon return to in-person schooling. Through these questions, we would seek to understand how COVID shaped their STEM trajectories and how different forms of social capital buoyed them afterward or made it more difficult for them to fully engage in in-person schooling. We would also include questions about how they understand their access to online instructional tools and the ways that they access and use such tools.

Much of the research on COVID-19 and its impacts on education tells the same story of the haves and have-nots. It is not a new story, but one brought into the spotlight due to the sheer scale and severity. But if we step back and think about *why* this is more extreme, it provides valuable lessons. While one lesson learned is that the digital divide is still prevalent in terms of access to remote education, educators also need to focus on the full range of factors that support students along their academic careers. Internal factors, such as students' sense of self-efficacy, and external factors, such as students' access to broadband internet, all play an important role in whether and how students progress along their STEM trajectories.

References

Bourdieu P (1986) Social space and symbolic power. Sociol Theory 7(1):14–25

Bourdieu P (2000) Pascalian meditations. Stanford University Press, Stanford

Bourdieu P, Passeron JC (1990) Reproduction in education, society and culture, 2nd edn. SAGE Publications

Burke M (2020) Thousands of Los Angeles high school students are not accessing online learning during school closures. EdSource. https://edsource.org/2020/thousands-of-los-angeles-high-school-students-are-not-accessing-online-learning-during-school-closures/627448#:~:text=About%2015%2C000%20high%20school%20students,Superintendent%20Austin%20Beutner%20said%20Monday

Dika SL, Singh J (2002) Social capital and academic achievement: findings from the national education longitudinal study. Res Sociol Educ Soc 13:273–298

Graham LJ (2020) Hysteresis and the sociological perspective: understanding the dynamics of social change. J Sociol 56(1):3–18. https://doi.org/10.1177/0001699320961814

Gross B, Opalka A (2020) Too many schools leave learning to chance during the pandemic. Center on Reinventing Public Education. https://files.eric.ed.gov/fulltext/ED605576.pdf. Accessed 14 March 2025

Hamilton J, Kim D (2022) Jennifer Hamilton: pandemic learning loss in U.S. high schools: a national examination of student experiences, a research update. In: Columbia academic commons, academic unit: northeast big data innovation hub, series: COVID-19 research lightning talks. https://doi.org/10.7916/vxqa-xq96

Hamilton LS, Grant D, Kaufman JH, Dilberti M, Schwartz HL, Hunter GP, Setodji CM (2020) COVID-19 and the state of K-12 schools: results and technical documentation from the spring 2020 American educator panels COVID-19 surveys. RAND Corporation. https://www.rand.org/pubs/research_reports/RRA168-1.html. Accessed 13 July 2024

Harris DN, Liu L, Oliver D, Balfe C, Slaughter S, Mattei N (2020) How America's schools responded to the COVID crisis. In: Technical report. National Center for Research on Education Access and Choice. https://www.edworkingpapers.com/sites/default/files/COVID%20Tech%20Report%20Merged%20Complete%20Final%20Working.pdf. Accessed 13 July 2024

Kim D, Hamilton J, McElvain C (2021) Remote learning in indigenous schools during COVID: final report for the Bureau of Indian Education. Washington DC

Kuhfeld M, Soland J, Tarasawa B, Johnson A, Ruzek E, Liu J (2020) Projecting the potential impacts of COVID-19 school closures on academic achievement. Educ Res 49(8):549–565

Lake R, Dusseault B (2020) Remote classes are in session for more school districts, but attendance plans are still absent. In: Center for reinventing public education. https://crpe.org/remote-classes-are-in-session-for-more-school-districts-but-attendance-plans-are-still-absent/. Accessed 14 March 2025

Liu N (2001) Social capital: a theory of social structure and action. Cambridge University Press, Cambridge

Makridis C, Wu C (2021) Ties that bind (and social distance): how social capital helps communities weather the COVID-19 pandemic. https://papers.ssrn.com/sol3/papers.cfm?abstract_id=3584581

Maltese AV, Tai RH (2011) Pipeline persistence: Examining the association of educational experiences with earned degrees in STEM among US students. Sci Educ 95(5):877–907

Map: Coronavirus and School Closures (2020) Education week. https://www.edweek.org/leadership/map-coronavirus-and-school-closures-in-2019-2020/2020/03

Perna LW (2004) Understanding the decision to enroll in graduate school: sex and racial/ethnic group differences. J Higher Educ 75(5):487–527

Saavedra A, Rapaport A, Silver D, Polikoff M, Garland M, Haderlein S (2020) Parents' perspectives on the effects of COVID-19 on K-12 education, April–July 2020. https://healthpolicy.usc.edu/evidence-base/parents-perspectives-on-the-effects-of-covid-19-on-k-12-education-april-july-2020/9

Sadler PM, Tai RH (2007) The two high-school pillars supporting college science. Science 317(5837):457–458

School Districts' Reopening Plans (2020) A snapshot. Education Week. https://www.edweek.org/leadership/school-districts-reopening-plans-a-snapshot/2020/07

Strand M, Lizardo O (2016) The hysteresis effect: theorizing mismatch in action. Theory Soc 45(1):1–27

Tai RH, Qi Liu C, Maltese AV, Fan X (2006) Planning early for careers in science. Science 312(5777):1143–1144

Tyson W, Lee R, Borman KM, Hanson MA (2007) Science, technology, engineering, and mathematics (STEM) pathways: high school science and math coursework and postsecondary degree attainment. J Educ Stud Placed Risk 12(3):243–270

Wang MT, Degol J (2013) Motivational pathways to STEM career choices: Using expectancy—value perspective to understand individual and gender differences in STEM fields. Develop Rev 33(4):304–340

Jennifer Hamilton is the Senior Vice President and Department Head of Education and Child Development at NORC at the University of Chicago. With over 30 years of research experience, she has utilized her methodological training to enhance the educational and economic outcomes of young Americans. Jennifer is a respected leader known for her work in generative AI as well as her work with policymakers, practitioners, and other education stakeholders ensuring that research is practical and addresses real-world needs. Her deep knowledge and methodological expertise ensure that all projects are conducted with the highest standards of rigor and efficiency. As principal investigator on numerous studies spanning from kindergarten to post-secondary education, Jennifer leads initiatives such as a project aimed at creating a research and development ecosystem to provide educators with high-quality instructional materials. She also oversees evaluations of civics education programs to assess their impact on high school students' civic learning and engagement. Before joining NORC, Jennifer worked to enhance the rigor and validity of over twenty 13 evaluations as a certified What Works Clearinghouse reviewer and managed various other education studies. Her ongoing motivation is to make a positive impact in education through research, collaboration, and a commitment to excellence.

Debbie Kim is a Senior Research Scientist in the Education and Child Development department at NORC at the University of Chicago. She has 20 years of educational research experience and has designed and led research projects from the federal to school levels. Debbie's work is deeply informed by institutional and organizational theory lenses. She is broadly interested in the ways that macro-level processes take form at the microlevel, and vice versa. Her research interests include policy-making and implementation, organizational change, decision making, and access. Throughout her work, she focuses on social inequities and the ways educational policies can support or hinder more equitable outcomes for youth. Debbie is principal investigator on a range of studies including studies of COVID's impact on high school students' science, technology, engineering, and mathematics (STEM) trajectories; the influence of STEM program participation in shaping organizational processes; and the efficacy of a workforce development program for young adults with disabilities. Debbie is a recognized expert in the field, with a forthcoming chapter on qualitative methodologies in a *Wiley Handbook of Research Methods* and publications in journals such as the *American Educational Research Journal* and *Teachers College Record*. She also serves as co-chair of the American Educational Research Association's Districts in Research and Reform Special Interest Group.

Career Confidence During the COVID-19 Pandemic Crisis: The Role of Individual Development Plans and Mentoring Support Among STEM Graduate Students

20

Chi-Ning Chang, Guan Saw, Cammie Justus-Smith, John Hui, and Ryan Culbertson

Abstract

The COVID-19 outbreak significantly disrupted graduate education in Science, Technology, Engineering, and Math (STEM) fields, particularly affecting in-person laboratory research and mentoring interactions. Mentoring support plays a vital role in preparing STEM graduate students for future careers. Since the early 2000s, STEM programs have increasingly promoted the use of Individual Development Plans (IDPs) to formalize mentorship and support students' academic and career development. The CHIPS and Science Act of 2022 reinforces this approach by advocating IDPs as essential tools to support professional development among STEM graduate students. However, the effectiveness of IDPs in enhancing mentoring and career outcomes during crises such as the COVID-19 pandemic remains unclear. This study offers insights into the role of IDPs during the pandemic by examining their relationship with mentoring support and career confidence with a nationwide sample of STEM graduate students. The findings reveal that while IDPs effectively promote mentor–mentee interactions and instrumental support, thereby benefiting career outcomes, they fall short in providing psychosocial support—a critical component of mentoring that fosters career confidence. Given the importance of psychosocial support in mentoring, this study underscores a more holistic approach to IDP

Research funded by

NSF Award #2031069—RAPID*: Collaborative Research: Electronic Mentoring to Address Challenges in Engineering Graduate Programs During the COVID-19 Pandemic;
NSF Award #2051263—RAPID*: Collaborative Research: Electronic Mentoring to Address Challenges in Engineering Graduate Programs During the COVID-19 Pandemic.

*Rapid Response Research Award

COVID Information Commons Video Presentation

The COVID Information Commons (CIC) Research Lightning Talk video presentation and individual multilingual transcripts for this project can be found in the Columbia Academic Commons (AC) (Saw 2020).

C.-N. Chang (✉) · C. Justus-Smith · J. Hui
Virginia Commonwealth University, Richmond, VA, USA
e-mail: changc10@vcu.edu

C. Justus-Smith
e-mail: justussmitcd@vcu.edu

J. Hui
e-mail: huij2@vcu.edu

G. Saw
Claremont Graduate University, Claremont, CA, USA
e-mail: guan.saw@cgu.edu

R. Culbertson
Texas Tech University, Lubbock, TX, USA
e-mail: ryaculbe@ttu.edu

F. D. Hudson (ed.), *The COVID Information Commons*, Sustainable Development Goals Series,
https://doi.org/10.1007/978-3-032-06125-6_20

design and mentoring practices, particularly during challenging periods.

Keywords

Career development · Mentoring · Individual development plan · COVID 19 · STEM graduate education

20.1 Goal of the COVID-19 Research

The COVID-19 pandemic, which began to emerge in early 2020, caused significant disruptions in higher education, particularly among STEM (Science, Technology, Engineering and Mathematics) graduate students. Universities worldwide suspended face-to-face classes and closed campuses, making it difficult for students to access study spaces, conduct research, attend conferences, and engage with faculty (Soria et al. 2020). Many STEM students reported delays in their expected graduation dates (Saw et al. 2020). The National Science Foundation's (NSF) Survey of Earned Doctorates confirmed a notable decline in STEM PhDs awarded between 2020 and 2021, marking the largest annual drop in the prior 40 years (Langin 2022).

Mentoring is crucial for the professional development of graduate students. Since the early 2000s, Individual Development Plans (IDPs) have been promoted in STEM programs to support structured mentor–mentee interactions, goal setting, and career planning (Hobin et al. 2012; Clifford 2002). Research has shown that IDPs enhance mentor–mentee relationships and professional development (Hobin et al. 2014; Vanderford et al. 2018; Chang et al. 2021). However, the effectiveness of IDPs in maintaining mentoring support and career outcomes during crises such as the COVID-19 pandemic remains unclear.

This chapter aims to provide evidence on the role of IDPs in fostering mentoring support and career confidence among STEM graduate students during the pandemic. By surveying a diverse, nationwide sample, this study seeks to understand the interrelationships between IDP usage, mentoring support, and career outcomes in the context of a global crisis. The findings aim to inform the design of resilient and comprehensive mentoring practices that better serve STEM graduate students in both regular and challenging circumstances, as well as in future crises.

20.1.1 The Individual Development Plans (IDPs)

The concept of IDPs was introduced in the early 1970s as a tool to help employees within the school system assess their skills and set career goals (Foster and Foster 1973). Since the early 2000s, IDPs have evolved into career planning tools that promote mentor–mentee interactions and professional development of students in higher education (Fuhrmann et al. 2011). IDPs now support STEM graduate students in self-assessment, exploring career paths, identifying career goals, and developing action plans to achieve those goals (Hobin et al. 2012; Clifford 2002).

Promoted by federal agencies such as the National Science Foundation (NSF) and the National Institutes of Health (NIH), professional organizations such as the American Association for the Advancement of Science (AAAS), American Chemical Society (ACS), and Federation of American Societies for Experimental Biology (FASEB), and recent legislation like the CHIPS and Science Act of 2022, IDPs have become integral to career development in STEM education. IDPs serve as actionable mentoring tools that support STEM graduate students in navigating their career paths. Commonly used IDP tools include myIDP by AAAS, ChemIDP by ACS, and tools developed by institutions like the University of California San Diego, University of Pennsylvania, University of Southern California, Stanford University's Office of Postdoc Affairs, and the University of Minnesota Medical School (Chang et al. 2023).

A limited but growing body of research has shown that students who use IDPs report greater clarity in their career paths and increased confidence in achieving their professional objectives (Chang et al. 2021; Thompson et al. 2020). By

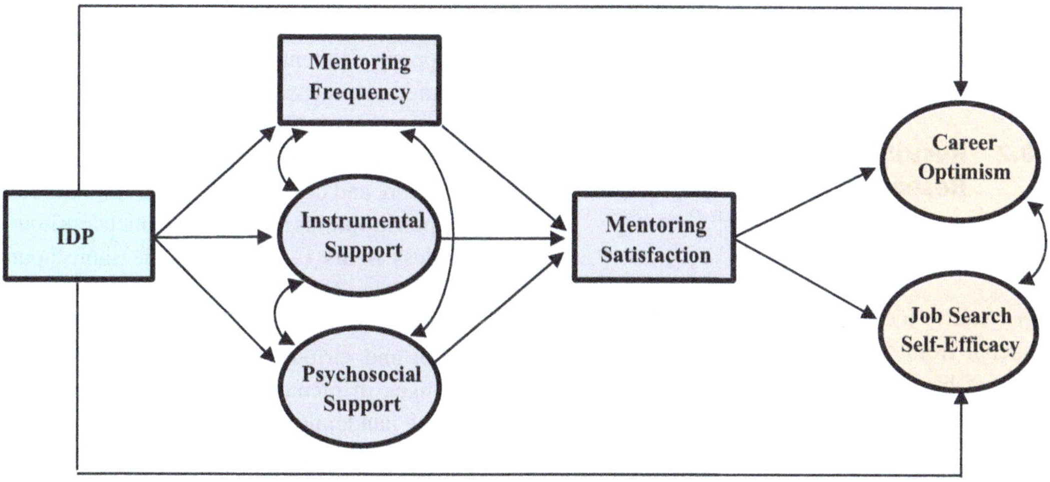

Fig. 20.1 A theoretical model of IDP-MIPO linking IDP, mentoring support and quality, and student outcomes. *Note* Blue = focal variable, the IDP usage; Purple = mentoring experience; Orange = career confidence; Oval = latent factor; Rectangle = observed variable

identifying skill gaps and providing a roadmap for acquiring the necessary competencies, IDPs empower students to take proactive steps toward their career aspirations. This is especially crucial during times of crisis, such as the COVID-19 pandemic, when traditional career development opportunities may be limited and students need additional support to navigate an uncertain job market. However, to date, little is known about the specific role of IDPs in maintaining and enhancing mentoring support and career outcomes during a crisis.

20.1.2 Mentoring Input-Process-Outcome (MIPO) Model

Although IDPs are valuable career development tools for STEM graduate students, their benefits are likely to be limited without strong mentoring support. The Mentoring Input-Process-Outcome (MIPO) model from the vocational psychology and career development literature suggests that the background characteristics of the mentee, mentor, or both (i.e., Inputs) affect the interaction and quality of mentoring (i.e., Processes), which in turn influences mentee outcomes (i.e., Outcomes) (Curtin et al. 2016; Eby et al. 2013). Building on the MIPO model and previous research (Chang et al. 2021; Saw et al. 2023), we propose a conceptual model of IDP-MIPO that connects IDP usage to mentoring processes and student career outcomes (see Fig. 20.1). Specifically, IDP usage is hypothesized to enhance student career motivational factors (i.e., career optimism, job search self-efficacy) directly, and indirectly through mentoring support (i.e., interaction frequency, instrumental support, and psychosocial support) and quality (i.e., mentoring satisfaction).

20.1.3 A National Survey

The goal of this research was to examine how IDP usage links to mentoring experiences and career confidence among STEM graduate students during the COVID-19 pandemic. We utilized data collected from a nationwide survey of STEM students during the COVID-19 outbreak funded by the NSF (Awards #2,051,263 and #2,031,069). In collaboration with deans and associate deans from colleges of engineering and science across the country, the 12–15-min survey was sent to STEM college students using the online platform Qualtrics on June 3–22, 2020. Informed consent was electronically obtained from all participants prior to their participation in the survey. In total,

4603 students from 157 higher education institutions in 41 states completed the survey.

20.2 Results of the COVID Research

20.2.1 Sample

This study focused on 755 STEM graduate students from 51 institutions in 26 states who reported having at least one mentor/advisor on campus in spring 2020. Of the total sample, 55.9% were master's students and 44.1% were doctoral students. Nearly half (48.9%) identified as male, 45.0% identified as female, and 6.1% identified as transgender, genderqueer, gender-nonconforming, other genders, or did not report. With respect to race/ethnicity: 45.4% identified as non-Hispanic White; 31.3% as non-Hispanic Asian; 14.0% as non-Hispanic Black, Hispanic, or non-Hispanic Native American; and 9.3% as other race/ethnicity, multiracial, or did not report their race. Approximately two-thirds of the respondents (64.6%) were United States (U.S.) citizens or permanent residents. Regarding mentoring experience: 73.5% of the respondents reported that their primary mentors were academic advisors or thesis/dissertation chairs; 13.8% were other faculty members; 5.2% were senior graduate students; 2.4% were staff members; and 5.1% were others. Among the respondents, 29.5% utilized an IDP with their mentors.

20.2.2 Measures

The key measures examined in this study included students' IDP status, interaction frequency, perceptions, and satisfaction with mentoring, career confidence, and demographic information. Table 20.1 provides descriptive statistics for all variables.

Students who reported having completed an IDP with their mentor were identified as the focal group of this study (yes = 1, no = 0).

Three mentoring process measures were examined: mentoring interaction frequency, perceived instrumental support, and perceived psychosocial support. Students were asked to compare the changes in interaction frequency with their primary mentor via face-to-face communication, video conferencing, email, phone, and social media before and during the COVID-19 outbreak. The response options ranged from much less hours (− 2), less hours (− 1), about the same hours (0), more hours (1), to much more hours (2). The total changes in interaction frequency, combining in-person and virtual methods, were defined as the changes in mentoring interaction frequency during the pandemic.

Perceived instrumental support and perceived psychosocial support were measured using two sets of four indicators each (Marie Taylor and Neimeyer 2009; Ortiz-Walters and Gilson 2005; Tenenbaum et al. 2001). Based on their experience before and during the pandemic, students rated various supports from their primary mentor on a 5-point Likert scale ranging from 1 ("much less support") to 5 ("much more support"). Cronbach's alphas, a measure of scale reliability, for perceived instrumental support and perceived psychosocial support were 0.85 and 0.89, respectively (Taber 2018).

Mentoring satisfaction, a proxy of mentorship quality, was assessed with a single item: "How satisfied were you with the support you received from your primary mentor during this past spring 2020 semester?" The response options ranged from 1 ("extremely dissatisfied") to 9 ("extremely satisfied").

Career confidence was measured by career optimism and job search self-efficacy. Career optimism, which refers to the tendency to expect the best career outcomes or to emphasize positive aspects of one's career since the COVID-19 outbreak, was assessed using two items on a scale from 1 ("strongly disagree") to 5 ("strongly agree") (Rottinghaus et al. 2005). Cronbach's alpha for career optimism was 0.85. Job search self-efficacy during the COVID-19 outbreak was measured using the three items. Based on the impact of the COVID-19 outbreak, students rated these items on a 5-point Likert scale from −2 ("much less confidence") to 2 ("much more confidence") (Manuti 2012). Cronbach's alpha for job search self-efficacy was 0.91.

Table 20.1 Summary statistics for STEM graduate students ($n = 755$)

Variables	Mean	S.D	Min	Max	Miss. (%)
Individual development plan (IDP)	0.30	0.46	0.00	1.00	0.00
Mentoring frequency					
Face-to-face	– 1.19	1.01	– 2.00	2.00	1.32
Video conferencing	0.60	1.26	– 2.00	2.00	1.06
Email	0.19	0.91	– 2.00	2.00	1.19
Phone	– 0.03	0.85	– 2.00	2.00	1.32
Social media	– 0.14	0.70	– 2.00	2.00	1.19
Instrumental support					
Finish my assignments/projects	0.06	0.78	– 2.00	2.00	1.99
Improve my writing skills	– 0.03	0.76	– 2.00	2.00	2.65
Prepare for my presentations	– 0.05	0.77	– 2.00	2.00	2.52
Explore my career options	– 0.09	0.79	– 2.00	2.00	2.52
Psychosocial support					
Discuss my concerns about academic projects	0.15	0.80	– 2.00	2.00	2.65
Pursue my learning interests	0.02	0.73	– 2.00	2.00	3.18
Work toward my career goals	0.02	0.75	– 2.00	2.00	2.78
Talk about my anxiety in career outlook	0.00	0.84	– 2.00	2.00	3.31
Mentoring satisfaction	7.06	2.13	1.00	9.00	8.08
Career Attitudes					
Career optimism					
More excited when I think about my career	2.65	1.03	1.00	5.00	0.26
More eager to pursue my career dreams	2.97	1.08	1.00	5.00	0.26
Job search self-efficacy—"Finding a job..."					
For which I am qualified	– 0.56	0.85	– 2.00	2.00	0.26
In a company/institution that I prefer	– 0.67	0.88	– 2.00	2.00	0.53
For which I am prepared	– 0.46	0.84	– 2.00	2.00	0.53
Demographics (Covariates)					
Male*	0.49	0.50	0.00	1.00	0.00
Female	0.45	0.50	0.00	1.00	0.00
Other gender/did not report—gender	0.06	0.24	0.00	1.00	0.00
White*	0.45	0.50	0.00	1.00	0.00
Asian	0.31	0.46	0.00	1.00	0.00
Black/Hispanic/Native American	0.14	0.35	0.00	1.00	0.00
Other race/did not report—race	0.09	0.29	0.00	1.00	0.00
Socioeconomic status (SES)	6.46	1.73	1.00	10.00	4.90
Non-disabled*	0.82	0.38	0.00	1.00	0.00
Disabled	0.14	0.35	0.00	1.00	0.00
Did not report disability status	0.04	0.20	0.00	1.00	0.00
US citizen/permanent resident*	0.65	0.48	0.00	1.00	0.00
International student	0.30	0.46	0.00	1.00	0.00
Did not report—citizenship status	0.05	0.22	0.00	1.00	0.00

(continued)

Table 20.1 (continued)

Variables	Mean	S.D	Min	Max	Miss. (%)
Age	27.30	5.61	18.00	73.00	4.37
PhD student	0.44	0.50	0.00	1.00	0.00
Master's student*	0.56	0.50	0.00	1.00	0.00

Note n = sample size; S.D. = standard deviation; Min. = minimum; Max. = maximum; Miss. = missing data
* = reference group

Several demographic variables were included as covariates in our models, including gender, race/ethnicity, socioeconomic status (SES), disability status, citizenship status, age, and degree level (see Table 20.1 for variable coding and descriptive statistics). The MacArthur Scale of Subjective Social Status was used to measure the students' SES. Students were given an image of a ladder with ten rungs and asked to choose from 1 to 10 the rung that best described their SES (Adler et al. 2000). Ten represented people with the most money, education, and best jobs, whereas one represented people with the least money, education, and worst jobs or no job.

20.2.3 Analytic Strategy

To test the proposed theoretical model of IDP-MIPO (Fig. 20.1), structural equation modeling (SEM) was employed. SEM allows for identifying the interrelationships among observed and latent variables simultaneously while accounting for measurement errors in the observed items. The analysis was conducted using *Mplus 8.4* statistical software. The measures in our SEM model were controlled for by the demographic covariates listed in Table 20.1. Missing data ranged from zero to a high of 8.08% for the variable of mentoring satisfaction. We used the full information maximum-likelihood (FIML) approach to handle the missing data and improve estimation (Mazza et al. 2015).

20.2.4 Analytic Results

The theoretical model of IDP-MIPO in SEM showed a good fit with the empirical data, root mean square error of approximation (RMSEA) = 0.035, comparative fit index (CFI) = 0.955, and standardized root mean squared error (SRMR) = 0.067 (Hu and Bentler 1999). Figure 20.2 presents the SEM results with statistically significant paths (solid lines), and the standardized path coefficients are denoted along the paths. The results indicated that students with an IDP reported interacting with their mentor more frequently and receiving a higher level of instrumental support from their primary mentor during the COVID-19 pandemic, compared with their peers without an IDP. However, using an IDP was not statistically predictive of psychosocial support from the primary mentor or career confidence, as measured by career optimism and job search self-efficacy during the COVID-19 crisis. Furthermore, the results revealed that both instrumental and psychosocial support from mentors were positively related to mentoring satisfaction, which in turn positively predicted students' career optimism and job search self-efficacy during the COVID-19 pandemic. Although mentoring interaction frequency was not statistically significantly associated with mentoring satisfaction, it was positively correlated with the other two critical mentoring factors—instrumental support and psychosocial support—during the COVID-19 crisis.

20.3 Recommendations Regarding Mitigation, Response, and Recovery from Future Pandemics

This study makes several theoretical, methodological, and practical contributions to the literature on IDPs, mentoring, career development, grad-

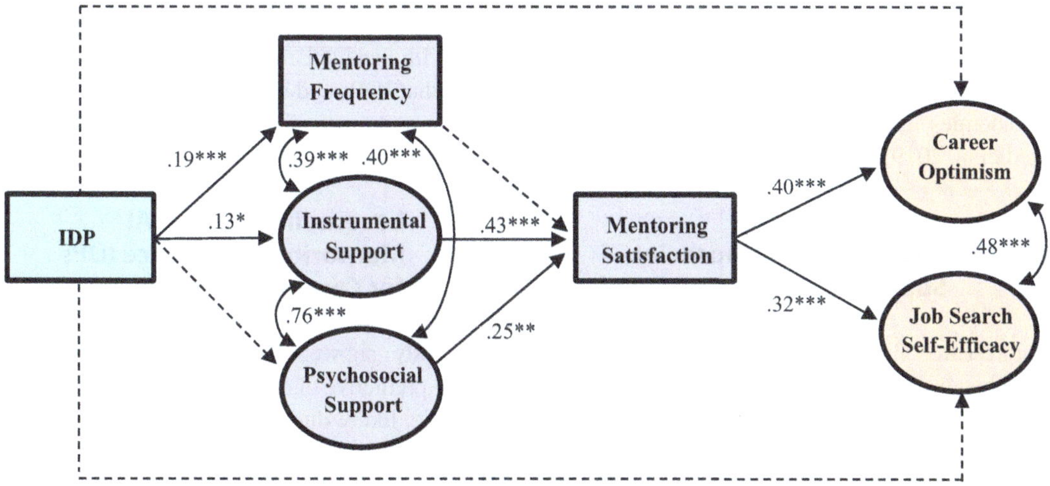

Fig. 20.2 SEM Results of interrelationships among IDP, mentoring support and quality, and career confidence of stem graduate students during the COVID-19 pandemic. *Note*: Blue = focal variable, the IDP usage; Purple = mentoring experience; Orange = career confidence; Oval = latent factor; Rectangle = observed variable. Values are standardized path coefficients. Dashed paths are not statistically significant. For reasons of clarity, all the covariates, factor loadings, and uniquenesses are not shown. All the factor loadings are over 0.700. R^2 for mentoring frequency = 0.08**; R^2 for instrumental support = 0.05*; R^2 for psychosocial support = 0.03*; R^2 for mentoring satisfaction = 0.45***; R^2 for career optimism = 0.25***; R^2 for job search self-efficacy = 0.21***. *$p < 0.05$, **$p < 0.01$, ***$p < 0.001$

uate STEM education, and crisis responses. At the theoretical level, it builds upon the literature on vocational psychology and career development to propose a novel theoretical framework of IDP-MIPO that links IDP usage, mentoring support, and student career outcomes. Our comprehensive theory-driven IDP-MIPO model is largely supported by empirical evidence. At the methodological level, this study is one of the first to measure and analyze the linkages of IDP usage with various dimensions of mentoring and career outcomes with a diverse, nationwide sample of STEM graduate students. Our findings have greater generalizability, improving on most prior work on IDP and mentoring that relied largely on a small sample from a single program or institution (Hernandez 2018). At the practical level, this study provides robust theoretical perspectives and empirical evidence demonstrating that IDPs effectively promote mentor–mentee interactions and instrumental support, thereby enhancing career-related outcomes. However, they fell short of providing psychosocial support. Based on our research findings, we offer the following recommendations to improve the support and development of STEM graduate students, especially during future pandemics or similar crises.

20.3.1 Promoting the IDP Usage and Enhancing Remote Mentoring Practices

The findings showed that students with IDPs reported more frequent interactions with their mentors and perceived a higher level of instrumental support. To effectively support students, especially during crises, institutions should encourage faculty and students to utilize IDPs and digital platforms for regular check-ins, meetings, and feedback sessions. Despite the increased interaction frequency, mentoring satisfaction was linked more to the quality of support than to the quantity of interactions. Creating structured communication schedules is crucial for providing ongoing and meaningful support and guidance. By establishing clear expectations and conducting regular check-ins through the usage of IDPs, mentors and mentees can maintain strong relationships even in remote settings. This approach ensures that both parties are aligned in their goals and can address any issue promptly, fostering a supportive and productive mentoring environment (Academies and of Sciences, Engineering, and Medicine, 2019).

20.3.2 Integrating Psychosocial Support in IDPs

Our research findings indicated that IDPs were effective in providing instrumental support but lacked emphasis on psychosocial support. This is consistent with recent research findings on the IDP tools used at minority-serving institutions (Chang et al. 2023). Psychosocial support, along with instrumental support, is critical for students' well-being, especially during crises (Saw et al. 2023). Institutions should broaden the scope of IDPs to include components addressing mental health, work-life balance, well-being, and burnout prevention (Chang et al. 2023). Training mentors on providing psychosocial support is also necessary, including recognizing signs of stress, anxiety, and burnout, and offering appropriate resources.

20.3.3 Developing Institutional Support for IDPs and Mentorship

The findings indicate that both instrumental and psychosocial support from mentors were positively related to mentoring satisfaction, which in turn positively predicted career optimism and job search self-efficacy. This underscores the need for institutional policies that bolster the implementation and effectiveness of IDPs and mentoring. Institutions should develop structured mentor training programs that focus on integrating IDPs into mentoring practices (Mehrubeoglu et al. 2023), and ensuring mentors provide both instrumental and psychosocial support. Flexible policies regarding deadlines and extensions, as an example, can accommodate the unique challenges faced during crises (Saw et al. 2020). Additionally, providing access to comprehensive resources such as virtual workshops and mental health support, along with mechanisms for regular feedback, can ensure that IDPs and mentoring programs remain effective and responsive to students' needs.

20.3.4 Leveraging Human-AI Mentoring to Enhance IDPs for Career Confidence

This study showed that IDP usage was not directly predictive of career confidence during the pandemic, likely due to the sudden and unprecedented challenges introduced by the crisis. The disruption caused by the pandemic may have overshadowed the potential benefits of IDPs, as students faced heightened uncertainty and stress. To improve IDPs for future crises, it is crucial to incorporate elements that specifically build resilience and adaptability. Our ongoing research highlights the potential of hybrid human-AI mentoring models (Chang et al. 2024). Human mentors play a key role by offering personalized guidance on crisis management, stress coping strategies, and continuous career planning despite uncertainties, which can help students maintain career confidence. Meanwhile, despite some concerns about using artificial intelligence (AI) (Chang et al. 2024), AI mentors (powered by large language models with internet access ability, e.g., GPT-4 and Gemini) could enhance IDP practices with timely and relevant feedback, real-time career information, and up-to-date job market trends, which are critical during crises. By combining the strengths of human and AI mentors, STEM graduate students could develop the confidence needed to navigate their career paths effectively, especially during times of uncertainty.

20.3.5 Strengthening Multicultural Mentoring Guidance

Effective mentoring is key to the success of IDP practices. Supporting underrepresented minorities in STEM is also a national priority (The White House 2021). Our prior study showed that most current IDP tools do not provide multicultural mentoring guidance (Chang et al. 2023). Successful mentorship requires faculty to recognize and respect their mentees' cultural beliefs and understand the identity-related obstacles they face (Byars-Winston et al. 2018). Creating specific guidelines for multicultural mentoring in implementing an IDP can help to address these challenges and support the growth of underrepresented minority students in STEM. Educating mentors on culturally aware and responsive mentorship practices is also essential. This training should help mentors to understand and address the unique challenges faced by students from diverse backgrounds, thereby fostering a more inclusive mentoring environment in STEM education.

References

National Academies of Sciences, Engineering, and Medicine (2019) The science of effective mentorship in STEMM. The National Academies Press, Washington, DC

Adler NE, Epel ES, Castellazzo G, Ickovics JR (2000) Relationship of subjective and objective social status with psychological and physiological functioning: preliminary data in healthy. White Women Health Psychol 19(6):586–592. https://doi.org/10.1037/0278-6133.19.6.586

Byars-Winston A, Womack VY, Butz AR, McGee R, Quinn SC, Utzerath E, Saetermoe CL, Thomas SB (2018) Pilot study of an intervention to increase cultural awareness in research mentoring: implications for diversifying the scientific workforce. J Clin Transl Sci 2(2):86–94. https://doi.org/10.1017/cts.2018.25

Chang CN, Justus-Smith C, Malagon-Palacios LJ, Hui J (2023) The use of the individual development plan at minority serving institutions. Front Educ 8:1–7. https://doi.org/10.3389/feduc.2023.1258273

Chang CN, Hui J, Justus-Smith C, Wang TW (2024) Navigating STEM careers with AI mentors: a new IDP journey. Front Artif Intell 7:1461137. https://doi.org/10.3389/frai.2024.1461137

Chang CN, Patterson CA, Vanderford NL, Evans TM (2021) Modeling individual development plans, mentoring support, and career preparedness relationships among Doctor of Philosophy (Ph.D.) trainees in the life sciences. F1000Research 10(626):1–10. https://doi.org/10.12688/f1000research.53705.2

Clifford PS (2002) Quality time with your mentor. The Scientist 16(19):59–60. https://www.the-scientist.com/quality-time-with-your-mentor-52734

Curtin N, Malley J, Stewart AJ (2016) Mentoring the next generation of faculty: supporting academic career aspirations among doctoral students. Res High Educ 57(6):714–738. https://doi.org/10.1007/s11162-015-9403-x

Eby LTdT, Allen TD, Hoffman BJ, Baranik LE, Sauer JB, Baldwin S, Evans SC (2013) An interdisciplinary meta-analysis of the potential antecedents, correlates, and consequences of protégé perceptions of mentoring. Psychol Bull 139(2):441–476. https://doi.org/10.1037/a0029279

Foster DM, Foster R (1973) A career development center: a model for school employee development. Santa Clara County Office of Education, San Jose, CA. https://eric.ed.gov/?id=ED083692

Fuhrmann CN, Hobin JA, Lindstaedt B, Clifford PS (2011) myIDP: science careers-individual development plan. https://myidp.sciencecareers.org/. Accessed on 31 July 2024

Hernandez PR (2018) Landscape of assessments of mentoring relationship processes in postsecondary STEMM contexts: a synthesis of validity evidence from mentee, mentor, and institutional/programmatic perspectives. Paper commissioned by the committee on the science of effective mentoring in STEMM. https://nap.nationalacademies.org/resource/25568/Hernandez%20-%20Landscape%20of%20Assessments%20of%20Mentoring.pdf

Hobin JA, Fuhrmann CN, Lindstaedt B, Clifford PS (2012) You need a game plan. Science. https://www.sciencemag.org/careers/2012/09/you-need-game-plan

Hobin JA, Clifford PS, Dunn BM, Rich S, Justement LB (2014) Putting PhDs to work: career planning for today's scientist. CBE—Life Sci Educ 13(1):49–53. https://doi.org/10.1187/cbe-13-04-0085

Hu LT, Bentler PM (1999) Cutoff criteria for fit indexes in covariance structure analysis: conventional criteria versus new alternatives. Struct Equ Model: Multidiscip J 6(1):1–55. https://doi.org/10.1080/10705519909540118

Langin K (2022) Pandemic led to historic drop in U.S. STEM Ph.D. graduates, new data suggest. Science. https://www.science.org/content/article/pandemic-led-historic-drop-u-s-stem-ph-d-graduates-new-data-suggest

Manuti A (2012) The meaning of working and career planning strategies: differences between high school and college students. Eur J Soc Sci 35(1):109–119. https://www.researchgate.net/publication/276277621

Marie Taylor J, Neimeyer GJ (2009) Graduate school mentoring in clinical, counselling, and experimental academic training programs: an exploratory study. Couns Psychol Q 22(2):257–266. https://doi.org/10.1080/09515070903157289

Mazza GL, Enders CK, Ruehlman LS (2015) Addressing item-level missing data: a comparison of proration and full information maximum likelihood estimation. Multivar Behav Res 50:504–519. https://doi.org/10.1080/00273171.2015.1068157

Mehrubeoglu M, Kelly K, Butler-Purry K, Richardson R, Walton S, Regisford G, Chaloo L, King S, Murphrey T, Burr E, Carson L, Coffin R, Moreira R, Carter C,

Challoo R (2023) Facilitating the transition of underrepresented minority STEM dissertators into academic careers: evolution of TxARM AGEP alliance model. J STEM Educ: Innov Res 24(1):51–61. https://jstem.org/jstem/index.php/JSTEM/article/view/2605
Ortiz-Walters R, Gilson LL (2005) Mentoring in academia: an examination of the experiences of protégés of color. J Vocat Behav 67(3):459–475. https://doi.org/10.1016/j.jvb.2004.09.004
Rottinghaus PJ, Day SX, Borgen FH (2005) The career futures inventory: a measure of career-related adaptability and optimism. J Career Assess 13(1):3–24. https://doi.org/10.1177/1069072704270271
Saw, G. (2020) Guan saw: collaborative research: electronic mentoring to address challenges in engineering graduate programs during the COVID-19 pandemic. Columbia Academic Commons, Academic Unit: Northeast Big Data Innovation Hub, Series: COVID-19 Research Lightning Talks. https://doi.org/10.7916/fe6e-se12
Saw GK, Chang C-N, Lin S, Hernandez PR, Culbertson R (2023) Disparities in mentoring and mental health problems of US college students in science and engineering during the COVID-19 pandemic. Educ Psychol 43(5):509–530. https://doi.org/10.1080/01443410.2022.2146055
Saw GK, Chang CN, Lomelí U, Zhi M (2020) Fall enrollment and delayed graduation among STEM students during the COVID-19 pandemic (NREED Data Brief. No 1). Network for Research and Evaluation in Education, Claremont, CA. https://par.nsf.gov/servlets/purl/10221725
Soria KM, Horgos B, McAndrew M (2020) Obstacles that may result in delayed degrees for graduate and professional students during the COVID-19 pandemic. https://escholarship.org/content/qt8d46b49s/qt8d46b49s_noSplash_675606d5773e4b84de21427c277d3e8d.pdf. Accessed on 31 July 2024
Taber KS (2018) The use of Cronbach's alpha when developing and reporting research instruments in science education. Res Sci Educ 48:1273–1296. https://doi.org/10.1007/s11165-016-9602-2
Tenenbaum HR, Crosby FJ, Gliner MD (2001) Mentoring relationships in graduate school. J Vocat Behav 59(3):326–341. https://doi.org/10.1006/jvbe.2001.1804
The White House (2021) Best practices for diversity and inclusion in STEM education and research: a guide by and for federal agencies. https://bidenwhitehouse.archives.gov/ostp/news-updates/2021/09/30/nstc-best-practices-for-diversity-inclusion-in-stem-education-research-a-guide-by-for-federal-agencies/. Accessed on 25 Mar 2025
Thompson HJ, Santacroce SJ, Pickler RH, Allen JK, Armer JM, Bakken S, Grey M (2020) Use of individual development plans for nurse scientist training. Nurs Outlook 68:284–292. https://doi.org/10.1016/j.outlook.2020.01.001
Vanderford NL, Evans TM, Weiss LT, Bira L, Beltran-Gastelum J (2018) A cross-sectional study of the use and effectiveness of the individual development plan among doctoral students. F1000Research 7(722):1–17. https://doi.org/10.12688/f1000research.15154.2

Chi-Ning Chang is an Assistant Professor of Research, Assessment, and Evaluation in Education at Virginia Commonwealth University (VCU). His research focuses on higher education mentoring, STEM education, and quantitative research methodology. His work has been supported by the National Science Foundation (NSF), the Institute of Education Sciences, and the U.S. Department of Health & Human Services. His recent research explores the potential synergy between artificial intelligence and human mentorship in enhancing individual development plans (IDPs) for STEM graduate students, particularly those from underserved backgrounds. This innovative work investigates how technology can complement traditional mentoring approaches to better support student career development in STEM. Chang's contributions to teaching, mentoring, and research have been recognized with the 2023 Charles P. Ruch Award for Excellence in Teaching and the 2024 Distinguished Early Career Faculty Award from VCU's School of Education.

Guan Saw is an Associate Professor in the School of Educational Studies at Claremont Graduate University. His research interests include educational inequality, STEM education and workforce, rural education, college access and success, mentorship, social capital, teacher victimization, health disparities, and research methodology. He is associate editor for *Handbook of the Sociology of Education in the 21st Century* (Springer, 2018) and an associate editor of *American Educational Research Journal* (AERJ; 2025-). Saw's research work has been funded by the National Science Foundation (NSF), Institute of Education Sciences (IES), and National Institute of Justice (NIJ). He received the National Rural Education Association (NREA) Best Research Award in 2024. He also served as a member of the National Academies of Sciences, Engineering, and Medicine (NASEM) Board on Science Education (BOSE) committee on K-12 STEM Education and Workforce Development in Rural Areas (2023–2025).

Cammie Justus-Smith is a doctoral student at Virginia Commonwealth University, specializing in research, assessment, and evaluation within the School of Education. Her research centers on Self-Directed Education, microschools, and critical methodologies, with a particular emphasis on fostering impactful research-practice partnerships. Cammie has collaborated on numerous projects exploring mentorship in higher education, deschooling, and program evaluation, applying diverse methodological approaches. Drawing on over a decade of K-12 teaching experience in secondary science education, she brings a practitioner's perspective to her scholarly work. A graduate of the University of Virginia, where she earned both a BA and MAT, Cammie is dedicated to advancing innovative and equitable educational practices through her research and advocacy.

John Hui is an advanced doctoral student at Virginia Commonwealth University and a cybersecurity professional. His research focuses on mentorship, mentor self-efficacy, and the social passage of expertise. His experiences as a first-generation immigrant and neurodivergent scholar shaped his perception of learning. With a late diagnosis of dyslexia, he learned how debilitating it can be to feel helpless in the classroom. Through the extraordinary mentors he found throughout his academic career, Hui found various strategies for mastery and learning. He advocates for diversified teaching strategies that will encourage students to perform at their best, working on descriptive and instrumental ways to bring these to practice. He has seen the tremendous positivity in motivation when people are given the right strategies to learn, the tools to empower themselves to be mentors, and the opportunity to pay it forward.

Ryan Culbertson, MA is the Executive Director of the American STEM Alliance (ASA), a Research Associate at Texas Tech University in the Department of Mathematics and Statistics, and a Research Fellow at Claremont Graduate University in the School of Educational Studies. Ryan has been in Science, Technology, Engineering, Math, and Medicine (STEMM) based educational support and development for over 15 years and has focused on developing and sustaining research to practice partnerships between educational practitioners, researchers, and scholars. His educational interest and research is focused in promoting equitable access to STEMM for underserved populations and providing relevant and intrinsically motivating education to improve individual and community happiness, health, and well-being through attainment of higher education, employment, and civil engagement.

The Importance of Novel Skill Learning and Growth Mindset in Adulthood: Insights from the COVID-19 Pandemic

21

Tania M. Rodriguez, Isadora Farias Lopes de Queiroz, Esra Kurum, Lilian Azer, Yena Kyeong, Leah Ferguson, Elizabeth Davis, Carla M. Strickland-Hughes, and Rachel Wu

Abstract

Learning new skills and having a growth mindset could be two key factors for adapting to changes. In particular, these factors may help buffer the negative effects of a pandemic, such as experiencing loneliness and decreased mental well-being from being socially isolated. Our research team conducted three longitudinal studies during the COVID-19 pandemic to investigate this idea. Through a survey across two time points over the first year of the pandemic, the first study found that worse subjective executive functioning during the COVID-19 pandemic predicted worse mental health across adulthood. However, learning new skills may have helped older adults adapt better to changes during the pandemic to increase mental health. The second study was a survey across three time points during two years of the COVID-19 pandemic. We found that having a growth mindset was generally beneficial for mental health, particularly for older adults who had engaged in an intense learning intervention prior to the start of the pandemic. The third study was a technology intervention conducted with low-income older adults in collaboration with a community

Research funded by

NSF/BCS* Award #2029575 - RAPID**: Older adults' learning and adaptation as resilience processes to counter social isolation during the COVID-19 pandemic.
*Division of Behavioral and Cognitive Sciences

**Rapid Response Research Award

COVID Information Commons Video Presentations

The COVID Information Commons (CIC) Research Lightning Talk video presentations and individual multilingual transcripts for this project can be found in the Columbia Academic Commons (AC) (Wu 2020; Wu and Azer 2022).

T. M. Rodriguez (✉) · I. F. L. de Queiroz · E. Kurum · L. Ferguson · E. Davis · R. Wu
University of California, Riverside, Riverside, CA, USA
e-mail: trodr020@ucr.edu

I. F. L. de Queiroz
e-mail: ifari006@ucr.edu

E. Kurum
e-mail: esrak@ucr.edu

L. Ferguson
e-mail: leah.ferguson@ucr.edu

E. Davis
e-mail: elizabeth.davis@ucr.edu

R. Wu
e-mail: rachel.wu@ucr.edu

L. Azer
University of California, Irvine, Irvine, CA, USA
e-mail: lazer001@ucr.edu

Y. Kyeong
National Taiwan Normal University, Taipei, Taiwan
e-mail: ykyeo001@ucr.edu

C. M. Strickland-Hughes
Colorado School of Mines, Golden, CO, USA
e-mail: cstricklandhughes@pacific.edu

F. D. Hudson (ed.), *The COVID Information Commons*, Sustainable Development Goals Series,
https://doi.org/10.1007/978-3-032-06125-6_21

organization. We found that digital independence predicted mental health outcomes. Taken together, our findings suggest that being open to and engaging in learning new skills, as well as having a growth mindset, may be beneficial to the long-term well-being of adults, especially older adults, when faced with drastic changes that require adaptation.

Keywords

Learning new skills · Growth mindset · Older adulthood · COVID-19 · Pandemic · Social isolation

21.1 Beneficial Outcomes of Novel Skill Learning

Novel skill learning, such as learning how to use new technological devices, occurs across the lifespan. In infancy and childhood, learners acquire basic skills for survival and navigating day-to-day life (e.g., how to walk, eat, speak, write, read (Berk 2022; Thompson 2001)). After infancy, individuals may also learn skills related to personal enjoyment and hobbies (e.g., a musical instrument (Li 2021)). In younger and middle adulthood, learners typically pursue skills for a particular career. Older adults learn new skills as they adapt to changes in the environment (e.g., learning to use a new device due to technological changes (Beier et al. 2015; Shultz and Adams 2007)) and within themselves (e.g., learning to use a new hearing aid due to hearing loss (Anstey and Low 2004; Papalia et al. 2007; Troen 2003)), as well as changes in their societal role (e.g., learning how to care for a sick spouse (Dow and Meyer 2010)). While biased societal expectations may lead some to think that learning new skills is something that primarily occurs and is most important earlier in lifespan, learning in older adulthood is just as necessary and crucial for many reasons. For example, we live in an ever-changing world that is always introducing new technologies, services, tools, and even illnesses (as we saw with the COVID-19 pandemic). Also, in order to remain employed in older adulthood, older workers must engage in continuous learning to stay up-to-date with developments within their workplace (Hennekam 2015). Learning new skills such as using new technological devices also enable older adults to independently use online platforms for finances, health care, and requesting services (e.g., transportation or food delivery (Charness and Boot 2009; Harvey et al. 2021)). It also can help increase work prospects (Maurer 2001) and social connectivity (Sen et al. 2021). As such, adults, especially older adults, can learn new skills to be able to keep up with and adapt to the changes that are constantly occurring within one's environment to avoid being dependent on others.

Besides the need for novel skill learning for adaptation to remain functionally independent (Nguyen et al. 2020), learning new skills also has been shown to increase cognitive abilities. Prior interventions that have taught older adults new skills, such as photography, technology, playing an instrument, painting, a new language, etc., have shown that participants experience improvements across different cognitive domains (e.g., episodic and working memory, processing speed, cognitive control (Bugos and Wang 2022; Bugos et al. 2007; Chan et al. 2016; Leanos et al. 2023; Noice et al. 2004; Park et al. 2014)). For example, Bugos and colleagues (Bugos et al. 2007) conducted an intervention that taught older adults how to play a new musical instrument, where the researchers offered older adult participants in the experimental group individualized piano instruction for six months. In this intervention, results indicated that, in comparison to the no-treatment control group, the experimental group showed significant improvements across two neuropsychological tests. Similarly, in a more recent study, older adults were placed in a piano training condition, a computer-assisted cognitive training condition, or a no-treatment control group for 16 weeks (Bugos and Wang 2022). Participants in the piano and computer-assisted cognitive training groups showed enhancements in working memory and processing speed. Those in the piano training condition also showed greater verbal fluency skills than the other two groups, suggesting a cognitive advantage of learning to play a new musical instrument. Learning new skills also has benefits related to improved well-being (Field 2009; Narushima et al. 2013; Perkins and Williamon 2014; Shapira et al. 2007). For instance, a study investigating the impact that learning to make music had on older adults found that participants that engaged in learning music experienced an increase in well-being and an enhancement in health-promoting behaviours (Perkins and Williamon 2014). Another study examining the impact of Internet training found that learning to use a computer and the Internet provided older adults with a sense of empowerment and improved well-being (Shapira et al. 2007).

Prior to the pandemic, Wu and colleagues conducted a learning intervention to test the cognitive benefits of learning multiple new skills simultaneously in older adulthood. In the three-month learning intervention, older adults learned three or more skills simultaneously (e.g., drawing, painting, a new language, how to use an iPad, photography, and music composition (Leanos et al. 2023)). Weekly classes were held on the University of California, Riverside's campus in addition to weekly motivational discussion sessions on growth mindset, motivation, neuroplasticity, and successful aging. The findings from the intervention showed that within six to eight weeks of the intervention, the cognitive abilities (i.e., working memory and cognitive control) of the older adult participants increased to middle-aged levels on average (30 years younger), and eventually younger adult levels (50 years younger) one year after the end of the intervention (Ferguson et al. 2023).

After confirming the feasibility of and potential benefits from this learning intervention, we aimed to conduct another wave of the intervention with more participants to test the optimal dosage of frequency and variety of learning content, funded by a National Science Foundation Faculty Early Career Development (NSF CAREER) Award to R. Wu. Unfortunately, the COVID-19 pandemic hit and prematurely shut down this new wave of the intervention, compelling our research team to investigate a more pressing matter, the impact of the pandemic on individuals, especially older adults. The goal of this chapter is to highlight three longitudinal studies that our research team conducted during the COVID-19 pandemic in Southern California with diverse adults. Across these studies, we assessed the role of novel skill learning, cognition, and having a growth mindset on navigating the negative effects of the pandemic. We first describe the negative impact that the COVID-19 pandemic had on individuals that led us to our specific research aims. Next, we describe the main findings and implications from each study. Lastly, we conclude with realizations, recommendations, and future directions based on our findings.

21.2 The Impact of the COVID-19 Pandemic and Our Research Goals

The COVID-19 pandemic and associated physical distancing restrictions had a negative impact on the overall well-being of adults as reflected by the increase in social isolation and perceived loneliness, anxiety, and depressive symptoms (Fancourt et al. 2021; Hwang et al. 2020; Krendl and Perry 2021; Tyrrell and Williams 2020). The psychological distress experienced by individuals was also fostered by the fear of being exposed to the virus among other secondary pandemic-related stressors, such as financial insecurity and the scarcity of resources (Fitzpatrick et al. 2020; Holingue et al. 2020; Zheng et al. 2021). Moreover, the grief of losing loved ones during these complicated times was exacerbated by the inability of individuals to say goodbye and attend memorial/funeral services, which contributed to negative mental health outcomes (Hernández-Fernández and Meneses-Falcón 2022; Menzies et al. 2020; Murphy 2020; Wang et al. 2022). During the pandemic, there was also a lot of uncertainty and misinformation alongside conspiracy theories that deepened fear, anxiety, and depression among adults (Coninck et al. 2021; Rocha et al. 2021).

Given the concern of mental health deterioration observed in adults as a result of the pandemic, we became interested in identifying potential factors that could promote resilience against situational mental health issues. In relation to our prior work, as described earlier, learning new skills has been found to improve well-being (Field 2009; Narushima et al. 2013; Perkins and Williamon 2014; Shapira et al. 2007). Studies conducted prior to the pandemic have also revealed that older adults with lower executive functioning (which includes cognitive control and working memory among other cognitive processes (Goldstein et al. 2014)) have higher depressive symptoms and loneliness, as well as lower well-being (Carrasco et al. 2020; Markova et al. 2017; Zlatar et al. 2018). Based on these findings, one of our research aims during the pandemic was to examine the role of cognitive ability and novel skill learning in predicting mental health outcomes (perceived loneliness, depressive symptoms, and overall well-being) among diverse younger to older adults across two time points (Study 1 (Azer et al. 2025)). In particular, given that our prior learning intervention (Leanos et al. 2023) increased older adults' executive functioning even one year after the end of the intervention (Ferguson et al. 2023), we wanted to investigate whether having participated in the intervention predicted older adults' mental health outcomes.

Another factor that has been identified as having positive implications for mental health outcomes is a growth mindset (Ortiz Alvarado et al. 2019; Zeng et al. 2016). Growth mindset is the belief that one's abilities and attributes can be developed and changed through dedication and effort (Dweck 2006). A growth mindset can specifically help foster resilience in the face of challenges and setbacks (Brown 2015; Yeager and Dweck 2012). A growth mindset may be specifically important when needing to learn new skills in order to adapt to new situations. Within the context of the pandemic, having a growth mindset predicts employing more positive coping strategies that lead to better mental health (Zhang et al. 2023). Thus, we wanted to investigate whether a growth mindset was predictive of mental health outcomes (depression levels, well-being, and adjustment of daily tasks) among diverse younger to older adults across three time points during two years of the COVID-19 pandemic. Considering the learning intervention conducted in the lab prior to the pandemic (Leanos et al. 2023), which included cultivating a growth mindset for learning new skills, we also wanted to examine whether having participated in the learning intervention resulted in an increased growth mindset and led to more positive mental health outcomes in older adults (Study 2 (Kyeong et al. 2024)).

The pandemic also brought on the need for individuals to adapt to a "new normal" of social distancing restrictions for a prolonged period of time in order to navigate their day-to-day life. Part of the adaptation and adjustment process included becoming familiar with and utilizing technology to make video calls to engage in social interactions (e.g., Zoom), attend virtual meetings and appointments (e.g., telehealth), and order services online (e.g., food or grocery delivery apps) to avoid contact with others and exposure to the virus. Having access to and knowing how to use technology was crucial to adapt to the transition from in-person to online interactions for these purposes. In particular, utilizing technology during the COVID-19 pandemic was viewed as a potential avenue to help attenuate some of the negative outcomes related to social isolation (Rodrigues et al. 2022). To be able to get the benefits of technology usage, many were forced to learn new technology-related skills. However, the pandemic highlighted that there were certain groups who were not able to feasibly adapt to new technology during these times due to the digital divide (Litchfield et al. 2021). The digital divide refers to the inequities in technology access and literacy that tend to primarily affect specific groups, such as older adults and those with a low income (Atske and Perrin 2021; Mitchell et al. 2019; Smith 2014; Vogels 2021). Thus, in collaboration with a community organization, we aimed to increase technology access and training for low-income older adults via a two-month technology intervention. We assessed the impact of this technology intervention on cognitive function, well-being, loneliness, depression, and growth mindset (Study 3 (Ferguson et al. 2024)).

21.3 Study 1: Skill Learning and Subjective Executive Function

The first study our laboratory conducted during the pandemic was executed in two parts (Study 1 (Azer et al. 2025)). Part 1 examined the role of subjective executive functioning and novel skill learning in predicting mental health outcomes among younger to older adults. Part 2 investigated the same question but specifically focused on a sample of middle-aged and older adults who had participated in the aforementioned learning intervention before the pandemic that yielded an increase in objective cognitive functioning (Leanos et al. 2023). The mental health outcomes included perceived loneliness, depressive symptoms, and general well-being. It was hypothesized that higher subjective executive functioning and higher levels of novel skill learning would predict better mental health. The data were collected via an online survey at two time points across the first year of the pandemic (2020–2021) and were analysed using linear mixed-effects regression models for each mental health outcome variable (while controlling for emotional closeness and sociodemographic variables). Emotional closeness referred to how emotionally close the adults felt to the people living with them.

The findings from this study showed that lower subjective executive functioning predicted lower mental health across all outcome measures for both parts of the study. These results mirror the literature before and during the pandemic showing the negative impact of perceived low cognitive functioning on mental health (Carrasco et al. 2020; Markova et al. 2017; Zlatar et al. 2018). Findings also revealed that experiencing emotional closeness during the pandemic had a positive impact on well-being. This finding aligns with the work that has shown that close relationships can help foster positive mental health outcomes (Zhang et al. 2023; Sarason et al. 2001). Furthermore, part 2 of the study revealed a significant difference in the mental health of middle-aged and older adults over 58 based on time spent on learning a new skill or behaviour. In particular, during the second time point, participants who reported spending more time learning a new skill or behaviour had lower loneliness and depressive symptoms. Similarly, prior studies have found that engagement in leisure activities (including learning new things) can positively impact mental health (Field 2009; Narushima et al. 2013; Perkins and Williamon 2014; Shapira et al. 2007; Poelke et al. 2016; Teh and Tey 2019). These results suggest that engaging in learning new skills could serve as a protective factor against negative mental health outcomes. Interestingly, results for part 2 of the study also demonstrated that the participants who had previously participated in the learning intervention were more likely to experience poorer levels of well-being during the pandemic in contrast to the participants who did not partake in the intervention. A potential reason for this finding is that those who participated in the intervention were perhaps more susceptible to the negative mental health consequences of the physical distancing restrictions (e.g., cancelation of in-person learning opportunities such as classes) imposed by the pandemic.

21.4 Study 2: Skill Learning and Growth Mindset

The second study our team conducted during the pandemic examined whether having a growth mindset predicted mental health outcomes among adults (from younger to older adulthood) and assessed whether having participated in the prior learning intervention from the lab that fostered an increased growth mindset led to more positive mental health outcomes among older adults (Study 2 (Kyeong et al. 2024)). The outcomes included depression levels, subjective well-being, and adjustment of daily tasks (which refers to adjusting tasks within one's daily routine during the pandemic based on the physical distancing restrictions). It was hypothesized that a higher level of growth mindset would predict better mental health, and that the older adults who participated in the learning intervention would demonstrate better mental health outcomes than those who did not. The data were collected via an online survey across three time points during the first two years of the pandemic (2020–2022). Linear mixed-effects regression models were conducted to test the effect of growth mindset on the mental health outcomes.

The results showed that a higher endorsement of a growth mindset predicted lower depression levels, higher well-being, and greater adjustment of daily tasks. Additionally, middle-aged and older adults who participated in the intervention displayed a growth mindset that was linked to an increase in well-being more so than for the adults who did not participate in the intervention. Finally, an interplay between intervention status and time was identified in that, over time, the intervention group experienced an increase in well-being as opposed to the group that did not participate in the intervention, who faced a decrease in well-being. These results highlight how a growth mindset can help foster resilience and induce a buffering effect against negative mental health outcomes during difficult times.

21.5 Study 3: Technology Learning and Digital Independence

The third study our lab conducted during the pandemic examined the impact of a two-month community-based technology intervention that trained low-income older adults to use a new tablet (Study 3 (Ferguson et al. 2024)). This pilot study was conducted in collaboration with the Riverside County Office on Aging (OoA), which is a department that provides systems of care (e.g., programs and services) to help improve the quality of life of vulnerable older adults, especially those with a low-income and disabilities. The OoA wanted to assess the feasibility and effectiveness of their program entitled "Senior Learning and Accessibility Technology Evaluation" (SLATE), which was a project that was created and executed during the pandemic for teaching low-income older adults how to use new technological devices and programs. Some of the technological-related skills being taught to participants, aside from learning to use and navigate the tablet device itself, were setting up and using email accounts, video calling, accessing online resources and benefit portals, using online health care services, and ordering groceries online. The outcomes of interest for this study included cognitive function, well-being, loneliness, depression, and growth mindset. It was hypothesized that each of these outcomes would increase for participants by the end of the program based on previous learning interventions, including the one conducted in our laboratory (Chan et al. 2016; Leanos et al. 2023). Linear mixed-effects models were used to analyse the data for each outcome measure.

The results from this intervention indicated that there were no significant improvements across all of the outcome measures in relation to learning technology. A potential reason for these findings is that the sample size was small ($N = 19$), as it was a pilot study that primarily tested the feasibility of the program. Nonetheless, analyses revealed that functional digital independence played an important role in predicting depression and well-being levels. Higher dependency on others for digital matters at pretest predicted lower well-being scores at post-test. These findings suggested that having more independence in navigating technology can potentially benefit an individual's mental health, similar to other findings (Cheshmehzangi et al. 2022; Fang et al. 2018). For depression, there was also a significant interaction between digital independence and age, where out of two participants of the same age, the one with lower digital independence was expected to have a higher depression score. Also, out of two participants with the same score on the digital independence measure, the older individual would be expected to have a higher depression score. It was also the case that the higher the digital independence score at pretest, the higher the depression scores at posttest. This finding is contrary to previous studies that have found that higher use of technology (e.g., the Internet) is linked to lower depression among older adults (Cotten et al. 2014), although other studies have shown that it may not always increase well-being (Bong et al. 2019; Dickinson and Gregor 2006; Elliot et al. 2014). Our results for this study should be interpreted with caution given that, aside from the small sample size, there could be other factors that have an influence on the mental health of individuals that were not taken into account in this study, such as physical activity and health (Callow et al. 2020; Ohrnberger et al. 2017). Lastly, an important factor that could have contributed to our findings is that participants voiced that they had a difficult time understanding the questions and their format of number scales, leading to inaccurate responses from individuals. Therefore, qualitative methods might be a better fit for this population (e.g., interviews) when investigating the impact of the intervention, and such methods should be employed in future work for greater accuracy (Rodriguez et al. 2025).

21.6 Realizations, Recommendations, and Future Directions

The findings from the three studies described above that we conducted during the pandemic taught us the following. From the first study, we learned that subjective ratings of cognitive abilities and learning new skills are important predictors of mental health outcomes. From the second study, we learned that a growth mindset has overall beneficial predictive effects on depression, well-being, and adjustment of day-to-day tasks. And from the third study, we learned that digital independence is important for depression and well-being outcomes. Altogether, our findings suggest that engaging in novel skill learning, as well as having a growth mindset, may be advantageous for mental health and adaptation to environmental changes for adults, especially older adults.

Before the pandemic, our research was focused on answering the question of whether aging individuals should engage in novel skill learning and have a growth mindset to increase cognitive abilities. However, this question is no longer the appropriate question. Instead, the question should be that given that adults need to learn new skills to adapt to changing environments (just like younger learners), as we have seen with the pandemic and emerging technology, how can learning be optimized for adults? How can we best encourage, support, and provide opportunities for all adult individuals to continuously engage in learning new skills?

Our recommendations for a smoother navigation of future pandemics would involve: (1) moving away from the perception that learning is meant for younger populations, (2) increasing awareness of the importance of learning new skills especially for older adults to adapt, and (3) generating more learning programs and providing more learning resources that are tailored to sociodemographic factors that make up different segments of the adult population. As we embrace the belief that learning new skills is important across the lifespan, society can help promote lifelong learning to its fullest potential. A key aspect of changing the minds of individuals is in part to widely disseminate the benefits that come from continued skill learning engagement and clarify the necessity of doing so for the better adaptation and adjustment to novel situations. Aside from disseminating the importance of engaging in novel skill learning, it is also key for policymakers and community leaders to come together to create more learning programs and to provide more free or low-cost resources that are accessible to adults from all backgrounds. New learning programs must take into account that individuals will face different barriers to learning engagement and, thus, they should tailor programs to address such barriers appropriately and ensure that it is feasible for all to learn new skills (Rodriguez et al. 2022). Some examples of barriers we have previously identified in our work, specifically with underserved older populations, include language, financial, time, motivation, transportation, and health-related barriers (Rodriguez et al. 2025).

Our lab is currently further investigating the impact of and how to best support engagement of novel skill learning among diverse older populations, such as low-income, non-English speaking, racial/ethnic minoritized, and immigrant older adults. We are creating connections and partnerships with community organizations to offer learning opportunities to older adults to supplement existing support and resources provided to them. We are also interested in reskilling older adults that wish to remain in the workforce due to their inability to retire or desire to continue working. Future work should aim to generate more programs to prepare individuals with the tools to engage in learning, not just for learning's sake, but also to help address potential needs when faced with new challenging situations. Providing appropriate resources and equipping adults of all ages with a growth mindset, which involves having the confidence that they will be able to learn to successfully adapt to changes, will be a powerful step toward better preparing everyone for future pandemics and other circumstances that require considerable and rapid adaptation.

References

Anstey KJ, Low LF (2004) Normal cognitive changes in aging. Aust Fam Physician 33(10):783–787

Atske S, Perrin A (2021) Home broadband adoption, computer ownership vary by race, ethnicity in the U.S. Pew Research Center. https://www.pewresearch.org/short-reads/2021/07/16/home-broadband-adoption-compute. Accessed on 22 Aug 2024

Azer L, Kürüm E, Ferguson L, Davis EL, Zhang W, Strickland-Hughes CM, Wu R (2025) (under review) Subjective executive functioning and skill learning during the COVID-19 pandemic associated with perceived loneliness, depressive symptoms, and well-being. PLOS Ment Health 2(7):e0000372

Beier ME (2015) The aging workforce and the demands of work in the 21st century. In: Finkelstein LM, Truxillo DM, Fraccaroli F, Kanfer R (eds) Facing the challenges of a multi-age workforce: a use-inspired approach. Routledge/Taylor & Francis Group, New York, pp 108–133

Berk LE (2022) Infants, children, and adolescents, 9th edn. SAGE Publications, California

Bong WK, Bergland A, Chen W (2019) Technology acceptance and quality of life among older people using a TUI application. Int J Environ Res Public Health 16(23):4706. https://doi.org/10.3390/ijerph16234706

Brown R (2015) Building children and young people's resilience: lessons from psychology. Int J Disaster Risk Reduct 14(2):115–124. https://doi.org/10.1016/j.ijdrr.2015.06.007

Bugos JA, Wang Y (2022) Piano training enhances executive functions and psychosocial outcomes in aging: results of a randomized controlled trial. JS Gerontol Ser B 77(9):1625–1636. https://doi.org/10.1093/geronb/gbac021

Bugos JA, Perlstein WM, McCrae CS, Brophy TS, Bedenbaugh PH (2007) Individualized piano instruction enhances executive functioning and working memory in older adults. Aging Ment Health 11(4):464–471. https://doi.org/10.1080/13607860601086504

Callow DD, Arnold-Nedimala NA, Jordan LS, Pena GS, Won J, Woodard JL, Smith JC (2020) The mental health benefits of physical activity in older adults survive the COVID-19 pandemic. Am J Geriatr Psychiatry 28(10):1046–1057. https://doi.org/10.1016/j.jagp.2020.06.024

Carrasco P, Crespo D, Pedrero-Pérez E, Peña M (2020) Subjective cognitive decline: mental health, loneliness, pain and quality of life: Poblational study. J Aging Sci 8(1):218. https://www.researchgate.net/publication/339029388_Subjective_Cognitive_Decline_Mental_Health_Loneliness_Pain_and_Quality_Of_Life_Poblational_Study

Chan MY, Haber S, Drew LM, Park DC (2016) Training older adults to use tablet computers: does it enhance cognitive function? Gerontologist 56(3):475–484. https://doi.org/10.1093/geront/gnu057

Charness N, Boot WR (2009) Aging and information technology use: potential and barriers. Curr Dir Psychol Sci 18:253–258. https://doi.org/10.1111/j.1467-8721.2009.01647.x

Cheshmehzangi A, Zou T, Su Z (2022) The digital divide impacts on mental health during the COVID-19 pandemic. Brain Behav Immun 101:211–213. https://doi.org/10.1016/j.bbi.2022.01.009

De Coninck D, Frissen T, Matthijs K, d'Haenens L, Lits G, Champagne-Poirier O et al (2021) Beliefs in conspiracy theories and misinformation about COVID-19: comparative perspectives on the role of anxiety, depression and exposure to and trust in information sources. Front Psychol 12. https://doi.org/10.3389/fpsyg.2021.646394

Cotten SR, Ford G, Ford S, Hale TM (2014) Internet use and depression among retired older adults in the United States: a longitudinal analysis. J Gerontol B Psychol Sci Soc Sci 69(5):763–771. https://doi.org/10.1093/geronb/gbu018

Dickinson A, Gregor P (2006) Computer use has no demonstrated impact on the well-being of older adults. Int J Hum Comput Stud 64(8):744–753. https://doi.org/10.1016/j.ijhcs.2006.03.001

Dow B, Meyer C (2010) Caring and retirement: crossroads and consequences. Int J Health Serv 40(4):645665. https://doi.org/10.2190/HS.40.4.e

Dweck CS (2006) Mindset: the new psychology of success. Random house, New York

Elliot AJ, Mooney CJ, Douthit KZ, Lynch MF (2014) Predictors of older adults' technology use and its relationship to depressive symptoms and well-being. J Gerontol B Psychol Sci Soc Sci 69(5):667–677. https://doi.org/10.1093/geronb/gbt109

Fancourt D, Steptoe A, Bu F (2021) Trajectories of anxiety and depressive symptoms during enforced isolation due to COVID-19 in England: a longitudinal observational study. Lancet Psychiatry 8(2):141–149. https://doi.org/10.1016/S2215-0366(20)30482-X

Fang Y, Chau AK, Wong A, Fung HH, Woo J (2018) Information and communicative technology use enhances psychological well-being of older adults: the roles of age, social connectedness, and frailty status. Aging Ment Health 22(11):1516–1524. https://doi.org/10.1080/13607863.2017.1358354

Ferguson L, Sain D, Kürüm E, Strickland-Hughes CM, Rebok GW, Wu R (2023) One-year cognitive outcomes from a multiple real-world skill learning intervention with older adults. Aging Ment Health 27(11):2134–2143. https://doi.org/10.1080/13607863.2023.2197847

Ferguson L, Kürüm E, Rodriguez TM, Nguyen A, Lopes de Queiroz IF, Lee J, Wu R (2024) Impact of community-based technology training with low-income older adults. Aging Ment Health 28(4):638–645. https://doi.org/10.1080/13607863.2023.2256271

Field J (2009) Good for your soul? Adult learning and mental well-being. Int J Lifelong Educ 28(2):175–191. https://doi.org/10.1080/02601370902757034

Fitzpatrick KM, Harris C, Drawve G (2020) Fear of COVID-19 and the mental health consequences in America. Psychol Trauma Theory Res Pract Policy 12(S1):S17–S21

Goldstein S, Naglieri JA, Princiotta D, Otero TM (2014) Introduction: a history of executive functioning as a theoretical and clinical construct. In: Goldstein S, Naglieri JA (eds) Handbook of executive functioning. Springer, New York, pp 3–12. https://doi.org/10.1007/978-1-4614-8106-5_1

Harvey PD, Forero DB, Ahern LB, Tibiriçá L, Kallestrup P, Czaja SJ (2021) The computerized functional skills assessment and training program: sensitivity to global cognitive impairment, correlations with cognitive abilities, and factor structure. Am J Geriatr Psychiatry 29:395–404. https://doi.org/10.1016/j.jagp.2020.08.019

Hennekam S (2015) Career success of older workers: the influence of social skills and continuous learning ability. J Manag Dev 34(9):1113–1133. https://doi.org/10.1108/JMD-05-2014-0047

Hernández-Fernández C, Meneses-Falcón C (2022) I can't believe they are dead. Death and mourning in the absence of goodbyes during the COVID-19 pandemic. Health Soc Care Community 30(4):e1220–e1232. https://doi.org/10.1111/hsc.13530

Holingue C, Kalb LG, Riehm KE, Bennett D, Kapteyn A, Veldhuis CB et al (2020) Mental distress in the United States at the beginning of the COVID-19 pandemic. Am J Public Health 110(11):1628–1634. https://doi.org/10.2105/AJPH.2020.305857

Hwang TJ, Rabheru K, Peisah C, Reichman W, Ikeda M (2020) Loneliness and social isolation during the COVID-19 pandemic. Int Psychogeriatr 32(10):1217–1220. https://doi.org/10.1017/S1041610220000988

Krendl AC, Perry BL (2021) The impact of sheltering in place during the COVID-19 pandemic on older adults' social and mental well-being. J Gerontol B Psychol Sci Soc Sci 76(2):e53–e58. https://doi.org/10.1093/geronb/gbaa110

Kyeong Y, Kürüm E, Sheffler P, Ferguson L, Davis EL, Strickland-Hughes CM, Wu R (2024) The implications of growth mindset for depression, well-being, and adjustment over 2 years during the COVID-19 pandemic. PLOS Mental Health 1(7):e0000182. https://doi.org/10.1371/journal.pmen.0000182

Leanos S, Kürüm E, Strickland-Hughes CM, Ditta AS, Nguyen G, Felix M, Yum H, Rebok GW, Wu R (2023) The impact of learning multiple real-world skills on cognitive abilities and functional independence in healthy older adults. J Gerontol Ser B 75(6):1155–1169. https://doi.org/10.1093/geronb/gbad053

Li H (2021) Piano education of children using musical instrument recognition and deep learning technologies under the educational psychology. Front Psychol 12:705116. https://doi.org/10.3389/fpsyg.2021.705116

Litchfield I, Shukla D, Greenfield S (2021) Impact of COVID-19 on the digital divide: a rapid review. BMJ Open 11(10):e053440. https://doi.org/10.1136/bmjopen-2021-053440

Markova H, Andel R, Stepankova H, Kopecek M, Nikolai T, Hort J, Thomas-Antérion C, Vyhnalek M (2017) Subjective cognitive complaints in cognitively healthy older adults and their relationship to cognitive performance and depressive symptoms. J Alzheimer's Dis JAD 59(3):871–881. https://doi.org/10.3233/JAD-160970

Maurer TJ (2001) Career-relevant learning and development, worker age, and beliefs about self-efficacy for development. J Manag 27:123–140. https://doi.org/10.1177/014920630102700201

Menzies RE, Neimeyer RA, Menzies RG (2020) Death anxiety, loss, and grief in the time of COVID-19. Behav Chang 37(3):111–115. https://doi.org/10.1017/bec.2020.10

Mitchell UA, Chebli PG, Ruggiero L, Muramatsu N (2019) The digital divide in health-related technology use: the significance of race/ethnicity. Gerontologist 59(1):6–14. https://doi.org/10.1093/geront/gny138

Murphy K (2020) Death and grieving in a changing landscape: Facing the death of a loved one and experiencing grief during COVID-19. Health Soc Care Chaplain 8(2):240–250. https://doi.org/10.1558/hscc.41578

Narushima M, Liu J, Diestelkamp N (2013) The association between lifelong learning and psychological well-being among older adults: implications for interdisciplinary health promotion in an aging society. Act Adapt Aging 37(3):239–250. https://doi.org/10.1080/01924788.2013.816834

Nguyen C, Leanos S, Natsuaki MN, Rebok GW, Wu R (2020) Adaptation for growth via learning new skills as a means to long-term functional independence in older adulthood: insights from emerging adulthood. Gerontologist 60(1):4–11. https://doi.org/10.1093/geront/gny128

Noice H, Noice T, Staines G (2004) A short-term intervention to enhance cognitive and affective functioning in older adults. J Aging Health 16(4):562–585. https://doi.org/10.1177/0898264304265819
Ohrnberger J, Fichera E, Sutton M (2017) The relationship between physical and mental health: a mediation analysis. Soc Sci Med 195:42–49. https://doi.org/10.1016/j.socscimed.2017.11.008
Ortiz Alvarado NB, Rodriguez Ontiveros M, Ayala Gaytan EA (2019) Do mindsets shape students' well-being and performance? J Psychol 153(8):843–859. https://doi.org/10.1080/00223980.2019.1631141
Papalia DE, Sterns HL, Feldman RD, Camp CJ (2007) Adult development and aging, 3rd edn. McGraw-Hill
Park DC, Lodi-Smith J, Drew L, Haber S, Hebrank A, Bischof GN, Aamodt W (2014) The impact of sustained engagement on cognitive function in older adults: the synapse project. Psychol Sci 25(1):103–112. https://doi.org/10.1177/0956797613499592
Perkins R, Williamon A (2014) Learning to make music in older adulthood: a mixed-methods exploration of impacts on wellbeing. Psychol Music 42(4):550–567. https://doi.org/10.1177/0305735613483668
Poelke G, Ventura MI, Byers AL, Yaffe K, Sudore R, Barnes DE (2016) Leisure activities and depressive symptoms in older adults with cognitive complaints. Int Psychogeriatr 28(1):63–69. https://doi.org/10.1017/S1041610215001246
Rocha YM, De Moura GA, Desidério GA, De Oliveira CH, Lourenço FD, de Figueiredo Nicolete LD (2021) The impact of fake news on social media and its influence on health during the COVID-19 pandemic: a systematic review. J Public Health 31:1007–1016. https://doi.org/10.1007/s10389-021-01658-z
Rodrigues NG, Han CQY, Su Y, Klainin-Yobas P, Wu XV (2022) Psychological impacts and online interventions of social isolation amongst older adults during COVID-19 pandemic: a scoping review. J Adv Nurs 78(3):609–644. https://doi.org/10.1111/jan.15063
Rodriguez TM, Sheffler P, Ferguson LE, Rebok GW, Wu R (2022) Cognitive and functional improvement via novel skill learning for low-income minoritized middle-aged and older adults. Prev Sci 24:926–935. https://doi.org/10.1007/s11121-022-01413-0
Rodriguez TM, Neri X, Wu R (2025) (Under review) Identifying learning barriers, engagement, and interests among low-income Latino older adults. J Aging Health 37(3–4 suppl):53S–65S
Sarason BR, Sarason IG, Gurung RAR (2001) Close personal relationships and health outcomes: a key to the role of social support. In: Sarason BR, Duck S (eds) Personal relationships: implications for clinical and community psychology. Wiley, pp 15–41
Sen K, Prybutok G, Prybutok V (2021) The use of digital technology for social wellbeing reduces social isolation in older adults: a systematic review. SSM—Population Health 17:101020. https://doi.org/10.1016/j.ssmph.2021.101020
Shapira N, Barak A, Gal I (2007) Promoting older adults' well-being through Internet training and use. Aging Ment Health 11(5):477–484. https://doi.org/10.1080/13607860601086546
Shultz KS, Adams GA (eds) (2007) Aging and work in the 21st century. Lawrence Erlbaum Associates, New Jersey
Smith A (2014) Older adults and technology use. Pew Research Center. https://www.pewresearch.org/internet/2014/04/03/older-adults-and-technology-use/. Accessed on 20 Aug 2024
Teh JKL, Tey NP (2019) Effects of selected leisure activities on preventing loneliness among older Chinese. SSM—Population Health 9:100479. https://doi.org/10.1016/j.ssmph.2019.100479
Thompson RA (2001) Development in the first years of life. Future Child 11(1):21–33. https://doi.org/10.2307/1602807
Troen BR (2003) The biology of aging. Mt Sinai J Med 70(1):3–22
Tyrrell CJ, Williams KN (2020) The paradox of social distancing: Implications for older adults in the context of COVID-19. Psychol Trauma Theory Res Pract Policy 12(S1):S214–S216. https://doi.org/10.1037/tra0000845
Vogels EA (2021) Digital divide persists even as Americans with lower incomes make gains in tech adoption. Pew Research Center. https://www.pewresearch.org/short-reads/2021/06/22/digital-divide-persists-even-as-americans-with-lower-incomes-make-gains-in-tech-adoption/. Accessed on 21 Aug 2024
Wang H, Smith-Greenaway E, Bauldry S, Margolis R, Verdery AM (2022) Mourning in a pandemic: the differential impact of COVID-19 widowhood on mental health. J Gerontol Ser B 77(12):2306–2316. https://doi.org/10.1093/geronb/gbac085
Wu R (2020) Older adults' learning and adaptation as resilience processes to counter social isolation during the COVID-19 pandemic. Columbia Academic Commons, Academic Unit: Northeast Big Data Innovation Hub, Series: COVID-19 Research Lightning Talks. https://doi.org/10.7916/dxyz-sp91
Wu R, Azer L (2022) Older adults' learning and adaptation as resilience processes to counter social isolation during the COVID-19 pandemic. Columbia Academic Commons, Academic Unit: Northeast Big Data Innovation Hub, Series: COVID-19 Research Lightning Talks. https://doi.org/10.7916/wsfs-gy91
Yeager DS, Dweck CS (2012) Mindsets that promote resilience: when students believe that personal characteristics can be developed. Educ Psychol 47(4):302–314. https://doi.org/10.1080/00461520.2012.722805
Zeng G, Hou H, Peng K (2016) Effect of growth mindset on school engagement and psychological well-being of Chinese primary and middle school students: the mediating role of resilience. Front Psychol 7:1873. https://doi.org/10.3389/fpsyg.2016.01873
Zhang S, Li Y, Ren S, Liu T (2023) Associations between undergraduates' interpersonal relationships and mental health in perspective of social network analysis. Curr Psychol 42(3):2059–2066. https://doi.org/10.1007/s12144-021-01629-3

Zheng J, Morstead T, Sin N, Klaiber P, Umberson D, Kamble S, DeLongis A (2021) Psychological distress in North America during COVID-19: the role of pandemic-related stressors. Soc Sci Med 270:113687. https://doi.org/10.1016/j.socscimed.2021.113687
Zlatar ZZ, Muniz M, Galasko D, Salmon DP (2018) Subjective cognitive decline correlates with depression symptoms and not with concurrent objective cognition in a clinic-based sample of older adults. J Gerontol B Psychol Sci Soc Sci 73(7):1198–1202. https://doi.org/10.1093/geronb/gbw207

Tania M. Rodriguez (ORCID: 0000-0002-8147-7796) completed her BA in Psychology at the California State University, Fullerton in 2021. She currently is a PhD candidate at the University of California, Riverside in the Department of Psychology. She works under the mentorship of Dr. Rachel Wu in the Cognitive Agility across the Lifespan via Learning and Attention (CALLA) laboratory. Tania has experience conducting qualitative research studies with low-income, minority older adult populations. Her research interests include understanding cognitive functioning and barriers to resources that benefit cognition among minoritized older adults, especially those with low income. In particular, she aims to help underserved populations improve their cognition, functional abilities, and well-being via learning new skills by providing them with the tools and opportunities to engage in learning. Utilizing mixed methods, her dissertation work is focused on executing a community-based learning intervention tailored for minority older adults with a low income that will address barriers to learning among this population.

Isadora Farias Lopes de Queiroz completed her BA in Psychology at Hollins University. She is now a graduate student at the University of California, Riverside in Developmental Psychology under the mentorship of Dr. Rachel Wu. In the Cognitive Agility across the Lifespan via Learning and Attention (CALLA) laboratory, her research work consists of investigating low-cost highly accessible novel skill learning interventions for older adults in order to improve their cognitive abilities, motivation to learn, and in turn, overall well being. She aims to optimize the learning environment for older adults to mitigate or prevent cognitive decline and promote healthy aging. Currently, her Master's degree thesis consists of optimizing older adults' learning environment through teaching them new skills (board games) and having a discussion time on healthy aging topics in a community environment. Her intended area of focus is underrepresented populations, specifically individuals with a Latinx background. In her free time, Isadora enjoys the outdoors, playing the guitar, and reading.

Esra Kurum is an Associate Professor in the Department of Statistics at the University of California, Riverside (UCR). She earned her PhD in Statistics from the Pennsylvania State University and was a Postdoctoral Associate in the Department of Epidemiology of Microbial Diseases at Yale University before joining UCR. Her research focuses on developing statistical methods that can simultaneously analyse longitudinal, spatial, and survival data while capturing dynamic patterns—such as temporal and spatial dynamics—within complex data sets. Dr. Kurum has collaborated with researchers across a broad range of disciplines, such as aging, stem cell biology, and medical specialties including infectious disease, nephrology, and rheumatology. Her work has been supported by extramural grants from the National Institutes of Health and the National Science Foundation. She is the recipient of a Fulbright Scholarship, a National Science Foundation (NSF)/Howard Hughes Medical Institute (HHMI) Scientific Teaching Fellowship, and the Kenneth Rothman Epidemiology Prize.

Lilian Azer is a postdoctoral researcher in the department of Neurobiology and Behavior at the University of California, Irvine. She is a cognitive psychologist whose research focuses on normal age-related cognitive decline, and Alzheimer's disease and related dementias. Her current research involves establishing normative data and validating a screening tool to detect hippocampal-related memory decline in older adults across diverse communities. Lilian earned her PhD in Psychology: Cognition and Cognitive Neuroscience from the University of California, Riverside in 2023 where she studied the interaction between cognitive and motor function. Specifically, her research investigated the effects of effortful physical exertion on working memory in younger and older adults. Before completing her PhD, she received a master's degree in clinical psychology from California State University, Dominguez Hills where she researched the relationship between cardiovascular health, mental health, and cognitive functioning in tobacco users.

Yena Kyeong is a developmental psychologist whose research focuses on emotional socialization across cultural contexts, as well as the associations between parenting practices and children's academic and socioemotional adjustment. She is particularly interested in how cultural values shape parent-child dynamics across the lifespan. Her work also explores non-cognitive factors, such as mindset and grit, and their role in mental health and developmental outcomes. She earned her master's degree in Counselling Psychology from Ewha Womans University in South Korea and her PhD in Developmental Psychology from the University of California, Riverside. Following her postdoctoral training at the National University of Singapore and Nanyang Technological

University, she will join the Department of Child and Family Science at National Taiwan Normal University in 2025 as an Assistant Professor.

Leah Ferguson is a current PhD student studying Developmental Psychology at the University of California, Riverside. Her research interests combine developmental, community, and cognitive psychology to create a holistic approach to cognitive aging research. Trained in cognitive engagement interventions, lifespan development, and longitudinal designs, her research contributes to the field's understanding of real-world skill learning in older adulthood and its influence on cognitive performance, mental health, and well-being. Her program of research aims to develop collaborative community-based programs that expand learning opportunities to improve quality of life and cognitive performance for older adults, while providing accessible preventative practices for all ages. She strives to incorporate individual needs, abilities, and experiences into the development of community-based cognitive engagement interventions by applying mixed-method approaches. She aims to encourage resource development, education accessibility, and policy changes through cognitive aging research.

Elizabeth Davis (ORCID: 0000-0003-2599-4390) is a Professor of Psychology at the University of California, Riverside (UCR). She is a developmental psychologist whose research focuses on cognition-emotion interactions in childhood, how children learn to regulate their negative emotions and stress, the psychobiology of emotional development, and the sociocultural contexts (like the parent-child relationship and broader environment) that shape child development. She earned her doctorate in Psychology and Social Behavior from the University of California, Irvine in 2009 and completed postdoctoral training in Psychology at the Pennsylvania State University from 2009-2012. Since joining the faculty at UCR in 2012, her research has primarily considered links between childhood experiences of emotion and emotion regulation for subsequent health and well-being, psychopathology, and learning. Her expertise in affective science, psychophysiology and psychobiology, and social contexts of development can be leveraged to inform novel theoretical and empirical approaches to developmental science.

Dr. Carla M. Strickland-Hughes (ORCID: 0000-0001-8446-8708) is a social-cognition researcher trained in experimental and intervention methodologies. She completed her PhD in Psychology at the University of Florida in 2017. Before joining the Trefny Innovative Instruction Center at the Colorado School of Mines (January 2025), she was a faculty member in the Psychology Department at the University of the Pacific (August 2017-January 2025, promoted to Associate Professor in September 2023). Her research program aims to enhance memory in adulthood via training and improvement on non-ability factors, such as personal beliefs. For example, her research has demonstrated the power of one's level of confidence in her memory ability. Relatedly, because pervasive negative age stereotypes can threaten adults' cognitive performance, her research aims to improve attitudes about old aging held by young and old alike. She is also interested in studying self-regulated learning and metacognition in adulthood, applied to higher education and aging.

Dr. Rachel Wu (ORCID: 0000-0002-7486-697X) is an Associate Professor of Psychology at University of California, Riverside. She completed her PhD in Psychology at Birkbeck, University of London (2011). She is a developmental cognitive neuroscientist investigating how we learn and attend across the lifespan. Her latest research focuses on optimizing learning in older adults based on child development research. She has shown that with a supportive learning environment (e.g., an undergraduate curriculum), older adults' cognitive abilities can be enhanced to levels similar to younger adults, while developing useful functional skills, such as for staying in the workforce. Dr. Wu has received more than two dozen awards and grants, including a Rising Star award from the Association for Psychological Science and an NSF CAREER award. Thus far, her research program has resulted in over 50 publications. She was in a rock band as a violinist for seven years (2003–2010) and also is a professional painter. Every few years, she learns to master new skills to gain a better understanding of the difficulties in being an adult learner.

Part VI
COVID and Public Mobility

Measuring Mobility and Social Mixing to Inform Pandemic Prediction and Response

22

Peyton Luiz, Jason R. Gantenberg, Samuel F. Rosenblatt, Guixing Wei, Aditya Khanna, Bryn C. Loftness, and Mark N. Lurie

Abstract

This chapter focuses on the need for fine-grained social mixing and mobility data to devise mitigation strategies and more accurately predict pathogenic spread before and during pandemics. We explore the availability of different data types and discuss our efforts to collect individual-level indoor movement data to measure patterns of social interaction in a congregate setting. To do so, we deployed a dense Bluetooth network and developed a smartphone-based app to record the signal strength and connection duration of Bluetooth beacons and Wi-Fi routers. Over a two-week study period in November 2023, we recorded 135 million data points from 122 unique participant devices. We outline ways signal strength data can be used to calculate a participant's place in space and time, allowing us to map the social interactions of study participants. We argue that these types of data are necessary to make informed decisions when planning targeted interventions in congregate settings, which can aid in our ability to predict and prevent future pandemics.

Research Funded by

NSF Award #2154941—PIPP* Phase I: Mobility Analysis for Pandemic Prevention Strategies (MAPPS).

*Predictive Intelligence for Pandemic Prevention

COVID Information Commons Video Presentation

The COVID Information Commons (CIC) Research Lightning Talk video presentation and individual multilingual transcripts for this project can be found in the Columbia Academic Commons (AC) (Lurie 2023).

P. Luiz (✉) · A. Khanna · M. N. Lurie
Brown University School of Public Health, Providence, RI, USA
e-mail: peyton_luiz@brown.edu

A. Khanna
e-mail: aditya_khanna@brown.edu

M. N. Lurie
e-mail: Mark_Lurie@brown.edu

J. R. Gantenberg · G. Wei
Brown University, Providence, RI, USA
e-mail: jason_gantenberg@brown.edu

G. Wei
e-mail: guixing_wei@brown.edu

S. F. Rosenblatt
University of Vermont, Burlington, VT, USA
e-mail: samuel.f.rosenblatt@uvm.edu

B. C. Loftness
Biobe and University of Vermont, Burlington, VT, USA
e-mail: Bryn.Loftness@uvm.edu

F. D. Hudson (ed.), *The COVID Information Commons*, Sustainable Development Goals Series,
https://doi.org/10.1007/978-3-032-06125-6_22

Keywords

Mobility · Social mixing · Contact · Smartphone · Bluetooth · Wi-Fi · Network epidemiology · Modeling · Congregate settings · Ethics · Social science · COVID-19

22.1 The MAPPS Project

In 2021, the U.S. National Science Foundation (NSF) launched the Predictive Intelligence for Pandemic Prevention Program to support interdisciplinary research tackling grand challenges in infectious disease pandemics. Under this program, we created the Mobility Analysis for Pandemic Prevention Strategies (MAPPS) Project to address the continued underutilization of mobility and social mixing data in designing pandemic response strategies. We assembled a diverse team representing public health, epidemiology, computational modeling, engineering, mathematics, data and computer science, network science, ethics, and the social sciences. The MAPPS Project has four aims: (1) develop new technologies to measure social interactions and mobility; (2) refine public health analytics, modeling, and forecasting; (3) translate research into tailored policies for congregate settings; and (4) create a framework for ethical practices when collecting sensitive personal information.

22.1.1 *Motivating Use Cases*

The MAPPS Project responds to the large body of research identifying detailed mobility and social mixing data as key to effective pandemic response. Throughout the COVID-19 pandemic, various countries, including South Korea, Japan, New Zealand, and China, demonstrated the effectiveness of utilizing social mixing and telecommunications data to curb pathogen transmission (Usherwood et al. 2021; Specht et al. 2022; Alessandretti 2021; Ibrahim 2020; Firth et al. 2020; Oliver et al. 2020; Rosenblatt et al. 2020; Bedson et al. 2021; Parsons and Worden 2021; Xu et al. 2020). Specific uses of mobility data included contact tracing apps, which allowed for higher contact tracing coverage than would have been possible manually (Lee and Kim 2021), and evaluation of the effectiveness of policies such as travel restrictions and social distancing mandates, enabling policymakers to allocate resources and efforts accordingly (Kraemer et al. 2020; Oh et al. 2021). Many epidemiologists and other experts have argued that data-driven non-pharmaceutical interventions are the only means to control pandemics without transmission-blocking measures like vaccination (Ferretti et al. 2020; Chung et al. 2021; The DELVE Initiative 2020; Perra 2021). Regions that did not utilize mobility and mixing data during the COVID-19 pandemic had limited non-pharmaceutical alternatives apart from severe interventions, such as widespread lockdowns, which saved lives but resulted in well-documented adverse social and economic impacts (Adams-Prassl et al. 2022; Allen 2022; Onyeaka et al. 2021; Piper 2020; Althouse et al. 2023).

22.1.2 Challenges in Acquiring and Utilizing Mobility and Social Mixing Data

Acquiring and utilizing fine-grained mobility and social mixing data presents numerous technical and ethical challenges. Individual-level interaction data have historically been expensive to generate, leading researchers to rely on a small number of datasets (e.g., the Copenhagen Networks Study (Stopczynski et al. 2014), the National Longitudinal Study of Adolescent Health (Add health 2024), Reality Mining (Eagle and Pentland 2006), SocioPatterns' primary school temporal network (Gemmetto et al. 2014; Stehlé et al. 2011), including datasets with known limitations (Christakis and Fowler 2013). Harmonization issues further complicate matters, as data from different sources cannot be easily combined or integrated (McClain et al. 2023; Wnuk et al. 2020; Zhang et al. 2020). Statistical and causal inferences made from synthetic data have their own limitations

(Tchetgen Tchetgen and VanderWeele 2012; Halloran and Hudgens 2016). Estimating the direct and indirect effects of interventions in the presence of contagion requires detailed data on the exposure and outcome status of those in the interconnected population. Depending on the contagion of interest, confounding variables may be impossible to account for, even with complete data on the network (Shalizi and Thomas 2011). The MAPPS Project explores the potential for Bluetooth and Wi-Fi data to increase precision and lower data collection costs.

Acceptability plays a crucial role in acquiring and utilizing data for pandemic prevention. The burden of contributing data may affect population members' willingness to participate in collection efforts (Stopczynski et al. 2014; Smyth et al. 2021). Additionally, data collection may be unacceptable in certain settings due to concerns about privacy loss and data misuse. For instance, the harms of disclosures in South Korea included invasions of individual privacy, reidentification, and loss of business at locations where infected individuals visited (Park et al. 2020). Balancing privacy and public safety in the context of pandemic prevention continues to be complicated by emerging technologies and evolving notions of digital autonomy (Zhang et al. 2020; Fahey and Hino 2020; Ribeiro-Navarrete et al. 2021; Bengio et al. 2020).

22.1.3 Applications of Mobility Data During the COVID-19 Pandemic

Mobility data are critical for pandemic prediction yet remain underutilized. Early in the SARS-CoV-2 pandemic, Kraemer et al. emphasized the importance of using real-time mobility data to study COVID-19's spread and the effect of control measures in China (Kraemer et al. 2020). Building on the five challenges outlined by Oliver et al. (Oliver et al. 2020), our systematic review of how mobility data were employed to respond to COVID-19 from 2019 through 2023 identified four persistent issues hindering utilization: (1) lack of unified standards for recording and distributing data; (2) opacity in third-party data collection processes; (3) unequal and waning data accessibility; and (4) limited policy-applicable work evaluating interventions.

First, there are no unified standards or protocols for developing and reporting mobility metrics, making it difficult to compare findings across studies. Second, the absence of transparency in data production by such third parties collecting mobility data muddies our understanding of the effect that human movement has on the spread of COVID-19 in different environments. Approximately 70% of the articles we reviewed utilized mobility data provided by large tech companies, social media platforms, and mobility data vendors (Go and Wei 2024). Many of these datasets do not include demographic information, citing privacy concerns, which makes it unclear who is represented in a given dataset (Kraemer et al. 2021). Third, despite calls for action by researchers, the global disparity in mobility data accessibility has not improved since the early days of the pandemic. Research is concentrated in the United States (U.S.), China, and the European Union, locales with various publicly available mobility datasets. Noticeably, many data collection and reporting efforts were phased out by 2022 as the perceived threat of COVID-19 waned, with several frequently cited datasets pausing updates or revoking researcher access entirely. The data disparity hampers efforts to develop comprehensive strategies for controlling or mitigating pandemic threats (Budd et al. 2020). Finally, processes for evaluating mobility-based interventions could be further developed to meet the needs of policymakers. Evaluating interventions and predicting outcomes require a substantial multidisciplinary knowledge base and coordinated infrastructure. Of the studies reviewed, 82% are limited to descriptive analysis (Go and Wei 2024), which tends to be less actionable from a policymaking point of view. The impact of descriptive analysis and intervention evaluation is magnified when data standardization and sharing are prioritized (Alessandretti 2021; Oliver et al. 2020; The DELVE Initiative 2020; Kraemer et al. 2021).

In the post-pandemic era, we still face significant challenges in effectively using mobility data for public health strategies. Despite the public pressure that led to progress in mobility data collection and availability during the early days of the COVID-19 pandemic, integration into public health strategies remained fragmented due to data disparities, expertise bottlenecks, privacy concerns, standardization issues, and other logistical hurdles. Since the pandemic has declined in the public eye, this goal of the data-backed precise intervention slips farther from reach as data sources become less available. Further research must address the pivotal need for fine-grained, epidemiologically relevant mobility and interaction data collected in a manner that protects the privacy of the individuals represented in the datasets.

22.2 Goals of the MAPPING@Brown Study

With the MAPPING@Brown study, the MAPPS team piloted, at an unprecedented scale, a multimodal location data collection system within a congregate setting. MAPPING@Brown investigated the acceptability and viability of a secure, community-driven indoor positioning system to track the near-constant location of participating students, faculty, and staff at the Brown University School of Public Health (SPH) in Providence, Rhode Island. Our goal was to see whether and how we could use various modalities, including Bluetooth and Wi-Fi data, (1) to localize individuals in the Brown University SPH building and (2) to learn about the contact networks that arise among faculty, students, and staff as they move throughout the school, while maintaining buy-in from the community and protecting the privacy of participants. Later in this chapter, we discuss potential applications for the collected data, focusing on their use in network epidemiology. Network epidemiology combines the biology and statistics of epidemiology with the theory of complex networks to improve understanding of infectious disease dynamics (Keeling and Eames 2005; Morris 2004; Meyers 2006). This subfield models the effects of human behavior and social contact on contagious processes by treating inferred networks as a substrate for disease spread (Meyers 2006; Kenah and Robins 2007).

22.2.1 Study Design

Twelve months of preparatory work, including literature reviews, focus group discussions, and expert collaboration, informed the MAPPING@Brown study's design and implementation. We aimed to capitalize on the ubiquity of Wi-Fi networks and smartphones equipped with Bluetooth Low-Energy (BLE) radios without jeopardizing user privacy. To avoid gathering individual-level data on non-participants, we decided against using passively collected institutional data, such as university ID card swipe logs and university Wi-Fi logs, because Brown community members did not actively consent to such information being used in research. We set rigorous standards for informed consent by making information and resources accessible to potential participants through our website, printed materials, and community forums (Lurie et al. 2023a). We prioritized participant autonomy in the data collection process, allowing app users to enable and disable scanning with a simple toggle and request retroactive data erasure during the study period. The consent process, demographic survey, and app-based Bluetooth and Wi-Fi modules represent separate data streams, as Fig. 22.1 depicts. After collection, all data sources were stored as separate tables in a highly secure computing environment.

22.2.2 Methodology

Throughout fall 2022, MAPPS team members engaged with Brown students, faculty, and staff to assess the acceptability of a study like MAPPING@Brown on campus and to garner feedback about privacy and other concerns. Activities included presentations to select undergraduate and graduate classes, presentations and

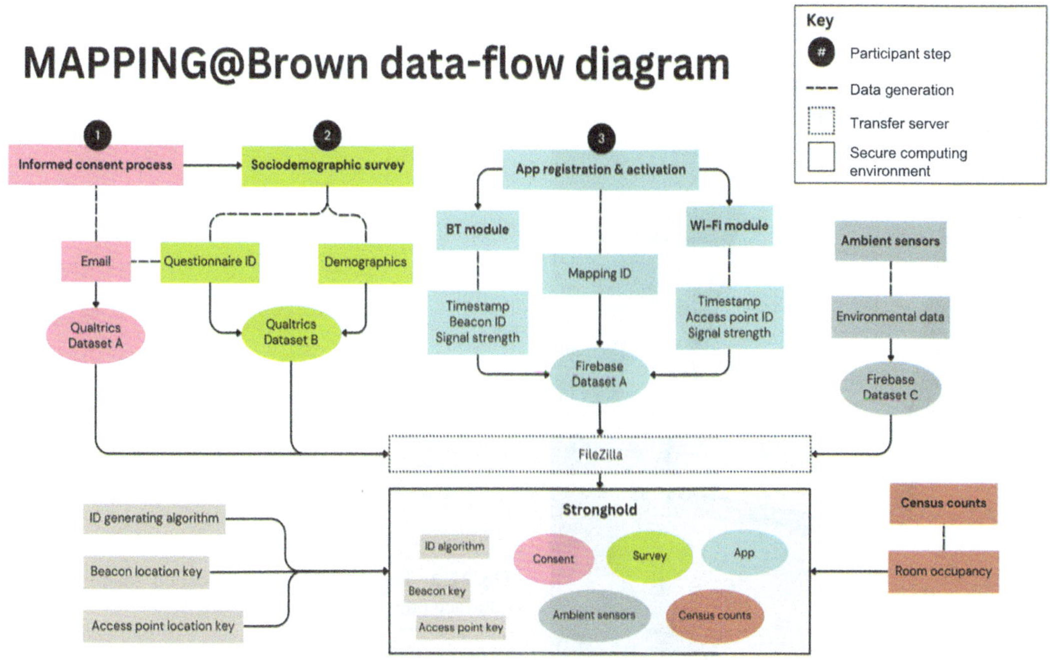

Fig. 22.1 Separation of the consent, demographic, app-based Bluetooth (BT) and Wi-Fi, environmental, and census data streams during collection and storage

discussions in faculty meetings, focus groups with students and staff, and deep engagement with the University and SPH leadership. We also established a community advisory board (CAB) of faculty, staff, and students focused on detailed discussions about privacy and study methodology. The CAB reviewed and suggested improvements to early prototypes of the app and provided insight and guidance on recruitment.

We developed the proprietary MAPPING@Brown smartphone application (see Fig. 22.2) with public health graduate students, upper-level engineering students, and the Brown Center for Computation and Visualization (CCV). Students in the Pandemics in Global Perspective and Designing Human-Centered Technology classes collaborated to develop the first prototype. The students worked with MAPPS Project staff to review previous indoor positioning and contact tracing work, design a proprietary app-based data collection tool, and present their proposal to research app developers from CCV, who would ultimately make the product a reality. Important prototype features retained in the final app included a simple and intuitive user interface, upfront presentation of privacy and ethics information, and easily accessible resources about the MAPPING@Brown study and data security.

The MAPPING@Brown app recorded Wi-Fi and Bluetooth signals within the SPH building. Equivalent apps were available to Android and iPhone users after completing the web-based consent process and sociodemographic survey. App-based data collection involved two systems: (1) Brown's existing network of Wi-Fi access points and (2) a dense network of Bluetooth beacons designed and installed by the research team. These beacons were small BLE transmitters that continuously broadcast a unique identifier to in-range smartphones, which recorded this information and the strength of the received signal. Beacon assembly and installation functioned as an educational and research exercise, with students outside of the core research team invited to help with large-scale assembly and deployment, thereby offering an opportunity to learn more about indoor positioning systems,

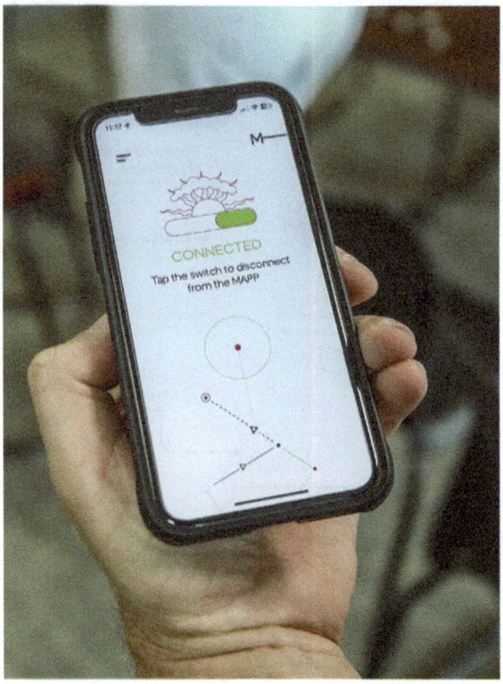

Fig. 22.2 MAPPING@Brown

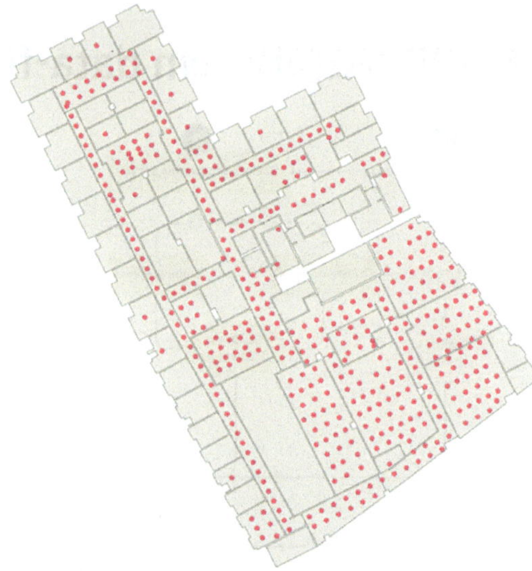

Fig. 22.3 Diagram of beacon location smartphone app interface on one floor of the Brown SPH

the MAPPING@Brown study, and the science of mobility data collection. With the help of several dozen volunteers, including undergraduate and graduate students, visiting researchers, and MAPPS co-investigators, we built and installed 2121 beacons in a grid pattern across the common spaces of eight floors (e.g., Fig. 22.3) in the SPH. Each beacon was suspended eighteen inches below the ceiling to decrease the distance to participants' phones.

When activated, the app scanned for Bluetooth beacons on the order of every 1–4 s (depending on the phone's operating system, battery life, and location settings) and recorded the beacon ID, Received Signal Strength Indicator (RSSI), and timestamp for each signal detected. Directly after a scan that detected BLE beacons, the app checked whether a participant's phone was connected to Brown Wi-Fi and, if so, recorded the access point ID number, Wi-Fi RSSI, and timestamp. Study participants were asked to activate data collection on their mobile phone whenever they entered SPH on weekdays during the two-week study period. During hours of peak foot traffic, MAPPS team members were available in the lobby to recruit participants, troubleshoot technical issues, and remind people of the ongoing study. Data collection was tied to the physical access point and beacon infrastructure at SPH. Thus, no data were collected once participants left the building and/or deactivated the app.

Finally, we collected environmental measures in key classroom locations during the study period. We measured carbon dioxide (CO_2) levels and temperature, which we plan to explore as potential proxy measures for room occupancy. These observations can be assessed against manual census counts taken periodically in the same locations.

22.3 Results

22.3.1 Study Acceptability

We received a largely positive response to the MAPPING@Brown project pitch in our eight focus group discussions (FGDs). Participants included eight staff members, three graduate students, and 20 undergraduate students—a total of 31 participants. When asked if they would

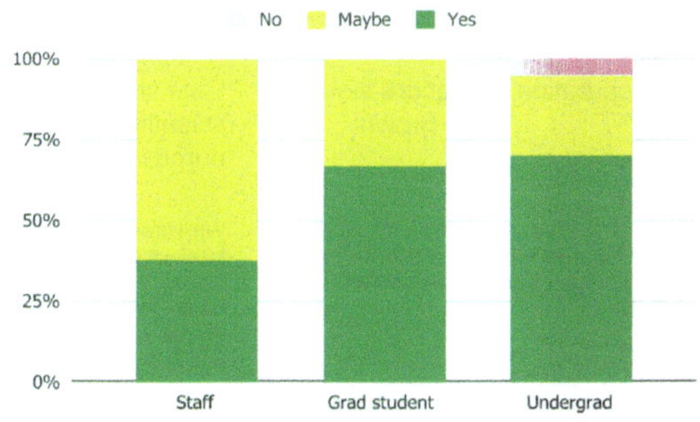

Fig. 22.4 Willingness to participate in MAPPING@Brown by role

enroll in a study like MAPPING@Brown, 19 participants said yes, 11 said they would need more details to decide, and one said no. Staff were least likely to commit, with 38% saying yes and 62% undecided. Graduate and undergraduate students responded affirmatively at similar rates (67% and 70% yes, respectively). Informal surveys of undergraduate and graduate students, taken during class visits with the goal of informing and recruiting potential study participants, yielded similar proportions. The one respondent who said they would not enroll was an undergraduate (see Fig. 22.4).

The most cited reason for agreeing to enroll was the belief that this research could benefit society. The intrinsic merit of public health research was mentioned by about half of staff and students. When asked if they had any concerns about participating in the study, 29% of respondents raised privacy concerns, and 19% raised concerns about phone functionality. Staff and graduate students were more likely than undergraduates to say they had concerns about reidentification risk and who would have access to their data. Regarding phone functionality, three staff and three undergraduate students said they were worried about the effect the study app might have on their phone's battery life and processing speed. Overall, staff and graduate students tended to be more cautious about data privacy, while most undergraduates reported being unconcerned about MAPPING@Brown jeopardizing their privacy. Five undergraduates across three FGDs expressed the view that their data were already "out in the world" and might as well be used for a "good cause" like pandemic prevention research (Luiz et al. 2024). Our findings suggest that undergraduates are more likely to hold strong views, both positive and negative, on the importance of data privacy.

22.3.2 Workshop on Privacy and Ethics in Pandemic Data Collection and Processing

In January 2023, MAPPS convened a week-long workshop on Privacy and Ethics in Pandemic Data Collection and Processing in collaboration with the Institute for Computational and Experimental Research in Mathematics (ICERM), an NSF-funded research institute at Brown University. Fifty experts and students from across the country attended. Through presentations, breakout sessions, and panel discussions, their contributions informed a detailed draft of the study methodology and a framework of ethical standards to guide the MAPPING@Brown study (Lurie et al. 2023b). The final set of privacy principles, shown in Fig. 22.5, is rooted in data ethics and reflects many of the provisions of the European Union's General Data Protection Regulation (GDPR).

22.3.3 Planned Data Analysis

The MAPPING@Brown study was recently completed at the time of writing in August 2024, and we are still in the early stages of data analysis and model development. We intend to use the data collected to hone methods for identifying minimally intrusive and maximally effective pandemic policies in local settings. Algorithms will be developed to calculate participants' spatial and temporal positions from the Wi-Fi and BLE signaling data. We hope this positional data will be the basis for detailed and flexible agent-based outbreak

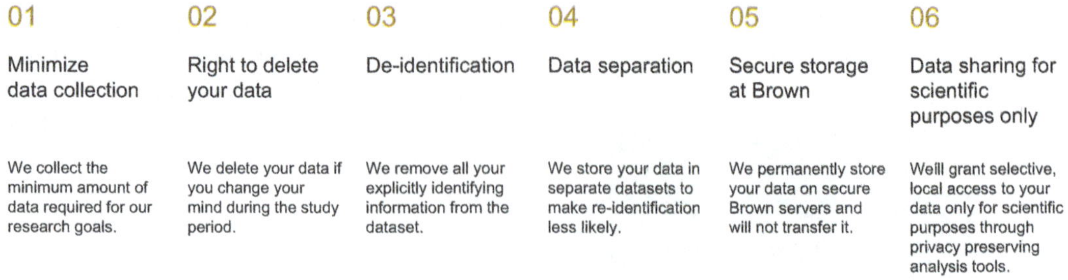

Fig. 22.5 Excerpt of a poster depicting the MAPPING@Brown study's privacy principles

modeling. Our envisioned modeling tool will have customizable network, environmental, and epidemiological parameters, allowing us to gain insights into social interaction patterns and disease spread dynamics under different conditions. For instance, model parametrization could simulate introducing a range of pathogen types, including those spread via direct human-to-human contact, like a respiratory virus transmitted via droplets; via indirect contact, like an airborne pathogen; or via fomites, like many bacterial and viral infections. Agent behavior and environmental conditions will be customizable to test the effect of outbreak interventions like room occupancy restrictions, mask mandates, or introducing Corsi-Rosenthal Boxes for air purification (Porto et al. 2022).

MAPPING@Brown will help us determine whether BLE and Wi-Fi are effective for capturing epidemiologically relevant networks and will prime us to consider these and other technologies that might be more efficient and/or accurate in the future. We are especially interested in developing data solutions for decision support in congregate settings, which tend to be densely populated and highly connected, making them hotspots for disease transmission and prime targets for intervention. The easiest and least expensive solution is to use existing infrastructure, typically Wi-Fi in our settings of interest. MAPPING@Brown and future MAPPS Project studies seek to determine when Wi-Fi is sufficient to inform decision-making and when Bluetooth is a necessary add-on. Similarly, we are interested in assessing the value added by environmental measures such as CO_2 and temperature readings.

The next steps for the MAPPS Project involve implementing data collection efforts like MAPPING@Brown in other settings. The framework developed in the MAPPING@Brown study lays the foundation for future campus-wide deployments of real-world mobility data collection studies. This framework could also be extended to other highly connected congregate settings, such as long-term care facilities (e.g., nursing homes and rehabilitation centers) or public transportation hubs (e.g., airports and bus stations). Through collaborations with researchers specializing in wearable technology, sensor physics, location management, information theory, signal processing, healthcare, database management, and other fields, data collection and analysis techniques stemming from the MAPPING@Brown study will continue to aid the development of high-resolution, privacy-preserving measurement of human movement and disease spread. The MAPPS Project aims to make decision support tools based on realistic modeling accessible to institutional leaders, empowering them to make informed public health policy choices in collaboration with their communities.

22.4 Recommendations for Pandemic Mitigation

Understanding social mixing and human mobility is crucial for identifying patterns of pathogen spread and designing effective mitigation strategies (Buckee et al. 2020). Effective outbreak preparedness hinges on detecting circulating pathogens, monitoring infection trends, forecasting spread, and swiftly coordinating public and private entities to implement population-level interventions, including measures to modify contact and mobility patterns. Leveraging high-resolution data analytics can enhance disease modeling and inform targeted interventions (Moghadas et al. 2009; Lal et al. 2022). Multi-modal data, in particular, can help us construct more accurate decision-making tools for congregate settings (Siddle et al. 2022; Petros et al. 2022; Dandekar et al. 2021).

22.4.1 Strengthening Predictive and Preventive Epidemiological Models

When developing policies to minimize exposure to contagion, public health and healthcare policymakers make assumptions regarding major intervention targets. The tools that we currently have at our disposal in 2024 include ones that target network mixing, the environment, or individual behavior. Interventions often aim to alter (1) the frequency, mix, proximity, and duration of human interactions, (2) the density and distribution of pathogens within spaces where this interaction occurs, and (3) the amount of pathogen deposited into the environment by infected individuals. Restricting travel, limiting the size of gatherings, cohorting, and prohibiting visits to healthcare facilities are examples of interventions that aim to slow disease transmission by altering network mixing, thus reducing the frequency and duration of exposure. Individual (e.g., vaccination, masking) and environmental (e.g., filtration, ventilation) interventions similarly seek to lower the disease burden by reducing the transmission probability.

Against the backdrop of infectious disease forecasting's growing popularity in recent years, the SARS-CoV-2 pandemic spurred widespread interest in the practical and theoretical applications of modeling (Ray and Reich 2018; Reich et al. 2019a, 2019b; Biggerstaff et al. 2018; Bannick et al. 2021). Numerous models forecasting hospitalizations and mortality (Cramer et al. 2022; Fox et al. 2022; Reich et al. 2022; Klein et al. 2023) and estimating the impacts of non-pharmaceutical interventions on disease incidence and death (Lorig et al. 2021; Zhang et al. 2023) have helped policymakers make difficult decisions about local and regional policies. However, most models failed to integrate high-resolution data on mobility, human interaction, and behavior, focusing instead on meso- and macro-level disease dynamics within cities, across continents, or globally (Moran et al. 2016; Achterberg et al. 2022; Shakeel et al. 2021; Wang et al. 2022). Uncertainty regarding basic epidemiological parameters, contact trends, and caseloads had the effect of complicating disease forecasting, simulation modeling, and cost–benefit analyses, sometimes resulting in policies that may not have optimally balanced trade-offs (Yan et al. 2021). These failures, both real and perceived, opened public health officials and researchers to criticism that they disregarded the economic, educational, and psychological impacts of stay-at-home orders and "lockdowns" (Ferguson et al. 2020; Fraser et al. 2004; Li et al. 2020).

Besides uncertainty around model parameters, model assumptions also play a key role in widening the gap between our current capabilities and those necessary for determining optimal interventions. Epidemiological analyses used for policy decision-making could benefit from integrating lower-level models (e.g., network and individual-based models) to complement or, in some cases, replace deterministic, averaged, and coarse-grained compartmental approaches (Garnett et al. 2011; Ferrante 2023). We posit that inferring contact structure is a viable and often favorable option since this approach can: (1) organically account for the uncertainty contained in mobility data; (2) identify key behavioral or infrastructural intervention points; and (3) integrate with other data streams on local context-dependent factors like ventilation, or traditional epidemiological data such as infection status, genomics or contact-tracing.

22.4.2 Prioritizing Congregate Settings in Pandemic Response

Congregate settings are densely populated, highly connected communities where outbreak risk is high—as is the potential for pandemic prevention impact. Unraveling transmission dynamics in congregate settings may be key to understanding and intervening on drivers of community transmission. First, significant superspreading events often occur in congregate settings, where close physical and social proximity facilitates a spike in cases. Identifying and limiting these events can prevent spillover into the broader community. Second, congregate settings may serve as microcosms of a larger population and provide appropriate environments for studying mobility, social mixing, and pathogen dissemination. Third, decision-makers in congregate settings often can revise policies and implement them more or less immediately, suggesting the potential for rapid response to emerging pathogens. During the exponential growth phase of an outbreak, pathogen spread usually outpaces the rollout of top-down interventions from governmental bodies like stay-at-home orders and lockdowns. Countering contagion at this stage requires rapid response and coordination of the type that institutions and facilities may be better equipped to achieve because of their small size. The observation that individuals and businesses voluntarily altered their behaviors in response to forecasts and epidemiologic data during the COVID-19 pandemic (Ferguson et al. 2020; Fraser et al. 2004; Li et al. 2020) suggests that bottom-up changes altering mobility and contact networks could reduce transmissions more quickly than large-scale interventions.

22.4.3 Future Directions

We suggest exploring college campuses and long-term care facilities (LTCFs) as grounds for pandemic mitigation research. College campuses and LTCFs often represent transmission hotspots during epidemic events (Petros et al. 2022; Ferguson et al. 2020). In particular, LTCFs like nursing homes and rehabilitation centers tend to experience high morbidity and mortality rates from infectious diseases due to preexisting health vulnerabilities (Blumberg et al. 2022; Fisman et al. 2020; Lau-Ng et al. 2020). For instance, in the U.S., more than half of all deaths during the first year of the SARS-CoV-2 pandemic occurred in LTCFs (Lau-Ng et al. 2020). Coupling modules governing detailed behavioral and environmental conditions can produce rich models capable of significantly improving our understanding of disease dynamics in these settings, and suggest ways to reduce transmission while minimizing disruption to everyday operations.

Ensuring that forecasting models and decisions that utilize them are robust in parameterization, assumptions, and data uncertainties is a necessary step of all modeling for public health when real decisions are on the line. Methods to protect against uncertainties include sensitivity analysis and identification of interventions that are consistently effective across a broad range of model assumptions and parameterizations (O'Gara et al. 2023). Yet, computational limitations and bias in uncertainty limit how effective these methods can be without the proper data and a framework for their integration into models. Network epidemiology provides a modeling framework that allows for the integration of high-resolution contact data to reduce uncertainty, especially when aided by Bayesian approaches to inferring network structure (Young et al. 2021), and in turn provides an opportunity to ground the study of network science more firmly in empirical evidence, which spurs advances in theory. We propose validating a network epidemiology framework to provide probabilistic forecasts grounded in evidence-based models of human behavior and disease transmission.

Integrating data-derived contact structures in epidemiological models offers significant advantages by addressing uncertainties in mobility data and epidemiological models. We can elucidate various contextual factors and construct more accurate forecasting models by pinpointing critical intervention points (e.g., in behavior or infrastructure) and combining various multimodal data sources. The outcome of these efforts will also offer researchers and decision-makers real-world data from which to plan and execute targeted interventions to help prevent or mitigate viral spread. Addressing these issues will require new technology, cross-domain and multi-level partnerships, and further collaborative motivation to contextualize human interactions and socialization patterns.

Acknowledgements We are grateful to Drs. Anna Lysyanskaya, Julia Netter, Sohini Ramachandran, Kimani Toussaint, and Thomas A. Trikalinos for their guidance and support as co-principal investigators on the MAPPS Project. We thank our fellow team members, Bradford Roarr and Galen Winsor from the Center for Computation and Visualization and Marta G. Wilson-Barthes from the International Health Institute, whose collaboration and expertise were essential to our work. We acknowledge the contributions of Nelson Anangwe, Lucas Blackmore, Julia Braeunig, Mallory Go, Zhijun Liu, Emily Ma, Maria Pieruccini, Yatharth Sharma, Xilin Wang, the members of our community advisory board, and the many volunteers who helped make the MAPPING@Brown study a success. We appreciate the support of the National Science Foundation and the Brown University School of Public Health, which made this research possible. Lastly, special thanks are due to our study participants for their time, effort, and enthusiasm around the topic of pandemic prevention.

References

Achterberg MA, Prasse B, Ma L et al (2022) Comparing the accuracy of several network-based COVID-19 prediction algorithms. Int J Forecast 38:489–504. https://doi.org/10.1016/j.ijforecast.2020.10.001

Adams-Prassl A, Boneva T, Golin M, Rauh C (2022) The impact of the coronavirus lockdown on mental health: evidence from the United States. Econ Policy 37:139–155. https://doi.org/10.1093/epolic/eiac002

Add health. https://addhealth.cpc.unc.edu/. Accessed 29 Jul 2024

Alessandretti L (2021) What human mobility data tell us about COVID-19 spread. Nat Rev Phys 4:12–13. https://doi.org/10.1038/s42254-021-00407-1

Allen DW (2022) Covid-19 lockdown cost/benefits: a critical assessment of the literature. Int J Econ Bus 29:1–32. https://doi.org/10.1080/13571516.2021.1976051

Althouse BM, Wallace B, Case BKM et al (2023) The unintended consequences of inconsistent closure policies and mobility restrictions during epidemics. BMC Glob Public Health 1:28. https://doi.org/10.1186/s44263-023-00028-z

Bannick MS, McGaughey M, Flaxman AD (2021) Ensemble modelling in descriptive epidemiology: burden of disease estimation. Int J Epidemiol 49:2065–2073. https://doi.org/10.1093/ije/dyz223

Bedson J, Skrip LA, Pedi D et al (2021) A review and agenda for integrated disease models including social and behavioural factors. Nat Hum Behav 5:834–846. https://doi.org/10.1038/s41562-021-01136-2

Bengio Y, Janda R, Yu YW et al (2020) The need for privacy with public digital contact tracing during the COVID-19 pandemic. Lancet Digit Health 2:e342–e344. https://doi.org/10.1016/S2589-7500(20)30133-3

Biggerstaff M, Johansson M, Alper D et al (2018) Results from the second year of a collaborative effort to forecast influenza seasons in the United States. Epidemics 24:26–33. https://doi.org/10.1016/j.epidem.2018.02.003

Blumberg S, Lu P, Kwan AT et al (2022) Modeling scenarios for mitigating outbreaks in congregate settings. PLoS Comput Biol 18:e1010308. https://doi.org/10.1371/journal.pcbi.1010308

Buckee CO, Balsari S, Chan J et al (2020) Aggregated mobility data could help fight COVID-19. Science 368:145–146. https://doi.org/10.1126/science.abb8021

Budd J, Miller BS, Manning EM et al (2020) Digital technologies in the public-health response to COVID-19. Nat Med 26:1183–1192. https://doi.org/10.1038/s41591-020-1011-4

Christakis NA, Fowler JH (2013) Social contagion theory: examining dynamic social networks and human behavior. Stat Med 32:556–577. https://doi.org/10.1002/sim.5408

Chung S-C, Marlow S, Tobias N et al (2021) Lessons from countries implementing find, test, trace, isolation and support policies in the rapid response of the COVID-19 pandemic: a systematic review. BMJ Open 11:e047832. https://doi.org/10.1136/bmjopen-2020-047832

Cramer EY, Huang Y, Wang Y et al (2022) The united states COVID-19 forecast hub dataset. Sci Data 9:462. https://doi.org/10.1038/s41597-022-01517-w

Dal Porto R, Kunz MN, Pistochini T et al (2022) Characterizing the performance of a do-it-yourself (DIY) box fan air filter. Aerosol Sci Technol 56:564–572. https://doi.org/10.1080/02786826.2022.2054674

Dandekar R, Henderson SG, Jansen HM et al (2021) Safe blues: the case for virtual safe virus spread in the long-term fight against epidemics. Patterns 2:100220. https://doi.org/10.1016/j.patter.2021.100220

Eagle N, Pentland A (2006) Reality mining: sensing complex social systems. Pers Ubiquitous Comput 10:255–268. https://doi.org/10.1007/s00779-005-0046-3

Fahey RA, Hino A (2020) COVID-19, digital privacy, and the social limits on data focused public health responses. Int J Inf Manage 55:102181. https://doi.org/10.1016/j.ijinfomgt.2020.102181

Ferguson N, Laydon D, Nedjati Gilani G et al (2020) Impact of non-pharmaceutical interventions (NPIs) to

reduce COVID19 mortality and healthcare demand. Imperial College COVID Response Team
Ferrante P (2023) Editorial: lessons and policy consequences of mathematical modeling in relation to ongoing pandemics. Front Public Health 11:1281493. https://doi.org/10.3389/fpubh.2023.1281493
Ferretti L, Wymant C, Kendall M et al (2020) Quantifying SARS-CoV-2 transmission suggests epidemic control with digital contact tracing. Science 368:eabb6936. https://doi.org/10.1126/science.abb6936
Firth JA, Hellewell J, Klepac P et al (2020) Using a real-world network to model localized COVID-19 control strategies. Nat Med 26:1616–1622. https://doi.org/10.1038/s41591-020-1036-8
Fisman DN, Bogoch I, Lapointe-Shaw L et al (2020) Risk factors associated with mortality among residents with coronavirus disease 2019 (COVID-19) in long-term care facilities in Ontario Canada. JAMA Network 3:e2015957. https://doi.org/10.1001/jamanetworkopen.2020.15957
Fox SJ, Lachmann M, Tec M et al (2022) Real-time pandemic surveillance using hospital admissions and mobility data. Proc Natl Acad Sci 119:e2111870119. https://doi.org/10.1073/pnas.2111870119
Fraser C, Riley S, Anderson RM, Ferguson NM (2004) Factors that make an infectious disease outbreak controllable. Proc Natl Acad Sci USA 101:6146–6151. https://doi.org/10.1073/pnas.0307506101
Garnett GP, Cousens S, Hallett TB et al (2011) Mathematical models in the evaluation of health programmes. Lancet 378:515–525. https://doi.org/10.1016/S0140-6736(10)61505-X
Gemmetto V, Barrat A, Cattuto C (2014) Mitigation of infectious disease at school: targeted class closure vs school closure. BMC Infect Dis 14:695. https://doi.org/10.1186/s12879-014-0695-9
Go M, Wei G (2024) Navigating mobility data landscape in the post-pandemic era
Halloran ME, Hudgens MG (2016) Dependent happenings: a recent methodological review. Curr Epidemiol Rep 3:297–305. https://doi.org/10.1007/s40471-016-0086-4
Ibrahim NK (2020) Epidemiologic surveillance for controlling Covid-19 pandemic: types, challenges and implications. J Infect Public Health 13:1630–1638. https://doi.org/10.1016/j.jiph.2020.07.019
Keeling MJ, Eames KTD (2005) Networks and epidemic models. J R Soc Interface 2:295–307. https://doi.org/10.1098/rsif.2005.0051
Kenah E, Robins JM (2007) Second look at the spread of epidemics on networks. Phys Rev E 76:036113. https://doi.org/10.1103/PhysRevE.76.036113
Klein B, Zenteno AC, Joseph D et al (2023) Forecasting hospital-level COVID-19 admissions using real-time mobility data. Commun Med 3:25. https://doi.org/10.1038/s43856-023-00253-5
Kraemer MUG, Yang C-H, Gutierrez B et al (2020) The effect of human mobility and control measures on the COVID-19 epidemic in China. Science 368:493–497. https://doi.org/10.1126/science.abb4218
Kraemer MUG, Scarpino SV, Marivate V et al (2021) Data curation during a pandemic and lessons learned from COVID-19. Nat Comput Sci 1:9–10. https://doi.org/10.1038/s43588-020-00015-6
Lal A, Ashworth HC, Dada S et al (2022) Optimizing pandemic preparedness and response through health information systems: lessons learned from Ebola to COVID-19. Disaster Med Public Health Prep 16:333–340. https://doi.org/10.1017/dmp.2020.361
Lau-Ng R, Caruso LB, Perls TT (2020) COVID-19 deaths in long-term care facilities: a critical piece of the pandemic puzzle. J Am Geriatr Soc 68:1895–1898. https://doi.org/10.1111/jgs.16669
Lee U, Kim A (2021) Benefits of mobile contact tracing on COVID-19: tracing capacity perspectives. Front Public Health 9:586615. https://doi.org/10.3389/fpubh.2021.586615
Li R, Pei S, Chen B et al (2020) Substantial undocumented infection facilitates the rapid dissemination of novel coronavirus (SARS-CoV-2). Science 368:489–493. https://doi.org/10.1126/science.abb3221
Lorig F, Johansson E, Davidsson P (2021) Agent-based social simulation of the Covid-19 pandemic: a systematic review. J Artif Soc Soc Simul 24:5. https://doi.org/10.18564/jasss.4601
Luiz P, Blackmore L, Go M et al (2024) MAPPING@Brown focus group discussions: response analysis and summary of findings
Lurie M (2023) Mobility analysis for pandemic prevention strategies (MAPPS). Columbia Academic Commons, Academic Unit: Northeast Big Data Innovation Hub, Series: COVID-19 Research Lightning Talks. https://doi.org/10.7916/7rb5-hz95
Lurie MN, Netter J, Luiz P (2023) MAPPS frequently asked questions. In: Mobility analysis for pandemic prevention strategies. https://www.mappsproject.com/faq. Accessed on 25 July 2024
Lurie M, Wilson-Barthes M, Luiz P, Gantenberg J (2023) Report of proceedings and recommendations from the workshop on privacy and ethics in pandemic data collection and processing
McClain C, Faverio M, Anderson M, Park E (2023) How Americans view data privacy. Pew Research Center, Internet, Science & Tech
Meyers L (2006) Contact network epidemiology: bond percolation applied to infectious disease prediction and control. Bull Am Math Soc 44:63–86. https://doi.org/10.1090/S0273-0979-06-01148-7
Moghadas SM, Pizzi NJ, Wu J, Yan P (2009) Managing public health crises: the role of models in pandemic preparedness. Influ Other Respir Viruses 3:75–79. https://doi.org/10.1111/j.1750-2659.2009.00081.x
Moran KR, Fairchild G, Generous N et al (2016) Epidemic forecasting is messier than weather forecasting: the role of human behavior and internet data streams in epidemic forecast. J Infect Dis 214:S404–S408. https://doi.org/10.1093/infdis/jiw375
Morris M (2004) Network epidemiology: a handbook for survey design and data collection, 1st edn. Oxford University, PressOxford

O'Gara D, Rosenblatt SF, Hébert-Dufresne L et al (2023) TRACE-omicron: policy counterfactuals to inform mitigation of COVID-19 spread in the United States. Adv Theory Simul 6. https://doi.org/10.1002/adts.202300147

Oh J, Lee H-Y, Khuong QL et al (2021) Mobility restrictions were associated with reductions in COVID-19 incidence early in the pandemic: evidence from a real-time evaluation in 34 countries. Sci Rep 11:13717. https://doi.org/10.1038/s41598-021-92766-z

Oliver N, Lepri B, Sterly H et al (2020) Mobile phone data for informing public health actions across the COVID-19 pandemic life cycle. Sci Adv 6:eabc0764. https://doi.org/10.1126/sciadv.abc0764

Onyeaka H, Anumudu CK, Al-Sharify ZT et al (2021) COVID-19 pandemic: a review of the global lockdown and its far-reaching effects. Sci Prog 104:368504211019854. https://doi.org/10.1177/00368504211019854

Park S, Choi GJ, Ko H (2020) Information technology-based tracing strategy in response to COVID-19 in South Korea-privacy controversies. JAMA 323:2129–2130. https://doi.org/10.1001/jama.2020.6602

Parsons TL, Worden L (2021) Assessing the risk of cascading COVID-19 outbreaks from prison-to-prison transfers. Epidemics 37:100532. https://doi.org/10.1016/j.epidem.2021.100532

Perra N (2021) Non-pharmaceutical interventions during the COVID-19 pandemic: a review. Phys Rep 913:1–52. https://doi.org/10.1016/j.physrep.2021.02.001

Petros BA, Paull JS, Tomkins-Tinch CH et al (2022) Multimodal surveillance of SARS-CoV-2 at a university enables development of a robust outbreak response framework. Med 3:883-900.e13. https://doi.org/10.1016/j.medj.2022.09.003

Piper K (2020) The devastating consequences of coronavirus lockdowns in poor countries. Vox

Ray EL, Reich NG (2018) Prediction of infectious disease epidemics via weighted density ensembles. PLoS Comput Biol 14:e1005910. https://doi.org/10.1371/journal.pcbi.1005910

Reich NG, Brooks LC, Fox SJ et al (2019) A collaborative multiyear, multimodel assessment of seasonal influenza forecasting in the United States. Proc Natl Acad Sci 116:3146–3154. https://doi.org/10.1073/pnas.1812594116

Reich NG, Lessler J, Funk S et al (2022) Collaborative hubs: making the most of predictive epidemic modeling. Am J Public Health 112:839–842. https://doi.org/10.2105/AJPH.2022.306831

Reich NG, McGowan CJ, Yamana TK, et al (2019) Accuracy of real-time multi-model ensemble forecasts for seasonal influenza in the U.S. PLOS Comput Biol 15:e1007486. https://doi.org/10.1371/journal.pcbi.1007486

Ribeiro-Navarrete S, Saura JR, Palacios-Marqués D (2021) Towards a new era of mass data collection: assessing pandemic surveillance technologies to preserve user privacy. Technol Forecast Soc Change 167:120681. https://doi.org/10.1016/j.techfore.2021.120681

Rosenblatt SF, Smith JA, Gauthier GR, Hébert-Dufresne L (2020) Immunization strategies in networks with missing data. PLoS Comput Biol 16:e1007897. https://doi.org/10.1371/journal.pcbi.1007897

Shakeel SM, Kumar NS, Madalli PP et al (2021) COVID-19 prediction models: a systematic literature review. Osong Public Health Res Perspect 12:215–229. https://doi.org/10.24171/j.phrp.2021.0100

Shalizi CR, Thomas AC (2011) Homophily and contagion are generically confounded in observational social network studies. Sociol Methods Res 40:211–239. https://doi.org/10.1177/0049124111404820

Siddle KJ, Krasilnikova LA, Moreno GK et al (2022) Transmission from vaccinated individuals in a large SARS-CoV-2 delta variant outbreak. Cell 185:485-492.e10. https://doi.org/10.1016/j.cell.2021.12.027

Smyth JM, Jones DR, Wen CKF et al (2021) Influence of ecological momentary assessment study design features on reported willingness to participate and perceptions of potential research studies: an experimental study. BMJ Open 11:e049154. https://doi.org/10.1136/bmjopen-2021-049154

Specht I, Sani K, Loftness BC et al (2022) Analyzing the impact of a real-life outbreak simulator on pandemic mitigation: an epidemiological modeling study. Patterns (n y) 3:100572. https://doi.org/10.1016/j.patter.2022.100572

Stehlé J, Voirin N, Barrat A et al (2011) High-resolution measurements of face-to face contact patterns in a primary school. PLoS ONE 6:e23176. https://doi.org/10.1371/journal.pone.0023176

Stopczynski A, Sekara V, Sapiezynski P et al (2014) Measuring large-scale social networks with high resolution. PLoS ONE 9:e95978. https://doi.org/10.1371/journal.pone.0095978

Tchetgen Tchetgen EJ, VanderWeele TJ (2012) On causal inference in the presence of interference. Stat Methods Med Res 21:55–75. https://doi.org/10.1177/0962280210386779

The DELVE Initiative (2020) Data readiness: lessons from an emergency. The Royal Society

Usherwood T, LaJoie Z, Srivastava V (2021) A model and predictions for COVID-19 considering population behavior and vaccination. Sci Rep 11:12051. https://doi.org/10.1038/s41598-021-91514-7

Wang P, Zheng X, Liu H (2022) Simulation and forecasting models of COVID-19 taking into account spatio-temporal dynamic characteristics: a review. Front Public Health 10.1033432. https://doi.org/10.3389/fpubh.2022.1033432

Wnuk A, Oleksy T, Maison D (2020) The acceptance of Covid-19 tracking technologies: the role of perceived threat, lack of control, and ideological beliefs. PLoS ONE 15:e0238973. https://doi.org/10.1371/journal.pone.0238973

Xu B, Gutierrez B, Mekaru S et al (2020) Epidemiological data from the COVID-19 outbreak, real-time case information. Sci Data 7:106. https://doi.org/10.1038/s41597-020-0448-0

Yan Y, Malik AA, Bayham J et al (2021) Measuring voluntary and policy-induced social distancing behavior during the COVID-19 pandemic. Proc Natl Acad Sci 118:e2008814118. https://doi.org/10.1073/pnas.2008814118

Young J-G, Cantwell GT, Newman MEJ (2021) Bayesian inference of network structure from unreliable data. J Complex Netw 8:cnaa046. https://doi.org/10.1093/comnet/cnaa046

Zhang B, Kreps S, McMurry N, McCain RM (2020) Americans' perceptions of privacy and surveillance in the COVID-19 pandemic. PLoS ONE 15:e0242652. https://doi.org/10.1371/journal.pone.0242652

Zhang W, Liu S, Osgood N et al (2023) Using simulation modelling and systems science to help contain COVID-19: a systematic review. Syst Res Behav Sci 40:207–234. https://doi.org/10.1002/sres.2897

Peyton Luiz is a Master of Public Health student concentrating in Causal Inference Epidemiology at Brown University School of Public Health and a graduate research assistant with BWell, Brown University's health promotion office. She previously earned her Bachelor of Science in Global Health at Georgetown University, where she was a Global Health Institute Fellow. Her recent work managing the Mobility Analysis for Pandemic Prevention Strategies (MAPPS) Project focuses on the ethical use of spatial data for outbreak decision-making. She is interested in applying similar techniques to study climate-related displacement and environmental health disparities.

Jason R. Gantenberg is a Research Scientist in the Center for Evidence Synthesis in Health and an Assistant Professor of Practice in the Department of Epidemiology at Brown University. His primary research interests include infectious disease transmission, computational modeling, causal inference, and applied machine learning. He is also interested in social/sexual networks and complex systems more generally. Dr. Gantenberg's Ph.D. dissertation focused on using agent-based models to study gonorrhea transmission among men who have sex with men. He has conducted epidemiologic research on respiratory syncytial virus, HIV/STI, and influenza. As a lead researcher on the Mobility Analysis for Pandemic Prevention Strategies (MAPPS) Project, Dr. Gantenberg implements Bluetooth and Wi-Fi-based positioning systems to study contact networks and disease propagation in indoor environments.

Samuel F. Rosenblatt brings expertise in network science, scientific computing, and modeling to a range of interdisciplinary projects from computational social science to agent-based contagion models. His Ph.D. dissertation presented network-based approaches to developing epidemic control interventions under constraints such as limited resources, missing data, and uncertainty. Sam is invested in scientific communication and outreach, serving as the Outreach Officer for the Society of Young Network Scientists (SYNS). In his free time, Sam is an avid podcast listener and enjoys board games and travel.

Dr. Guixing Wei serves as Senior Research Scientist within the Spatial Structures in the Social Sciences (S4) and the Population Studies and Training Center at Brown University. He also holds the position of Senior Research Scientist at Brown University Health (formerly Lifespan) and serves as the Associate Director of the Lifespan Cancer Institute Center for Performance Measurement. Dr. Wei earned his Ph.D. in Geographic Information Science (GIS) from Texas State University and brings over a decade of expertise in GIS and spatial analytics to his work. His research focuses on spatial epidemiology, with a particular emphasis on cancer epidemiology, health disparities, and human mobility and travel behavior modeling. Dr. Wei's work makes substantial contributions to understanding the complex spatial and social determinants of health outcomes, offering valuable insights for public health interventions and policy development.

Aditya Khanna (ORCID: 0000–0003-1049–3374) is an Associate Professor of Behavioral and Social Sciences at the Brown University School of Public Health. Dr. Khanna is a computational social scientist specializing in using network analysis, agent-based modeling and predictive analytics to inform policy initiatives to achieve health equity. He received his Ph.D. in Quantitative Ecology and an MS in Statistics from the University of Washington. He completed postdoctoral training in Global Health at the University of Washington and in Medicine (Infectious Diseases) at the University of Chicago. Dr. Khanna integrates data from diverse sources to create representations of public health systems which guide policy reform and the implementation of biobehavioral interventions. His research has mostly focused on the prevention of new HIV infections among sexual and gender minorities, particularly persons of color. More recently, he has expanded his portfolio to incorporate substance use behaviors, incarceration, breast cancer screening, and COVID-19.

Bryn C. Loftness (ORCID: 0000–0003-4597–0783) is the CEO and Co-Founder of Biobe. Currently, she is a Ph.D. candidate in the Complex Systems and Data Science program at University of Vermont and a National Science Foundation (NSF) Graduate Research Fellow. She works in the Neurobotics Lab and M-Sense Research Group at the University of Vermont Complex Systems Center developing machine learning and digital phenotyping algorithms. The data she works with are primarily biological and behavioral signals derived from on-body and off-body sensors, health, and/or contextual records. She also works remotely with the Sabeti Lab through the Broad Institute, specializing in behavioral phenotyping and digital epidemiology projects. In this role, she collaborates significantly with the Mobility Analysis for Pandemic Prevention Strategies (MAPPS) Center at Brown University. Bryn engages in a wide variety of research across her collaborations, with an overall vision for developing novel computational methods supporting the betterment of community, physical, and mental health across populations. Bryn most prominently contributes to evolving cutting edge research broadening the availability, objectivity, and precision of early childhood mental health screening across pre-adolescent populations.

Mark N. Lurie, Ph.D. (ORCID: 0000–0002-3153–1715) is a Professor of Epidemiology at the Brown University School of Public Health with an honorary appointment at the University of Cape Town, South Africa. A native South African, Dr. Lurie earned his master's degree in African History from the University of Florida and his Ph.D. from Johns Hopkins University School of Public Health. His areas of expertise include HIV/AIDS, STIs, silicosis, and tuberculosis in sub-Saharan Africa. He has studied the role of migration in the spread of HIV, examined the evidence for concurrency as a major driver of the HIV epidemic, and examined the impact of antiretroviral therapy on HIV epidemic dynamics in the region. Since the COVID-19 pandemic, Lurie has focused on developing novel methods for measuring human social interaction through his National Science Foundation (NSF)-funded Mobility Analysis for Pandemic Prevention Strategies (MAPPS) study. In addition, the MAPPS project aims to develop mathematical models of disease transmission that make maximum use of social interaction data to inform policy decisions in the face of new pathogens.

Democratizing Pedestrian Dynamics in Infection Modeling Applications

23

R. Goodenough, Y. Wu, R. Pahle, J. Coffey, S. Namilae, and A. Srinivasan

Abstract

Pedestrian dynamics simulates the movement of individuals in a crowd. It has powerful societal applications, such as in steering public health policy and the design of the built environment. It can suggest fine-tuned interventions that mitigate infection risk without disrupting human activities. Broad use of pedestrian dynamics in such applications is hindered by the need for contributions from domains where experts lack substantial computing background. We discuss our solution through cyberinfrastructure developments, which ameliorates this problem by providing (i) a domain-specific language (DSL) in which experts could provide input that gets translated to C++ code, (ii) a modular framework in which researchers could contribute models, (iii) a recommender system to enable users to determine suitable models, and (iv) a dynamically load balanced parallel parameter sweep engine to examine "what-if" scenarios. We then show results using pedestrian dynamics, which suggest effective public health interventions for air travel that can substantially reduce infection risk without reducing passenger traffic. The significance of this chapter lies in describing a solution

Research funded by

NSF Award #1931511—Collaborative:Elements:RUI*:Cyberinfrastructure for Pedestrian Dynamics-Based Analysis of Infection Propagation Through Air Travel;
NSF Award #1931483—Collaborative:Elements:Cyberinfrastructure for Pedestrian Dynamics-Based Analysis of Infection Propagation Through Air Travel;

NIH Award 1R15LM013382-01A1—A data analytics framework for the application of pedestrian dynamics to public health.

*Research in Undergraduate Institutions

COVID Information Commons Video Presentation

The COVID Information Commons (CIC) Research Lightning Talk video presentation and individual multilingual transcripts for this project can be found in the Columbia Academic Commons (AC) (Srinivasan and Namilae 2020).

R. Goodenough (✉) · J. Coffey · A. Srinivasan
University of West Florida, Pensacola, FL, USA
e-mail: rtg13@students.uwf.edu

J. Coffey
e-mail: jcoffey@uwf.edu

A. Srinivasan
e-mail: asrinivasan@uwf.edu

Y. Wu · S. Namilae
Embry-Riddle Aeronautical University, Daytona Beach, FL, USA
e-mail: WUY3@my.erau.edu

S. Namilae
e-mail: namilaes@erau.edu

R. Pahle
New York University, New York, NY, USA
e-mail: pahle@nyu.edu

F. D. Hudson (ed.), *The COVID Information Commons*, Sustainable Development Goals Series,
https://doi.org/10.1007/978-3-032-06125-6_23

to enable interdisciplinary contributions to complex pedestrian dynamics applications in public health applications.

Keywords

Pedestrian dynamics · Domain-specific language · Recommender system · Infection modeling · COVID-19 · High performance computing

23.1 Goals of the Research

Pedestrian dynamics is a mathematical modeling approach to simulate the trajectories of individuals in a crowd (Helbing et al. 2000). It has a variety of applications of societal importance. For example, it is used to identify interventions to reduce infection risk, as local movement and interaction patterns of individuals congregated in public locations drive the spread of proximity-based diseases, such as COVID-19 (Islam et al. 2021). It is also used to design the built environment for efficient and safe movement of people (Wei-Guo et al. 2006).

A large class of pedestrian dynamics simulations use social force models. These pose human movement as an ordinary differential equation (ODE) and simulate the trajectory of each human using a differential equation solver. Such models are the focus of this work, although the solution proposed here applies to other types of models too. The movement of pedestrians in real-world situations also depends on human behavior that is not captured by social force models.

A challenge here is that simulations in decision-making contexts require input on human behavior from experts in domains where computational expertise is lacking. For example, behavioral scientists may provide insights into aspects of human behavior that are not captured by the pedestrian dynamics model, such as airplane passengers yielding right of way to those in rows ahead of them while disembarking. Public health decision-makers may wish to input different policy choices whose impacts need to be examined, as illustrated in Fig. 23.1. Such users would also use the tools without expertise in all domains that contribute to the model. There is, therefore, a need for a cyberinfrastructure that enables interdisciplinary contributions and use of pedestrian dynamics for infection spread modeling. While ease of use is an important factor, so is computational efficiency, for the following reasons: (1) the movement of pedestrians depends on human behavior, which is inherently variable. In order to use it for prescriptive purposes, it is essential to simulate a variety of possible scenarios, ideally with a variety of models, so that decisions are robust to inherent variations in human behavior. This is computationally intensive and high performance computing (HPC) resources are required (Chunduri et al. 2018); and (2) the descriptions of human behavior go into the inner loops of the simulation code, which makes their computational performance critical.

Viral Infection Propagation during Air travel (VIPRA) cyberinfrastructure (VIPRA) aims to achieve this, thus promising a transformative impact on democratizing the use of pedestrian dynamics in applications critical for public health. In particular, it provides a domain-specific language (DSL) for human behavior in which experts could provide input that gets translated to C++ code, a modular framework in which researchers could contribute models, a recommender system to enable users to determine suitable models and model parameters, and a dynamically load balanced parallel parameter sweep engine to examine "what-if" scenarios for different possible human responses to interventions (Islam et al. 2023).

We describe this cyberinfrastructure (CI) below, with a discussion of the balance between ease of use and performance. Important components of the CI are shown in Fig. 23.2. We also show results that identify suitable interventions to reduce infection risk in air travel.

23.1.1 Pedestrian Dynamics Simulations

Social force models are a popular choice for pedestrian dynamics and capture many emergent behaviors exhibited by crowds. A simplified view

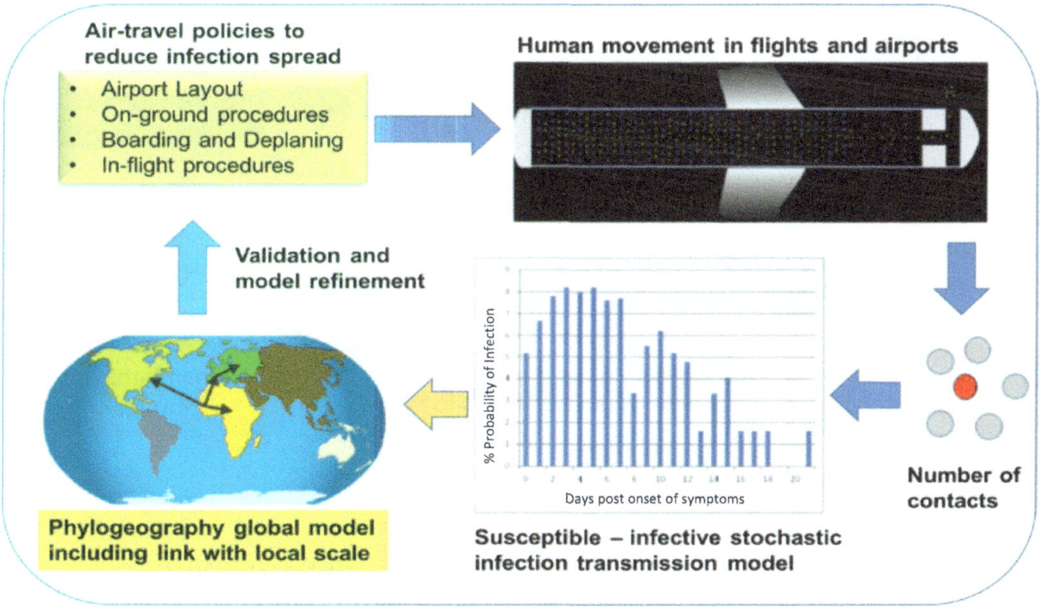

Fig. 23.1 Example workflow for using pedestrian dynamics to identify interventions to reduce infection spread

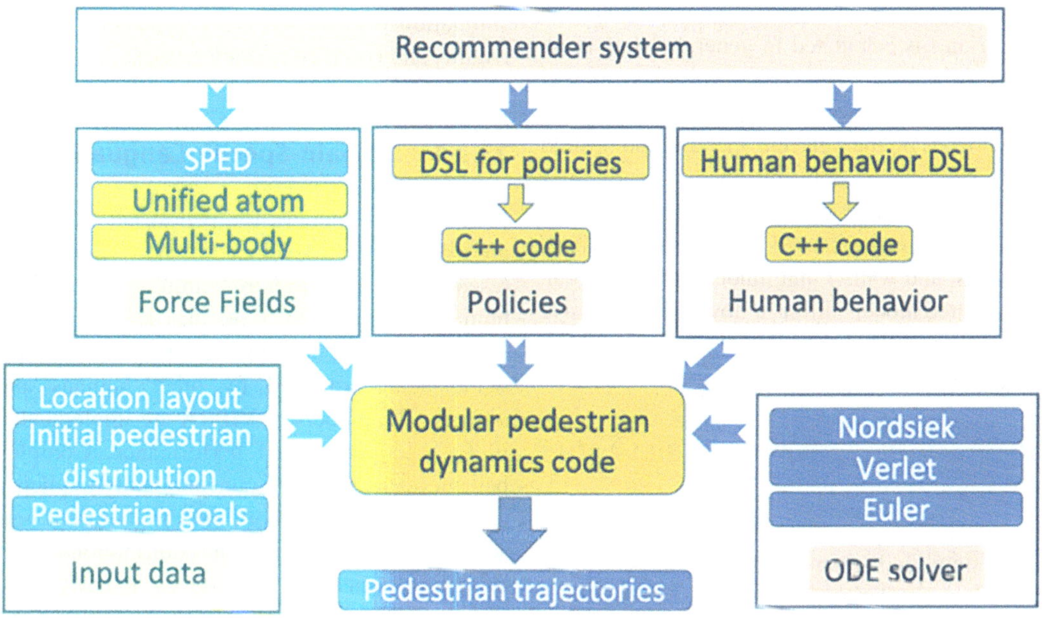

Fig. 23.2 VIPRA cyberinfrastructure framework

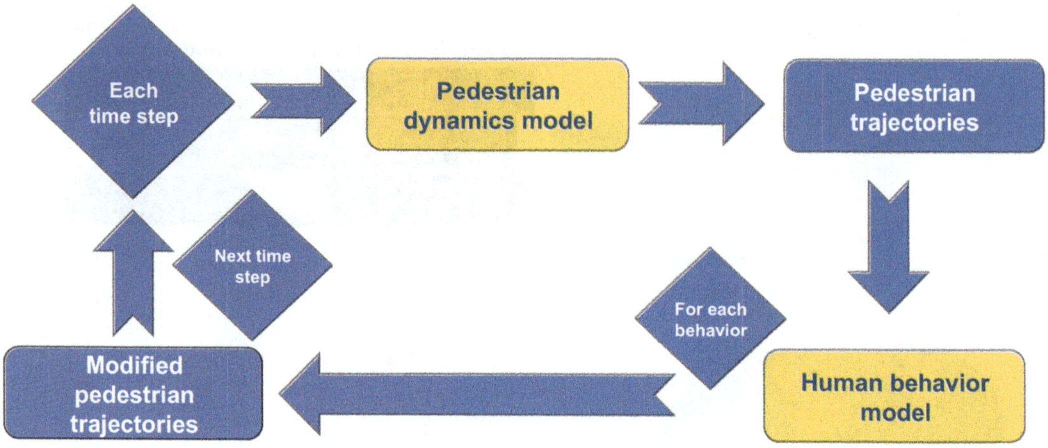

Fig. 23.3 Simplified view of pedestrian dynamics simulations

of pedestrian dynamics simulations is shown in Fig. 23.3. These models borrow their idea from molecular dynamics; each pedestrian is modeled as a particle, with internal and external forces that act upon it.

Equation 23.1 shows how the force f_i on each pedestrian i is calculated in general (Islam et al. 2021). At time t, a pedestrian with mass m_i and reaction time τ_i experiences a propulsive force that depends on their current speed v_i and desired speed v_{0i}, which motivates them toward their goal. This propulsive force is counteracted by repulsive forces f_{ij} caused by other pedestrians, obstacles, and walls j that hinder i from its goal. Social force models differ in how these repulsive forces are modeled. These are typically sparse, being non-zero only in the proximity of i.

$$f_i = m_i(d\boldsymbol{v}_i/dt) = (m_i/\tau_i)(\boldsymbol{v}_{0i}(t) - \boldsymbol{v}_i(t)) + \sum_{i \neq j} f_{ij} \tag{23.1}$$

A simplified pseudocode for a pedestrian dynamics simulation is provided in Algorithm 23.1 (see Sect. 23.2.1), which omits details for clarity. Our actual implementation also includes additional features, such as path finding, collision detection, and its resolution, which are omitted for ease of exposition. This is further explained in Sect. 23.2 Cyberinfrastructure Components.

23.2 Cyberinfrastructure Components

The cyberinfrastructure includes four key components: (i) domain-specific language, (ii) modular simulation framework, (iii) parameter sweep engine, and (iv) recommender engine.

23.2.1 Domain-Specific Language

Mathematical models for pedestrian dynamics are typically developed by engineering researchers and capture simple features of human behavior, such as the desire of individuals to move toward their respective goals, while being impeded by other humans and objects. The models in VIPRA are implemented in C++ to yield good performance. Complex behavioral features are conventionally explicitly coded. This impedes both the simulation of new scenarios and inclusion of contributions from behavioral scientists and policy experts, who may lack programming expertise. Furthermore, such behavior is used in the inner loop of the simulations. Therefore, coarse-scale tools for integrating components are inadequate to deliver good performance.

The VIPRA Behaviors DSL allows contributors to specify human behaviors for use in pedestrian dynamics, without requiring an understanding of programming or pedestrian dynamics models. Behaviors are written in a descriptive, YAML-like language. Behaviors are broken into components, namely: Conditions, Actions, Events, Locations, and Selectors, summarized as follows.

- Conditions are used to determine if a behavior should be applied to a pedestrian.
- Actions are used to modify the pedestrian's state, such as changing their velocity or position. Actions can be triggered by a wide range of factors, such as other pedestrians' proximity and state, internal state, location, and events occurring.
- Events are used to trigger behaviors, such as a pedestrian being injured, a fire starting, or an announcement.
- Locations are used to give context to behaviors, because some behaviors are only applicable in certain locations. For example, a pedestrian in a plane would stop to grab their luggage when they are in the aisle, and not when they are in a window seat.
- Selectors are used to choose which pedestrians exhibit a given behavior, thus capturing the behavioral heterogeneity of human populations.

These components are then combined to describe a behavior.

Figure 23.4 is an example illustrating the action for a behavior named "Luggage_Grabber," which specifies an airplane passenger's action in taking luggage from an overhead bin. It is invoked when a passenger enters the "Aisle" region, causing the pedestrian to stop for 10–20 s. Some behaviors are already included in the VIPRA human behavior model library, such as grabbing luggage when disembarking a vehicle, running from a fire, fighting, and helping injured pedestrians walk. Simple behaviors can also be composed into more complex ones.

The behaviors are interpreted, with each of the behavior components mentioned above being implemented as a C++ *functor*, that is, a callable object. This approach was chosen to allow for

```
Action ( Luggage_grabber ):
Condition : enter @Aisle
Response : set velocity to {0}
Duration : random 10 – 20 seconds .
```

Fig. 23.4 Example luggage grabber DSL snippet

researchers to quickly iterate on their behaviors without recompiling code. Behaviors are specified by their contributors in Behavior Files.

During simulation initialization, Behavior Files are parsed by ANTLR (ANTLR) and produce an abstract syntax tree (AST) loaded into a "Behavior Builder." This builder constructs Behavior objects that contain and manage these component functors (Fig. 23.5). These runtime objects' structures match the grammar almost 1-to-1. For example, Actions hold their Conditions; these Conditions are evaluated and fire the action when they are met, then call the Response object. This can be seen as a very coarse threaded code interpreter (Bell 1973).

After construction, each behavior is orchestrated by the behavior model and used in line 2.c.i of Algorithm 23.1. This provides the flexibility of letting users provide new models, while not requiring any hard coding of changes to behaviors. Most of the overhead occurs by adding it to the initialization step, Line 1 of Algorithm 23.1, thus reducing the overhead in loop 2.

23.3 Algorithm 1: Pedestrian Dynamics Simulation

1. Initialization with input: Pedestrian initial states, pedestrian goals, fixed objects' positions
2. While t ≤ maximum time *or* a terminating event has not happened
   a. for each pedestrian *i*
      i. Compute f_i using Eq. 23.1
   b. for each pedestrian $_i$

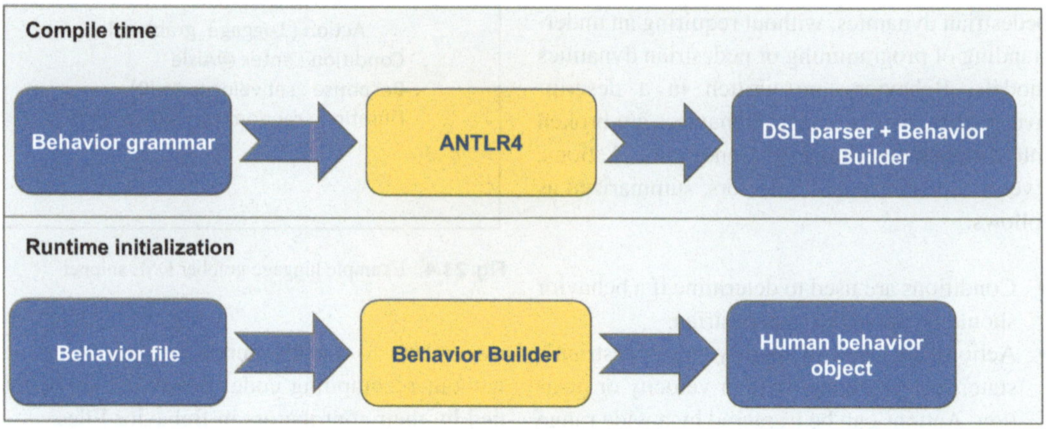

Fig. 23.5 Human behavior DSL initialization

   i. Compute acceleration $a_i = f_i/m_i$
   ii. Numerically integrate a_i to obtain v_i and position x_i
c. for each pedestrian $_i$
   i. Update v_i and position x_i based on human behavior (which may override the social force computation, and may depend on human response to interventions)
d. Output $x_i(t)$ and $v_i(t)$ for each pedestrian i
e. Increment t

23.3.1 Modular Simulation Framework

The VIPRA Cyberinfrastructure Framework, illustrated in Fig. 23.2, is written in C++ to ensure high performance (Driesen and Hölzle 1996). The framework is modular, and different types of pedestrian dynamics models can be contributed by pedestrian dynamics experts for use in Algorithm 23.1 line 2.a.1 by implementing the desired functional forms for the f_{ij} term of Eq. 23.1. Different types of human behaviors can be used in Algorithm 23.1 line 2.c.i either by directly coding it in C++ or specifying it using the DSL described above. Algorithm 23.1 line 2.b.ii is a differential equation solver that currently uses the Euler method. But this can be replaced with higher order schemes. Functionality is also provided for common needs, such as pathfinding, nearest neighbor computation, and collision avoidance (Sadeghi Lahijani et al. 2020).

The simulation configuration, module parameters, and simulation output are, by default, in a JSON format. This enables coarse-scale interfacing with other models, such as infection spread models, through a RESTful API. An example of this is (Chainbuilder), a web application that allows for the creation of workflow pipelines for use in decision-making.

23.3.2 Parameter Sweep Engine

Functionality is provided to enable parameter sweep over the model parameters to generate many possible scenarios to evaluate. For example, variability in human behavior is modeled by representing sources of uncertainty as model parameters. A parameter sweep then generates a variety of scenarios. Interventions that are effective under the most possible scenarios can be considered robust (VIPRA). Users can provide ranges for the parameters, and a parameter sweep engine then generates parameters within these ranges. We include two options, (a) low discrepancy and (b) pseudorandom sweeps. These have different theoretical and empirical tradeoffs related to ensuring good load balance, efficient coverage of the parameter space, and convergence checks (VIPRA). The engine then runs simulations of the parameter sweep in parallel, with each simulation itself being sequential. Load imbalance typically arises, which is handled automatically through a supervisor-worker based greedy dynamic load balancing implementation which assigns a task to the next available core. Functionality is provided for users to also change models without compilation, which can help with multi-model ensemble simulations.

23.3.3 Recommender System

The VIPRA Recommender System (VRS) (Coffey et al. 2024) aims to enable (i) users to identify useful models given the computational resources available to them and the application scenario they intend to simulate, and (ii) contributors to provide a broad set of information to help users select the right models among the ones recommended by VRS. VRS is currently capable of providing assistance with identification of: (i) local pedestrian dynamics modeling capabilities in a variety of enclosed spaces, (ii) epidemiological modeling capabilities of disease spread on the basis of pedestrian motion, (iii) global simulators of disease spread, and (iv) resources and references that elaborate modeling capabilities.

The goal of the development of VRS was to combine concepts from knowledge sharing (White and Lutters 2007; Alotaibi et al. 2013; Chen 2021) with ideas pertaining to recommender systems (Elsweiler et al. 2016) to broaden and democratize use of specialized modeling tools. Development to date has produced a hierarchically organized prototype system organized with an overall view of the capabilities at the top and affordances to tunnel down through global, local to global, and local modeling capabilities. At the bottom of the hierarchy are the available individual modeling tools.

Users who are new to modeling can utilize a Quickstart capability where they can view modeling scenarios currently included in VRS. At the global modeling level (at the country or intercontinental level), they currently can choose between two capabilities that model disease spread. At the local level, they can choose from alternatives including airplane boarding and deplaning, movement in concourses, movement in security lines, general group movement in a corridor, and others. Epidemiological models can be superimposed over many of the local pedestrian modeling capabilities.

The current system provides editing capabilities to those granted the privilege. Edits include adding a new individual model, a new set of parameters with recommended values for a particular application of a model, adding resources such as scholarly or white papers pertaining to a model, contact information for model developers, etc. Edit capabilities include the ability to modify or delete any component. Figure 23.6 shows snapshots of the VRS website.

The current system has undergone a preliminary usability study of basic ability to use VRS. The goal of the study was to determine how readily an end user could navigate around VRS, find a relevant model for a simulation, find resources elaborating the workings of a model, and to find contact information for a model's creator(s). Results of the preliminary study guided modifications to enhance usability. The usability of the editing capabilities has not yet been studied because current initiatives are toward the goal of semi-automating new model inclusion. Additional work will likely bring important changes to the current manual entry of new model information.

23.4 Performance Results

To test the scaling of the parameter sweep engine, a parameter sweep was performed on CPU-only nodes of the Perlmutter supercomputer at the National Energy Research Scientific Computing Center (NERSC) in Berkeley, California. The code was compiled on the GNU g++ compiler with −O3 and interprocedural optimizations. Parallel execution was performed through Open MPI 5.0.0.

The disembarkation of an Airbus A320 airliner with 144 passengers was simulated using the Constrained Linear Movement Model (CALM) (Sadeghi Lahijani et al. 2020), with behaviors modeling passengers grabbing their luggage and waiting for those in front of them

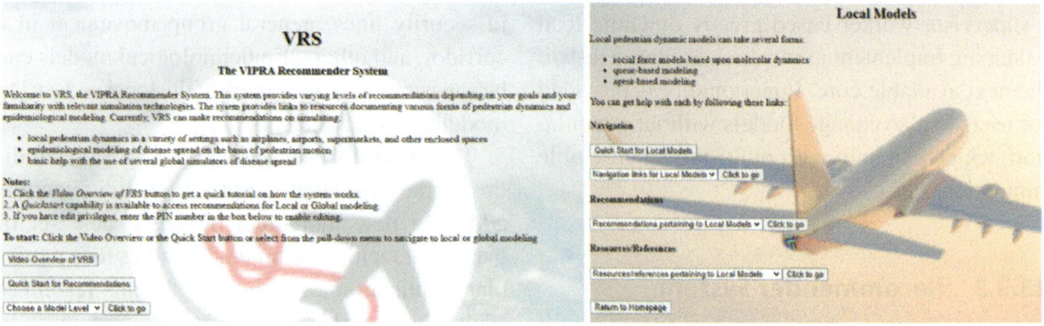

Fig. 23.6 Snapshots of VRS system

Fig. 23.7 Weak scaling results

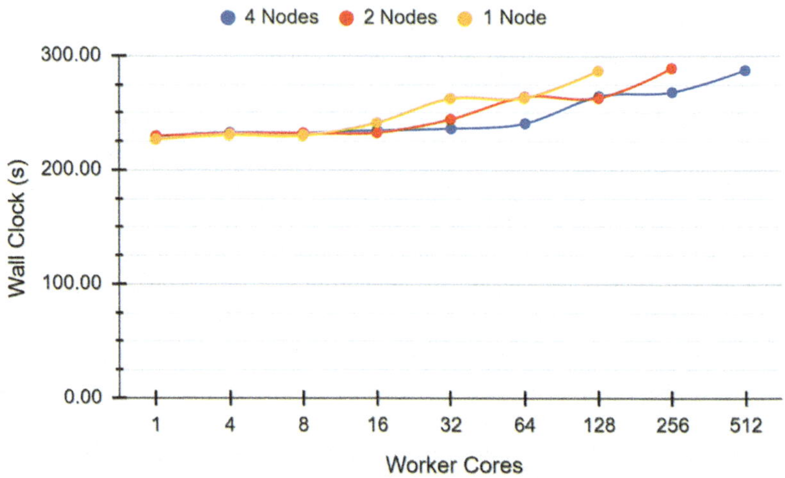

to leave their seats before they proceed down the aisle. The parameter sweep explored average pedestrian shoulder lengths, masses, reaction times, and desired speeds. The weak scaling results are plotted in Fig. 23.7 with 20 simulations per core. (Note that production runs would require thousands to hundreds of thousands of simulations.) Compute runs were repeated five times and the lowest time is reported to eliminate the impact of noise due to causes such as operating system (OS) jitter. While weak scaling is reasonably good, we can see that the time is not constant. This does not arise from communication or Input/Output (IO) overheads, which are less than 1%. The load is fairly well balanced, and is not the main cause of the performance degradation. Memory contention is the primary cause of loss in performance. The Perlmutter CPU-only nodes hold two AMD EPYC 7763 CPUs, each with 64 physical cores. Each CPU comprises four Non-Uniform Memory Access (NUMA) domains of 16 cores each. When the number of cores used in a NUMA domain increases, there is loss in performance. For example, there is a 10% loss with 15 processors and another 10% for 16. This problem can be ameliorated using more nodes with fewer cores used per node, as seen in Fig. 23.7.

23.4.1 Application to Infection Propagation

Pedestrian Dynamics models can be directly utilized for analyzing the contacts between pedestrians, which can indicate the potential for infection spread. The contact minutes in a given trajectory set can be used to evaluate the policies and procedures that reduce infection spread. We performed parameter sweeps to evaluate the contact minutes in different boarding procedures, particularly those introduced newly in response to the COVID-19 pandemic (Islam et al. 2021). Our results indicated that the new boarding procedures, such as back-to-front, further worsened infection risk by increasing social proximity, due to the increased assembling of travelers in aisles. Boarding the window seat before the aisle seat, and reducing luggage stowing time can ameliorate this. While such insights can be garnered by using pedestrian dynamics alone, combining pedestrian movement models with infection spread models can be used to model past infection spread events and suggest strategies for reducing infection spread.

We utilized the pedestrian dynamics models in conjunction with infectious disease spread models to analyze the spread of COVID-19 during air travel (Namilae et al. 2017, 2022). These models leveraged the results of pedestrian dynamics that capture population mixing behaviors and are a function of the environment's layout, the behavioral preferences of people, and proxemic behavior of walking groups, which determine an individual's path of movement. Pedestrian dynamics provides the trajectories of people by outputting the position of each person every Δt seconds. If we know the positions of infective and susceptible persons at a given time, then we can estimate the virus dose that each susceptible person is exposed to in a small time interval using dose–response models. The combined model can be used to simulate past infection spread events and evaluate procedures that reduce infection spread.

The dose–response infection spread model (Namilae et al. 2022) consists of the following two components: (i) quantification of the amount of pathogens ingested (infection dose) by a susceptible individual while in close vicinity of an infectious individual, with the dose monotonically decreasing with distance, and (ii) computation of the differential risk of successful transmission during a certain event using infection dose. The primary mechanisms captured in the model include the duration and intensity of the effective contact with an infectious individual and the temporal distribution of distance between individuals, which are obtained from passenger seating and movement during air travel.

We sum the virus dose over all time intervals to find a normalized, unitless measure V_n of total dose received by the nth susceptible person over the duration of the simulation. There are numerous dose–response relationships available to estimate infection probability from the dose (Smieszek 2009; Brouwer et al. 2017). The commonly used exponential model, beta-Poisson model, and Weibull model yield the infection probability P_n for the nth susceptible person as given in Eqs. 23.2–23.4, respectively, where b_1, b_2, q_1, and q_2 are statistical parameters describing the infectivity of the pathogen. P_n is summed over all passengers to yield the expected number of infections over the flight.

$$P_n = 1 - \exp(-V_n) \tag{23.2}$$

$$P_n = 1 - \left(1 + \frac{V_n}{b_2}\right)^{b_2} \tag{23.3}$$

$$P_n = 1 - \exp(-q_1 V_n^{q_2}) \tag{23.4}$$

The viral load will decay with distance, and we assume a threshold d_0 beyond which the viral load is zero. We note that the viral load decreases with distance and that in a short time interval, the exposure is proportional to the time of exposure. We incorporate these insights into a functional form that gives the dose $V_{n,t}$ on susceptible person n

from M infective persons in the tth time step as shown in Eq. 23.5:

$$V_{n,t} = k.\Delta t. \sum_{m=1}^{M} \left(1 - \frac{d_{n,m}}{d_0}\right)^{\alpha}, d_{n,m} < d_0 \tag{23.5}$$

Here, $d_{n,m}$ is the distance between the nth and m^{th} passengers at time t. V_n is obtained by summing $V_{n,t}$ over all the time steps. The parameter κ is a measure of the dose a person is exposed to per unit time while α controls how quickly the virus concentration drops with distance. The model parameters κ, α, d_0, q_1, and q_2 are unknown. We fit these parameters in the above model based on a superspreading event in a restaurant in Guangzhou, China early in the COVID-19 pandemic (Lu et al. 2020). This situation involves no movement, which makes parameter estimation easier than it would be otherwise. The obtained values for these parameters, along with fitting errors (within a threshold of 10%), are listed in Table 23.1. Details of the infection context and the corresponding parameterization are described in (Namilae et al. 2022) and (Wu et al. 2024). To compare the dose levels quantified using three models, we normalized the unit dose (per unit time) to the exponential model, as listed in Table 23.1. We observed that the beta-Poisson model quantified a lower level of dose in parameterization compared to the exponential model, whereas the Weibull model obtained a significantly higher level of dose.

Each model encapsulates different assumptions on the infection transmission mechanism. In a new infection outbreak, the exact mechanism is often unknown. Consequently, in the context of prescriptive modeling, it is critical to generate a variety of scenarios, using different models, so that the interventions selected are robust to the uncertainty in the underlying mechanism. Due to model simplicity concerns, we use the exponential model in the validation of our approach, as it does not involve any statistical parameters. In subsequent comparison of dose–response model applicability and for mitigation policy recommendations, all three models are utilized in a hypothetical analysis of infection spread in a generic two-aisle aircraft configuration.

Masks function by reducing viral emission near the infected individual and by decreasing the effective distance of viral transmission. To model these effects, we adjusted parameters κ and d_0. N95/FFP2 masks, with approximately 97% filtration efficiency, substantially lower viral emission (κ) compared to unmasked conditions (Srinivasan et al. 2020). For cloth masks, we applied a 50% filtration rate (Clapp et al. 2021). Additionally, mask use reduces the distance traveled by respiratory droplets, which we model by setting d_0 to 1.7 m for all mask types (Dbouk and Drikakis 2020).

We studied infection spread in two flights that had detailed information on in-flight COVID-19 seating arrangements (Fig. 23.8). A London flight (Fig. 23.8a) is used to study the impact of passenger movement on infection risk when masks are not used (Khanh et al. 2020; Chen et al. 2020). The ten-hour flight had 16 crewmembers and 201 passengers onboard. One infected passenger, referred to as an index passenger, was located in the business class, resulting in 11 secondary infections in the first-class and two in

Table 23.1 Model parameters obtained from a superspreading event

Model parameterization	Exponential model	Beta-poisson model	Weibull model
α	2.5	1.6	0.53
β	0.15	1.45	1.5
r_{thr} (meter)	3.5	3.5	3.5
Infectivity parameters		$b_1 = 2.3$ $b_2 = 52.3$	$q_1 = 0.0016$ $q_2 = 1.487$
Parameterization error (%)	7.5	2.5	3.3
Dose per min	1	0.770	3.155

the economy cabin. Mask usage was not mandated on this flight, and its use was only sporadic. A Singapore flight (Fig. 23.8b) is used to examine the impact of mask wearing (Yang et al. 2020; Github project site). The total flight duration was five hours, with 321 passengers on board. Mask usage was mandatory for this flight. All infected cases were wearing the masks, although the mask type is not known (Yang et al. 2020; Github project site). There were two infected cases on the plane, sitting far apart. An additional 14 passengers tested positive within the 14 days quarantine. Some of these passengers could have been exposed outside of the flight (Yang et al. 2020), and the number of inflight infections is estimated as 12 (Github project site). Once the model is validated, we utilized it to examine counterfactuals regarding mask wearing and social distancing.

23.4.2 Infection Modeling Results

We estimate infection risk in the United Kingdom (UK) flight in order to verify that the model yields reasonable results in the absence of masks, and to examine the role of passenger movement in explaining superspreading patterns. The expected number of cases from the simulations, 12.86, is close to the observed 13 cases. In addition, it also explains the two cases observed in the economy section.

We next examine the impact of passenger movement. Simulations without passenger movement estimate around nine secondary infections in the Business class cabin, which is not far off from the observed eleven cases. But it yields no infections in the premium economy and economy cabins. Adding inflight movement results only in expected 0.4 cases in the premium economy and economy cabins. Lastly, including the simulation of boarding and deplaning results in 4.4 cases, 3.3 in business class cabin and 1.1 in economy cabin. This suggests that boarding/deplaning contributes significantly to the infection spread during flights. Note that the combined expected number of cases for the flight is less than the scenarios of no movements and boarding/deplaning only combined, because the infection from different stages of flight are not mutually exclusive. Figure 23.8a illustrates the infection risk probabilities throughout the UK flight. Areas with high infection risk are mainly constrained within the business class cabin, aligning well with the empirical data. The presence of lower infection risks in the economy cabin is explained by the pedestrian movement.

We now examine the validity of the infection model addressing mitigation policies, using the Singapore flight, where masks were mandated. The specific type of masks used is not known. We assume a regular cloth mask rather than N95/FFP2. The estimated number of cases from the simulations, including passenger movement, is 10.7. This matches well with the 12 secondary cases reported to be from the flight (Github project site). If everyone had used FFP2 or N95 masks for the entire duration of the flight, then the model indicates that there would be 2.3 secondary infections. If there had been no mask usage, then there would be 55 secondary infections.

Figure 23.8b shows the risk profile as a function of the seat. The high-risk regions highly cluster around the index cases. Furthermore, in the area of the index case sitting in the front, a "U-shaped" gradient profile of infection risk presents, along the aircraft aisle. This may reflect the captured movements of the index or other passengers passing by. Otherwise, if the infection spread is mainly caused by the close proximity with the index, the profile in this area should be close to the pattern present in the business cabin of the UK flight as shown in Fig. 23.8 a. Overall, the use of pedestrian dynamics models along with the infection modeling explains the in-flight infection spread with high resolution.

23.5 Summary and Recommendations

The efficiency of masks varies depending on the mask type, manufacturing quality and facial-fitting. These factors affect the filtration efficiency. Moreover, various distance thresholds can result from the diverse design and fitting of masks. A parameter sweep by varying the mask efficiency

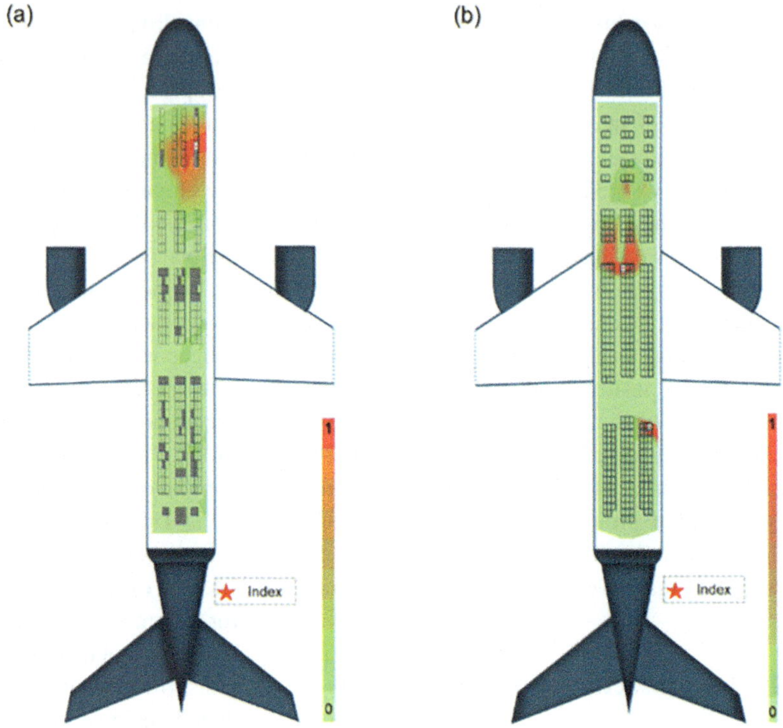

Fig. 23.8 Modeling results of the distribution of secondary infections in the, **a** UK flight and, **b** Singapore flight for the duration of the flight

and distance threshold from a no mask case to a well-fitting N95 mask provides a clearer understanding of the role of dose and mask usage in infection spread. Therefore, we performed hypothetical analysis of mitigation efficiency of masks in a generic two-aisle airplane, with a duration of two hours to simulate common regional flights, with model parameter choices to simulate a superspreading scenario in the absence of masks. Without any masks used in such scenarios, simulations estimate 52.6, 52.7, and 54.8 cases of 157 passengers, using exponential, beta-Poisson, and Weibull models, respectively. Subsequently, we systematically incorporate the mitigating effect through infection-related parameters, i.e., the distance threshold and filtration efficiency. Figure 23.9 shows the variation of mitigation efficiency compared to the superspreading in the airplane configuration using exponential, beta-Poisson and Weibull models. All three models captured a consistent reduction in infections with mask usage, with well-fitting N95 masks resulting in close to 100% reduction. Lower quality masks exhibit a limited mitigation efficiency in exponential and beta models, as shown in Figs. 23.9a and 23.9b. Comparatively, the reduction in number of infections varies more rapidly with improving mask quality in the Weibull model, as shown in Fig. 23.9c. This suggests that the Weibull model is more sensitive to the reduction of mitigating parameters and thereby is not suitable for prescriptive modeling where a conservative modeling result is preferred for mitigation policy recommendations. In transportation scenarios (e.g., air travel) policy development for infection mitigation needs to be conservative to account for extreme events like superspreading. Therefore, one would use more conservative models like the exponential model to develop mitigation policies. Moreover, an exponential model is computationally efficient for implementing large parameter sweeps.

The widespread application of pedestrian dynamics models has been limited due to the technical expertise requirements. To address this, we developed a user-friendly platform that allows

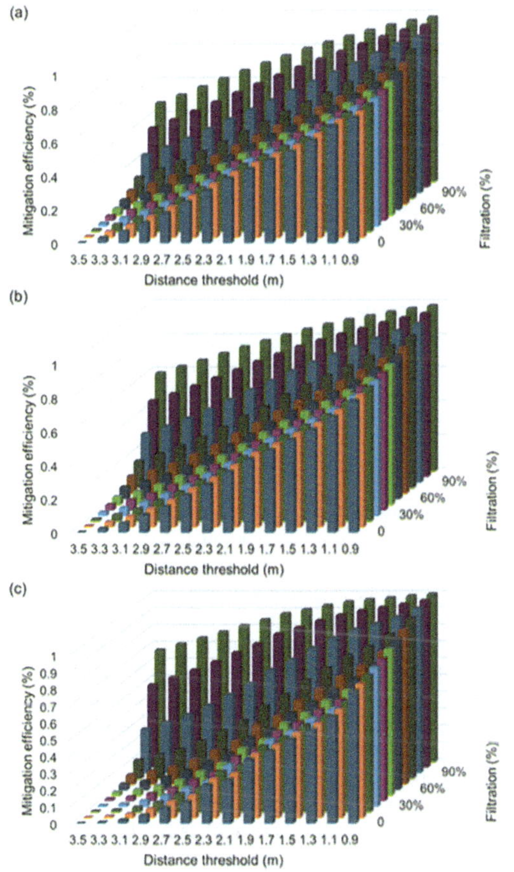

Fig. 23.9 Mitigation efficiency in airplane configuration using, **a** exponential, **b** beta-Poisson and, **c** Weibull models

researchers without extensive computing knowledge to contribute and utilize these models. Our platform includes tools for creating and sharing models, selecting appropriate models, and running large-scale simulations. By applying an early prototype of this platform to pedestrian dynamics in air travel, we identified potential public health interventions to significantly reduce infection risk without impacting passenger volume.

This VIPRA model is available for use in similar situations for the COVID pandemic as it evolves and continues, and in future pandemic and viral scenarios, to mitigate infectious spread and viral transmission. Potential users can access the code and example case studies through the (Github project site).

References

Alotaibi H, Crowder R, Wills G (2013) Investigating factors for knowledge sharing using web technologies. In: Proceedings of the 13th international conference on knowledge management and knowledge technologies, pp 1–4. https://doi.org/10.1145/2494188.2494230

ANTLR. www.antlr.org

Bell JR (1973) Threaded code. Commun ACM 16(6):370–372. https://doi.org/10.1145/362248.362270

Brouwer AF, Weir MH, Eisenberg MC, Meza R, Eisenberg JN (2017) Dose-response relationships for environmentally mediated infectious disease transmission models. PLoS Comput Biol 13(4):e1005481. https://doi.org/10.1371/journal.pcbi.1005481

Chainbuilder. www.chain-builder.net

Chen J, He H, Cheng W, Liu Y, Sun Z, Chai C, Chen Z (2020) Potential transmission of SARS-CoV-2 on a flight from Singapore to Hangzhou, China: an epidemiological investigation. Travel Med Infect Dis 36:101816. https://doi.org/10.1016/j.tmaid.2020.101816

Chen M (2021) Exploration in recommender systems. In: Proceedings of the 15th ACM conference on recommender systems, pp 551–553.https://doi.org/10.1145/3460231.3474601

Chunduri S, Ghaffari M, Lahijani MS, Srinivasan A, Namilae S (2018) Parallel low discrepancy parameter sweep for public health policy. In: 2018 18th IEEE/ACM international symposium on cluster, cloud and grid computing (CCGRID). IEEE, pp 291–300. https://doi.org/10.1109/CCGRID.2018.00044

Clapp PW, Sickbert-Bennett EE, Samet JM, Berntsen J, Zeman KL, Anderson DJ, Bennett WD (2021) Evaluation of cloth masks and modified procedure masks as personal protective equipment for the public during the COVID-19 pandemic. JAMA Intern Med 181(4):463–469. https://doi.org/10.1001/jamainternmed.2020.8168

Coffey JW, Pahle R, Srinivasan A, Namilae S, Scotch M (2024) A knowledge sharing approach to foster interdisciplinary pedestrian dynamics and epidemiological modeling research and practice. In: International congress on information and communication technology. Springer Nature Singapore Singapore, pp 313–324. https://doi.org/10.1007/978-981-97-3556-3_25

Dbouk T, Drikakis D (2020) On respiratory droplets and face masks. Phys Fluids 32(6). https://doi.org/10.1063/5.0015044

Driesen K, Hölzle U (1996) The direct cost of virtual function calls in C++. In: Proceedings of the 11th ACM SIGPLAN conference on object-oriented programming, systems, languages, and applications, pp 306–323. https://doi.org/10.1145/236337.236369

Elsweiler D, Ludwig B, Said A, Schaefer H, Trattner C (2016) Engendering health with recommender systems.

In: Proceedings of the 10th ACM conference on recommender systems, pp 409–410. https://doi.org/10.1145/2959100.2959203
Github project site: https://github.com/vipra-uwf/VIPRA
Helbing D, Farkas I, Vicsek T (2000) Simulating dynamical features of escape panic. Nature 407(6803):487–490. https://doi.org/10.1038/35035023
Islam T, Lahijani MS, Srinivasan A, Namilae S, Mubayi A, Scotch M (2021) From bad to worse: airline boarding changes in response to COVID-19. R Soc Open Sci 8(4):201019. https://doi.org/10.1098/rsos.201019
Islam T, Srinivasan A, Namilae S (2023) Evaluation of parameter sweeps for computationally efficient infection risk analysis using pedestrian dynamics. In: 2023 IEEE aerospace conference. IEEE, pp 1–10. https://doi.org/10.1109/AERO55745.2023.10115639
Khanh NC, Thai PQ, Quach HL, Thi NH, Dinh PC, Duong TN, Mai LTQ, Nghia ND, Tu TA, Quang N, Quang TD, Nguyen TT, Vogt F, Anh DD (2020) Transmission of SARS-CoV 2 during long-haul flight. Emerg Infect Dis 26(11):2617–2624. https://doi.org/10.3201/eid2611.203299
Lu J, Gu J, Li K, Xu C, Su W, Lai Z, Zhou D, Yu C, Xu B, Yang Z (2020) COVID-19 outbreak associated with air conditioning in restaurant, Guangzhou, China, 2020. Emerg Infect Dis 26(7):1628–1631. https://doi.org/10.3201/eid2607.200764
Namilae S, Derjany P, Mubayi A, Scotch M, Srinivasan A (2017) Multiscale model for pedestrian and infection dynamics during air travel. Phys Rev E 95(5):052320. https://doi.org/10.1103/PhysRevE.95.052320
Namilae S, Wu Y, Mubayi A, Srinivasan A, Scotch M (2022) Identifying mitigation strategies for COVID-19 superspreading on flights using models that account for passenger movement. Travel Med Infect Dis 47:102313. https://doi.org/10.1016/j.tmaid.2022.102313
Sadeghi Lahijani M, Islam T, Srinivasan A, Namilae S (2020) Constrained linear movement model (CALM): simulation of passenger movement in airplanes. PLoS ONE 15(3):e0229690. https://doi.org/10.1371/journal.pone.0229690
Smieszek T (2009) A mechanistic model of infection: why duration and intensity of contacts should be included in models of disease spread. Theor Biol Med Model 6:1–10. https://doi.org/10.1186/1742-4682-6-25
Srinivasan A, Mubayi A, Akman O (2020) Should N95 respirators be recommended for the general public. Lett Biomath 7(1):143–152
Srinivasan A, Namilae S (2020) Leveraging new data sources to analyze the risk of COVID-19 in crowded locations. Columbia Academic Commons, Academic Unit: Northeast Big Data Innovation Hub, Series: COVID-19 Research Lightning Talks. https://doi.org/10.7916/d8-0bvw-pd06
VIPRA. https://www.cs.fsu.edu/vipra/
Wei-Guo S, Yan-Fei Y, Bing-Hong W, Wei-Cheng F (2006) Evacuation behaviors at exit in CA model with force essentials: a comparison with social force model. Physica A 371(2):658–666. https://doi.org/10.1016/j.physa.2006.03.027
White KF, Lutters WG (2007) Structuring cross-organizational knowledge sharing. In: Proceedings of the 2007 ACM international conference on supporting group work, pp 187–196. https://doi.org/10.1145/1316624.1316651
Wu Y, Namilae S, Srinivasan A, Mubayi A, Scotch M (2024) Parametric analysis of SARS-CoV-2 dose-response models in transportation scenarios. PLoS ONE 19(6):e0301996. https://doi.org/10.1371/journal.pone.0301996
Yang N, Shen Y, Shi C, Ma AHY, Zhang X, Jian X, Qian G (2020) In-flight transmission cluster of COVID-19: a retrospective case series. Infect Dis 52(12):891–901. https://doi.org/10.1080/23744235.2020.1800814

Rolland Goodenough is a Computer Science master's student with a focus on computational modeling and simulation. His research is centered on creating the Viral Infection Propagation Through Air-Travel (VIPRA) framework for pedestrian dynamics, enabling the detailed study of how people move and interact in various environments, from urban spaces to evacuation scenarios. As part of their work, Rolland developed a domain-specific language (DSL) designed to describe pedestrian behaviors in a flexible and intuitive way. This language allows researchers and practitioners to define complex movement patterns and interactions, making it easier to create and test realistic pedestrian simulations. The framework provides a foundation for analyzing how human behaviors affect the motion of pedestrians. Rolland is also interested in high performance computing (HPC) and its applications across various fields. He is particularly drawn to the challenges of building efficient systems that can process large amounts of data and run complex simulations. This interest aligns with his passion for solving computational problems at scale and exploring how HPC can support advancements in areas like modeling, simulation, and data analysis.

Yuxuan Wu recently earned his Ph.D. in Aerospace Engineering from Embry-Riddle Aeronautical University. His research focuses on particle dynamics modeling, including pedestrian dynamics and the discrete element method. He has applied these models to study the impact of pedestrian movement on the spread of infectious diseases and to simulate powder behavior during additive manufacturing processes. He has been awarded a Dwight D. Eisenhower Transportation Fellowship.

Robert Pahle is a senior research scientist at New York University (NYU) Research Technology. He has a PhD from Arizona State University. His focus areas are on computer mediated reality infrastructure and applications. Among other things he works on the development of NYU's new High Speed Research Network. Previously he worked at Arizona State University where he was the technical director for the decision theater and later faculty in geography within the GeoDa center. One of his significant milestones in technology development has been the creation of an innovative framework—ChainBuilder—which is capable of integrating disparate models and simulations to holistically interact with and visualize them. In addition, he worked with geographic information system (GIS) databases and the parallelization of GIS analytics.

John Coffey holds a BS in Psychology from the College of William and Mary, and a BS in Systems Science, a MS in Computer Science, and EdD with a Computer Concentration from the University of West Florida (UWF). Coffey joined the Institute for Human and Machine Cognition (IHMC) in 1990, while he was a graduate student at UWF. He was the principal programmer of NUCES (Nuclear Cardiology Expert System), the original project for which IHMC was funded. For many years after graduation, John divided his time between the Computer Science Department at UWF, where he held the rank of Professor until semi-retirement, and IHMC. He has worked on projects with NASA Glenn Research Center, several agencies in the U.S. intelligence community, Chief of Naval Aviation and Training, Boeing, Tennessee Valley Authority Nuclear, the Electrical Power Research Institute, Palo Alto Research Center, and others. He now teaches one class for UWF and continues to work with the VIPRA research team. His research interests spanned knowledge elicitation and representation for knowledge management. He also explored advanced educational software technologies, emphasizing customizable delivery and assessment systems integrated with student models.

Sirish Namilae is a Professor of Aerospace Engineering at Embry-Riddle Aeronautical University. He earned a Ph.D. from Florida State University. His research interests are in the areas of particle dynamics, multiscale modeling, and composite materials. Dr. Namilae's research has been supported by the National Science Foundation (NSF), Department of Transportation (DOT), and National Institutes of Health (NIH). He is a fellow of the Royal Aeronautical Society, Associate Fellow of the American Institute of Aeronautics and Astronautics (AIAA), and a Fulbright Senior Research scholar.

Ashok Srinivasan is the William Nystul Eminent Scholar Chair and Professor of Computer Science at the University of West Florida. He earned his PhD in Computer Science from the University of California, Santa Barbara (UCSB), in 1996 and conducted postdoctoral research at the University of Illinois at Urbana-Champaign. Subsequently, he held faculty positions at the Indian Institutes of Technology Bombay, UCSB, and Florida State University. He has served as a Fulbright Senior Research Scholar and as a Program Director at the National Science Foundation (NSF). His research focuses on supercomputing for public health policy, high performance computing, numerical methods, and scientific software. Dr. Srinivasan's work has been supported by NSF, the Department of Energy, and the Department of Defense. He has received multiple Best Paper and Best Poster awards at international conferences. His research has attracted global attention, with coverage by more than 300 news outlets worldwide, and has been cited in testimony to the U.S. Congress.

24 Resilience in Crisis: Strengthening Food Supply Networks During the COVID-19 Pandemic

Matthew Albrecht, Jeffrey C. Johnson, Tracy Van Holt, Craig Carter, and George Perrett

Abstract

The COVID-19 pandemic exposed critical vulnerabilities within food supply networks, causing widespread disruptions throughout social and economic systems. Through mixed-methods analysis of independent restaurants in Gainesville, Florida, this research examines how these establishments navigated unprecedented operational challenges, focusing on both systemic adaptations and human relationships within supply networks. Our semantic network analysis of the interviews identifies the co-occurrence of themes associated with operational responses and community relationships. At the same time, survey data demonstrates specific strategic adaptations including menu modifications (65% of establishments), delivery service adoption (80%), and outdoor dining expansion (49% increase). Our study addresses a crucial gap in understanding how local independent restaurants maintain operational continuity during systemic disruptions. While existing research has explored supply network resilience in large-scale operations, the unique challenges and adaptations of independent restaurants during the pandemic provide valuable insights into resilience-building at the community level. Our findings reveal the critical interplay between vertical supply relationships and horizontal peer networks in fostering resilience. The study's findings contribute to both theoretical frameworks of network resilience and practical strategies for enhancing supply network adaptability. By examining responses to COVID-19 disruptions, we identify actionable approaches for strengthening food system resilience against future challenges, whether arising from pandemics, natural

Research funded by

NSF Award #2032312—RAPID* Collaborative: Networks and Spatial Dynamics of the US Food Supply Chain amid the COVID-19 Pandemic.
*Rapid Response Research Award

COVID Information Commons Video Presentation

The COVID Information Commons (CIC) Research Lightning Talk video presentation and individual multilingual transcripts for this project can be found in the Columbia Academic Commons (AC) (Holt 2021).

M. Albrecht (✉) · J. C. Johnson
University of Florida, Gainesville, FL, USA
e-mail: matthewalbrecht@ufl.edu

J. C. Johnson
e-mail: johnsonje@ufl.edu

T. Van Holt · G. Perrett
New York University, New York, NY, USA
e-mail: vanholt.tracy@gmail.com

G. Perrett
e-mail: gp77@nyu.edu

C. Carter
Arizona State University, Tempe, AZ, USA
e-mail: craig.carter@asu.edu

F. D. Hudson (ed.), *The COVID Information Commons*, Sustainable Development Goals Series,
https://doi.org/10.1007/978-3-032-06125-6_24

disasters, economic instability, or geopolitical tensions. These insights demonstrate how local business adaptations can inform broader strategies for maintaining critical infrastructure during periods of systemic stress.

Keywords

Supply chain resilience · COVID-19 pandemic · Supply chain management · Horizontal supply networks · Food service industry

24.1 Contemporary Supply Networks, the COVID-19 Pandemic, and Resilience

The COVID-19 pandemic revealed fundamental vulnerabilities in food supply networks, triggering cascading disruptions that profoundly impacted local food systems and communities. Restaurants across the United States (U.S.) experienced unprecedented systemic shock, with widespread closures, labor shortages, health safety concerns and shifts in consumer behavior rapidly altering operational norms (Jribi et al. 2020; Rizou et al. 2020). These operational disruptions were exacerbated by factory closures and transportation restrictions (Ivanov 2020). The simultaneous transformation of supply and demand patterns fundamentally forced restaurants to reimagine their supply relationships and operational models. The sudden surge in demand for essential goods and the network's inability to meet that demand underscored the fragility of modern supply chains (Pettit et al. 2020). These vulnerabilities illuminated the critical role of restaurant resilience in maintaining food security and economic stability. Modern food supply networks have undergone a profound transformation, reflecting broader shifts in global economic and social systems. This evolution is characterized by an increasing consolidation of retail power structures and the systematic industrialization of supply chain processes (Saitone and Sexton 2017; MacDonald 2017). These changes reshaped operational frameworks and the human relationships that underpin food distribution. Systemic transformation has prioritized efficiency through standardization and repetition, creating highly uniform processes that boost productivity but may reduce adaptability. Just-in-time inventory and global sourcing exemplify this focus on cost-cutting and efficiency (Christopher and Peck 2004). Before the COVID-19 pandemic, restaurant supply networks typically operated on predictable patterns of demand, established supplier relationships, and standardized delivery schedules. While effective under stable conditions, these models face increasing criticism for weakening supply network resilience and endangering food security (Tendall et al. 2015).

Food supply network resilience is multifaceted involving the capacity over time of a food system to withstand, adapt, and recover from disruptions while maintaining continuous operations and food availability (Tendall et al. 2015). It encompasses dimensions such as robustness, redundancy, resourcefulness, and rapidity (Ponomarov and Holcomb 2009; Ivanov and Dolgui 2020). Building these resilience capabilities presents a significant challenge as the necessary investments in redundancy and flexibility often conflict with operational efficiency objectives under normal conditions (Pettit et al. 2020). The global nature of many food supply networks further complicates resilience efforts, as disruptions in one region can have cascading effects worldwide.

Previous studies on supply network resilience have explored various dimensions of how food networks respond to disruptions. For instance, during natural disasters such as hurricanes and earthquakes, supply chains have demonstrated both vulnerabilities and strengths in their ability to maintain operations (Ivanov and Dolgui 2020). Economic crises, such as the 2008 financial recession, have also tested the resilience of supply networks, revealing critical weaknesses in financial liquidity and demand forecasting (Jüttner and Maklan 2011). Furthermore, geopolitical conflicts and trade wars expose network weaknesses associated with political instability and regulatory changes (Bednarski et al. 2023).

While existing research has significantly advanced our understanding of supply network resilience, several gaps remain. For instance, (Christopher and Peck 2004) provide a comprehensive framework for resilience but primarily focus on theoretical constructs without extensive empirical validation. Ivanov and Dolgui (2020) highlight the impacts of economic crises but do not fully explore the interplay between different types of disruptions, such as how economic and health crises might compound each other's effects. This chapter addresses these gaps through empirical analysis of adaptive strategies employed by independent restaurants during the COVID-19 pandemic. The integration of insights from various crisis types provides a comprehensive perspective on the multifaceted nature of supply chain disruptions and the strategies required for effective mitigation.

24.2 Horizontal and Vertical Supply Networks

The development of resilient supply networks depends on the interplay between vertical and horizontal networks. These complementary network structures create a complex adaptive system supporting both operational efficiency and innovative problem-solving capacities. Understanding how these networks interact and evolve during systemic stress provides crucial insights into building more resilient food systems.

Vertical food networks comprise the relationships between different supply chain stages, linking producers, processors, distributors, and retailers (Gellynck and Kühne 2010). These networks facilitate the efficient flow of goods, services, and information up and down the supply network. For example, a vertical network might include the relationships between farms, food processors, distributors, and restaurants. The effectiveness of vertical networks depends on clear communication between stakeholders and the establishment of protocols and policies for managing expectations and relationships.

Horizontal food networks involve collaborations among entities operating at the same level of the supply chain (Gellynck and Kühne 2010). These peer-to-peer relationships focus on knowledge sharing, resource pooling, and collective problem-solving. For example, membership in restaurant associations allows restaurants to share information about supply chain disruptions and collaborate on alternative delivery methods, such as curbside pickup or online ordering platforms. Horizontal networks utilize social capital to enable stakeholders to share best practices, coordinate responses to market changes, and pool resources to address common challenges (Burt 2004). These collaborations foster collective learning and innovative responses to systemic challenges.

During crises like the COVID-19 pandemic, the interaction between vertical and horizontal networks becomes particularly significant, revealing the essential role of multidimensional relationships in organizational adaptation. Vertical networks contribute to operational efficiency and responsiveness to market demands, while horizontal networks foster innovation and collective adaptation to challenges. This synergy between network structures is essential for maintaining both operational continuity and community food security. These network structures create a complex adaptive system that supports both operational resilience and innovative problem-solving capacities. Understanding and strengthening these relationships represents a crucial component in enhancing food system resilience against future disruptions.

24.3 Background and Sampling

Data collection began in March 2020 and continued through June 2021. The research took place in Gainesville, Florida, an ideal community to conduct COVID-19 supply network research as the city's restaurant sector underwent sudden and dynamic change amidst statewide surges in COVID-19 cases. The city is home to the University of Florida (UFL); as such, the city's population waxes and wanes seasonally depending on numerous factors, including the presence of students as well as university-related activity (i.e., student graduation, sporting events, professional conferences) and non-university activity (i.e., concerts).

Gainesville hosts diverse independent restaurants, ranging from small family-owned establishments to larger, more upscale venues. This diversity allows for a comprehensive examination of various business models and adaptive strategies. The city's restaurant industry reflects broader national trends in the food service sector, making it an ideal location to study the impacts of the pandemic and the adaptive responses of supply networks.

Data collection involved ethnographic key informant interviews followed by a survey comprised of close-ended and open-ended questions. A purposive sample design was used to select independent restaurants (Johnson 1990). Independent restaurants, characterized by lean inventories and reliance on daily deliveries of fresh ingredients, represent a crucial yet vulnerable segment of the food service sector, contributing approximately 4% to the U.S. national gross domestic product (GDP) and generating over $863 billion in annual sales (National Restaurant Association 2019). To be considered for inclusion in this study, restaurants had to meet a set of criteria developed to emphasize specific aspects of the restaurant industry. Restaurants had to meet each of the following characteristics to be considered for data collection: (1) was not a "chain restaurant" (e.g., McDonalds, Texas Roadhouse), (2) must have dinner service, defined as having service after 6 p.m., and (3) must use servers during dinner service (no counter service). A list of 50 eligible restaurants was built from Gainesville Chamber of Commerce data. Each eligible restaurant was contacted three times, once by email, once by phone, and once in person, resulting in the participation of twenty total independent restaurants. Only five of 50 eligible restaurants closed before the study concluded, with one closure among the 20 interviewed.

24.4 Data Collection

Key informant interviews were conducted with six restaurant owners to gain insider knowledge about the Gainesville restaurant supply network and its response to the COVID-19 pandemic. These interviews focused on several key areas, including challenges faced during the pandemic, adaptive strategies employed, relationships within the network, changes in product sourcing, specific product shortages and any COVID-19-related topics informants deemed relevant. These interviews occurred from March 2020 through May 2020.

After completing and analyzing key informant interviews, a mixed-method survey was designed and developed to explore salient topics identified during key informant interviews. Twenty owner-operators and general managers ultimately responded and were surveyed. Restaurant owners and managers were explicitly asked questions about supplier networks and restaurant operations, including changes to their menu, the price and availability of products, and staffing alterations. Informants also discussed the usage of delivery services during the pandemic, accounting for the specific services used and the percentage of income these services accounted for before and during the pandemic. Survey data collection spanned from June 2020 until June 2021.

24.4.1 Data Analysis

The key informant interviews were recorded and transcribed. Following transcription, key informant interviews were analyzed using an adaptive grounded theory approach based on inductive theme analysis (Charmaz and Belgrave 2012). The process began with open line-by-line coding to identify initial themes (Corbin and Strauss 1990). As analysis progressed, themes were compared across interviews, and a codebook was developed and refined (MacQueen et al. 1998). Emerging themes were linked to create a theoretical model of informant experiences during the pandemic (Glaser and Strauss 2017; Strauss and Corbin 1994). This iterative process allowed continuous refinement of the theoretical model as new data were collected and analyzed. These key informant interviews in Gainesville were used to create a mixed-methods survey.

Semantic network analysis was used to elucidate the complex challenges faced by independent restaurants during the early months of the COVID-19 pandemic. A semantic network is a structured representation of relationships between concepts, depicted as nodes connected by edges that signify associations (Bernard et al. 2016). The links among themes reflect the degree to which these themes co-occur across restaurant interviews.

The resulting transcribed interviews were coded for themes yielding an informant by theme $n \times m$ two-mode matrix. The $n \times m$ matrix was transformed into an $n \times n$ one mode matrix using the sums of cross-products (i.e., overlap or co-occurrence) method for the themes across informants (Everett and Freeman 2002). Cell values x_{ij} in the matrix reflect the number of thematic overlaps shared by informants i and j. We categorized network connections into two levels: core strength connections with overlaps greater than 59% (Fig. 24.1) representing high-frequency co-occurrences, and moderate strength connections (Fig. 24.2) representing less frequent but still meaningful thematic relationships with overlaps greater than 39%.

This approach allowed for visualization and analysis of the multifaceted nature of the pandemic's impact on the restaurant industry. Our network consisted of nodes representing themes relevant to restaurants' pandemic experiences, while edges illustrated thematic overlaps vis-a-vis informant's themes. The results provided a comprehensive view of the industry's initial pandemic experience, offering insights into critical areas requiring support. Synthesizing complex data into a visually interpretable format facilitated a deeper understanding of the challenges faced and informed recommendations for policymakers and industry stakeholders.

The quantitative data from our mixed-methods survey was systematically analyzed to complement the qualitative findings. We calculated descriptive statistics (means, frequencies, and percentages) to quantify key operational adaptations during the pandemic. This analytical process included computing averages for delivery service utilization, revenue attribution changes, and physical space transformations. We tabulated the proportion of establishments implementing specific strategies and categorized pricing responses into distinct approaches. This methodology allowed us to measure the magnitude of operational changes while validating patterns identified in our semantic network analysis. By triangulating these mixed-methods data sources, we developed a comprehensive understanding of both statistical trends and contextual meanings of pandemic-driven adaptations in restaurant supply networks.

24.5 Semantic Network Results

Our semantic network analysis identified co-occurring challenges and adaptations described by restaurant owners during the pandemic. In this analysis, core strength nodes (Fig. 24.1) represent central themes with high co-occurrence across key informant interviews, serving as primary anchors that link various challenges and adaptations. Moderate strength nodes (Fig. 24.2) provide important contextual detail by elaborating on core themes. While these moderate strength nodes may have fewer co-occurrences, they remain vital for understanding the complete ecosystem of restaurant adaptations during the pandemic. It is important to note that while the network edges illustrate frequent co-occurrence of themes, they do not imply a direct relationship between these concepts; rather, they reflect the statistical association in how restaurant owners discussed these issues.

We identified several central thematic nodes, shown in Fig. 24.1, that informants consistently mentioned in the same interview: "COVID risk staff," "distribution," "mask-wearing," and "planning," among others. Each of these themes

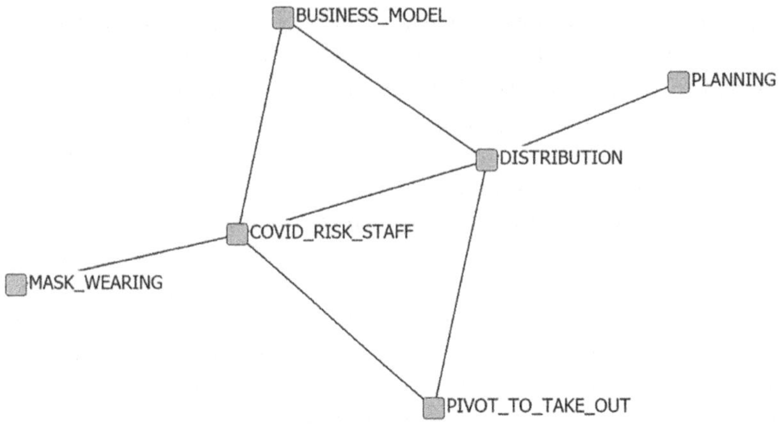

Fig. 24.1 Core strength semantic network of key informant interviews. Each node is a theme that was identified in the six key informant interviews. Network ties indicate thematic co-occurrence in greater than 59% of interviews

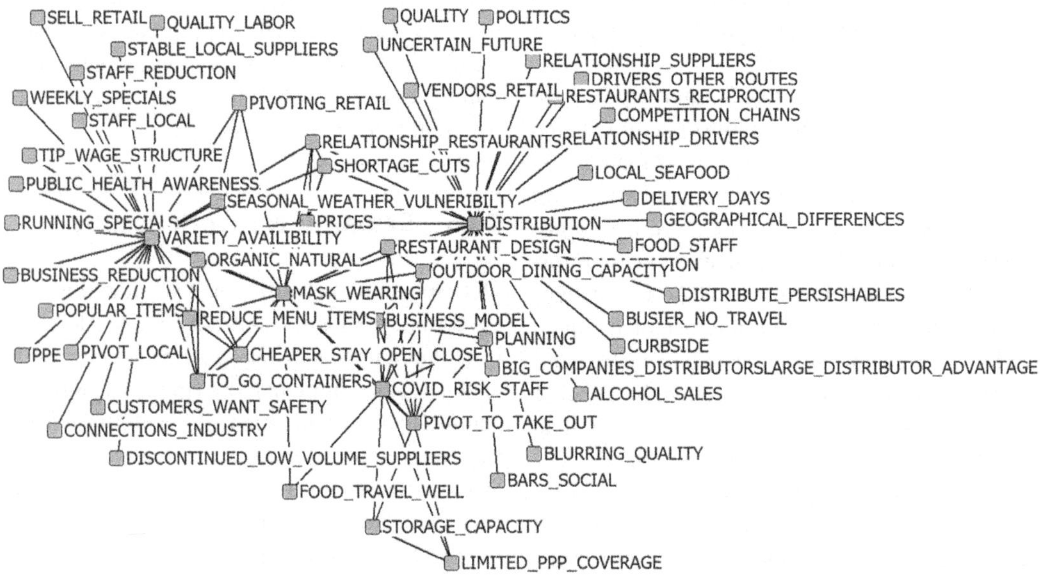

Fig. 24.2 Moderate strength semantic network of key informant interviews. Each node is a theme that was identified in the six key informant interviews. Network ties indicate thematic co-occurrence in greater than 39% of interviews

co-occurred frequently with other concepts in the interviews while also frequently appearing alongside a myriad of moderate strength themes, shown in Fig. 24.2.

In our analysis, restaurant owners frequently discussed "business model" alongside various themes, including "outdoor dining capacity," "mask-wearing," and "planning." These themes appeared together in discussions about protecting staff and streamlining operations when restaurant owners were talking about distribution challenges and health concerns. Our semantic network revealed patterns where multiple themes co-occurred in restaurant owners' descriptions of their adaptation strategies. For example, "pivoting to takeout" potentially influenced "distribution" by altering supply needs (to-go containers, food that travels well, masks, and other personal protective equipment (PPE)), which in turn affected "COVID risk staff" through changes in work patterns. Restaurant owners integrated these changes back into their business models, creating dynamic, evolving systems of challenges and responses.

We identified that the "COVID risk staff" theme frequently co-occurred with "distribution" in our semantic network, indicating these two themes were discussed in the same interview and possibly indicate a relationship between health concerns and supply network issues. The co-occurrence of this theme connection underscores how health risks to restaurant employees and throughout the supply network contributed to disruptions in food distribution. Interestingly, both central nodes are indirectly connected, meaning that they did not co-occur in many interviews, through two other prominent themes: "business model" and "pivoting to takeout." These indirect connections revealed the adaptive strategies employed by restaurants in response to the dual challenges of staff safety and distribution problems.

Our analysis identified numerous themes that frequently appeared in restaurant owners' discussions of their adaptations, including outdoor dining capacity, mask-wearing (PPE), other staff and customer protection measures, and pivoting to takeout. These adaptation themes commonly co-occurred with discussions about staff risk, distribution challenges, and business model changes. For example, when restaurant owners discussed "pivoting to takeout," they often mentioned reduced staff-customer interactions, changes to food delivery approaches, and adjustments to operational procedures.

Informants frequently discussed multiple aspects of the supply network together, suggesting they experienced these elements as interconnected. When describing supplier challenges, such as supplier relationships and product scarcity, restaurant owners commonly mentioned distribution issues in the same context. These discussions often occurred alongside conversations about business model adjustments, including takeout services and outdoor seating arrangements. Restaurant owners also described how their adaptations related to their ongoing supplier interactions.

Through semantic network visualization of co-occurring themes, we illustrated the complex, we illustrated the complex, multifaceted nature of the pandemic's impact on restaurants. The results demonstrate how challenges in one area rapidly propagated through the entire operational ecosystem, necessitating holistic, adaptive responses. The analysis underscores restaurateurs' intricate decision-making processes, balancing staff safety, supply chain management, and business viability through interlinked adaptive strategies. This analysis offers valuable insights into the complex decision-making processes and adaptive strategies employed by restaurateurs in response to the multifaceted challenges posed by the COVID-19 pandemic. The semantic network analysis revealed co-occurring themes of operational challenges and adaptive responses in interviews. These patterns provide essential context for understanding the specific strategies captured in our survey data, particularly how restaurants balanced competing pressures across multiple operational domains.

24.6 Survey Results

Through our semantic network analysis, we identified co-occurring themes of challenges and adaptations that shaped restaurant decision-making during the early pandemic months. These findings provide essential context for interpreting the survey responses by highlighting the underlying structural relationships between key themes, such as "COVID risk staff," "distribution," "business model," and "pivoting to takeout."

The network's key themes illustrate how restaurants navigated these pressures in real time. For instance, our analysis demonstrated that restaurant owners viewed pivoting to takeout not merely as an operational shift but as a strategic response to intertwined factors, including distribution challenges, staff safety concerns, and financial survival. This is reflected in the survey data, where 80% of restaurants (16 out of 20) adopted third-party delivery, with ten relying solely on third-party services and six using a combination of third-party delivery and in-house takeout. Restaurants on average used 2.28 delivery services allowing customers multiple options, while increasing the complexity of vertical networks. This added complexity was necessary for maintaining operational continuity. The significance of these relationships is underscored by the increased proportion of overall income attributed to these services. Before the pandemic, takeout/delivery accounted for about 9% of independent restaurant income; during the pandemic, this figure rose to 23%, with some restaurants reporting a three to fivefold increase. The widespread adoption of third-party delivery services (80% of restaurants), despite significant financial implications, reveals how survival pressures overwhelmed traditional cost–benefit considerations. This adaptation represents a fundamental shift in business models, where maintaining revenue streams took precedence over profit margin optimization. As one owner noted, "At those rates (delivery service rates), you're working for free." This willingness to sacrifice profit margins for operational continuity challenges traditional operational theories that emphasize margin protection, suggesting a more complex decision-making framework during crises. The increase in delivery revenue from 9 to 23% of total income underscores the magnitude of this operational pivot, while the average use of 2.28 delivery services per restaurant indicates efforts to diversify distribution channels. Despite this challenge, adapting their operational models to accommodate greater reliance on takeout and delivery became crucial for restaurant survival, creating greater network complexity as new relationships emerged and existing ones underwent alterations.

As restaurants pivoted themselves to takeout and delivery options and food availability strained, they adapted their menus to meet these challenges. Most restaurants (13 out of 20) streamlined operations by removing certain menu items to reduce food waste. They prioritized takeout-friendly options that met evolving customer behavior, "Our fish dishes didn't travel well, by the time they arrive they're soggy, so, we dropped those. Plus, it's harder to find decent cuts at reasonable prices, it's expensive and inconsistent." At the same time, eight restaurants developed new menu items citing similar rationale for menu item removal, including catering to changing customer health preferences, adapting to delivery constraints, and capitalizing on emerging trends. Restaurant owners and managers strategically changed their menus to optimize offerings, balance efficiency, and maintain customer interest during unprecedented challenges. The co-occurrence of themes identified in the semantic network analysis underscores how decisions, such as reducing menu items, were not made in isolation but were shaped by broader supply chain issues, including product price and availability, health concerns to staff, and shifting customer expectations.

The complexity, uncertainty, and unfamiliarity associated with supply networks during the pandemic led to a divergence in pricing responses. Nine out of 20 restaurants raised menu prices, citing a need to maintain profitability amid higher operating costs. Conversely, eight restaurants kept prices stable to retain customers, motivated by a desire to demonstrate customer loyalty and deliver value. The remaining three restaurants decreased prices strategically to attract more customers. These varied responses underscore the complex decision-making processes restaurant owners and managers faced in balancing viability, customer retention, and market positioning.

As restaurants adapted to changing customer preferences with menu modifications and delivery integration, they also had to reconsider their physical dining spaces to accommodate safety concerns. Government-imposed indoor seating limits challenged existing operational models and knowledge bases, while simultaneously customer health preferences underwent significant change with restaurants estimating that only 51% of customers felt comfortable dining indoors. To address these shifts in customer preference, many restaurants pivoted their operations to allow for more outdoor seating. The average outdoor seating capacity increased from 31 seats to 46.2 seats. Meanwhile, indoor seating was reduced from an average of 104.7 seats per restaurant to 78.3 seats. Outdoor seating represented an important adaptation for many restaurants, allowing them to offset indoor dining restrictions, while also catering to new customer preferences. However, the capacity for restaurants to add outdoor seating varied. Thirteen out of 20 reported minimal difficulty in adding outdoor seating, while seven faced significant difficulty largely stemming from space limitations, including four who were unable to add any outdoor seating. These empirical findings demonstrate how restaurants' adaptive strategies emerged from complex interactions between operational constraints and network relationships. This interplay between practical adaptations and theoretical frameworks offers important insights for understanding supply network resilience during crises.

24.7 Network Adaptation and Decision-Making During Crisis

Our findings extend (Christopher and Peck 2004) framework of supply chain resilience by examining how local organizations navigated compound disruptions in both operational and physical dimensions. The observed adaptations offer insights into how theoretical principles of network resilience manifest in practice, particularly as organizations balance competing pressures while maintaining operational continuity. The COVID-19 pandemic severely disrupted established decision-making models in restaurant networks, which typically rely on predictable patterns and stable operational norms. Our semantic network analysis revealed that restaurant owners developed organizational adaptations with their concerns about "COVID risk staff," "distribution," and "business model" anchoring their response strategies. These co-occurring themes manifested in both operational and physical adaptations, as evidenced by our empirical findings, where 80% of restaurants adopted third-party delivery services and 65% modified their menus in response to systemic pressures.

The transformation of physical space further illustrates this complex adaptive process. Average outdoor seating capacity increased from 31 to 46.2 seats, while indoor seating decreased from 104.7 to 78.3 seats, reflecting both regulatory requirements and shifting customer preferences. The varying ability to implement these changes, with 65% of establishments successfully integrating outdoor dining while 35% faced significant constraints, demonstrates how physical infrastructure influences adaptive capacity within broader network resilience.

Traditional reliance on historical data and seasonal patterns became particularly problematic, as demonstrated by significant shifts in revenue sources from 9% delivery/takeout pre-pandemic to 23% during the pandemic. This disruption created an environment of profound uncertainty, requiring restaurants to develop new operational models while simultaneously reconfiguring their physical spaces. Restaurant owners and managers reported that multiple simultaneous challenges amplified their impact. Our analysis showed that when they pivoted to takeout, this decision influenced their distribution needs, staff safety concerns, and spatial utilization patterns, creating complex feedback loops within the business model.

Restaurant owners and staff demonstrated remarkable adaptive capacity through this multi-faceted transformation. Our data shows that while 13 out of 20 restaurants streamlined menus and reconfigured spaces, eight developed new menu items reflecting both constraints and opportunities presented by the crisis. These adaptations were discussed alongside various factors that frequently co-occurred in our semantic network analysis, including staff safety, distribution challenges, spatial constraints, and evolving customer preferences. This integrated response to operational and physical challenges extends our understanding of how organizations build resilience through the dynamic reconfiguration of both network relationships and material resources

The varying capacity for adaptation among establishments is particularly evident in their ability to reconfigure physical spaces and operational systems, revealing important insights into organizational resilience during the COVID-19 pandemic. These findings suggest that effective resilience strategies must account for both network relationships and physical infrastructure constraints, expanding traditional frameworks to encompass the material dimensions of organizational adaptation.

24.8 Development of Network Relationships

Our survey data shows a transformation of customer relationships during the pandemic revealed intricate adaptations within both vertical and horizontal networks, fundamentally reshaping traditional service delivery models. Our empirical analysis demonstrates that restaurants developed complex intermediary partnerships, averaging 2.28 delivery service relationships per establishment, creating new layers of network connectivity while striving to preserve direct customer engagement. Restaurant owners and managers transformed their operational structures most prominently by widely adopting third-party delivery services with 80% of surveyed restaurants (16 out of 20) implementing these partnerships despite significant concerns about profit margin erosion and operational complexities.

The integration of delivery intermediaries substantially reconfigured vertical network relationships, introducing new connections between restaurants and their customers while simultaneously increasing operational complexity. These changes required restaurants to develop new coordination mechanisms and information flows, fundamentally altering how value was created and delivered within the network. The shift from direct customer interaction to mediated service experiences necessitated significant adaptations in traditional service models, as restaurants balanced the benefits of expanded delivery capability against the challenges of maintaining meaningful customer relationships through digital intermediaries.

These structural adaptations illustrate how crisis events can catalyze fundamental changes in network architecture, requiring organizations to balance operational efficiency with relationship maintenance in increasingly complex service environments. While the adoption of multiple delivery partnerships enhanced distribution capability and provided operational redundancy through service provider diversification, it also introduced new vulnerability points in customer relationship management. These vulnerabilities might include fragmented customer data across platforms limiting unified service delivery, reduced direct communication channels, inconsistent quality control across delivery providers potentially damaging brand reputation, and diminished ownership of customer relationships as third-party platforms mediate these crucial connections. The added reliance on third-party delivery highlights the dynamic nature of network resilience, where adaptations intended to address immediate operational challenges can lead to longer-term structural changes in how organizations connect with and serve their communities.

As these new network configurations emerged, they raised important questions about the human dimensions of supply chain resilience and customer relationship management. As these intermediary relationships become more embedded in operational models, understanding their impact on network stability, service quality, and organizational adaptability becomes increasingly critical. The experiences of restaurant owners and managers reveal how technological solutions, while providing operational resilience, can fundamentally alter the social fabric of service relationships. This tension between technological efficiency and human connection emerges as a central challenge in developing truly resilient service networks.

These findings suggest that effective network resilience strategies must consider not only operational efficiency but also the preservation of meaningful human relationships within increasingly digitized service environments. The transformation of customer relationships during the pandemic demonstrates how crisis-driven adaptations can reshape both the structural and social dimensions of service networks, requiring careful attention to both technological integration and relationship maintenance in building sustainable resilience strategies.

24.9 Theoretical Contributions and Practical Implications

Our findings both confirm and extend existing theoretical frameworks on supply chain resilience while revealing new insights about organizational adaptation during systemic disruption. The semantic network analysis reveals patterns that align with (Christopher and Peck 2004) framework of supply chain agility in local food systems, particularly through the frequent co-occurrence of themes related to operational challenges and adaptive responses. These adaptations reflect not just operational changes but fundamental transformations in how organizations maintain community connections during crisis.

The research particularly enriches (Ivanov and Dolgui 2020) work on adaptability during economic crises by demonstrating how local organizations navigate compound disruptions. Our semantic network analysis demonstrates the nature of these challenges as perceived by restaurant owners, who reported that changes in one area (such as "pivoting to takeout") created ripple effects across multiple operational domains. Since we specifically asked restaurant owners about these domains in our interviews, these thematic co-occurrences were somewhat anticipated. The survey data quantifies these adaptations: restaurants averaged 2.28 delivery service partnerships while simultaneously managing significant changes in seating capacity, where indoor seating decreased from 104.7 to 78.3 seats while outdoor seating increased from 31 to 46.2 seats. These findings demonstrate how organizations balance multiple competing pressures while maintaining operational continuity.

For restaurant owners and suppliers, our findings underscore the importance of developing flexible, multidimensional business models that can adapt to systemic disruption. The varied pricing strategies observed in our study demonstrate the complex decision-making required to balance business viability with community needs. Successful adaptation depends not only on operational flexibility but also on the ability to maintain and strengthen both vertical and horizontal network relationships.

Despite their impact on profit margins, our survey data shows that restaurant owners and managers adopted third-party delivery services, transforming their customer relationships in ways that suggest how crises can accelerate systemic change. Our data shows that delivery/takeout revenue increased from 9 to 23% during the pandemic, highlighting the need for businesses to develop robust digital capabilities while maintaining community connections. Key strategies identified in our analysis include implementing flexible delivery systems, developing outdoor dining capabilities where possible, and maintaining product adaptability to navigate future disruptions.

For policymakers, our findings demonstrate the critical need for supportive frameworks that enhance system-wide resilience. The varied ability of restaurants to adapt their physical spaces highlights how structural constraints can limit adaptive capacity. Future policy frameworks should address these disparities while supporting the food service industry's overall adaptability and resilience. This includes developing flexible regulatory approaches to accommodate rapid operational changes while maintaining essential health and safety standards.

24.10 Conclusion

The COVID-19 pandemic revealed fundamental vulnerabilities in U.S. food supply networks while simultaneously demonstrating remarkable adaptive capacity within the independent restaurant industry. Our research extends existing theoretical frameworks by showing how local organizations navigate compound disruptions through the dynamic reconfiguration of both operational systems and network relationships.

The observed adaptations show how resilience principles manifest in practice, particularly in how organizations balance competing pressures while maintaining community connections during periods of systemic stress.

The demonstrated importance of both vertical and horizontal relationships, combined with the need for physical and operational flexibility, suggests that resilience-building requires attention to both structural capabilities and relational resources. Organizations must develop adaptive strategies that account for both the technical and human dimensions of service delivery, as evidenced by restaurants' efforts to balance operational efficiency with relationship maintenance in increasingly complex service environments.

As we move beyond pandemic-driven strategies, several critical research directions emerge. Future studies should examine: (1) the long-term sustainability of pandemic-era adaptations, particularly the integration of digital intermediaries in traditional service relationships; (2) the role of digital transformation in enhancing supply network resilience while preserving meaningful human connections; and (3) the development of metrics for assessing adaptive capacity in local food systems. Additionally, comparative studies across different geographic regions and restaurant types could provide valuable insights into the generalizability of these findings.

Beyond the immediate context of restaurant supply networks, this research illuminates broader patterns in how organizations and communities adapt to systemic disruptions. The observed interplay between operational adaptation and relationship maintenance offers valuable insights for understanding organizational resilience across various sectors facing similar challenges in an increasingly uncertain environment. These lessons prove particularly relevant as organizations continue to navigate the complex balance between technological efficiency and human connection in building truly resilient service networks.

References

Bednarski L, Roscoe S, Blome C, Schleper MC (2023) Geopolitical disruptions in global supply chains: a state-of-the-art literature review. Prod Plan Control 1–27. https://doi.org/10.1080/09537287.2023.2286283

Bernard HR, Wutich A, Ryan GW (2016) Analyzing qualitative data: systematic approaches. SAGE publications

Burt RS (2004) Structural holes and good ideas. Am J Sociol 110(2):349–399. https://doi.org/10.1086/421787

Charmaz K, Belgrave L (2012) Qualitative interviewing and grounded theory analysis. SAGE Handb Interview Res Complex Craft 2:347–365. https://doi.org/10.4135/9781452218403

Christopher M, Peck H (2004) Building the resilient supply chain. Int J Logist Manag 15(2):1–14. https://doi.org/10.1108/09574090410700275

Corbin JM, Strauss A (1990) Grounded theory research: procedures, canons, and evaluative criteria. Qual Sociol 13(1):3–21. https://doi.org/10.1007/bf00988593

Everett MG, Freeman LC (2002) UCINET for windows: software for social network analysis. Analytic Technologies, Harvard, MA

Gellynck X, Kühne B (2010) Horizontal and vertical networks for innovation in the traditional food sector. Int J Food Syst Dyn 1(2):123–132. https://doi.org/10.18461/ijfsd.v1i2.124

Glaser B, Strauss A (2017) Discovery of grounded theory: strategies for qualitative research. Routledge. https://doi.org/10.4324/9780203793206

Van Holt T (2021) RAPID collaborative: networks and spatial dynamics of the US food supply chain amid the COVID-19 pandemic. Columbia Academic Commons, Academic Unit: Northeast Big Data Innovation Hub, Series: COVID-19 Research Lightning Talks. https://doi.org/10.7916/4zrc-q685

Ivanov D (2020) Predicting the impacts of epidemic outbreaks on global supply chains: a simulation-based analysis on the coronavirus outbreak (COVID-19/SARS-CoV-2) case. Transp Res Part E Logist Transp Rev 136:101922. https://doi.org/10.1016/j.tre.2020.101922

Ivanov D, Dolgui A (2020) Viability of intertwined supply networks: extending the supply chain resilience angles towards survivability. A position paper motivated by COVID-19 outbreak. Int J Prod Res 58(10):2904–2915. https://doi.org/10.1080/00207543.2020.1750727

Johnson JC (1990) Selecting ethnographic informants, vol 22. SAGE Publications, Incorporated. 1991-97110-000

Jribi S, Ben Ismail H, Doggui D, Debbabi H (2020) COVID-19 virus outbreak lockdown: what impacts on household food wastage? Environ Dev Sustain 22(5):3939–3955. https://doi.org/10.1007/s10668-020-00740-y

Jüttner U, Maklan S (2011) Supply chain resilience in the global financial crisis: an empirical study. Supply Chain Manag Int J 16(4):246–259
MacDonald JM (2017) Consolidation, concentration, and competition in the food system. Econ Rev 01612387
MacQueen KM, McLellan E, Kay K, Milstein B (1998) Codebook development for team-based qualitative analysis. Cam J 10(2):31–36. https://doi.org/10.1177/1525822x980100020301
National Restaurant Association (2019) 2019 Restaurant industry factbook
Pettit TJ, Croxton KL, Fiksel J (2020) Ensuring supply chain resilience: development and implementation of an assessment tool. J Bus Logist 34(1):46–76. https://doi.org/10.1111/jbl.12009
Ponomarov SY, Holcomb MC (2009) Understanding the concept of supply chain resilience. Int J Logist Manag 20(1):124–143. https://doi.org/10.1108/09574090910954873
Rizou M, Galanakis IM, Aldawoud TM, Galanakis CM, Galanakis CM (2020) Safety of foods, food supply chain and environment within the COVID-19 pandemic. Trends Food Sci Technol 102:293–299. https://doi.org/10.1016/j.tifs.2020.06.008
Saitone TL, Sexton RJ (2017) Concentration and consolidation in the US food supply chain: the latest evidence and implications for consumers, farmers, and policymakers. Econ Rev 01612387
Strauss A, Corbin J (1994) Grounded theory methodology: an overview. 1994-98625-016
Tendall DM, Joerin J, Kopainsky B, Edwards P, Shreck A, Le QB, Six J (2015) Food system resilience: defining the concept. Glob Food Sec 6:17–23. https://doi.org/10.1016/j.gfs.2015.08.001

Dr. Matthew Albrecht is a cultural anthropologist and Postdoctoral Researcher at the University of Florida Bureau of Economic and Business Research, Survey Research Lab. His research explores the complex relationships between organizational structures, human systems, and adaptive capacity during periods of disruption. He combines rigorous mixed-methods approaches with ethnographic insight to examine how organizations navigate systemic challenges while maintaining operational resilience. Dr. Albrecht's research portfolio includes studies on social sentiment modeling of infectious disease transmission, climate change adaptation among small-scale farmers in Jamaica, and, most recently, restaurant supply chain resilience during the COVID-19 pandemic. His methodological expertise spans quantitative analysis, social network modeling, and qualitative inquiry, allowing for nuanced examination of both formal and informal organizational networks. He holds a PhD in Cultural Anthropology from the University of Florida and a Master's degree in Development Practice with a certificate in Global Health and Development from the University of Arizona.

Jeffrey C. Johnson, Ph.D. is Professor of Anthropology at the University of Florida. He was a former Program Manager with the Army Research Office (IPA) where he started the basic science research program in the social sciences. He has conducted extensive long-term research, supported by the National Science Foundation (NSF), comparing group dynamics and the evolution of social networks of over-wintering crews at the American South Pole Station, with those at the Polish, Russian, Chinese, and Indian Antarctic Stations. In related research, he has studied aspects of team cognition and social networks on success in simulated space missions. He has published extensively in anthropological, sociological, biological, aerospace, and marine science journals and was the founding editor of the *Journal of Quantitative Anthropology*, co-editor of the journal *Human Organization*, the author of *Selecting Ethnographic Informants* (Sage, 1990), and co-author of *Analyzing Social Networks* (Sage 2018), *Analyzing Social Networks Using R* (Sage 2023), and *Analyzing Social Networks*, 3ed. (2024). He is the recipient of the 2014 Simmel Award for recognition of his work for the study of social networks.

Tracy Van Holt is the head of Sustainable Systems Lab, and an interdisciplinary researcher and data scientist with a Ph.D. in Interdisciplinary Ecology, specializing in social science research methods, AI tools, network science, and spatial analysis. With expertise in human behavior, sustainable supply chains, conflict dynamics, and landscape change, she integrates AI-driven approaches with social and environmental research. She leverages surveys, machine learning, natural language processing, and predictive modeling to uncover patterns in sustainability, human behavior, and ESG decision-making. Her work bridges disciplines to develop ethical, data-driven solutions that enhance decision-making across business, health, and environmental systems.

Dr. Craig Carter is the John G. and Barbara A. Bebbling Professor of Supply Chain Management in the W.P. Carey School of Business at Arizona State University. He is also an Advising Editor and Editor Emeritus of the Journal of Supply Chain Management and Fearon Fellow at CAPS Research. Professor Carter has worked in the areas of transportation and logistics with Ryder Systems, Hechinger Company, and the U.S. Department of Transportation. Additional field-based procurement research with over 100 Fortune 1000 size firms in the U.S. and Germany includes work with CAPS Research and McKinsey & Company. His research focuses on sustainable supply chain management and the often overlapping area of behavioral decision-making phenomena within supply management contexts. Dr. Carter's research has been cited over 25,000 times and has appeared in the leading supply chain management journals, including the *Journal of Supply Chain Management*, *Journal of Business Logistics*, *Journal of Operations Management*, *Production and Operations Management*, and *Decision Sciences Journal*. He serves as a Senior Editor or Associate Editor for the *Journal of Business Logistics*, *Decision Sciences Journal*, and *Journal of Operations Management*.

George Perrett is an applied statistician working at the intersection of social science and statistics. He has a Bachelor of Arts degree in Psychology from the University of Wisconsin-Madison, a Master of Public Health degree in Health Behavior and Health Education from the University of Michigan, and a Master of Science degree in Applied Statistics from New York University. Currently, he is pursuing a doctoral degree in Statistics and Computational Social Science at New York University.

Index

F. D. Hudson (ed.), *The COVID Information Commons*, Sustainable Development Goals Series,
https://doi.org/10.1007/978-3-032-06125-6

The manufacturer's authorised representative in the EU is Springer
Nature Customer Service Centre GmbH, Europaplatz 3, 69115 Heidelberg,
Germany. If you have any concerns regarding our products, please
contact ProductSafety@springernature.com

Printed and bound by CPI Group (UK) Ltd, Croydon, CR0 4YY
16/09/2026
02217783-0012